CONTENTS

TRIBOLOGY SERIES, 8

INDUSTRIAL TRIBOLOGY

The Practical Aspects of Friction, Lubrication and Wear

edited by

MERVIN H. JONES
Department of Mechanical Engineering, University College of Swansea, Swansea, U.K.

and

DOUGLAS SCOTT
Consultant, Editor of "Wear", Secretary of The Institution of Engineers and Shipbuilders in Scotland, Glasgow, U.K.

ELSEVIER SCIENTIFIC PUBLISHING COMPANY
AMSTERDAM –OXFORD –NEW YORK 1983

ELSEVIER SCIENTIFIC PUBLISHING COMPANY
Molenwerf 1,
P.O. Box 211, 1000 AE Amsterdam, The Netherlands

Distributors for the United States and Canada:

ELSEVIER SCIENCE PUBLISHING COMPANY INC.
52, Vanderbilt Avenue
New York, N.Y. 10017

Library of Congress Cataloging in Publication Data
Main entry under title:

Industrial tribology.

(Tribology series ; 8)
Includes bibliographical references and indexes.
1. Tribology. I. Jones, Mervin H., 1939-
II. Scott, Douglas, 1916- . III. Series.
TJ1075.I48 1983 621.8'9 82-24248
ISBN 0-444-42161-0 (U.S.)

ISBN 0-444-42161-0 (Vol. 8)
ISBN 0-444-41677-3 (Series)

Printed in The Netherlands

ACKNOWLEDGEMENT

The editors gratefully acknowledge the assistance of Mr. G. Williams and Mr. J. Thomas of the lubrication department of British Steel Corporation, Port Talbot and Dr. G. Thomas of the Extra Mural Department, University College, Swansea, who have been actively associated with the annual Seminars on Industrial Tribology since their inception.

Acknowledgement is also due to visiting lecturers who generously gave their time in developing the Seminar to its present successful format. These include R. Gronbech - Davey United, J. Bathgate - David Brown Gears Ltd., P. Gadd - NAML Gosport, N.W. Morris - Farvalube Ltd., D. Hatton - Shell International, Dr. E.T. Jagger - Angus Seals Ltd., and Dr. D.J. Haines - British Aircraft Corporation.

The editors extend their gratitude to the typing expertise of Mrs. M.A. Williams and Mrs. P.T. Hancock who have so expertly produced this camera ready copy.

FOREWORD

Some eleven years ago Mr. Ronald Dale, British Steel Corporation, Port Talbot, visited the College to propose the introduction of a course in Tribology for Steelworks Design Staff. This proposal was readily accepted because preparations for the establishment of the Tribology Centre were already well advanced.

The first course was rather ambitious insofar as it was of two weeks duration, the first week being devoted to fairly heavy theory whilst more practical material was introduced during the second week. A particular feature was the inclusion of projects based on contemporary problems at the steelworks. These projects were allocated to small teams and, on the final day, the leader of each team had to report to the whole course on the solution arrived at by his team.

The demand for repeat courses was strong and they have been run at annual intervals ever since. However, the pattern of instruction has been modified from year to year as the result of questionnaires completed by course members and in response to representations by employers. Thus the present course is of shorter duration and is orientated towards practice from the outset, theory only being introduced when required to point the way towards the solutions of practical problems.

One of the objects of the first course, to contrive the maximum degree of interchange of information between participants, has been retained and developed, particularly since the course is now recruited from a number of industries besides the steel industry and indeed from several countries.

The term 'Tribology', defined in the Chambers Twentieth Century Dictionary as follows:- a science and technology embracing all subjects involved when surfaces in contact move in relation to each other (Greek tribein, to rub, and logos, speech, a discourse) had then been only recently introduced to emphasize the multi-disciplinary nature of the study of bearing system. The term 'Lubrication' which had been used previously was considered to be inadequate because it focussed attention on one element only of the bearing system - notably the lubricant to the exclusion of other factors such as the material of construction of the interacting elements.

The need for the introduction of the new term was not necessitated by any failure to develop the subject from the research and development point of view but rather to draw attention to a body of knowledge which was not thought to be sufficiently applied within industry at that particular time. The development of Tribology has always been related to advances in the State of the Art of

Engineering. Thus, during the nineteenth century, progress was dominated by the railway and the steamship. The hydrodynamic theory of lubrication was developed to explain Beachamp Towers experiments on bearings used on the Metropolitan Railway, and Michell's tilting pad thrust bearings were considered to be very suitable for ships' propeller shafts. The thirties this century were devoted to the aircraft and automobile; the forties to the gas turbine. Much of the early work of N.E.L. and N.A.S.A. Cleveland was directed to the problems of high-speed bearings. Then came atomic energy. The origin of our sister institution, The National Tribology Laboratory at Risley, can be attributed to this need and, more recently, space technology has introduced a whole range of new problems particularly related to operation in rarified atmospheres.

If one has to decide on today's special character it is the urge for greater productivity mainly achieved by automatic and otherwise capital-intensive equipment. This has focussed attention on the importance of reliability and of reduced maintenance, which call for the utmost refinement of tribological design. Means for monitoring the condition of machines so as to anticipate failure and to rationalise maintenance have now assumed the utmost importance. The tribology of the manufacturing industry will probably be the most important growth area of our subject during the next decade.

Notwithstanding the output of present day research schools in Tribology, of which there are a growing number, the body of knowledge which has been built up during the past century provides powerful tools for any engineer who wishes to improve a product or practice and it is hoped that the following volume will point the way to the application of sound tribology in many industries.

F.T. Barwell
University College of Swansea

1 TRIBOLOGY IN PERSPECTIVE

D. SCOTT, Consultant, Editor of Wear

1.1 INTRODUCTION

As our technological civilization expands, material and energy conservation is becoming increasingly important. Wear is a major cause of material wastage, so any reduction of wear can effect considerable savings. Friction is a principal cause of energy dissipation and considerable savings are possible by improved friction control. Lubrication is the most effective means of controlling wear and reducing friction. Thus tribology, which is the science and technology of friction lubrication and wear, is of considerable importance in material and energy conservation. The history of this relatively new science which is concerned with problems that have always presented man with a challenge has been recorded [1], and the fundamentals reviewed [2].

1.2 IMPACT OF TRIBOLOGY

Since the publication of the Lubrication Report [3] there has been an increasing awareness throughout industry of the subject of tribology. In the UK the National Centre for Tribology and Industrial Units of Tribology have been set up to provide advice to industry on the utilisation of existing knowledge. These are now viable establishments operating as contract research organisations selling their services at commercial rates. Over thirty universities, polytechnics and technical colleges have incorporated courses on various aspects of tribology into their syllabuses. A basic tribology module [4] for undergraduate mechanical engineering courses has been drawn up. Tribology is an elective subject for the higher national certificate (H.N.C.) in engineering in the United Kingdom and a tribology content is included in some committee for national academic awards (C.N.N.A.) courses. Post-graduate research in tribology, leading to higher degrees is carried out at several universities; three have chairs in tribology. Various courses and training programmes are also available to industry.

Tribology is now recognized universally and President Carter of U.S.A. [5] declared it to be a generic technology underlying many industrial sections and the prospectus for an Industrial Tribology Institute at Rensselaer Technology Center has been presented [6].

Numerous papers on tribology are published annually and many report research directed towards a better understanding of the fundamental principles governing interacting surfaces. Unfortunately, most of the information provided is not suitable for direct use by designers and engineers as research workers generally find it more convenient to express results in terms of non-dimensional parameters rather than as the specific data required for design purposes. A tribology handbook [7] has been produced with the object of providing information to industry in a form that is readily accessible and understood by engineering designers, draughtsmen and works engineers. A synoptic journal [8] has been introduced to reduce time spent in literature perusal.

There is a steady growth in the formation of Tribology Societies on an international scale.

1.3 ECONOMIC ASPECTS OF TRIBOLOGY

The Lubrication Report [3] estimated, within an error of twenty-five per cent, that an amount exceeding five hundred million pounds per annum can be saved in the civilian sector of the UK economy by improvements in education and research in tribology. Such improvements are significant, not merely in cost savings, but are crucial to technological progress and have doubly significant implications for the economic well-being of the nation and the reputation of its engineering products.

The ASME Research Committee on Lubrication in their "Strategy for Energy Conservation through Tribology" [9] reported the magnitudes of energy conservation that can potentially be obtained in the four major areas of road transportation, power generation, turbo machinery and industrial processes through progress in tribology. The estimated 11 per cent total savings in annual US energy consumption is equivalent to some sixteen billion US dollars by an expenditure in research and development of an estimated twenty-four million dollars.

A techno-economic study [10] concluded that the application of tribological principles and practices can effect national energy savings of considerable magnitude in the United Kingdom, in the areas covered which comprise the major parts of 87% of energy consumption. These savings are estimated at £468 to £700 million per annum.

Erosion can be expensive and it has been reported [11] that the ingestion of dust clouds can reduce the lives of helicopter engines by as much as 90 per cent; local stall can be caused by removal of as little as 0.05 mm of material from the leading edges of compressor blades. In pneumatic transportation of material through pipes, the erosive wear at bends can be up to fifty times more than that in straight sections. Even wood chips can cause such wear [12]. Analyses of the failure of boiler tubes indicate that about one third of all occurrences were due to erosion [13].

Although abrasive wear is useful to shape and polish engineering components, its unwanted occurrence is probably the most serious industrial wear problem. In the agricultural industry as many as forty per cent of the components replaced on equipment have failed by abrasive wear [14].

The wear of tools used for cutting metals is of considerable importance to the economics of the engineering industry. It was estimated in 1971 [15] that forty billion dollars was spent in the USA on the machining of metal parts. In the UK about twenty million carbide cutting tools are used per year at a cost of fifty million pounds.

Several estimates have been made on the cost of friction and wear. Jost [16] stated that friction and wear in the USA accounted for an expenditure of one hundred billion dollars per annum. A Committee of the Ministry of Research and Technology of F.R.G. [17] estimated that friction and wear caused a national economic waste of ten billion DM per annum of which about fifty per cent is due to abrasive wear. Rabinowicz [18] has estimated that about ten per cent of all energy generated by man is dissipated in friction processes.

Tribological failures are invariably associated with bearings and to illustrate the costs which can be involved it has been reported [19] that a simple bearing failure in a fully integrated steel mill can lead to a total shut down which at full output rate may cost one hundred and fifty to three hundred pounds per minute. A similar bearing failure on a modern generator set could involve the Central Electricity Generating Board in a loss of one to twenty pounds sterling per minute till the set was again operational. A similar bearing failure in the USA has been quoted to cost twenty-five thousand dollars per day [20]. It has been reported [21] that the total cost of wear for a US naval aircraft amounted to two hundred and forty three dollars per flight hour.

1.4 MECHANISMS OF WEAR

Progress in wear control can be aided by a better understanding of the mechanisms by which it occurs. Research workers have tended to isolate and study specific wear mechanisms such as adhesion, abrasion, erosion and fatigue. Such research has generally been directed towards the study of surfaces in relative motion, the changes brought about by their interaction and the effects of the lubricant and the environment present. Little attention has been given to the products of wear, that is to the debris generated. Recently, particle tribology [22] has allowed postulation of the mechanisms of their formation which together with refined techniques of surface investigation and the study of sub-surface changes aids the elucidation of the wear process.

Advances in understanding emerge only from a willingness to question accepted theories. Questioning of the theories of wear aided by refined investigation techniques are now stimulating the pioneering spirit. Suh's delamination theory of wear [23] is the typical example of recent progress.

Surface examination and wear particle analysis has led to the hypothesis [24] that interaction polishes the surfaces and creates a shear mix layer of short crystalline order of almost superductile material which spreads over the surface as first proposed by Beilby [25]. Repeated rubbing contact causes the shear mix layer to fatigue and characteristic particles flake off.

Further work is required to provide a more complete description of the surface behaviour of materials and the wider application of new theories must await the additional evidence. It may thus be possible to predict the wear rates of materials based on first principles and fundamental properties.

The application of a system analysis to wear problems is receiving considerable attention [26]. The complex nature of wear has delayed its investigation but it now appears that the era when wear was considered a branch of studies in friction and lubrication is coming to an end. The success of the first International Conference on Wear of Materials [27] established wear as a subject of international importance in its own right. A second [28] and a third International Conference [29] have been held and a fourth is planned. Microscopic aspects of wear are receiving attention [30] and calculation methods for friction and wear have been reviewed [31]. A state of the art review of wear is available [32].

1.5 SURFACE STUDIES

The frictional and wear behaviour of materials is greatly dependent upon the surface material and its topography. Surface interaction causes changes in these properties but detailed knowledge of happenings in the interface when wear is occurring is difficult to acquire. It has been usual to study surfaces at various stages of wear to postulate the sequence of events. Besides this procedure greater attention is now being given to the size, morphology and structure of wear particles as well as to the localised nature of damage to surface, interface and subsurface material. Several new tools are available for the study of surfaces at atomic level, notably Auger electron spectroscopy, x-ray photon electron spectroscopy, scanning ion spectroscopy and ion scattering spectroscopy which with complementary information from x-ray energy analysis in the scanning electron microscope and micro-probe analysis aid the tribological elucidation of surface phenomena [33,34].

Advances have been made in the application of statistical techniques to the characterisation of rough surfaces [35,36]. The entire statistical microgeometry of certain rough surfaces can now be completely described in terms of the number of peaks and mean line crossings counted on a single profile. These techniques are now being applied in tribology and it appears that in instrumentation three-dimensional mapping is now well established [37]. For the measurement, assessment and characterisation of very fine surfaces, a laser beam technique of light scattering appears potentially attractive for quality control purposes [38].

1.6 LUBRICATION

Since Reynolds [39] produced his equation following the pioneering work of Tower, (see [40]) the mathematical expression of the process of film formation between relatively moving surfaces has been fundamental to all lubrication theory. Equations have been derived and applied to the study of the various surface configurations used in practice, and the introduction of the high-speed digital computer allowed the simultaneous solution of Reynolds equation together with equations for the elastic deformation of the surfaces. Optical studies of elastohydrodynamic lubrication (EHL) films, infra-red temperature measurements and the elucidation of the response of viscous liquids to high frequency shear have greatly improved the understanding of elastohydrodynamic contacts. It is perhaps better to describe the lubricant in a highly loaded EHL contact as an elasto-plastic solid rather than as a viscous fluid. Based on the new understanding, a theory of EHL traction has been advanced [41] which may be applied to engineering components such as rolling bearings and variable-speed drives. The elastohydrodynamics of elliptical contacts has been applied to ball and roller bearing lubrication [42].

Progress in hydrodynamic lubrication appears to be centred on detailed developments rather than improved fundamental understanding. Work on boundary lubrication seems to be oriented towards specific problem areas such as elevated temperatures and hostile environments. Two centuries of study have failed to unravel completely the mysteries of lubrication problems most important to mankind, the mechanism of human joints. Following the tentative proposal of squeeze films [43] and the emphasis on the protective motion of the simple squeeze film [44] it is considered that the prospect of EHL is good but that the promising mode is squeeze film and not rolling sliding [45]. Collections of information are available on the tribology of natural and artificial joints [46] and the mechanical properties of biomaterials [47].

1.7 LUBRICANTS

When failure in service occurs it is commercially more acceptable to change the lubricant rather than the design. Thus research and development work is continuously directed towards improved lubricants, additives to impart or reinforce desirable properties and synthetic lubricants with unique properties. The more recent major developments in lubricant formulation appear to have been on cutting fluids, fire-resistant hydraulic fluids and synthesised hydrocarbon fluids. Although the latter may cost more than mineral-oil based products, experience indicates that they may give an overall cost saving.

Complications caused by lubricants lead to consideration of wear-resistant materials with good frictional properties which can operate without lubrication. Anti-pollution and conservation is placing emphasis on sealed, lubricated-for-life machinery using solid lubricants and surface treatments which lubricate. Under such conditions interfacial conditions become important. Plastics materials are receiving increased attention especially where chemical and thermal inertness are required. Polytetrafluoroethylene (P.T.F.E.) has become the standard solid lubricant in cryogenic applications. Its tendency to cold flow has been controlled by suitable reinforcement. Newer polymers are being increasingly used where high thermal stability is required. Metal film lubricants are now finding use and potential developments in solid lubricant technology may arise from composite solid-liquid lubricants to use the specific properties of each.

1.8 MATERIALS

The emergence of new design concepts is a major incentive for the development of wear-resistant materials and the acquisition of materials data. The thermal and stress problems associated with advanced tribo-engineering require high-strength, light-weight materials. Conventional materials have been improved by orthodox methods almost to the limit of their potential mechanical properties so that new types of materials such as composites, synthetic diamond and sapphire, new graphites and carbides, metal borides and nitrides which approach the hardness of natural diamond are being developed. To utilize their specific properties new design concepts are required as the substitution of such materials in existing designs can lead to problems and failures in service. Besides replacing metals, ceramics may be used as coatings to complement desirable metal characteristics with refractory properties, insulating and erosion, wear, oxidation and corrosion resistance.

In the field of plain bearings no major development of soft metal bearings appears likely in the immediate future as the possible alloys of all commercially feasible softer metals have been fully exploited. Available materials come close to utilising fully the potentiality of plain bearings of current designs and lubrication systems. The development of plastics bearing materials capable of being manufactured to and maintaining the close tolerances of metals could cause something of a revolution in the bearing field.

Plastics and their composites dominate the dry bearing scene mainly due to the availability of design and performance data [48,49]. A significant advance in fundamental understanding of the wear of plastics composites has been the recognition of the dominant role of the counterface metal [50]. Vacuum deposition techniques such as sputtering, ion-implantation, ion-plating and chemical vapour deposition (C.V.D) appear potentially attractive for solid film lubricant solutions to a wide range of dry bearing problems.

In the field of rolling bearing materials, methods such as reduction of gas content [51] and deleterious carbide segregation [52] are being developed to improve rolling contact fatigue resistance. Developments in high speed tool steel bearings have centred around a finer dispersion of carbides [53] and weight saving [54]. The use of higher than normal additions of alloying elements to provide marginally improved properties may not justify the increased cost. There appears to be a steady but unspectacular development of ceramic materials such as silicon nitride [55].

Brakes and clutches require to dissipate continually greater energies due to loads and speeds generally increasing and improved materials are constantly demanded to contend with more arduous duties and higher temperatures. Owing to the possible health hazard there is considerable pressure to replace asbestos, the most effective filler material for phenolic resins due to its fibrous nature and heat resistance. Sintered metal matrices are now used for severe duty applications but attempts to introduce other organic and inorganic materials have not yet succeeded in displacing conventional materials except in highly specialised fields. Concorde uses carbon composites against themselves and these materials with cheaper fibres and fillers may be a promising method of approach to the replacement of asbestos-filled phenolic resins.

1.9 SURFACE TREATMENTS

The material of engineering components must have structural characteristics to satisfy the design requirements and surface characteristics to contend with wear, fatigue and environmental effects. Surface coatings offer the best compromise to these requirements. Surface treatments are also attractive as an alternative to design or lubricant changes to combat wear in service [56,57].

Besides the conventional treatments recently developed, surface treatments involving thin surface films with specific properties are now finding increasing use and proving to be advantageous as wear-resistant coatings. The treatments include physical and chemical vapour deposition processes. The use of TiC coatings on sintered carbide cutting tools is a typical example of reducing tool wear and cutting costs [58]. Low temperature CVD processes and controlled nuclear thermo-chemical deposition are being developed to produce equiaxed grain material of exceptional hardness. Ion plating and ion implantation also appear to be finding use for specific applications.

1.10 COMPUTER-AIDED DESIGN

It is only by the combination of improved scientific undetstanding and its speedy industrial utilisation that rapid technological progress can be achieved. In the past, a major difficulty has been the delay or lack of feedback from industry and thus the delay in the time taken from the inception of a good idea to its fruitful application. To eliminate such delays increasing use is being made of the computer in design to enable almost instantaneous feedback. In the field of plain bearings, using a suitable programme the designer need only transmit information on bearing design to be informed of performance characteristics. In this way he can have instant feedback and make use of the latest research results without being an expert in the fields of tribology, computation or programming. In effect, he has at hand what amounts to a universal testing machine in which he can plan his design, test its characteristics, modify the design and again measure its characteristics, continuing the process until satisfied that he has the optimum design before committing himself to full-scale test, production or service. Materials selection for optimum performance by computer is now approaching rapidly.

1.11 MACHINERY CONDITION MONITORING

Economic pressures are causing the practice of withdrawing equipment from service at periodic intervals for inspection and maintenance to be replaced by failure prevention maintenance. Thus, means have been developed to determine the condition of machinery whilst in service and to detect any deterioration of performance so that remedial action can be taken before the breakdown point is reached. The monitoring technique chosen depends upon the specific information required and the cost of acquiring the information compared with the savings such information can effect.

There has been a gradual acceptance of vibration analysis although this has proved to be neither the simplest nor the most effective method to use. The problem of data interpretation usually creates the need for expensive trend analysis from a massive build-up of data.

The history of a wear process is recorded in the wear debris produced and magnetic plugs and spectrographic oil analysis (SOAP) are now extensively used to detect abnormal wear. The US Defense Department spends forty million dollars per year on oil analysis [20] to predict only certain types of failure in one power system, the aircraft gas turbine, to save twice this figure in terms of direct repair costs. Although these techniques have proved effective in providing warning of changes in a system, they have some disadvantages. SOAP provides a knowledge only of the quantity of metal in the lubricant but no information on the size or shape of the wear particles. Some damage has usually occurred when the magnetic plug picks up debris large enough for observation.

Ferrography [24,59] a convenient method for the isolation and analysis of wear particles has opened up a new dimension in wear detection and assessment in the form of particle tribology, [22]. Non-metallic particles can also be isolated from lubricants so that lubricant degeneration products can be identified to assess condition and performance. Recent developments [60] have enabled the adoption of ferrography to bio-engineering for the study of prosthesis joints, which should assist the development of improved materials and design of artificial implants. As arthritic joints are subjected to wear the analysis of aspirated synovial fluid appears potentially attractive for the study of wear rates, mechanisms and biological responses to wear in human joints. Ferrographic synovial fluid analysis should augment understanding of the etiology and pathogenesis of degenerative arthritis and provide a method for the diagnosis, documentation, prognostication and treatment of the disease. Some fifteen million USA citizens are afflicted with osteoarthritis [20]. The first International Conference on Ferrography has been planned [61].

1.12 CONCLUSIONS

In the short time since tribology was launched as a concept on its own, it has been described as the world's fastest growing applied science, as still in its infancy [62] and as a means of national wealth creation without commensurate capital investment [63]. So that tribology may quickly achieve maturity still greater use must be made of existing knowledge. Most industrial tribological problems can be solved satisfactorily by a logical systematic investigation of the problem and the application of existing knowledge. It thus appears that increased effort is required to disseminate knowledge in a readily understood form to effect greater energy, materials and manpower savings at a minimum cost. It is hoped that the subsequent chapters go some way towards doing this.

Future trends may be to experiment less but to measure and interpret more. Research may be justified only if it can provide information to allow industry to solve its immediate problems or can produce significant advances in technological progress. The environmentalists may influence tribologists by demanding reduced noise levels of mechanisms and the elimination of pollution and toxicity from lubricants. Lubricants generally do not wear out but become contaminated and so, from the point of view of the environment and conservation, reclamation, which is presently only practised if economical, will tend to become of major importance as we approach an era in which quality of life, safety and a clean environment may well be the motivating force behind technological innovation.

REFERENCES

1 Dowson,D. History of Tribology, 1979, Longmans, London.
2 Suh,N.P. and Saka,N. (Ed.), Fundaments of Tribology, 1980, M.I.T. Press, Cambridge, Mass., U.S.A.
3 Lubrication (Tribology) - Education and Research. A Report on the Present Position and Industry's Needs, 1966, HM Stationery Office, London.
4 A Basic Tribology Module, 1973, Dept. Trade and Industry, London.
5 White House Fact Sheet - The President's Industrial Innovation Activities, Oct. 1979, White House Press Secretary, U.S.A.
6 Industrial Tribology Institute - Prospectus, Sept. 1981, Rensselaer Polytechnic, Troy, N.Y.
7 Neale,M.J. (Ed.), Tribology Handbook, 1973, Butterworths, London.
8 Synoptic Journal, Inst. Mech. Engrs., London.
9 Strategy for Energy Conservation Through Tribology, 1978, ASME, N.Y.
10 Jost,H.P. and Schofield,J. Energy Saving Through Tribology - The James Clayton Lecture, Feb. 1981. I. Mech. Engrs., London.
11 Tilly,G.P., 8th ICAS Congress, Amsterdam, 1972.
12 Lehrke,W.D. and Nonnen,F.A., 1st Int. Conf. Protection of Pipes, Durham, 1975, Paper G2, BHRA, Cranfield.
13 Raask,E., Wear, 1968, 13, 301.
14 Richardson,R.C.D., Jones,M.P. and Attwood,D.G., Proc. Agric. Eng. Symp., 1967, Div. 2, Paper 26, Inst. Agric. Engrs., London.
15 Zlatin,L., 1st Int. Cemented Carbide Conf., Chicago, 1971, Paper 1071-918.
16 Jost,H.P. in Halling,J. (Ed.), Principles of Tribology, 1975, XII, Macmillan, N.Y.
17 Tribologie Res. Rep., T76-35, 1976, Ministry of Research and Technology, Zentralstelle fur Luft und Raumfahrtdokumentation und Information, Munich.
18 Rabinowicz,E., In Chynoweth,A. and Walsh,Wm. (Eds)., Materials Technology, 1976, p. 165 (Amer. Inst. Phys. Conf. Proc. No.32, N.Y.).
19 Braithwaite,E.R., Industrial Lubrication, 1969, 21, 241.
20 Ling,F.F., Proc. of the Tribology Workshop, 1974, 32, National Science Foundation, U.S.A.
21 Devine,M.J. (Ed.), Proc. of a Workshop on Wear Control to Allow Product Durability, 1977, Naval Air Development Centre, Warminster, PA.
22 Scott,D., Proc. Inst. Mech. Engrs., 1975, 189, 623.
23 Suh,N.P. Wear, 1977, 44, 1.
24 Scott,D., Seifert,W.W. and Westcott,V.C., Sci. Amer., 1974, 230, 88.
25 Beilby,G., Aggregation and Flow in Solids, 1921, Macmillan, London.
26 Czichos,H., Tribology - A Systematic Approach to the Science and Technology of Friction, Lubrication and Wear, 1979, Elsevier, Amsterdam.
27 Ludema,K. (Ed.), Wear of Materials, 1977, ASME, N.Y.
28 Ludema,K. (Ed.), Wear of Materials, 1979, ASME, N.Y.
29 Ludema,K. (Ed.), Wear of Materials, 1971, ASME, N.Y.
30 Georges,J.M. (Ed.), Microscopic Aspects of Adhesion and Lubrication, 1982, Elsevier, Tribology Series 7, Amsterdam.
31 Kragelsky,I.V., Dobychin,M.N. and Kombalov,V.S., Friction and Wear - Calculation Methods, 1982, Pergamon, Oxford.
32 Scott,D. (Ed.), Treatise on Materials Science and Technology, Vol.13, Wear, 1979, Academic Press, N.Y.
33 Buckley,D.H., Wear, 1978, 46, 19.
34 Buckley,D.H., Surface Effects in Adhesion, Friction and Wear, 1981, Elsevier, Tribology Series 5, Amsterdam.
35 Properties and Metrology of Surfaces, Proc. Inst. Mech. Engrs., 1967/68, 182, 3K, London.
36 Thomas,T.R. and King,M., Surface Topography in Engineering - A State of the Art Review and Bibliography, 1977, BHRA, Cranfield.
37 Thomas,T.R., (Ed.), Rough Surfaces, 1982, Longman, London.
38 Forsyth,I., and Scott,D. Characterisation of Micro-machined Mirror Surfaces, Wear, 1982, In Press.

39 Reynolds,O., Phil. Trans. Roy. Soc., 1886, 177, 157.
40 Cameron,A. Proc. Inst. Mech. Engrs., 1979, 193, Preprint No.25, London.
41 Johnson,K.L. and Tevazwark,J.L., Proc. Roy. Soc. A., 1977, 356, 215.
42 Hancock,B.T., and Dowson,D., Ball Bearing Lubrication - The Elasto-hydrodynamics of Elliptical Contacts, 1981, Wiley & Sons, N.Y.
43 Fein,R.S., Proc. Inst. Mech. Engrs., London, 1967, 181, (3J), 125.
44 Dowson,D., Proc. Inst. Mech. Engrs., London, 1967, 181, (3J), 45.
45 Higginson,G., Proc. Inst. Mech. Engrs., London, 1977, 191, Preprint 33/77.
46 Dumbleton,J.H., (Ed.), The Tribology of Natural and Artifical Joints, 1980, Elsevier, Tribology Series, 3, Amsterdam.
47 Hastings,G.W., and Williams,D.F. Mechanical Properties of Biomaterials, 1980, Wiley & Sons, N.Y.
48 A Guide to the Design and Selection of Dry Rubbing Bearings, 1976, E.S.D.U. Data Item, 76029.
49 Lancaster,J.K., Tribology Int., 1973, 6, 219.
50 Bartenev,G.M., and Lavrentev,V.V., Friction and Wear of Polymers, 1981, Elsevier, Tribology Series, 6, Amsterdam.
51 Scott,D. and McCullagh,P.J., Wear, 1973, 25, 339.
52. Scott,D. and Blackwell,J., Wear, 1975, 46, 273.
53 Scott,D. and Blackwell,J., Wear, 1978, 34, 149.
54 Scott,D., Tribology Int., 1976, 9, 261.
55 Scott,D., Wear, 1977, 43, 71.
56 Wilson,R.W., Proc. 1st Euro. Tribology Congress, 1975, p.165, I. Mech. Engrs., London.
57 Scott,D., Wear, 1978, 48, 283.
58 Hintermann,H.E., Wear, 1978, 48, 407.
59 Bowen,R., Scott,D., Seifert,W.W. and Westcott,V.C., Tribology Int., 1976, 9, 261.
60 Scott,D. and Westcott,V.C., Proc. Inst. Mech. Engrs., London, 1978, Preprint C42/78, 123.
61 First International Conf. on Ferrography, 1982 (Sept.), Univ. Swansea (In Press).
62 Eurotrib, '77, 1977, Bundesrepublik, Deutschland, Dusseldorf.
63 Jost,H.P., Tribology, 1978, 11, 34.

2 WEAR

D. SCOTT, Consultant, Editor of Wear

2.1 INTRODUCTION

Wear may be defined as the undesired displacement or removal of surface material, although under some circumstances, the initial stages of wear or mild wear which tends to smooth surfaces, may be beneficial for the running-in of mechanisms. The economic implications of wear cause concern in industry, as a reasonable life is required of mechanical equipment to cover capital and maintenance costs. Whilst, in many instances, wear may not place an absolute limit on the life of an investment, it certainly causes a great deal of expenditure on maintenance that must take place; such maintenance is costly in itself, but also costly in lost productivity whilst it is being carried out.

Although wear has for long been a subject of practical interest, fundamental knowledge of wear is sparse. This is due probably to the interdisciplinary nature of wear making it difficult to elucidate and the fact that wear has been accepted as inevitable and unavoidable and so mechanical part replacement technology has dominated wear control technology.

Progress in wear control and prevention can be made only after a better understanding of the mechanisms by which it occurs and of the controlling factors has been acquired.

2.2 THE WEAR PROCESS

Wear may take many forms depending upon surface topography, contact conditions and environment but, generally, there are two main types, mechanical and chemical. Mechanical wear involves processes which may be associated with friction, abrasion, erosion and fatigue. Chemical wear arises from surface attack by reactive compounds and the subsequent rubbing or breaking away of the reaction products by mechanical action. The different types of wear may occur singly, sequentially or simultaneously, but all wear phenomena centre on a common characteristic, an overstressing of the surface [1].

When two surfaces are in contact, the real area of contact is considerably less than the apparent area of contact being confined to a number of small areas where opposing high spots touch. Pressure in these areas will be high and the surface material deformed by the applied load until the contact area becomes sufficiently large to support the load. According to Bowden and Tabor [2], in the absence of an effective separating film a junction may be formed between the surfaces, and relative motion will cause the junction to be broken, resulting in the removal of material from one or the other of the surfaces. Ming Feng [3] considers that plastic flow occurs at contacting asperities so that mating surfaces conform in a pattern of ridges and grooves and giving rise to strain hardening. Mechanical interlocking prevents slip at the interface during relative motion and shear occurs in the softer subsurface material unaffected by strain hardening. The sheared off material may be in the form of debris or if the shearing process produces a sufficient rise in temperature, the wear products may become attached to one of the surfaces.

The adhesive wear theory as described by Archard [4,5] postulates the formation of wear particles at contacting asperities which are hemispherical in shape. Rabinowicz [6] considers that the ratio of surface energy to material hardness is an important factor in wear and may have some effect on wear particle size. Although wear equations [4,6] derived from these theories are consistent with experimental results, they do not account for the basic metallurgy of the materials and are based on several arbitrary assumptions. Seifert and Westcott [7,8] have demonstrated that rubbing wear particles take the form of thin flakes of metal with highly polished surfaces and are not hemispherical fragments generally proposed by adhesion theory. Suh and others [9,10] have proposed a delamination theory of wear based on the behaviour of dislocations at the rubbing surface, subsurface void and crack formation and the subsequent joining of cracks by shear deformation of the surface. The delamination theory predicts that the wear particle shape is thin flake-like sheets as opposed to the hemispherical shape proposed by the adhesion theory and that the surface layer can undergo large plastic deformation. Experimental results showing the process of wear sheet formation by delamination are claimed to substantiate the theory [11]. Suh has reported that bulk material hardness itself is not the controlling factor on wear and that the delamination theory satisfies the thermodynamic requirements of the frictional and wear behaviour of metals [12]. Westcott and others [8,13] suggest that surface interaction polishes the surfaces and creates a shear mix layer of short crystalline order of almost superductile material which spreads over the surface to effect the smooth nature of run-in surfaces as first proposed by Beilby [14]. Repeated rubbing causes the shear mix layer to become fatigued and rubbing wear particles flake off [13]. Such a state of equilibrium maintains benign wear, but disruption of this state can cause initiation of a more severe mode of wear [13]. A rheological mechanism of penetrative wear [15] has been proposed for the formation of plate-like debris. The study of subsurface effects during the sliding of metals supports the assumption of delamination [16]. Study of the wear behaviour of ultra high molecular weight (UHMW) polythene revealed a plastically flowed smooth surface and subsurface cracking which separated sheets of polymer from the wear track in the form of smooth surfaced platelets [17]. Hirst [18], Kragelskii [19] and Scott [20] have reviewed the subject of wear. Bickerman [21] has critically reviewed the theories of adhesion and frictional phenomena in sliding contact and Barwell [22] has reviewed the theories of wear and their significance for engineering practice.

2.3 SCUFFING

Under sliding conditions, the chief task of a lubricant is to allow relative motion between surfaces, with low friction and no damage. This can be achieved if the lubricant film is thick enough to keep the surfaces apart and hydrodynamic conditions prevail. If however, ideal conditions cannot be maintained, the surfaces will come into contact and wear or damage in the form of scuffing will occur, Fig.1. The metallographically changed, scuffed material, Fig.2, is considerably harder (varying from 300 to 850HV) than the original carbon

Fig.1. (x75) Scuffed steel surface.

Fig.2. (H= x110, V= x1100) Taper section through a scuffed surface.

steel (180HV). Electron microscopical investigation of the scuffed material [23,24] reveals that it is martensite and tempered martensite or trootstite, indicating that the material has been heated to above the austenitising temperature and rapidly cooled. Hydrocarbon lubricant breakdown in the contact zone [15] can lead to increase of carbon content and hardness due to diffusion of carbon and gases into the heated deformed material.

The complex mechanism of scuffing is difficult to elucidate as the process, by cumulative action, destroys evidence of its initial stages. To study scuffing and to follow the development of surface failure a crossed cylinder machine has been used [26] in which one cylinder is rotated and a mating cylinder, at right angles to it, is so traversed that the area of contact moves along the surface of both cylinders. Examination of the helical track round a tested cylinder reveals how surface damage builds up with test duration and load [27]. The bearing tracks on lubricated steel cylinders are distinguished by the increased optical reflectivity. As the load increases failure is initiated by short fine marks, followed by incipient scuffing and continuous failure.

Study of taper sections has helped considerably in the elucidation of the effects of rubbing action; the initial polishing seems to be achieved by smoothing of the grinding asperities by plastic deformation as suggested by Westcott [7,8] and Suh [9], Fig.3. Subsurface metallographic changes occur, Fig.3. Heat produced by rubbing action appears to temper the hard steel. Surface hardness may be preserved by rapid quenching from above the austenitising temperature by lubricant or bulk material. The white-etching, hard surface material usually contains cracks. Thermal softening may occur by conduction of heat in the subsurface area away from lubricant quenching action. Local heating, and subsequent quenching by lubricant or cold bulk metal may be sufficient to metallographically change and harden the surface of soft steel, so that both hard and soft steel when scuffed, and the scuffed material, develop a similar metallographic structure and hardness. The electron microscope has revealed the fine metallographic structure of rubbed material, the nature of the original ground surface and the smooth run-in surface and has revealed that the initiation of wear on a sub-microscopic scale develops in a similar manner to that observed on a larger scale. With light loads, the surfaces are smoothed by plastic deformation of the asperities, metallurgical transformations occur and scuffing appears to initiate from small scores. Within the scores on hard steel, features are visible supporting the Bowden and Tabor (2) mechanism of failure, Fig.4a. Plastic deformation and roughening of soft steel indicate over-stressing of the surface material which is not inconsistent with the ideas of Blok [1] and Ming-Feng [3], Fig.4b. Extreme pressure (e.p.) additives are used to prevent metal to metal contact between heavily loaded moving surfaces. Detailed examination of surfaces by the electron microscope has shown the build up of protective films and provides experimental evidence for the generally agreed action of these additives; reaction with surface metal to form adherent surface films with good boundary properties [27].

Fig.3a. (x80) Finely ground surface adjacent to run-in surface.

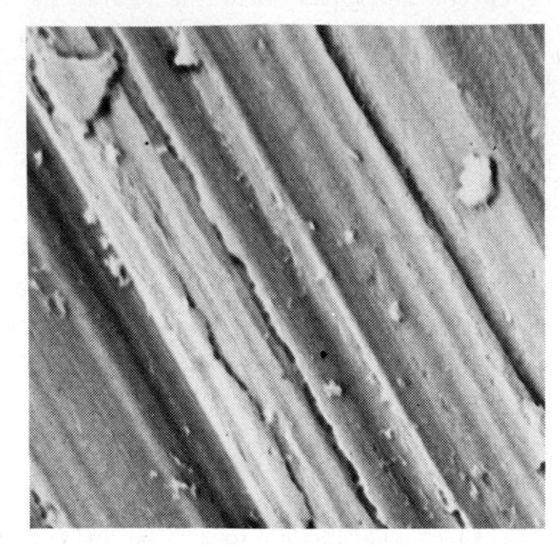

Fig.3b. (x7500) Finely ground surface.

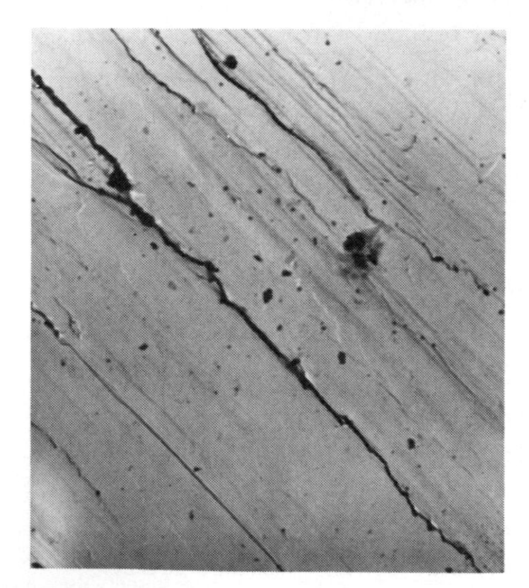

Fig.3c. (x7500) Run-in surface.

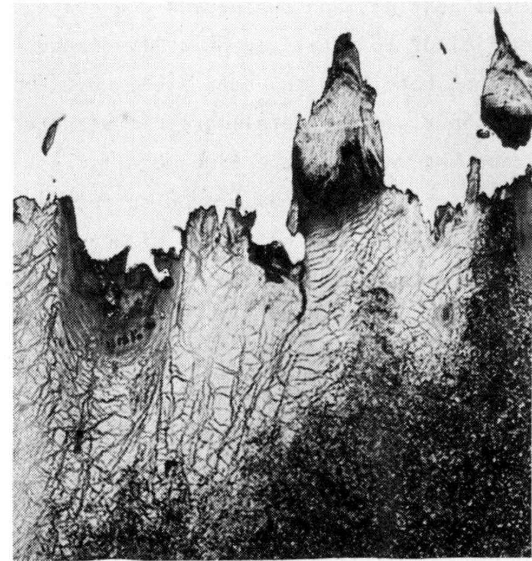

Fig.3d. (H= x100, V= x1100) Taper section through a scuffed surface.

Fig.4a. (x7500) Initiation of sliding wear on a hard steel surface.

Fig.4b. (x7500) Initiation of sliding wear on a soft steel surface.

2.4 ABRASIVE WEAR

From an economic point of view, abrasive wear caused by ploughing or gouging of a hard surface, hard particles or debris, against a relatively softer mating surface is probably the most serious single cause of wear in engineering practice. There are indications that abrasion, primarily a crude machining process, is related to indentation hardness and hence to static yield stress [28,31]. Evidence of extensive damage originating from a small particle of debris has been reported [32,34], Fig.5. A particle of hard brittle material may cause damage in a single pass through the area of minimum film thickness of a bearing. However, in so passing, it may be rendered ineffective due either to breakdown into smaller particles of dimensions smaller than the minimum oil film thickness or by being completely embedded in the softer of the mating materials. On the other hand a soft material particle may work harden on passage between relatively moving surfaces. In gouging the softer bearing material surface, the particle may, if the bearing material also work hardens, be only partially embedded in an equally hard surface area and become a source of further damage to the mating surface.

a. Large bearing machined by hard particle

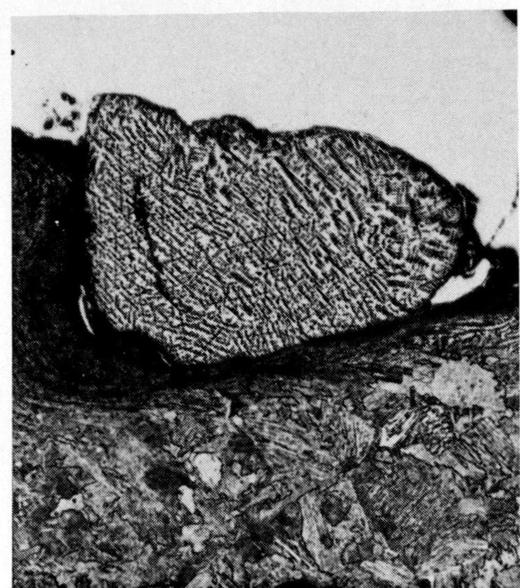

b. (x150) Section through the partially embedded hard particle

Fig.5. Wire-wool type bearing failure.

If abrasive particles are conveyed by a fluid stream the impact of the abrasive particle laden fluid will give rise to erosive wear of any interposed surface [35]. The extent and type of wear depends upon the impinging angle of the particles and the ductility of the surface.

2.5 FRETTING

Fretting is a specific form of wear which occurs when there is slight vibratory movement between loaded surfaces in contact and which manifests itself by pitting of the surfaces and the accumulation of oxidised debris, Fig.6. An electron microscopical study of the initiation of fretting [23] has provided support for the suggestion that at the outset it is no different from other forms of wear, but that the fine debris produced by the initial damage due to metal to metal contact and relative motion provides the starting point for a cumulative abrasive action [36,38]. The debris, being largely the oxide of the metals involved, occupies a greater volume than that of the metal destroyed and in a limited space, this can lead to a pressure build up and seizure. The form and extent of fretting damage depends on the chemical nature of the environment and on whether or not the debris can escape or is built up between the surfaces.

It has been suggested that the initial debris has a platelike form produced by a process of delamination [39].

The actual rate of wear may slow down if the debris acts as a buffer between the two surfaces. Thus a process which initiates as adhesive wear may change to abrasion and then the wear rate may slow down due to debris keeping the surfaces apart. The final failure may then be by fatigue fracture, crack initiation being effected by the stress raising role of fretting pits.

Fig.6a. (x7500) Initiation of fretting on hard steel.

Fig.6b. (x10,000) Fretting damage on a titanium implant.

2.6 FLUID AND CAVITATION EROSION

These wear mechanisms arise from the impact of fluids at high velocities. Fluid erosion damage caused by small drops of liquid can occur in steam turbines and fast flying aircraft through the impact of water droplets causing plastic depressions in the surface. As the fluid flows from the deformed zone it can cause shear deformation in peripheral areas and repeated deformation causes a fatigue type of damage by pitting and roughening of the surface. Cavitation erosion damage is caused by impact from the collapse of vapour or gas bubbles formed in contact with a rapidly moving or vibrating surface. The physical damage to metals is characterised by pitting suggestive of a fatigue origin. The ultimate resilience of material, measured as the energy that can be dissipated before appreciable deformation and cracking occurs, appears to be an important property of metals in cavitation resistance.

2.7 ROLLING CONTACT FATIGUE

The useful life of rolling elements is limited by surface disintegration pits or fracture being caused by a fatigue process dependent upon the properties of the material, the nature of the lubricant and the environment [40,42]. The phenomenon is characterised by the sudden removal of surface material or fracture due to repeated alternating stresses. The process has three phases, preconditioning of the material prior to crack initiation, crack initiation and crack propagation.

Rolling contact fatigue cracks initiate either at the surface and propagate into the material, or start below the surface in the area of calculated maximum Hertzian stress and propagate towards the surface depending upon prevailing circumstances. The propagation of surface cracks seems to be controlled by the nature of the lubricant and the environment [43]. The cracks, transverse to the rolling direction, Fig.7, propagate steadily into the material at an acute

Fig.7. (x15,000) Rolling contact fatigue cracks in En31 ball bearing steel.

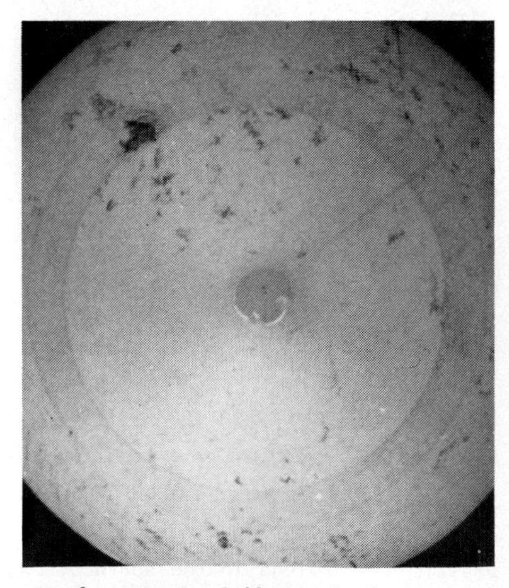

Fig.8. Single failure pit in a mineral oil lubricated bearing ball.

angle to the rolling direction, then influenced by the maximum shearing stresses, propagate parallel to the surface to detach surface material and form a pit, Fig.8. If the environment is deleterious, for example, if it leads to hydrogen embrittlement [44,45] the cracks may propagate rapidly, deep into the material, Fig.9, so that fracture ensues, Fig.10.

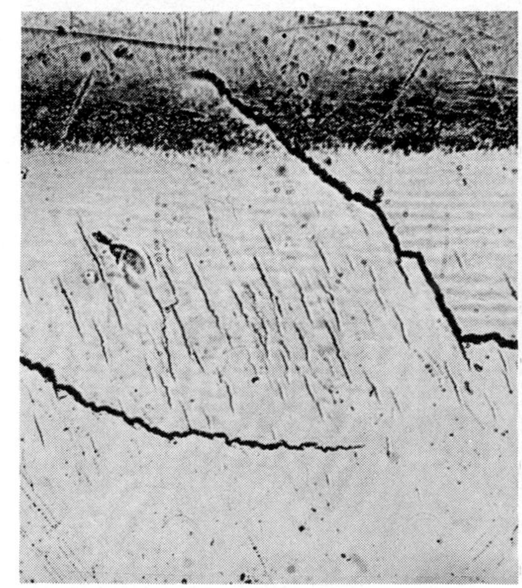

Fig.9. (x75) Rapid crack propagation in a non-flammable fluid lubricated bearing ball.

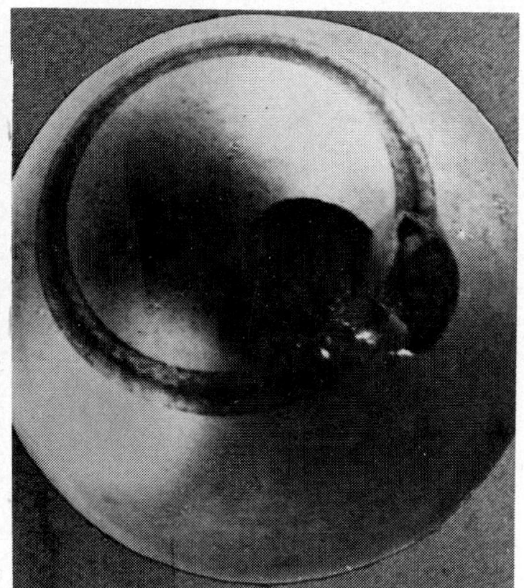

Fig.10. Fractured non-flammable fluid lubricated bearing ball.

Subsurface cracks initiate at depths associated with the region of calculated maximum Hertzian stress and propagate parallel to the surface to remove surface material, Fig.11. Crack initiation may be facilitated by brittle, non-metallic inclusions in the stressed region which crack, break the metallic continuity and act as stress raisers.

Owing to rolling and sliding action, mechanical and metallographic changes occur in the stressed surface and immediate subsurface material of rolling elements [41,42,46]. The structure of conventional En 31 ball bearing steel consists of finely dispersed carbide spheroids in martensite, Fig.12. The metallographically changed material of the surface layer is devoid of carbides as a result of high contact stresses and local high temperature flashes causing solution of the carbides followed by rapid quenching under pressure, Fig.13. Absorption of gases from lubricant breakdown may contribute to surface hardening and crack initiation. Stringer type carbides may form in the subsurface area of contact due to annihilation of the coarse carbides by plastic deformation, Fig.14. Sections transverse to the rolling direction may reveal the presence of localised areas of tempered martensite; cracks develop in such areas, Fig.15. There is a threshold stress level above which metallographic change

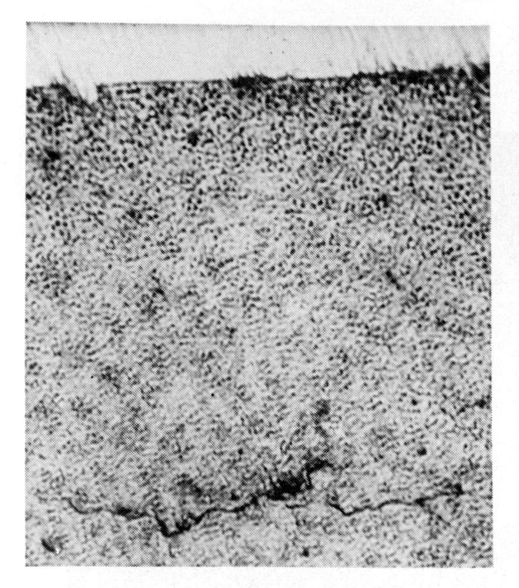

Fig.11. (x375) Subsurface crack in an En31 steel bearing ball.

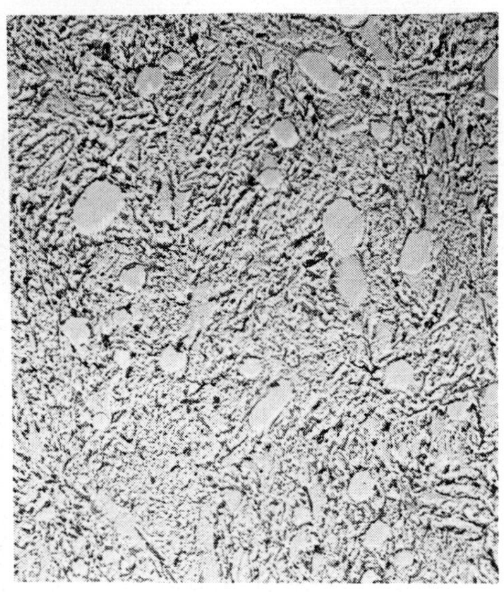

Fig.12. (x8000) Structure of En31 ball bearing steel.

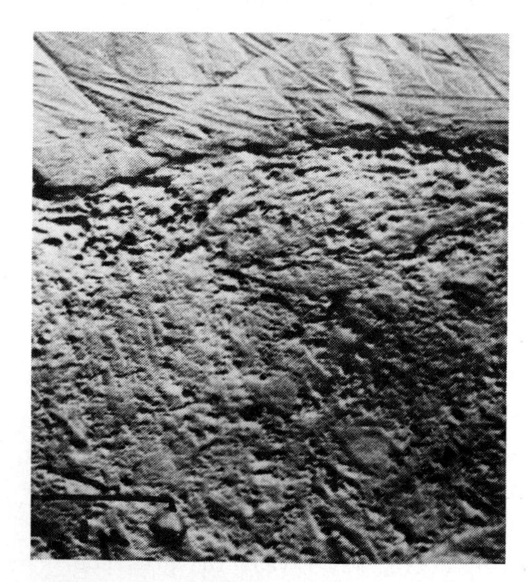

Fig.13. (x12,000) Deformed metallographically changed, spherical carbide free, surface material of a used bearing ball.

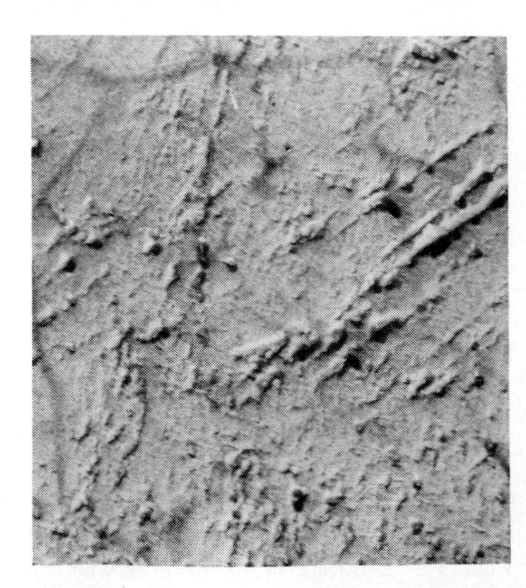

Fig.14. (x15,000) Stringer type carbides in subsurface metallographically changed ball bearing steel.

occurs, suggesting that the changes may be due to a yielding or plastic flow phenomenon rather than a tempering effect or be indicative of overload or a long duration of stressing.

Similar subsurface changes and associated cracks are found in sections of rolling elements parallel to the rolling direction together with associated elongated white etching areas of increased hardness, Fig.16. The extreme hardness of the white etching material may be due to a fine cell size or the almost colloidal dispersion of very fine carbides formed possibly by strain induced precipitation following solution of coarse carbides, Fig.17.

Fig.15. (x575) Tempered martensite in the subsurface area of rolling contact.

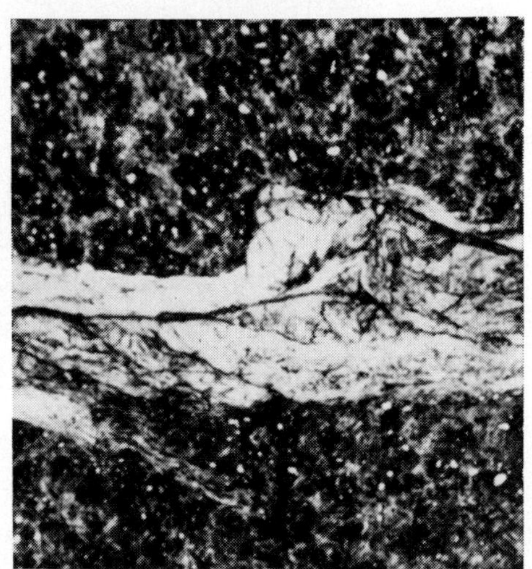

Fig.16 (x1100) Subsurface cracks and associated white etching material.

It appears that several different modes of rolling contact fatigue can cause cracks to nucleate and propagate independently at various rates; thus the phenomenon is greatly influenced by highly localised conditions. Whilst the general properties of the bulk material are important, specific aspects such as the steelmaking process, gas content and cleanliness are also equally important [48,52]. The nature of the lubricant and the environment can have a dominant effect on failure.

Material combination [53] and material lubricant combination require careful consideration to ensure satisfactory performance [54,55].

In rolling contact in the absence of a lubricant, failure occurs, not by the usual failure mechanism but by excessive wear limiting the useful life due to vibration and noise [56,57].

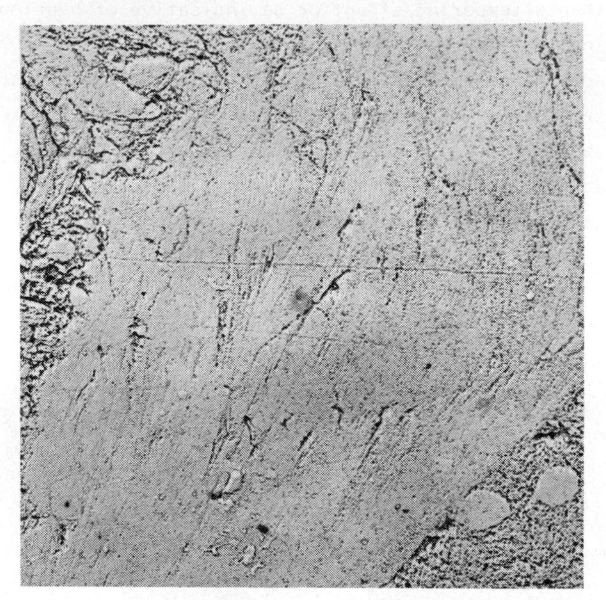

Fig.17. (x7000) Fine structure in subsurface white etching material of En31 steel.

2.8 WEAR DETECTION AND ASSESSMENT

One of the most difficult problems in engineering design is the prediction and assessment of possible wear. An equally important and difficult task is the detection of wear during the operation of machines. Simple methods of wear measurement such as the determination of changes in surface topography by stylus measurement and the determination of weight loss have disadvantages. Machinery must be dismantled for the measurements to be made and inaccuracies in weight may arise due to oxidation and absorption of lubricant.

As the history of the wear process is recorded in the wear debris produced [58] an attractive method of wear detection and assessment is contaminant analysis of the lubricant used. Lubricants in operating mechanisms may be conveniently checked for the detection of wear by spectrographic analysis although the method has some disadvantages [59] being in some instances relatively blind to large particles. A simple device such as a magnetic plug in an oil sump can collect ferrous debris and indicate wear of a moving part [60] but some serious damage may have occurred before debris large enough to be detected has been collected.

Wear particles are unique, having individual characteristics which bear evidence of the conditions under which they were formed [8,58,61]. Careful examination of the morphology and determination of the composition of wear particles can thus yield specific information concerning the surfaces from which they were produced, the mechanism of their formation and the operative wear mode in the system from which they were extracted.

Ferrography [7,8,62,63] is a technique developed to separate wear debris and contaminant particles conveniently from a lubricant for examination and analysis. The duplex Ferrograph analyser consists of two particle separators, a standard analyser and a direct reading (DR) Ferrograph. The DR Ferrograph is a simple instrument used to determine the amount and size distribution of wear particles in a lubricant sample from which significant numerical data can be derived [62, 64]. When successive lubricant samples yield constant density readings it may be concluded that the machine is operating normally and producing benign wear particles at a steady rate. A rapid increase in the quantity of particles and in particular in the ratio of large to small particles indicates the initiation of a more severe wear process. The use of a simple equation provides a single figure for a comparative severity of wear index. Full Ferrographic analysis using the bichromatic microscope [62,63], electron microscopy [62,65] and heating techniques [66] may be used to supplement the information from particles precipitated according to size on a prepared substrate by the analytical Ferrograph.

Particles generated by different wear mechanisms have characteristics which may be identified with the specific wear mechanism [8,61,62,67]. A wear particle atlas has been prepared [68]. Rubbing wear particles found in the lubricant of most machines have the form of platelets and indicate normal permissible wear, Fig.18. Cutting wear particles take the form of miniature spirals, loops and bent wires similar to cuttings from a machining operation, Fig.19. A concentration of such particles is indicative of a severe abrasive wear process.

The operative regimes of sliding wear can be classified by the particles produced. Six regimes which generate characteristic particles have been identified [13]. Free metal particles are produced in regimes 1,2 and 3 and these regimes may be recognised by particle size, ranging from the small particles of regime 2 associated with hydrodynamic lubrication, Fig.18, to large metallic particles of regime 3, Fig.20, which may vary in size up to 250μm when the shear mixed surface layer becomes unstable and localised adhesion occurs.

Three distinct particle types, laminar, spherical and chunks are associated with rolling bearing fatigue. Laminar particles are thin metal particles up to 50μm in major dimension containing holes formed in passage through the rolling contact. Such particles are generated throughout the life of the bearing. Their concentration increases with the onset of spalling. Spherical particles,

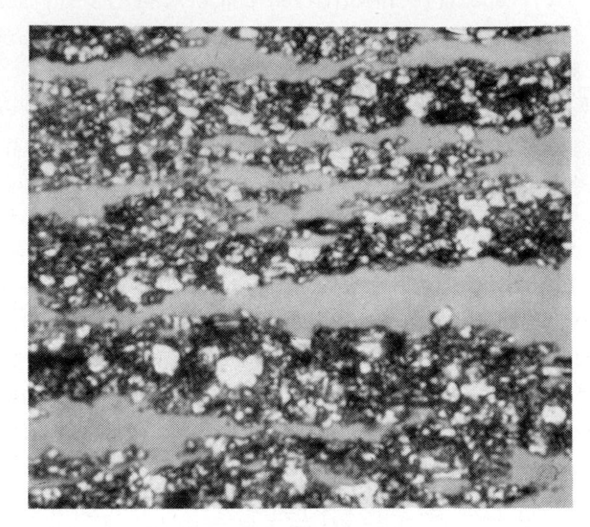

Fig.18. (x750) Rubbing wear particles and friction polymer.

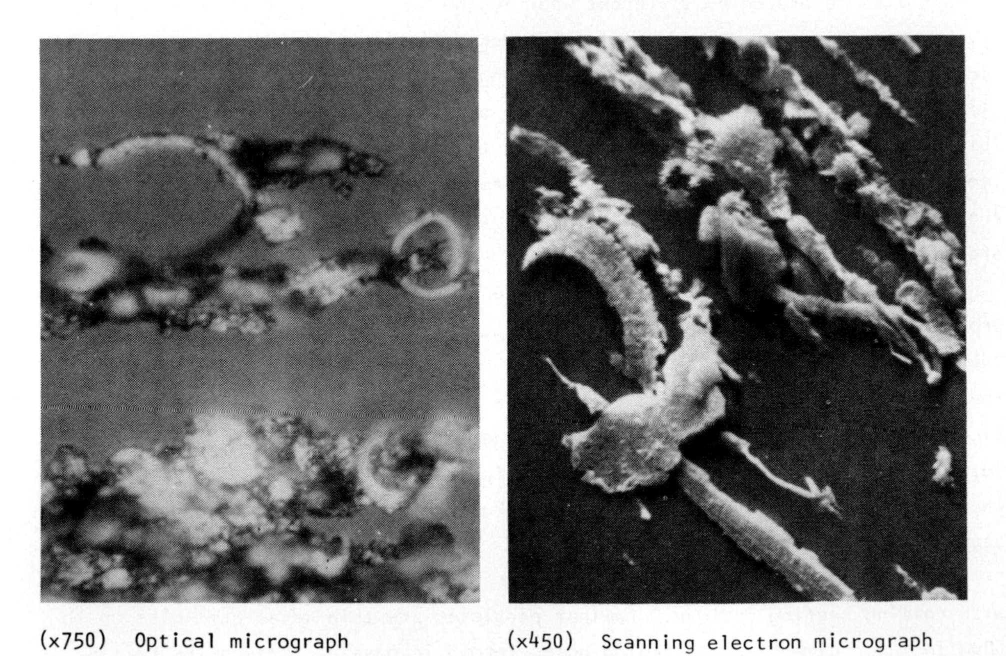

(x750) Optical micrograph (x450) Scanning electron micrograph

Fig.19. Cutting wear particles.

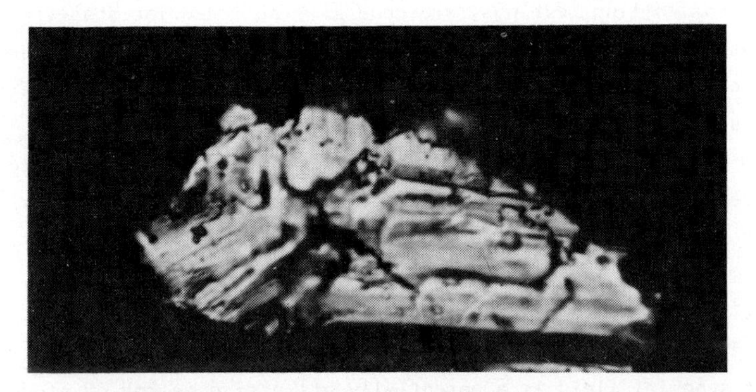

Fig.20. (x400) Large metallic wear particle.

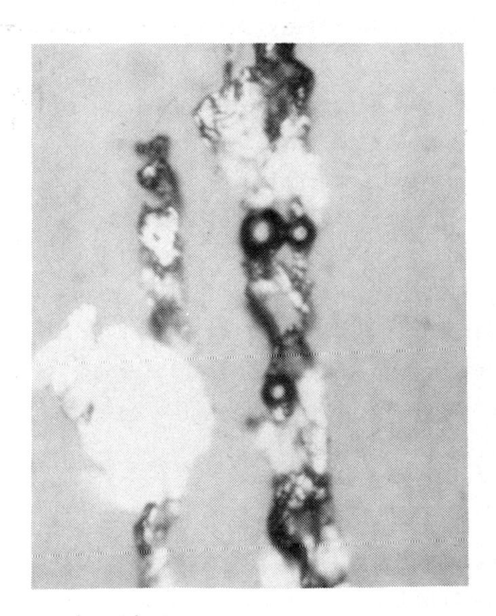

a. (x750) Optical micrograph

b. (x3,000) Scanning electron micrograph

Fig.21. Steel spherical particles.

Fig.21, are generated within a propagating fatigue crack and their detection gives warning of impending fatigue failure [69]. X-ray energy analysis in the scanning electron microscope can establish that they are composed of rolling bearing material. Fatigue chunks constitute the material removed by surface disintegration and pit formation.

Fatigue particles from a gear tooth although of similar dimensions differ from those from rolling bearings, are generally of irregular shape, free from holes and have a smooth surface. The number of particles increases as the fatigue process progresses. Larger fatigue chunks of gear material are indicative of surface deterioration by pitting.

Scuffing of gears causes an increase in the number of wear particles which tend to have a rough surface and an irregular shape. As the severity of scuffing increases, the larger particles produced have surface striations indicative of sliding action. Owing to the thermal effects of scuffing, particles may be partially oxidised with a range of temper colours.

As both arthritic and artificial joints are subjected to wear, Ferrographic analysis of synovial fluid is a potentially attractive method of studying the mechanisms and biological responses to wear in human joints [70].

2.9 CONCLUSIONS

In the present and foreseeable future world economic situation, material and energy conservation is becoming increasingly important. As wear is a major cause of material wastage, any reduction in wear can effect considerable savings in material and the energy necessary for their production. Thus increasing emphasis will be given to methods of wear control and prevention.

The complex mechanisms of wear, however, are not easily elucidated as the process by cumulative action obliterates evidence of the important initial stages of damage. Being an interdisciplinary subject, a multidisciplinary approach is required for the investigation of wear and the controlling factors to enable the most suitable design to be chosen embodying the best material and the correct lubrication to ensure minimum wear and satisfactory service performance from moving mechanisms.

REFERENCES

1 Blok,H., Engineering, London, 1952, 173(4502)594.
2 Bowden,F.P. and Tabor,D., "The Friction and Lubrication of Solids", 1950, Clarendon, Oxford.
3 Ming-Feng,I., J.Appl.Phys., 1952, 23(9)1011-1019.
4 Archard,J.F., J.Appl.Phys., 1953, 24(8)981-988.
5 Archard,J.F., Research, 1952, 5(8)395-396.
6 Rabinowicz,E., "Friction and Wear of Materials", 1966, J.Wiley, London.
7 Seifert,W.W. and Westcott,V.C., Wear, 1972, 21,27-42.
8 Scott,D., Seifert,W.W. and Westcott,V.C., Scient.Amer., 1974, 230(5)88-97.
9 Suh,N.P. and co-workers, Wear, 1977, 44,1-162.
10 Suh,N.P., Wear, 1973, 25,111-124.
11 Jahanmir,S., Suh,N.P. and Abrahamson,E.P., Wear, 1974, 28, 235-249.
12 Suh,N.P. and Sridharan,P., Wear, 1975, 34,291-299.
13 Reda,A.A., Bowen,E.R. and Westcott,V.C., Wear, 1975, 34,261-273.
14 Beilby,G., "Aggregation and Flow of Solids", Macmillan, London, 1921.
15 Bates,T.R., Ludema,K.C. and Brainard,W.A., Wear, 1974, 30,365-375.

16 Kirk,J.A. and Swanson,T.D., Wear, 1975, 35, 63-67.
17 Dumbleton,J.H. and Shen,C., Wear, 1976, 37, 279-289.
18 Hirst,W., Metall. Rev., 1965, 10, 145-172, I. Metals, London.
19 Kragelskii,I.V., "Friction and Wear", 1955, Butterworths, London.
20 Scott,D. (Ed.), "Treatise on Materials Science and Technology, 13, "Wear", 1979, Academic Press, N.Y.
21 Bickerman,J.J., Wear, 1976, 39, 1-14.
22 Barwell,F.T., in "Treatise on Materials Science and Technology, 13, "Wear", (Scott, D.Ed.) 1979, 1-83, Academic Press, N.Y.
23 Scott,D. and Scott,H.M., Proc. Conf. Lubrication and Wear, 1957, 609-612, Inst. Mech. Engrs., London.
24 Scott,D., Proc. Inst. Mech. Engrs., 1967, 181 (3L) 39-51.
25 Welsh,N.C., J.Inst. Metals, 1959, 88, 103-111.
26 Barwell,F.T. and Milne,A.A., Br. Pat. No. 732, 447, 1955.
27 Milne,A.A., Scott,D. and Macdonald,D., Proc. Conf. Lubrication and Wear, 1957, 735-741, Inst. Mech. Engrs., London.
28 Wright,K.H.R., Engineering, London, 1961, 191 (4956) 546-547.
29 Kruschov,M.M. and Babichev, , Akad. Nauk. SSR, 1960, 66-76. NEL Translation 893, National Engineering Laboratory, East Kilbride, Glasgow.
30 Kruschov,M.M., Wear, 1974, 28, 69-88.
31 Moore,M.A., Wear, 1974, 28, 59-68.
32 Barwell,F.T. and Scott,D., Proc. 4th Lubrication and Wear Convention, 1966, 277-297, Inst. Mech. Engrs., London.
33 Hother-Lushington,S., Proc. 4th Lubrication and Wear Convention, 1966, 243-252, Inst. Mech. Engrs., London.
34 Dawson,P.H. and Fidler,F., Proc. Inst. Mech. Engrs., 1965/66, 180, 513-530.
35 Engel,P.A., "Impact Wear of Materials", 1976, Elsevier, Amsterdam.
36 Wright,K.H.R., Proc. Inst. Mech. Engrs., 1952/53, IB(11) 556-574.
37 Godfrey,D. and Bailey,J.M., Lubric. Engng., 1954, 10, 155.
38 Waterhouse,R.B., "Fretting Corrosion", 1972, Pergamon, Oxford.
39 Waterhouse,R.B., in "Wear of Metals", 1977, 55, ASME, N.Y.
40 Scott,D., in "Fatigue in Rolling Contact", 1963, 103-115, Inst. Mech. Engrs., London.
41 Scott,D., in "Low Alloy Steels", 1965, 203-209, I.S.I., London.
42 Scott,D., Rolling Contact Fatigue in Wear, in (Scott,D. Ed.) "Treatise on Materials Science and Technology", 1979, 13, 321-361, Academic Press, N.Y.
43 Scott,D., Proc. Inst. Mech. Engrs. Conf. Lubrication and Wear, 1957, 463-468.
44 Grunberg,L. and Scott,D., J. Inst. Petrol., 1958, 44 (419), 406-410.
45 Grunberg,L., Scott,D. and Jamieson,D.T., Phil. Mag. 1963, 8(93) 1553-1568.
46 Scott,D., Loy,B. and Mills,G.H., Proc. Inst. Mech. Engrs., 1967, 181(31) 94-106.
47 Scott,D. and McCullagh,P.J. Wear, 1973, 24, 235-242.
48 Scott,D. and Blackwell,J., Proc. Inst. Mech. Engrs., 1964, 178, (3N) 81-89.
49 Scott,D. and Blackwell,J., Wear, 1978, 46, 273-279.
50 Scott,D. and Blackwell,J., Proc. Inst. Mech. Engrs., 1968, 182(3N) 239-242.
51 Scott,D. and McCullagh,P.J., Wear, 1975, 34, 222-237.
52 Scott,D. and McCullagh,P.J., Wear, 1973, 25, 339-344.
53 Scott,D., in "Rolling Contact Fatigue and Performance of Lubricants", (Tourret,R. and Wright,E.P. Eds.), 1977, 3-17, Heydon and Sons, London.
54 Scott,D. and Blackwell,J., Proc. Inst. Mech. Engrs., 1966, 180, (3K) 32-37.
55 Scott,D., Proc. Inst. Mech. Engrs., 1968, 182(3J) 116-123.
56 Scott,D., Proc. Inst. Mech. Engrs., 1967, 182(3A) 325-341.
57 Scott,D. and Blackwell,J., NEL Report 278, 1967, National Engineering Laboratory, East Kilbride.
58 Scott,D., Proc. Inst. Mech. Engrs., 1976, 189/75, 623-633.
59 Seifert,W.W. and Westcott,V.C., Wear, 1973, 23, 239-249.

60 Collacott,R.A., "Mechanical Fault Diagnosis and Condition Monitoring", 1977, Chapman and Hall, London.
61 Scott,D., Wear, 1975, 34, 15-22.
62 Bowen,E.R., Scott,D., Seifert,W.W. and Westcott,V.C., Tribology Int., 1976, 9(3) 109-115.
63 Scott,D. and Westcott,V.C., Proc. Eurotrib 77, 1977, Band 1, paper 70, 1-6.
64 Westcott,V.C., Naval Research Reviews, 1977 (March) 1-8, Office of Naval Research, Washington.
65 Scott,D. and Mills,G.H., in "Scanning Electron Microscopy", 1974, Part IV, 838-888, I.I.T., Chicago.
66 Barwell,F.T., Bowen,E.R. and Westcott,V.C., Wear, 1977, 44, 163-171.
67 Ruff,W. Wear, 1977, 42, 49-62.
68 Bowen,E.R. and Westcott,V.C., "Wear Particle Atlas", 1976, Foxboro/Transonics Inc., Mass, U.S.A.
69 Scott,D. and Mills,G.H., Wear, 1973, 24, 235-242.
70 Mears,D.C., Wear, 1978, 50, 115-126.

3 SELECTION OF BEARINGS

M.J. NEALE
Michael Neale and Associates Ltd.

3.1 INTRODUCTION

The selection of an appropriate type of bearing, for use in a particular application, is a decision that is usually made very early in a design process. At that stage, very detailed information on bearing performance is not usually necessary, and what is really required is broad guidance on the important characteristics of the various types. The information presented in this section ntended to meet this requirement.

3.2 BEARING TYPES

The basic function of bearings is to allow a load to be transmitted between two surfaces which are in relative motion. There are three main types of bearing as shown in Figure 1, and these are plain bearings, rolling bearings and flexures. In plain bearings the load is transmitted over a considerable area, while in rolling bearings the area actually in contact, and transmitting the load, is very small. The third type depends on the use of flexible components and is only suitable for oscillatory movement.

It can be seen from Figure 1 that there are five basic principles behind the operation of the various types and these are:-

(i) To permit the two surfaces to rub together and to arrange the surface properties so that seizure or excess friction does not occur and so that an acceptable rate of wear is obtained. In practice this is usually achieved by the choice of materials with suitable bulk properties or by the use of some form of surface coating, which may either be applied in advance or allowed to form in situ.

(ii) To keep the surfaces separated by a film of fluid, so that the relative movement can occur within the film. To do this the fluid must be maintained at a sufficient pressure to hold the surfaces apart against

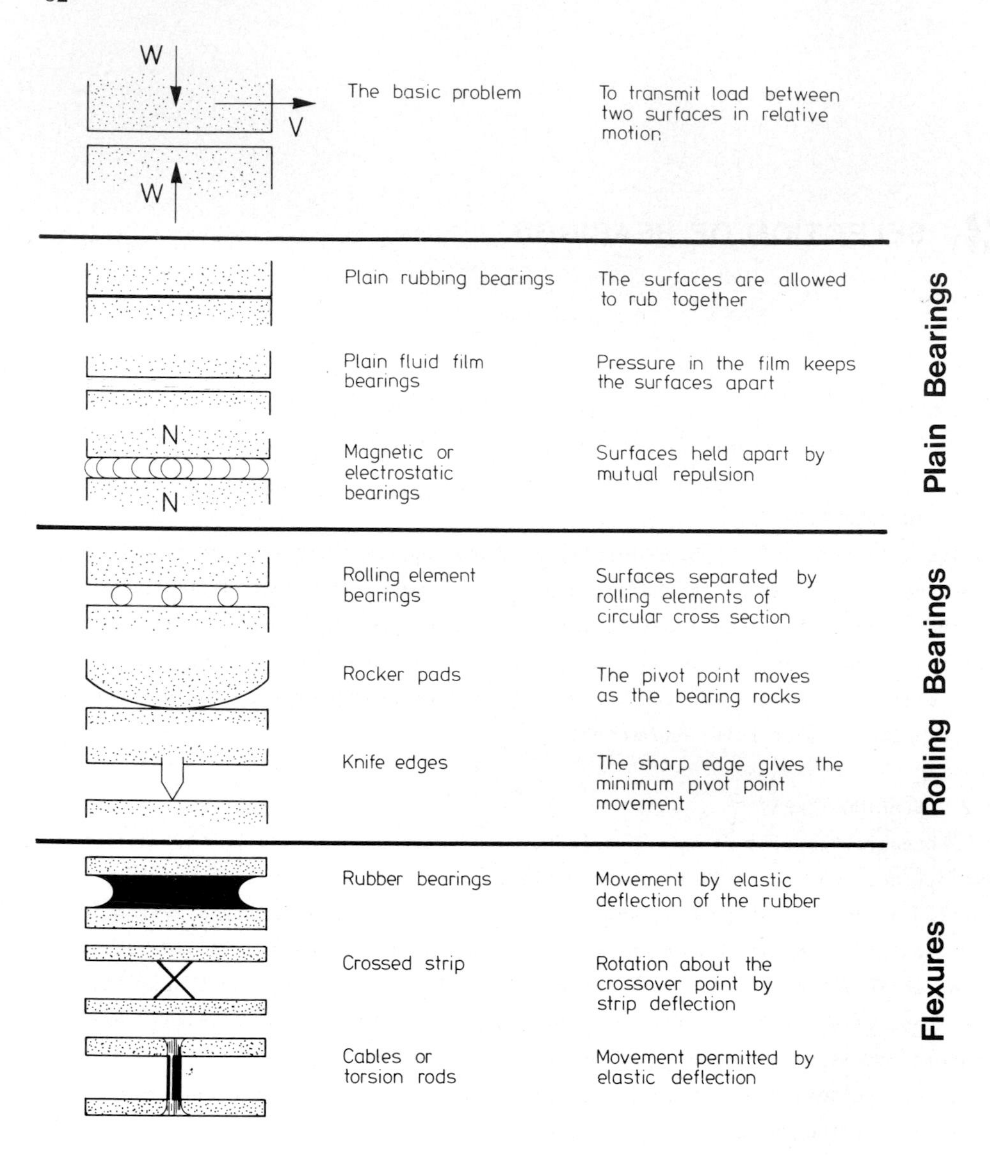

Fig.1 Various types of bearing

the applied load. This pressure may be obtained by feeding in from an external high pressure source, or may be generated within the film by relative movement of suitably shaped surfaces flooded with a viscous fluid.

(iii) One surface may be allowed to roll on the other. This, however, produces an interaction between the angular and translational movements of the adjacent components, which is determined by the shape of the rolling surfaces, and in practice may not always allow the required degrees of freedom. To overcome this problem the rolling surfaces are often permitted to slide as well as roll, such as in gear teeth, or the problem may be overcome by inserting a third element between the surfaces of the two original components as in a rolling element bearing.

(iv) To produce a repulsive force between the surfaces by magnetic or electrostatic means.

(v) To position a flexible member between the two components which can deflect to allow a relative oscillating movement to occur between them. Suitable members can be formed by filling most of the space between the surfaces with an adequate thickness of elastomeric material or by using thin connecting ligaments of higher strength materials.

3.3 PERFORMANCE OF VARIOUS TYPES OF BEARING

Since the ability to transmit a load with relative movement is the basic function of a bearing, a study of the relationship between the allowable load and speed for bearings of various sizes and types should provide a convenient standard for the comparison of their performance.

For the determination of load and speed characteristics of the various bearing types, the first information that requires to be established is the approximate shape of the relationship, so that a general comparison can be made between the various types. The basic forms of these relationships can be derived from the physical principles which govern the operation of the various bearing types. This is discussed below and the resulting performance curves applied to journal bearings compared in Fig.2.

3.3.1 Rubbing Bearings

In a bearing which operates by permitting the two surfaces to rub together, the physical limitations on performance are the risk of overheating and seizure and the possibility of excessive wear.

The generation of heat at the rubbing surfaces arises from the movement against the frictional resistance of the contact, and this heat has to be conducted away along heat-flow paths, of which the area will be approximately proportional to the projected bearing area A.

The risk of overheating at a sliding speed V is therefore approximately proportional to

$$\frac{\mu WV}{A} \text{ or } \mu PV$$

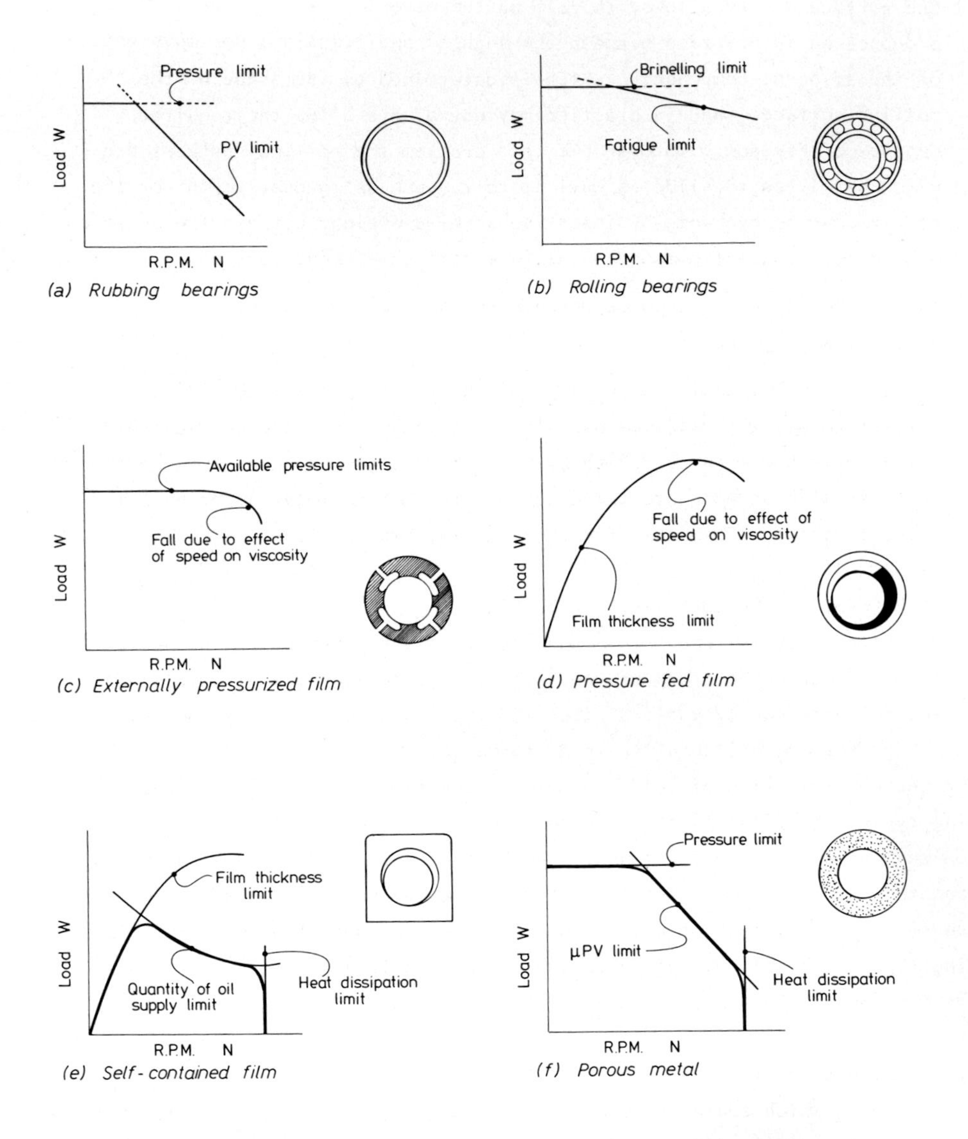

FIGURE 2 Performance of various types of journal bearing (log W plotted against log N)

When two surfaces are in rubbing contact, the volume of material worn from the rubbing surfaces after sliding a distance x with a load W is approximately proportional to Wx.

The wear volume is, however, of little significance in the performance of a bearing, and the depth wear rate is much more relevant as a design factor, since this is a measure of the rate at which slackness in the assembly is likely to be produced.

The depth wear rate will be approximately proportional to

$$\frac{Wx}{At} \text{ or } PV$$

Thus both the physical limitations to the performance of rubbing bearings indicate that the severity of the bearing operation is related to PV (bearing pressure x sliding speed). Fig.2a shows the shape of the load-speed curve for a rubbing type of journal bearing. This is a composite characteristic including a PV limit and also a maximum allowable pressure limit associated with the risk of fatigue or extrusion of the bearing material, which is frequently non-metallic.

3.3.2 Roller Bearings

Owing to the concentrated contact stresses in this type of bearing the ultimate limit on performance arises from the possible fatigue of the rolling elements or races, although in practice rolling bearings often fail as well from lack of proper lubrication, dirt contamination and unsuitable loading. If fatigue is taken as the performance limit it is reasonable to assume that the life of a rolling contact bearing measured in revolutions is proportional to $(1/W)^3$. Consequently for a given life $W \propto N^{-1/3}$ and this characteristic is shown in Fig.2b. As before, the actual maximum load is also limited by the capacity of the bearing to carry static load which in this case is limited by the resistance of the races to brinelling by the rolling elements.

3.3.3 Fluid Film Bearings

If these are of the externally pressurized or hydrostatic type, the load capacity is primarily dependent on the available supply pressure, and running speed has very little effect, although with liquid lubrication a slight fall in load capacity may occur at higher speeds owing to the reduction in lubricant viscosity caused by higher temperatures. The resulting load-speed characteristic is therefore as shown in Fig.2c.

With the hydrodynamic type of fluid film bearing, the load capacity increases with speed provided that the film is kept adequately supplied with lubricant, although there is normally a tendency with liquid lubricants for the load capacity to fall away at the highest speeds owing to the heating of the lubricant.

This gives a characteristic of the general shape shown in Fig.2d.

In order to keep the film fully supplied with lubricant, this has to be fed under pressure, which in turn requires some form of lubricant supply system. It is not always convenient or economical to have to include such a supply system and consequently many hydrodynamic fluid film bearings are used with self-contained lubricant supplies. In the larger sizes, the oil is generally contained in a sump below the shaft and lifted by a ring or disc, while in the smaller sizes it may be fed from an oil-soaked pad or retained in the bearing structure by making it porous. These methods of oil feeding have a lower efficiency than a pressure fed arrangement and result in the bearing operating with a load-carrying film of reduced circumferential length and therefore of reduced load capacity. It is also usually found with these systems that the volume of oil delivered per revolution of the shaft decreases with speed and consequently the load-carrying capacity also decreases with speed.

In addition to this effect, the absence of a lubrication system means that all the heat generated has to be dissipated directly to the surroundings, and with a maximum allowable temperature for a long life of mineral oil lubricants, places a limit on the maximum allowable speed.

These various effects result in load-speed characteristics for bearings of this type which are of the general shape shown in Figs.2e and 2f.

3.3.4 Flexible Members

Bearings of this type, which use either elastomers in shear or high tensile ligaments in bending to allow an oscillating motion to occur, are in many cases physically limited in performance by the fatigue strength of the material or its bonding on to the adjacent components. The stresses arise both from the deflection and from the applied load and are usually additive at some critical region of the assembly, with the result that the allowable loads tend to decrease with the permitted angle of movement. Blocks of elastomeric material used in compression do not have this additive stress condition, but in this case the allowable deflection is limited by the block thickness, which in turn limits the maximum load if the assembly is to be stable against excessive bulging or buckling. This can be overcome to some extent by the use of stiffening plates in the elastomer arranged to be parallel to the required direction of motion, but even when these are incorporated the maximum deflection is still limited by the amount that the load can be permitted to be offset.

It can therefore be generally assumed that the load-deflection characteristic of flexible bearings will generally show a decrease in allowable load as the deflection is increased.

3.4 SELECTION OF A SUITABLE BEARING

The characteristic relationships between load and movement for the various types of bearings can be compared with the bearing performance required in various applications and used as a guide to the selection of a suitable type of bearing.

Any possible application for a bearing in a machine or structure will have some form of characteristic relationship between the type of load to be carried and the movement to be allowed. In practice an important feature is whether the load and movement are nominally steady or whether they vary in some cyclic manner. This can give rise to four possible combinations of load and movement as indicated with examples in Table 3.1.

Probably the greatest number of bearing applications are of the unidirectional load and continuous movement type and in this category the way in which the bearing load varies with rotational speed, during the operation of the machine, is an important factor in bearing selection and design. For this reason the unidirectional loads have been sub-divided in Table 3.1 to draw attention to this situation.

Table 3.1 Examples of various types of load and movement patterns

Type of load	Type of movement	Examples
Unidirectional		
Constant	Continuous	Turbine journal bearings
Rising with speed	Continuous	Marine gearbox pinion bearings
Falling with speed	Continuous	Hydraulic motor bearings
Unidirectional	Oscillatory	Bridge support bearings Grinding machine tables
Multidirectional	Continuous	Piston-engine crankshaft bearings
Multidirectional	Oscillatory	Linkage bearings

3.4.1 Applications with Unidirectional Load and Continuous Movement

This is the first category of bearing applications listed in Table 3.1 and the various types of bearing which could be used are the rubbing, rolling element and fluid-film types. Fig.2 shows that these various types of bearing have quite different load-speed characteristics.

In fact, examples are given in Table 3.1 of various applications which also have several forms of relationship between load and speed, and ideally a bearing should be selected with a matching characteristic. This kind of approach to the problem indicates that turbine journal bearings should ideally be of the externally pressurized or hydrostatic type, while an application such as marine gearbox pinion bearing is particularly well suited to the simple type of hydrodynamic bearing. While this latter example is in line with current practice, turbines are not at present using the hydrostatic type of bearing except in those used to drive high-speed dental drills. This may be an example of a design situation where a particular type of machine is developed on the basis of a workable but not ideal form of bearing design and it then requires considerable commercial courage to make the necessary change on a vital component in a machine of such high capital value. From the economic point of view, however, hydrostatic bearings are particularly applicable in situations where a source of high pressure fluid is already available, and this is in fact the case with a steam turbine. Problems of erosion may occur if the steam is allowed to condense in the bearing clearance but this problem should not be insuperable.

Hydraulic motors were suggested as an example of an application in which the load decreased with increasing speed, although this characteristic tends to be common to any machine driven by a source of approximately constant power. For such an application, a rolling contact bearing would appear to be ideal in that it has a matching characteristic, as well as a low starting friction which is usually also required in this particular type of application. A rubbing type of bearing would also appear to be possible, but in fact the loads and speeds allowable with this type of bearing are considerably below the corresponding values for the rolling-contact type.

Although this technique of matching the form of the load-speed relationship of the application with that of various bearings is a useful guide in bearing selection, it is still necessary to consider the actual values of load and speed which can be carried by different sizes of bearings of the various types, in order to be certain that they are suitable.

A convenient way of doing this is to plot the performance of the various types of bearing on one diagram so that comparisons can be made, and if this is done on logarithmic axes, the whole span of engineering loads and speeds can be covered.

This technique can be used to show the performance of steadily loaded journal bearings with continuously rotating shafts, and Fig.3 shows the type of diagram which results.

Fig.3 is only intended to give broad guidance but it does show the general trends quite clearly. It indicates, for example, that fluid-film plain bearings are the best type to use at high speeds and that this is particularly true in the case of all shafts greater than about 100 mm diameter. This figure also indicates that the rubbing type of bearing is only really suitable for low

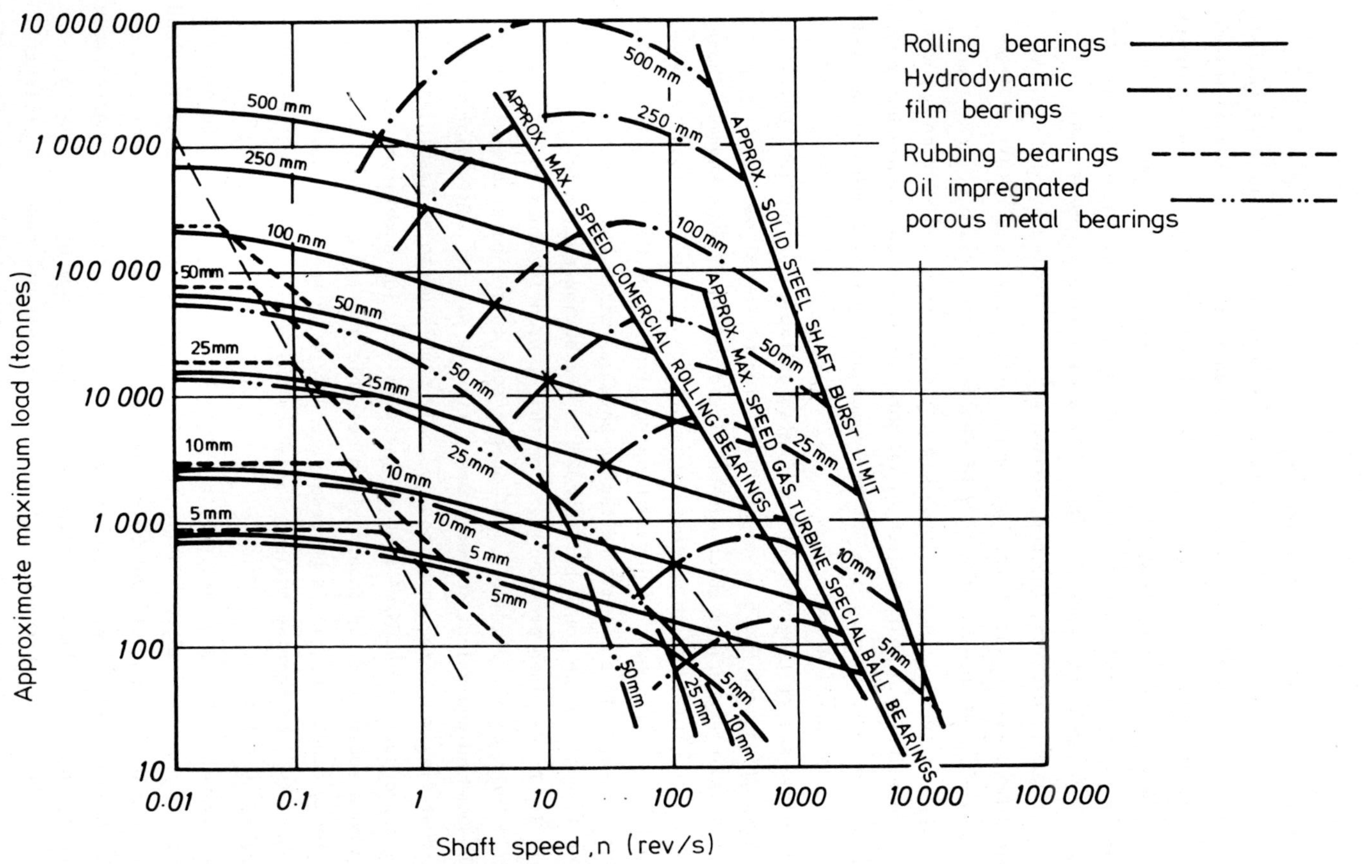

FIGURE 3 Indication of the performance of common types of journal bearings for shafts of various diameters

rotational speeds, and at anything more than one or two hundred revolutions per minute a rolling bearing is usually a better solution.

3.4.2 Applications with Oscillating Movement

The rubbing type of bearing is, however, particularly good for applications with oscillating movement, since it does not need to build up an oil film each time it starts to move, and with small angles of oscillation it cannot brinell in the same way as a rolling bearing can do under these circumstances.

It is therefore useful to consider the probable performance of the various types of rubbing bearing under conditions of oscillating movement. As in the case of continuous rotation, the performance will be limited by the maximum allowable load pressure and by the wear which will occur as a result of the rubbing movement. A convenient method of showing a comparison between the various types is to plot their allowable bearing pressure and sliding speed and this been done in Fig.4. The values given here are only approximate but should help to give an indication of the various types of material which can be used in any particular application.

The flexible member type of bearing is also suitable for application with oscillating motion, and Fig.5 gives an indication of the allowable maximum loading on flexible bearings which can be fitted within a given space. These bearings are commonly used in various types of linkage and the cross-sectional area available at the end of a lever which requires to be connected to another component is generally the limiting factor on size rather than the diameter of any connecting pin which could be regarded as the equivalent of the shaft in the applications examined previously. For this reason the load in Fig.5 is plotted in terms of the allowable bearing pressure on the outside diameter of rubber bushes and equivalent shaped crossed flexure pivots. This diagram indicates that flexible member bearings are only usable up to an absolute maximum equivalent pressure of 14 MN/m^2 2000 lb/in^2 with very low angular movements, and with much lower loads up to a maximum possible angular movement of 30 degrees. Outside these conditions, it is necessary to revert to the more conventional pin-and-bush type of bearing design, for which the material can then be selected as indicated by Fig.4.

It may appear from Fig.5 that rubber bushes are always likely to be superior to flexural ligament bearings. While this is true in terms of compressive load and allowable deflection, it must be remembered that flexural ligaments can be designed to have a much lower stiffness for a given load capacity compared to a rubber bush, and they will also accept a much wider range of environmental conditions. The data presented for flexural ligaments are based on crossed flexure pivots and it is important to remember that the design can be simplified to a much simpler single ligament, usually in tension, if the rotational centre

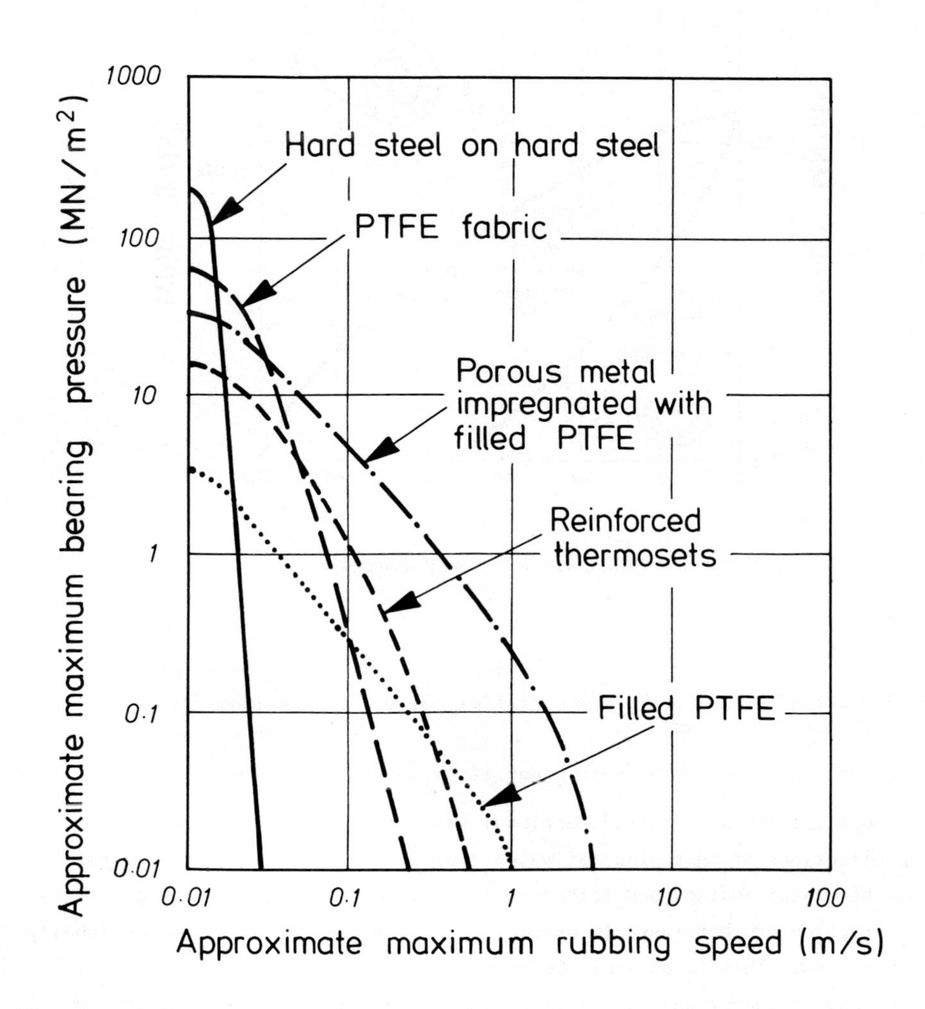

Fig.4 Indication of the performance of dry rubbing bearing with oscillating motion

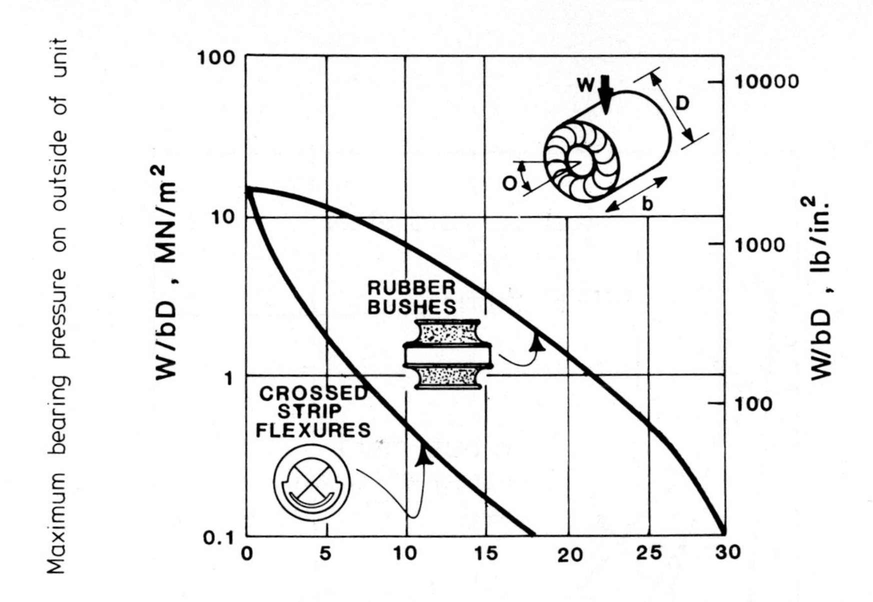

Fig.5 Indication of the performance of flexible member bearings with oscillating motion

of motion does not have to be kept under close control.

3.4.3 Applications with Multidirectional Load and Continuous Movement

In applications of this kind, of which typical examples are the crankshaft bearings of piston engines and reciprocating compressors, the bearing pressures used in practice are considerably greater than those applied to unidirectionally loaded bearings. This is possible because

(i) the load frequency changes in direction as well as in magnitude, and the shaft does not have time to squeeze the oil from the film to a sufficient extent to make metal-to-metal contact before the load reverses and lifts the shaft away again.

(ii) since various positions of the bearing and shaft surface carry the minimum film region in turn, the local thermal conditions are not as severe as in a steadily loaded bearing where one region of the bearing metal is continuously subjected to local heat input. As a result lower values of film thickness may be permitted for a given revolution speed and shaft size, provided that shaft surface finish is smooth enough.

The methods of calculating the performance of dynamically loaded bearings are essentially rather complex and the most comprehensive method of assessing their probable performance is to calculate the probable locus of the shaft with a computer.

From the point of view of very general design guidance, however, it can probably be appreciated that since much greater bearing pressures are allowable, the strength of the bearing metals becomes an important factor in bearing selection, together with the method of feeding oil to the bearings in order to ensure that the bearing clearances are kept as full of oil as possible in readiness for carrying the dynamic loads.

It is difficult to quote precise data for the allowable loads on various bearing materials since factors such as shaft deflection, efficiency of lubrication and the type of loading pattern have a considerable effect. Table 3.2 however, gives some broad guidance on the type of materials that are likely to be suitable.

Table 3.2 Possible materials for plain bearings in reciprocating machinery

Maximum bearing pressure		Possible bearing material
lb/in^2	MN/m^2	
Below 2500	17	Whitemetal
2500- 4000	17 - 28	Copper leads
2500- 5000	17 - 34	Aluminium tins
4000- 6000	28 - 41	Low tin lead bronzes
6000-10000	41 - 70	High tin lead bronzes
10000-15000	70 - 100	Phosphor bronzes

The stronger materials are harder than the weaker ones and it is therefore desirable to pick a material which is only just adequately strong enough in order to have the maximum possible conformability with the shaft and embeddability for dirt. With harder bearing materials it is also necessary to use a hard shaft, partly to maintain an adequate hardness differential with the bearing material and partly because the higher bearing pressures which have called for the harder material are also likely to result in lower film thicknesses and therefore greater dirt sensitivity.

As a very general guide, it is not usually necessary to harden the shaft at loads below 24 MN/m^2 (3500 lb/in^2) but generally desirable to do so at loads above 31 MN/m^2 (4500 lb/in^2).

Rolling element bearings can also be used for piston engine crankshaft bearings, although in most cases they offer no better performance than a fluid-film bearing, at the expense in larger engines of a rather more complicated design. Rolling bearings do have the particular merit of requiring much less oil supply than fluid-film bearings and this makes them a first choice for two-stroke engines of the petroil type, and for other small cheap engines where the opportunity they offer of eliminating the oil supply system results in a more commercial design.

Figures were reprinted by permission of the Council of the Institution of Mechanical Engineers from the Proceedings of 1967 International Conference on Lubrication and Wear.

4 DESIGN OF PLAIN BEARINGS
Use of Bearing Data Design Charts

F.A. MARTIN and D.R. GARNER
Research & Development Organisation
The Glacier Metal Company Limited.

4.1 INTRODUCTION

The design of hydrodynamic bearings can be a daunting task for the average machinery designer. With rolling element bearings the design process is very much biased towards bearing selection data, where manufacturers' catalogues give tabulated information on size, load, speed and life. For plain bearings it is not so easy to find such 'potted' information, and in the past the designer had either to become involved in the mathematical complexities of the hydrodynamics, or rely on the vagaries of rule-of-thumb guides. With the continual uprating of machinery, this latter approach is seldom adequate, but it can be difficult to obtain more detailed information in a readily digestible form.

Various techniques for analysing and checking the performance of plain journal bearings are presented here, so that the designer has the means to appreciate the effects of the various parameters on bearing performance and design, without the need for extensive mathematical manipulation.

4.1.1 Notation

b = axial bearing length (m)

c_d = bearing diametral clearance (m)

c_r = bearing radial clearance (m)

d = bearing diameter (m)

e = distance between bearing and shaft centres (m)

h_{min} = minimum oil film thickness (m)

H = power loss (W)

M = effective rotor mass at bearing (kg)

N = shaft rotational speed (rev/s)

N_L = rotational speed of load vector (rev/s)
P = bearing specific load W/bd (Pa or N/m^2)
Q = oil flow requirement (m^3/s)
W = applied load (N)
δ = journal misalignment over bearing length (m)
η = effective film viscosity (Ns/m^2)
θ_e = effective temperature (°C)
θ_{max} = maximum temperature (°C)
ρ = oil density (kg/m^3)
ω = angular velocity (rad/s)

Dimensionless Terms (in any consistent set of units eg. as given above)

load variable $W' = \frac{W}{bdN\eta}\left(\frac{c_d}{d}\right)^2$

turbulence variable $Y = \frac{d^2N^2\rho}{P}\left(\frac{d}{c_d}\right)$

eccentricity ratio $\varepsilon = e/c_r = 1 - h_{min}/c_r$

dimensionless critical mass $c_r\ \omega^2\ M/W$

4.2 JOURNAL BEARING DESIGN LIMITS

The total design process for most mechanical components, including plain bearings, involves many stages, with functional, economic and perhaps aesthetic aspects all needing to be considered. It is the first of these which normally involves the designer in most effort and which is to be considered in detail here. For convenience we will split this functional design stage into two:-

(i) ensuring that the bearing is capable of operating satisfactorily under the imposed conditions, and that it is neither too close to its limits of operation to endanger reliability nor so far away that it has penalties in over-design,

(ii) predicting the performance of the component as it affects the design of its associated parts or the overall system.

Initially, therefore, the limits of operation of bearings must be defined.

4.2.1 Limits of Operation

Consider a bearing of some given size and geometry, with a defined lubricant grade and feeding conditions. The limiting conditions of load and speed which this bearing can successfully accept are shown in Fig.1; these limits will now be considered in more detail.

4.2.1.1 Thin Oil Film Limit

The danger here is of metal-to-metal contact of the surfaces, with consequent severe wear (and perhaps overheating) leading to a breakdown in bearing operation. There is evidence to show that this contact occurs at a predicted film thickness which is a function of the surface roughness [1], which in turn is dependent upon machining process and bearing size. This has been considered in some detail in reference [2], and the conclusion reached that a realistic 'failure' value of the film thickness is given by the peak-to-valley surface finish (Rmax) of the journal, assuming that the surface finish of the bearing is of the same order.

An approximate correlation between Ra and Rmax values is also given in reference [2], enabling the information given in Fig.2 to be presented. The surface finishes shown here are representative of those which can be obtained by normal manufacturing methods. Since the Rmax values have been thought of as failure values, some additional factor must be applied if we are to specify safe values of oil film thickness. This factor has to allow for slight unintentional misalignment which may take place between shaft and bearing, and for dirt contamination in the oil supply. The allowable film thickness values shown in Fig.2 are a factor of three above the failure values, and have been found by experience to be acceptable. For very high standards of build and operation a factor of two can be satisfactory, but it is considered that the values shown

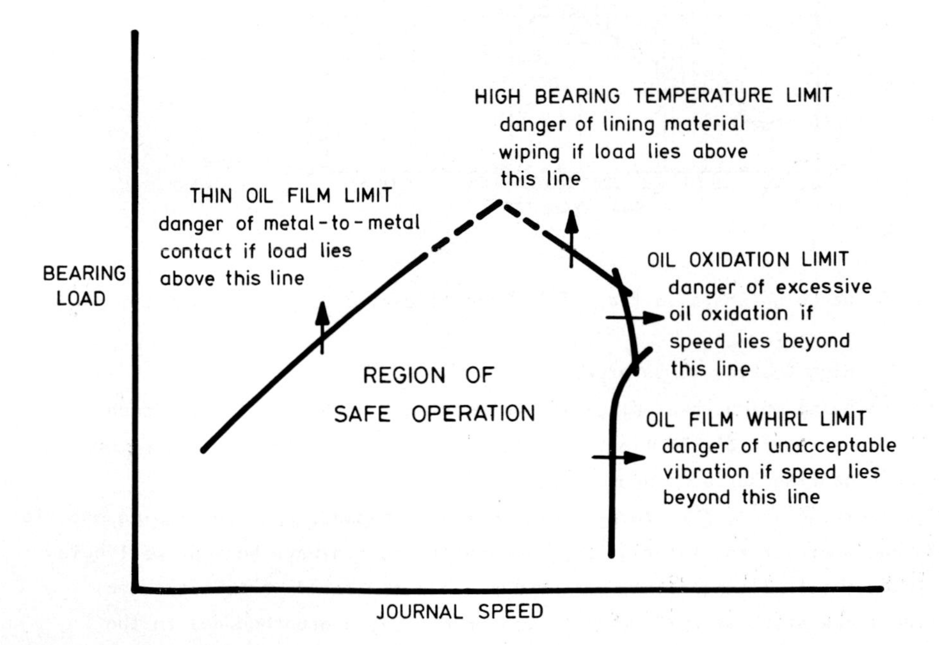

Fig.1 Limits of safe operation for hydrodynamic journal bearings

in Fig.2 should be used as a general guide to avoid working too close to failure limits. It must be emphasised that the relationship between Ra and Rmax is only approximate, as evidenced by the "spread of results" band on the figure, but it is adequate to show the order of reduction in the allowable film thickness that can be permitted by improving surface finish. To increase the operating oil film thickness, for given conditions of load and speed, the bearing size must be increased, or a thicker lubricant and/or a reduced inlet temperature used (change in oil supply pressure will usually have little effect). An increase in clearance may either increase or reduce the film thickness, depending upon the precise operating conditions.

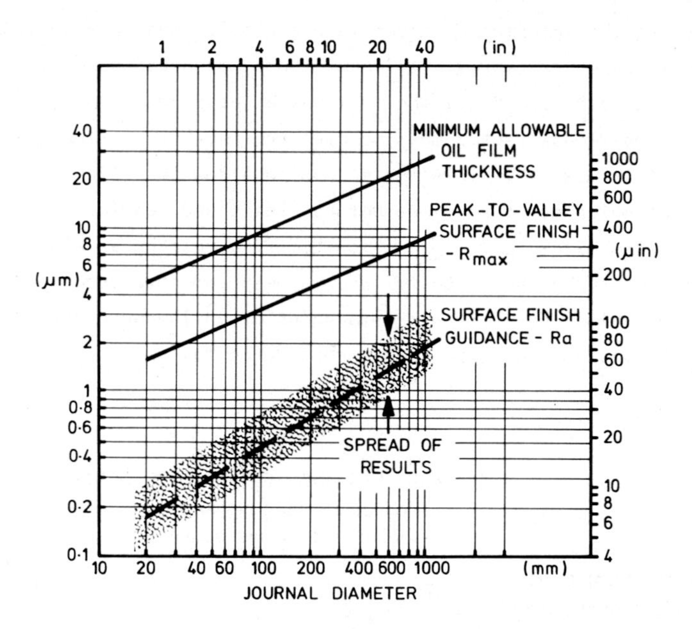

Fig.2 Guidance on shaft surface finish and allowable oil film thickness

4.2.1.2 High Bearing Temperature

The shearing which takes place in the bearing clearance space is 'seen' externally as power dissipation, and also as an increase in the temperature of shaft and bearing surfaces and of the lubricant.

The limiting acceptable temperature obviously depends upon the lining material used, but whatever the material the temperature must always be kept well below its theoretical melting point. For example, in tin based white metal the melting temperature of 232°C cannot even be closely approached due to the softening and subsequent plastic flow of the material which occurs at temperatures well below 200°C under the influence of hydrodynamic pressure. Booser et al. [3] observed limiting whitemetal temperatures in journal bearings as low as 130°C; however, since only calculated temperatures are available at the design stage, and current methods of estimation are known to be far from accurate, it is safer to lower the limit to about 120°C. In machines in which some dynamic loading can occur on top of the normal steady load, and where bearing fatigue is therefore a possible problem, it is customary to try and keep the temperature below 100°C. If higher temperatures have to be accommodated then the use of an aluminium tin or a copper lead material must be considered. The former of these, when containing about 40% tin, can be used at temperatures up to 150-160°C, and is almost equal to whitemetal in its ability to withstand seizure conditions and dirt contamination. Copper lead is much less forgiving in this respect, and ideally should have a soft, thin overlay plate to aid in bedding-in, but it can be used at temperatures of 200°C plus. It requires a hardened shaft (about 300 HV) and very good lubricant filtration.

It is worth noting that if a high temperature condition is present in a bearing there is usually nothing that can be done to the oil feeding conditions to improve the situation, apart from changing the oil grade. The oft-used 'palliatives' of increasing oil feed pressure or grooving area in order to force more oil through the bearing are usually not successful since they only reduce the bulk temperature of the oil passing through the bearing (see section 4.2.1.3). The bearing material temperature is controlled by the amount of lubricant passing through the active part of the oil film, and this is very insensitive to feeding conditions, provided that the bearing is not grossly starved. As a rough approximation, the maximum bearing temperature will be dropped by about a half of any decrease in oil inlet temperature, ie a 10°C reduction in inlet temperature may be expected to decrease the maximum bearing temperature by about 5°C. The bearing temperature may also generally be decreased by increasing the bearing size or clearance, or by using a thinner grade of oil.

One further point on bearing temperatures which is sometimes ignored is the necessity of considering the influence of machine temperatures on the bearings. If there can be appreciable heat soak along the shaft, or (less often) through the bearing housing, then the design must allow for adequate oil flow to deal with it. Additionally, in some machinery, the worst temperature conditions at the bearing can occur after shut down when heat soak raises temperatures well above peak running values, and the choice of material should then be dictated by conditions which are often not drawn to the attention of the bearing designer.

4.2.1.3 Oil Oxidation Limit

Straight mineral oils in a normal (oxygen containing) atmosphere can be rapidly oxidised at the order of temperatures that we have been discussing above. There is no precise 'go/no-go' limit for this process, rather the rate of degradation is a function of temperature [4]. Industrial mineral oils usually contain anti-oxidants which retard this process, but for commonly used turbine oils, for lives in the order of thousands of hours, it is necessary to restrict bulk temperatures of oil in tanks, reservoirs etc. to about 75-80°C. Thus a bulk drain temperature from a bearing at a higher value than this is unlikely to be acceptable.

As mentioned above, the drain temperature can usually be reduced by increasing oil supply pressure or providing bleed grooves in the bore, and thereby helping to avoid the oxidation limit. It is unfortunate that the bulk outlet temperature is often used to judge bearing performance, presumably because it is the easiest temperature to measure, since this juggling with supply conditions can radically alter drain temperatures without appreciably affecting material temperatures. The outlet temperature can be used as a long-term monitoring device (ie a temperature which has been steady at 70-75°C for months should not suddenly rise to 80-85°C) but it is not suitable, and may indeed be misleading, if used for 'setting up' oil feed conditions on a new machine.

4.2.1.4 Oil Film Instability

Under certain conditions, normally at low load and/or high shaft speeds, a self excited and self sustaining motion can occur in which the shaft centre precesses around the bearing centre at something slightly less than half shaft speed, typically 0.42-0.47 of shaft speed. Under these conditions the hydrodynamic action of the bearing is all but lost, and metal-to-metal contact can occur; in practice, if left for long periods, a fatigue type damage is produced due to the high temperatures generated. There can also be a large, and perhaps unacceptable, vibration transmitted through the machine. Guidance on the likelihood of instability in cylindrical bore bearings is given in Fig.3. This chart is strictly only applicable to rigid, simply supported shafts [5], but experience has shown it to give a fair guide for other systems, for example overhung rotors. It can be seen that if the operating eccentricity ratio is greater than about 0.8 then the bearing is stable. This is the reason why derating a bearing - adding extra grooving in the loaded region, using a thinner grade of oil etc. - can sometimes cure half speed whirl problems. Additionally, whatever the operating eccentricity ratio there is some value of the dimensionless critical mass below which the bearing will always be stable. In heavily loaded applications, such as gearboxes, the applied load is normally of a much higher magnitude than the shaft mass and instability, at any realistic speed, is not a problem. However there may well be some part load condition, for example a spin test, at which stability needs to be carefully checked.

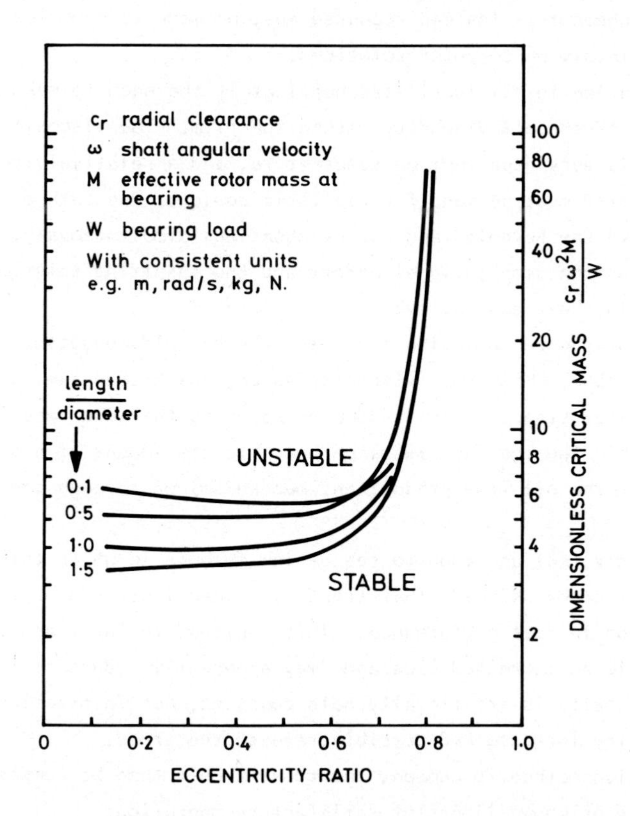

Fig.3 Oil film whirl instability of cylindrical bore journal bearings - Limiting dimensionless mass

4.2.2 Region of Safe Operation

The limits just defined serve to bound a region of safe operation. The simplest approach to design is merely to calculate such operating parameters as film thickness and temperatures and to compare them against the defined limits. If both the limits and the design point are plotted, in a form similar to Fig.1, then judgements on the degree of safety can easily be made.

There are various methods of tackling this, and the most common will now be described; the final design charts presented enable very rapid and accurate estimates of bearing operation to be made with minimal effort.

4.3 CALCULATION AND DESIGN PROCEDURES

The analysis of oil film conditions within a bearing, described by the Reynolds equation, is a very involved mathematical exercise. The equation is deceptively simple, but an analytical solution cannot be obtained for any but the simplest of arrangements. Instead recourse must be made to numerical techniques, usually involving computer solutions.

A further complication in oil lubricated bearings is the need to make a realistic estimation of the oil viscosity within the film. The viscosity of normal mineral oils is very dependent on temperature, and a relatively accurate assessment of the latter must be made for realistic design. One method of doing this is to solve the Reynolds and energy equations simultaneously. This considerably increases the computational effort and the resulting solutions are somewhat restricted in their generality.

Instead it is usually more convenient to solve the Reynolds equation assuming a single, global, effective viscosity value, which then can, and must, be determined at a later stage. This is done by equating the work done in shearing the oil within the bearing clearance space to the amount of heat carried away, both by the oil (the predominant mechanism) or through the bearing and housing.

It is, unfortunately, not uncommon to see design methods in which this heat balance is omitted, thereby assuming that trends in dimensionless groups accurately portray trends in real performance. This can lead to incorrect conclusions; for example an increased clearance may apparently reduce film thickness if the viscosity is artificially held constant, but in practice the operating viscosity may increase and possibly reverse the trend.

Whatever calculation method is chosen, the results must then be compared against the estimated or known limits of satisfactory operation.

Fig.4 illustrates the various stages in producing design data:-

Stage I is the solution of the Reynolds equation to give dimensionless data.
Stage II takes this data and determines a realistic oil viscosity to give specific answers.
Stage III uses the results from the previous two stages to produce 'easy-to-use' design charts.

4.3.1 Dimensionless Data - Stage I

Dimensionless design data have been published for the most commonly used bearing configurations [6,7,8,9] and Fig.5 shows a typical chart relating a load variable to a film thickness ratio for a cylindrical bore journal bearing with a steady (or pure rotating) load. This may be used for bearings with two axial oil feed grooves (loaded midway between them), with a single feed groove

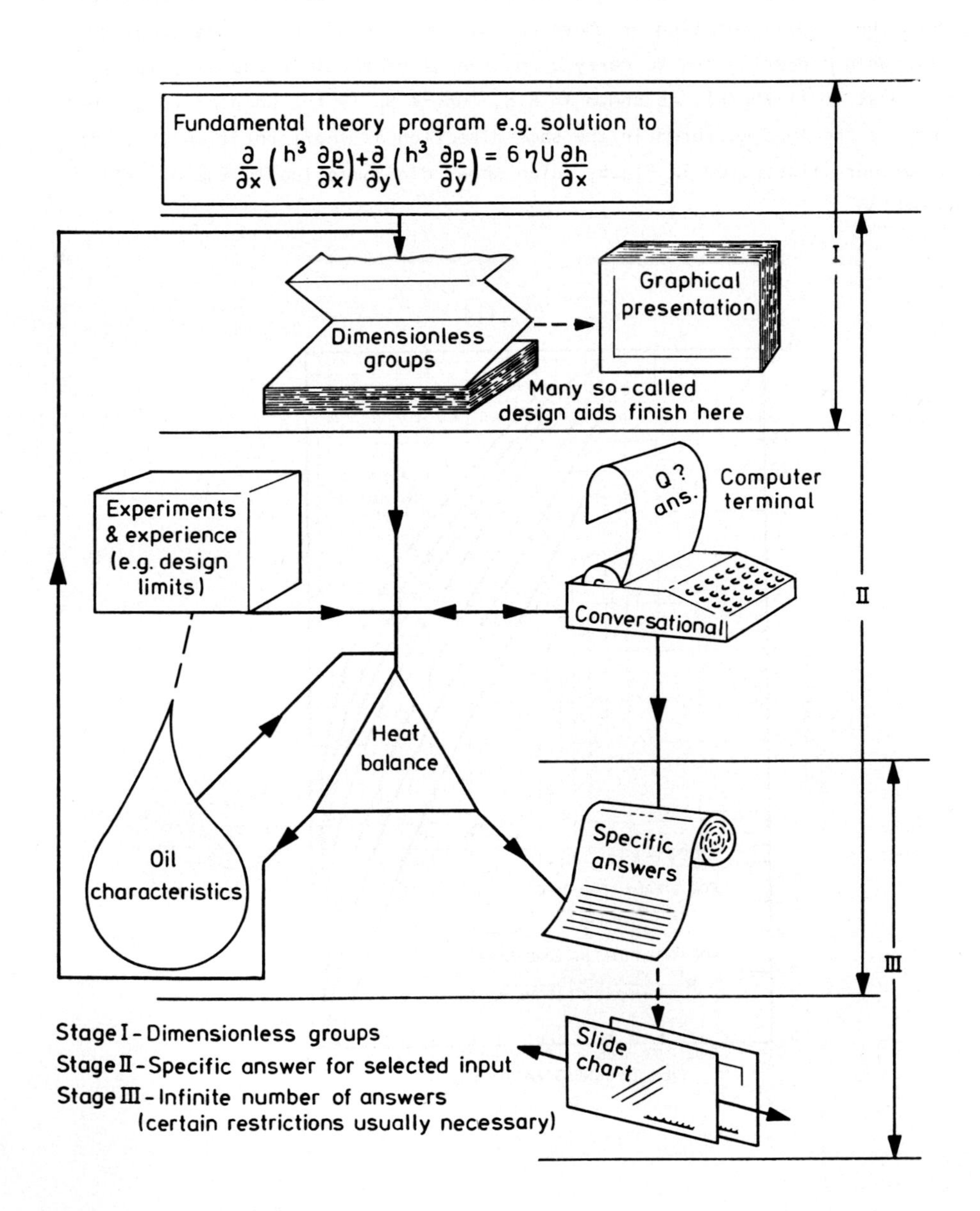

Fig.4 Stages in design

in the unloaded half or with a central circumferential groove; with the latter the length is that of one land and the load is half the total applied load. The central circumferentially grooved bearing is generally used in applications where the load is rotating or where an unknown load direction has to be catered for. When a bearing has to carry a rotating load the load capacity is proportional to $(1-2N_L/N)$, as shown in Fig.5 where N_L is the angular rotational speed of the load measured in the same direction as shaft rotation (N). This is further illustrated in Fig.6, which shows diagrammatically the oil film formation.

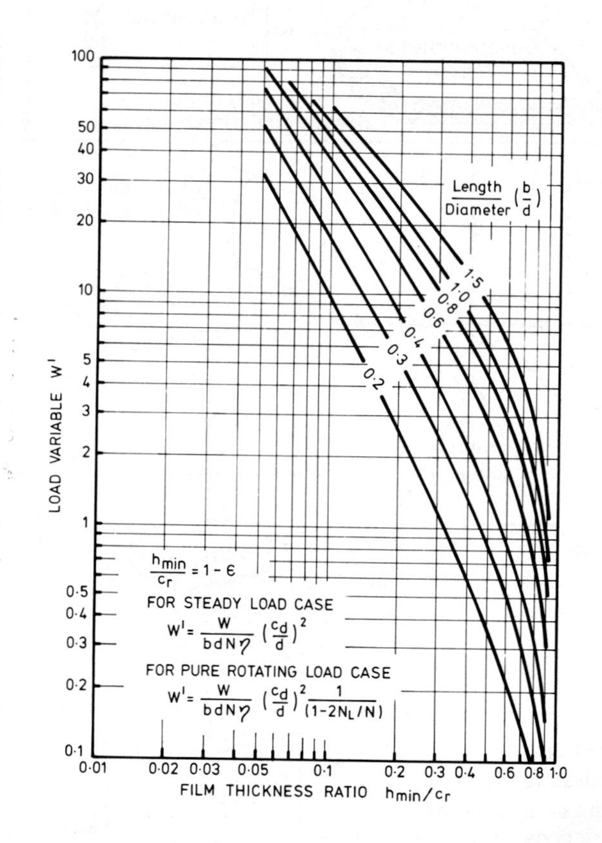

Fig.5 Bearing load capacity for steady and pure rotating load cases

An examination of dimensionless data can also be instructive where gross effects on performance are apparent. For instance the position of oil feed grooves in a steadily loaded bearing is very important, since they can seriously derate the load-carrying capability of the bearing. Such a derating effect is clearly demonstrated in Fig.7.

Case (a) shows an uninterrupted coverging oil film which generates a substantial hydrodynamic pressure to support the load (W).

Case (b) shows the same converging oil film as (a) but interrupted by an oil groove. This will only support a very reduced load because of the smaller integrated pressure.

Case (c) shows an uninterrupted film with the same load as (a) which needs a smaller oil film thickness to generate sufficient pressure.

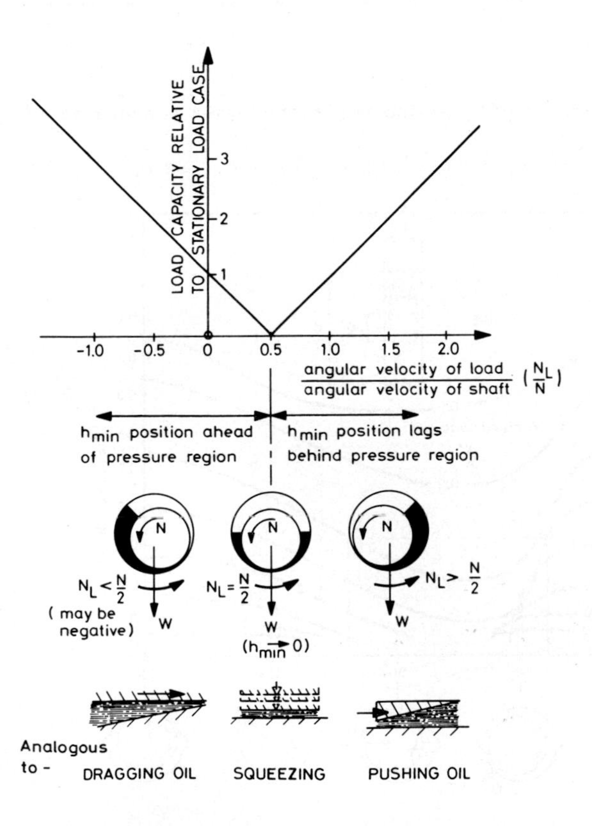

Fig.6 Film formation and relative load capacity under rotating load conditions.

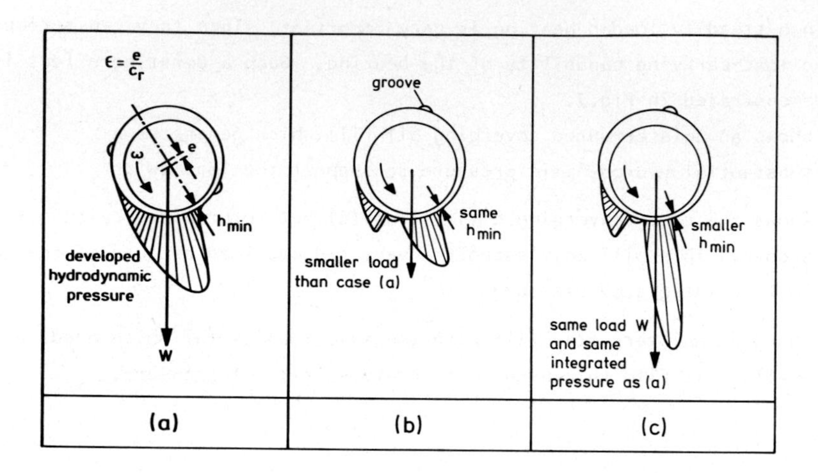

Fig.7 The derating effect of grooving positioned in a converging oil film

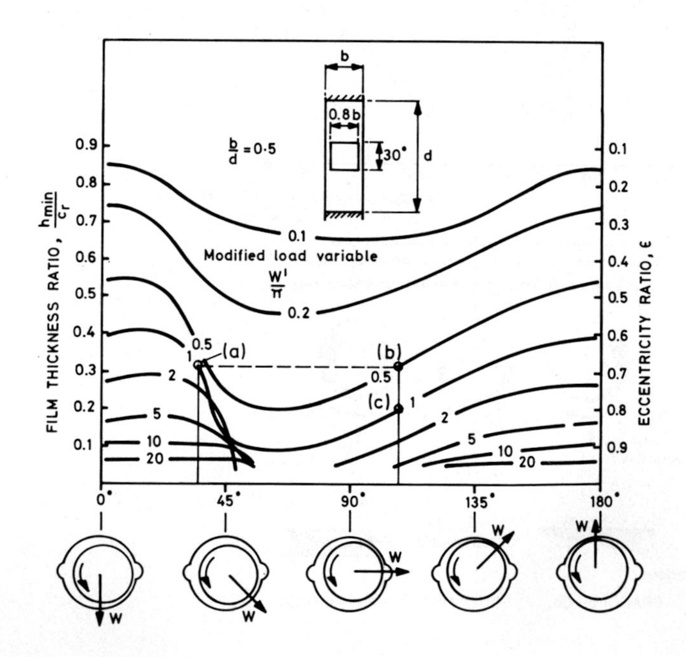

Fig.8 Variation of load capacity with direction of load for a bearing with two axial grooves

The dimensionless load and film thickness chart, Fig.8, shows more precisely the variation in load capacity with any load direction relative to the grooves. In this figure a bearing with a length equal to half the diameter is considered, having two axial grooves displaced 180° apart, each groove extending 30° around the bearing. These are typical groove arrangements.

With the load direction exactly between two grooves the load capacity shown in Fig.8 is the same as that in Fig.5 and variations in the angle within ± 20° cause little difference. As the load line gets near to a groove position the load capacity drastically reduces as seen by the reduction in the modified load variable (W'/π) for the same film thickness. It can be seen that the load variable is halved when going from point (a) to point (b) in Fig.8, where points (a), (b) and (c) can be identified with cases (a), (b) and (c) in Fig.7. Similarly for the same value of load parameter the film thickness for case (c) is much smaller than case (a).

It should be noted that the condition of loading directly into a groove does not normally produce the lowest oil film thickness; instead this occurs when the load is directed just before the groove. This is because the peak oil film pressure occurs just beyond the load line (for a steady load condition), and a groove in this position therefore more severely derates the bearing.

4.3.2 Design Procedures with Heat Balance - Stage II

The heat balance procedure can be carried out by hand calculation in a relatively straightforward manner, but is ideally suited to simple computer techniques. A particularly convenient and well documented procedure for cylindrical bore bearings is given in reference [9] which can form the basis of a computer programme [10]. The procedure [9] enables the influence of a very wide range of variables to be considered and is an extremely valuable design tool, especially when in the format of a conversational type computer programme.

4.3.3 Improvement in Design Aids - Stage III

Whilst the above arrangement is suitable for predicting bearing performance with 'single shot' answers, it does not provide a 'feel' for how near a particular design is to the various safe operating limits. Therefore, there is a need for stage III where the 'computed answers' for bearing cases are represented in design chart form. These enable the designer to answer such questions as:-

- Will this bearing operate safely?
- How near is it to the limits of operation?
- What changes can be made to improve the design?

When assessing safe limits of operation many factors are involved, for example the bearing data considered must include diameter, length, oil grade, load and speed. The various permutations for a range of these factors results in many thousands of cases to be considered. If such data were represented in conventional graphical form, a thick volume of unmanageable design aids would result. The 'slide chart' obviates the need for this and gives a direct and instant feel for the effect of changing any of the variables, thus bringing the power of the computer to the designers' desk top.

4.4 DESIGN PROCEDURE FOR CYLINDRICAL BORE BEARINGS

4.4.1 Method of Approach

The procedure detailed here presents design aids which allow for limiting conditions of operation and the prediction of bearing performance in cylindrical bore journal bearings which have two axial grooves spaced 180° apart. This type of bearing, with a steady load midway between the grooves, is one of the most common in use. All the tedious heat balance calculations have already been carried out by computer, based on the design procedure given in ESDU Data Item 66023 [9], and are consequently inherent in the design aids.

Comparison of experimental work with results from this theory indicated that whilst correlation was good for most variables, the predicted maximum temperature could, in many circumstances, be considerably in error. A new method was evolved based on considerable experimental evidence which relates the maximum bearing temperature to the calculated effective temperature [11].

The resulting procedure was used to calculate bearing performance over a wide range of operating conditions. However, to keep the problem to a manageable size it was decided that certain variables should be fixed throughout, to values commonly found in current practice. These were:-

(i) oil groove dimensions:
axial length = 0.8 of bearing length
circumferential width = 0.25 of bearing diameter.
(ie 150° between the edges of grooves, in top and bottom of bearing).
Grooves with 'square' ends

(ii) oil feed conditions at the bearing:
pressure 0.1 MPa (approx 15 lbf/in^2)
temperature 50°C

(iii) oil grades:

in order that the oil grade could be defined simply, a trend in viscosity-temperature characteristics typical of industrial mineral oils was assumed to apply throughout. The definition of an oil grade can then be made by quoting one viscosity at a certain (arbitrary) temperature; 40°C was used here to define the oil grade, this being the same basic reference as for ISO viscosity grades. Fig.9 shows the viscosity trends considered.

In a design problem the bearing diameter and operating conditions of speed and load are generally determined prior to the bearing design stage, since they usually form part of an overall system. For similar reasons the lubricating oil grade to be used is often also imposed leaving clearance and bearing length as the two dimensions still to be determined.

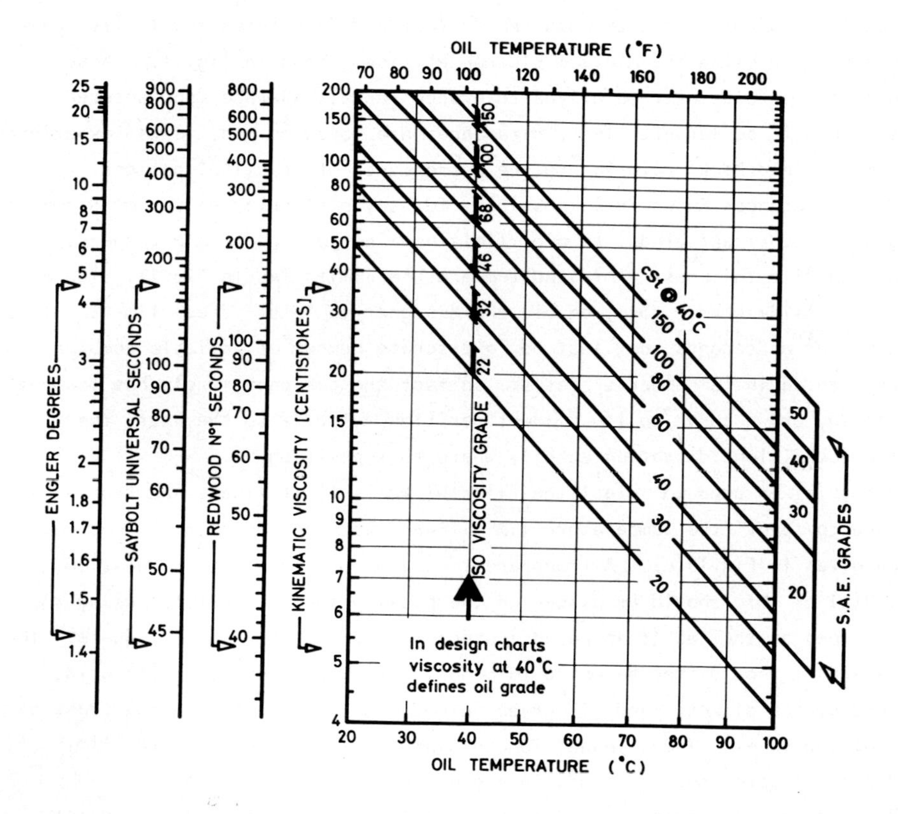

Fig.9 Viscosity ∿ Temperature characteristics for the range of oils considered in the design charts

Several authors have considered optimising on clearance [12,13] to give the largest possible oil film thickness in the bearing or the lowest power loss. Unfortunately these two optima do not coincide and a compromise solution is usually required. Indeed when the problem is treated purely as a mathematical exercise there is a danger of having excessively slack clearances for optimum power loss and, in some instances, very tight clearances for optimum film thickness, especially if the effect of change in clearance on viscosity is ignored.

4.4.2 Guidance for Safe Operation

4.4.2.1 Oil Film Thickness and Temperatures

It is appropriate to consider the tight clearance end of the tolerance range when examining conditions at high temperature. These tight clearances may also be used when making judgements on safe operation relating to small oil films since the variation of clearance throughout most practical tolerance ranges generally has little effect on oil film thickness. The design aids, Figs.10(a), (b) and (c), associated with both temperature and film thickness limits, have been developed using the minimum clearance values shown in Fig.11. These values relate to an extreme of the tolerance range; average manufactured clearance will be larger. These clearances are based on many years' experience with much 'feed-back' relating to the performance of practical designs.

With clearances known it is possible, for any particular values of diameter, length and lubricant grade, to plot limiting lines of load against speed, representing film thickness limit and temperature limits (as in Fig.1). The design aids, in 'slide chart' form, enable the designer to 'plot' these limiting lines in a matter of seconds, not just for a discrete number of variable combinations, but for an infinite number of cases. In fact the designer merely has to move an already drawn curve to its correct position relative to the axes, the bearing conditions dictating exactly where this position is.

Fig.10(a) is used to 'plot' the film thickness limit lines, and Fig.10(b) is used to 'plot' the temperature limit lines, both on the grid of load and speed given in Fig.10(c). A transparent version of the chart in question, Fig.10(a) or (b), should be placed on the backing sheet, Fig.10(c) using the guide lines on the top (transparent) sheet to keep the two sheets square. The transparency should then be moved to a position where a point in its grid, defined by the diameter and oil grade values, is coincident with the cross on the backing sheet. The relevant length/diameter line then shows the limit of load against speed for the conditions considered.

By using the two transparencies the relative position of a design point (defined by its speed and load values) to the limiting lines can be seen. If this point is within the limits then the bearing, under reasonable environmental

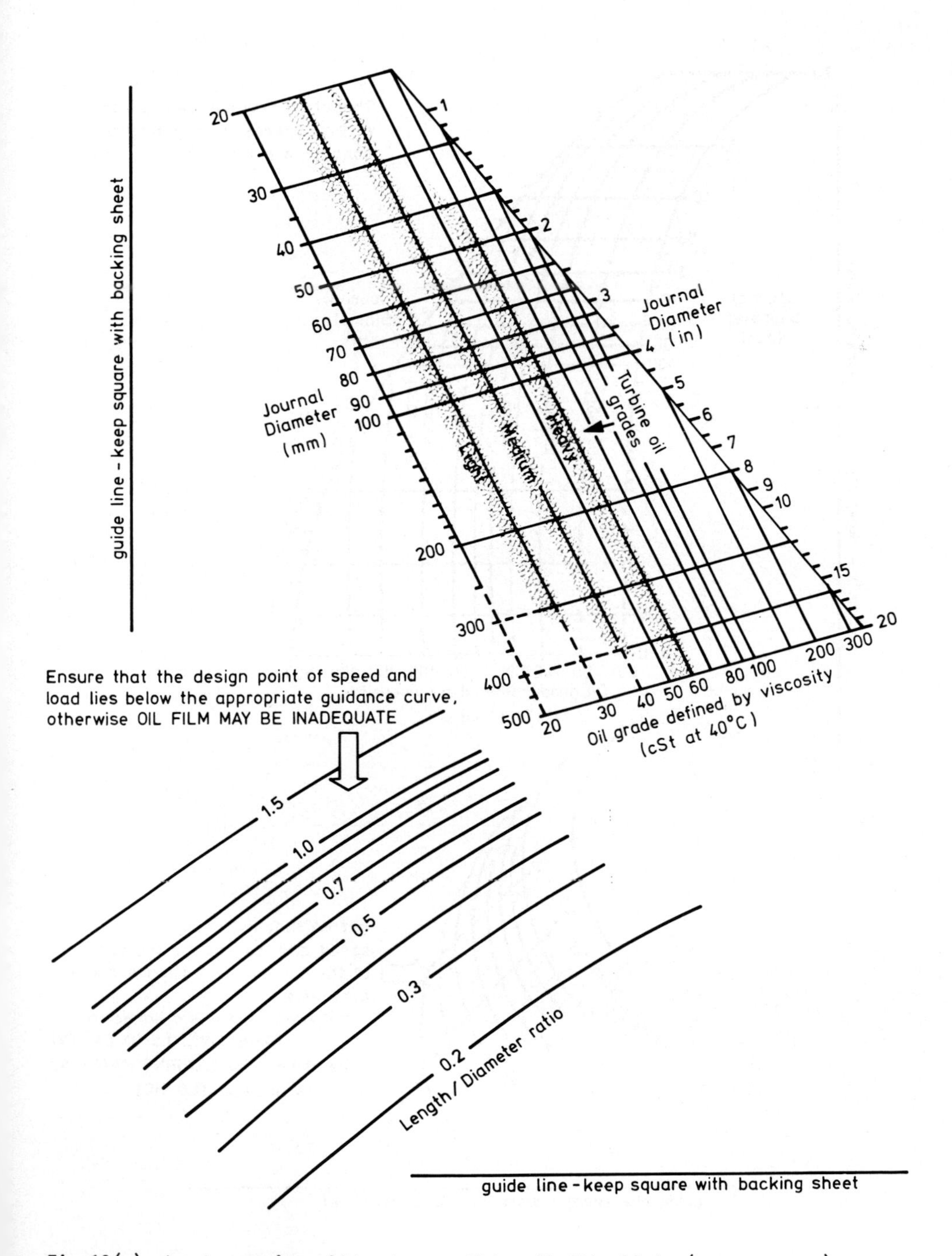

Fig.10(a) Load capacity slide chart: Thin oil film limit (transparency)

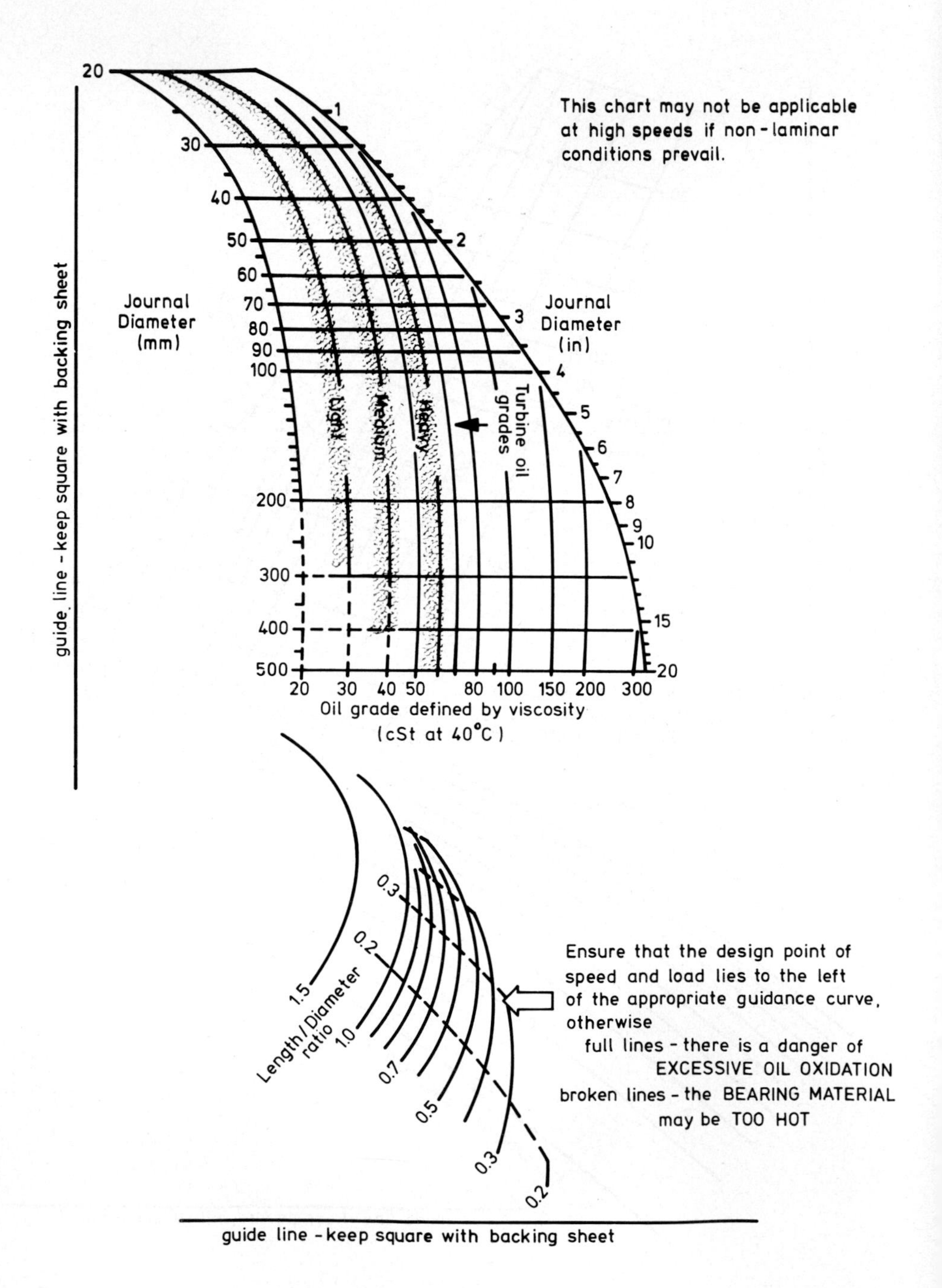

Fig.10(b) Load capacity slide chart: High temperature limits (transparency)

ERRATUM

F.A. Martin and D.R. Garner, Design of plain bearings:
Use of bearing data design charts, in: M.H. Jones and D. Scott (Editors), *Industrial Tribology — The Practical Aspects of Friction, Lubrication and Wear*, Elsevier, Amsterdam, 1983, pp. 45—79.

For technical reasons, the three parts of Fig. 10 received different treatment. This inadvertently resulted in the backing sheet (Fig. 10(c), on page 63) being reproduced in the book in an incorrect size. Figure 10(c) is given below in its proper size. We apologize to the authors of this chapter and to the readers for this mishap.

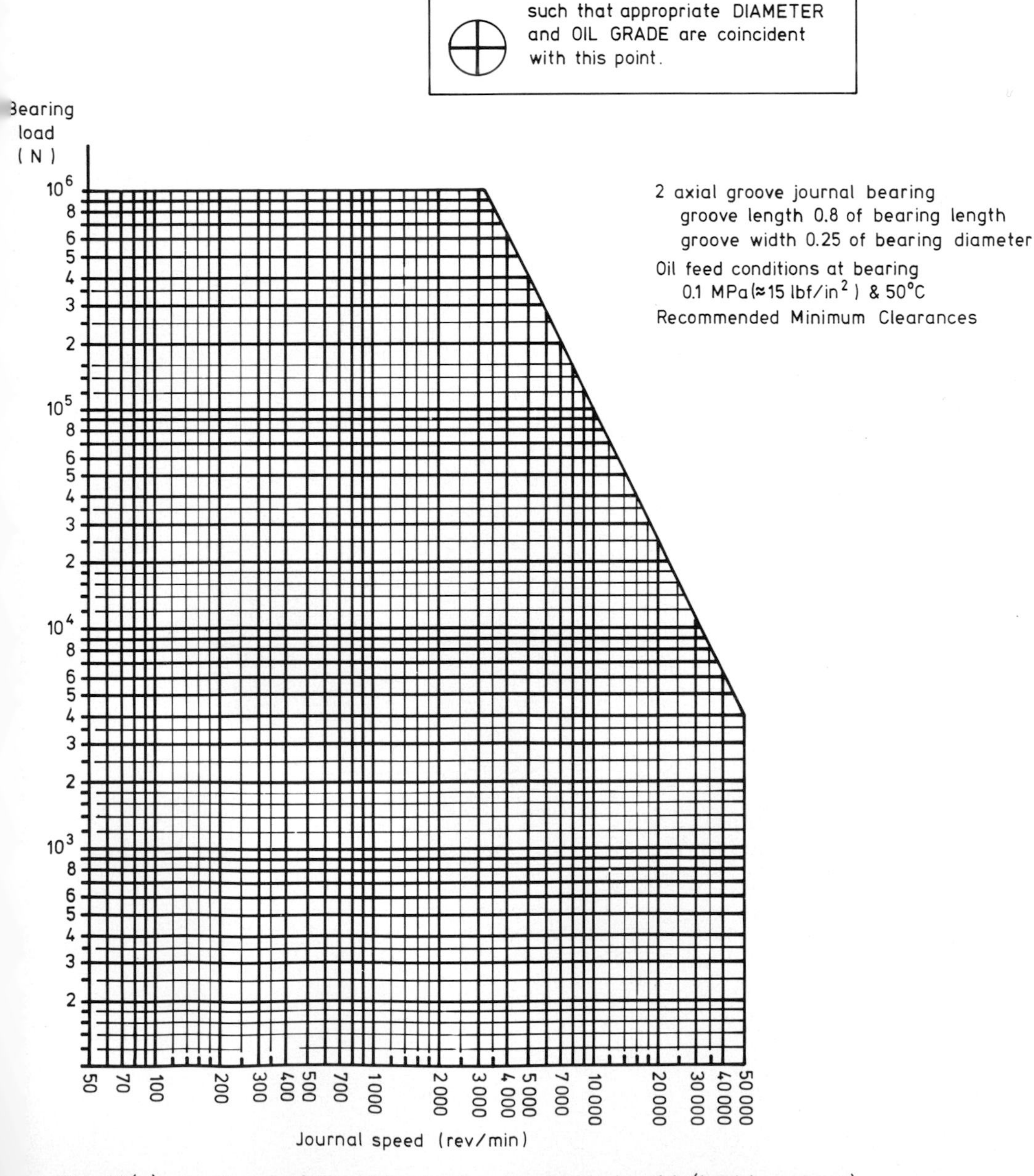

Fig.10(c) Load capacity slide chart: Load-speed grid (backing sheet)

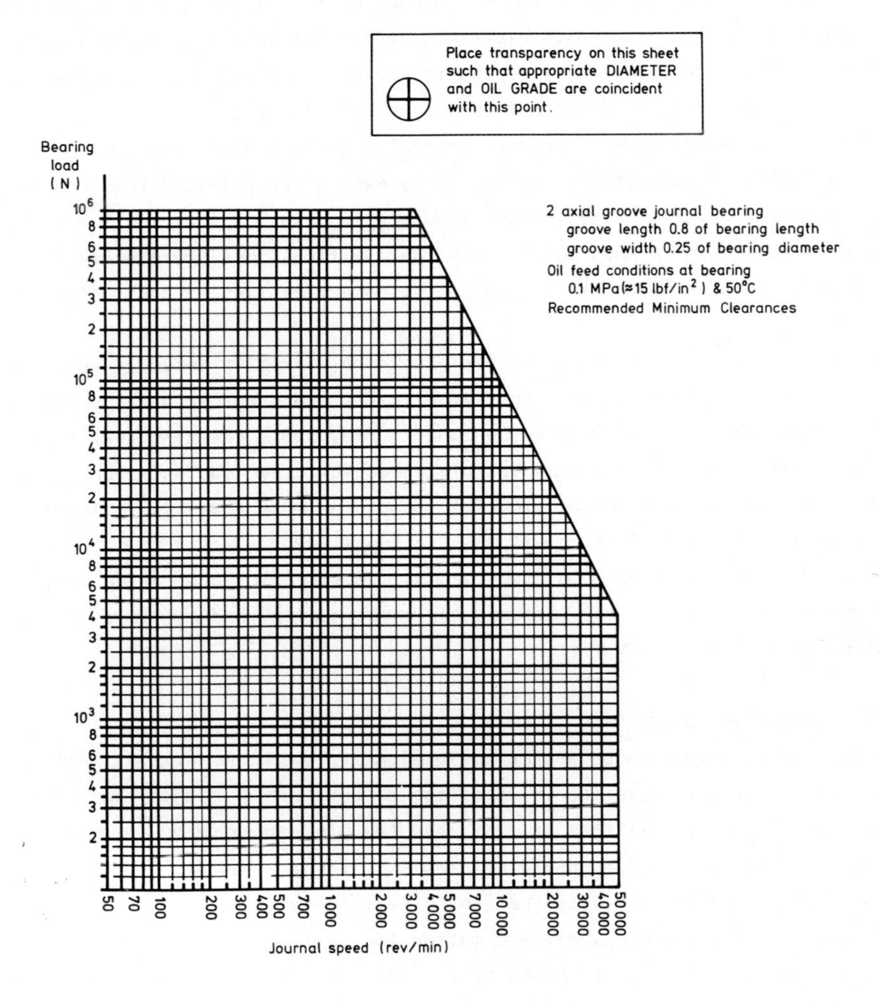

Fig.10(c) Load capacity slide chart: Load-speed grid (backing sheet)

conditions, should operate satisfactorily. If it is close to or outside the limits then the slide charts provide a quick method of determining which variable or variables can be changed to improve the design. For instance increasing the bearing length will raise the limit line for film thickness, but may worsen a high temperature situation. The latter effect was noted experimentally by Brown and Newman [14]. At these high speeds the charts demonstrate and quantify the advantage of thinner oils. If the use of thick oils is unavoidable then an increased clearance, above the values given by Fig.11, may help at high speeds. The slide charts are then not applicable, but a suitable design method is detailed in section 4.4.3.

Finally, it is advisable, wherever possible, to work away from the actual limiting lines; this allows an extra safety margin for unintentional adverse conditions such as small misalignment (see section 4.4.3), contaminated oil, etc. It is also advisable to work well within the limiting lines where there is a chance of having two modes of failure at the same time (ie. at the dotted apex in Fig.1).

Current practice is to quantify this additional limit in terms of specific load (load divided by projected area, W/bd). For hydrodynamic bearings it is usual to keep the specific load below about 4 MPa, still checking that film thickness and temperatures are acceptable. However, the load on a bearing at start-up or run down must be considerably lower, no more than 1 to 1.3 MPa otherwise high pressure jacking oil must be supplied.

Fig.12 has been developed to give the maximum specific load rating based on the thin oil film limit; this assumes the same geometry and oil feed conditions as considered in the slide chart.

4.4.2.2 Oil Film Instability

This is only likely to be a problem in lightly loaded bearings at high speeds. As a first check calculate the value of the dimensionless critical mass parameter shown in Fig.3. If this is less than the minimum value on the particular b/d curve, then the bearing should be stable. If not, then stable operation may still be predicted provided that the eccentricity ratio $(1-h_{min}/c_r)$ is large enough; section 4.4.3 gives a method for obtaining this value. The resistance to instability is raised by any external damping within the system, for example gear meshing, so that the limits can be unduly pessimistic.

If instability is a problem the bore profiling can often provide the solution; section 5. considers this in some detail.

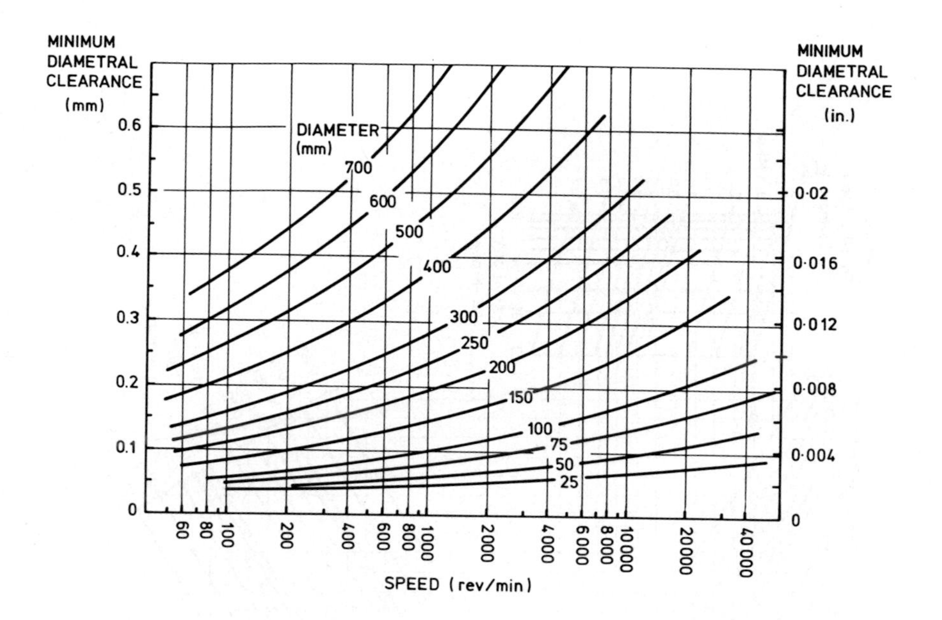

Fig.11 Recommended minimum clearance

4.4.3 Performance Prediction

Having determined that a given bearing is likely to operate satisfactorily, the designer then often needs to know the power loss and oil flow requirement of the bearing. As well as being of interest as far as the overall efficiency of a machine is concerned, these quantities have a direct influence on the lubricant supply system - the size of oil pump and supply lines, the need for coolers etc. This system must be capable of adequately supplying sufficient cooled oil (heat is the tangible form of bearing power loss) for any bearing within the spread of the manufacturing tolerance on clearance. When determining the oil pump capacity, the bearing flow at maximum possible clearance (within the tolerance range) should be considered since this has the maximum flow requirement. Maximum power loss on the other hand, important when considering the heat dissipation from the overall system, can occur anywhere within the tolerance range of clearance. The clearance has therefore been left as a variable in the prediction charts for power loss, oil flow and temperature. However, the clearance range should preferably still have a minimum value corresponding with Fig.11, since the slide charts (which inherently contain these clearances) can then be used to check for safe operation prior to predicting bearing performance.

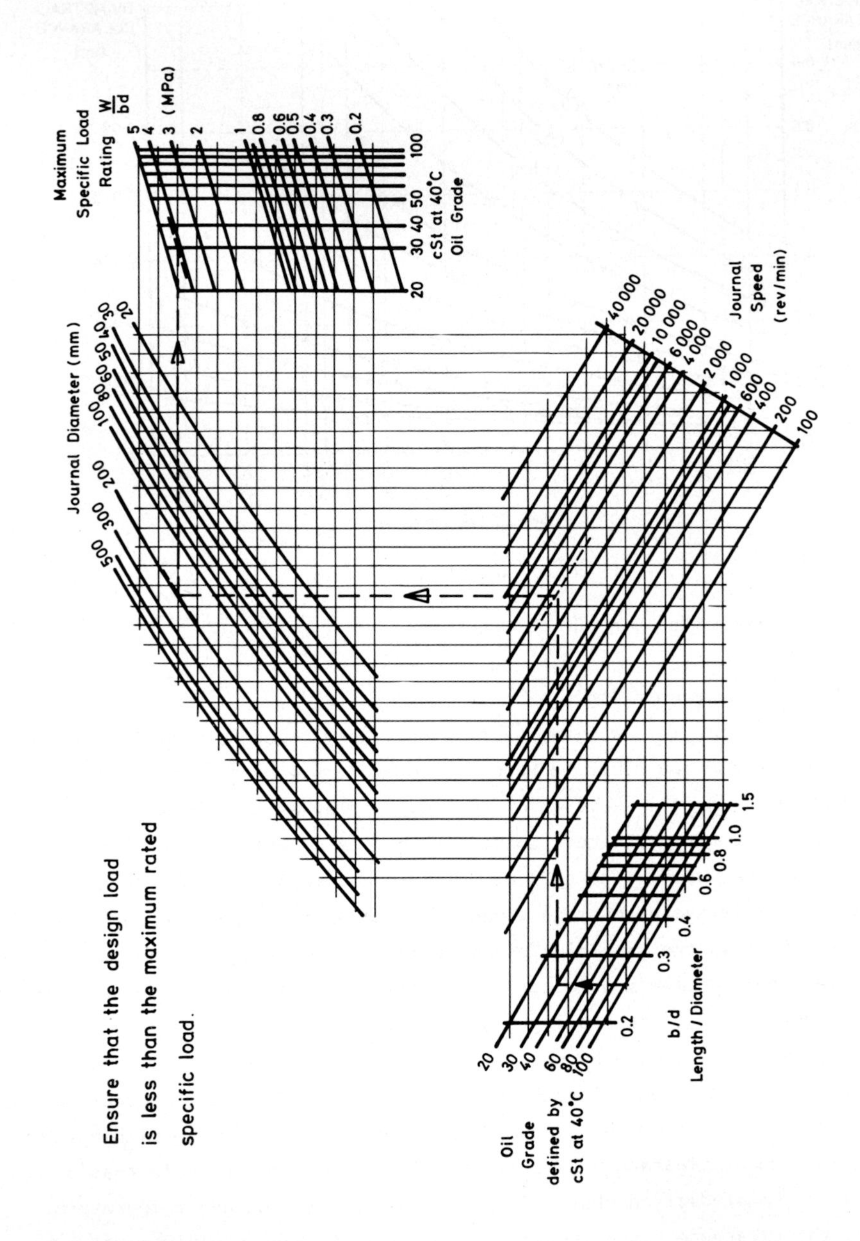

Fig.12 Maximum specific load rating for a two axial groove bearing based on thin oil film limit

Under some critical conditions the need to control the variations in film thickness, power loss or oil flow might regulate the range of the clearance tolerances, but usually predefined manufacturing details will impose a given clearance range which the designer must cater for adequately. The size of this tolerance range is mainly a function of economic considerations, and may therefore vary considerably. However, a practical guide is that the tolerance (mm) on diametral clearance lies within the range $(2.10^{-6}x\ dia)^{1/3}$ to $(5.10^{-6}xdia)^{1/3}$ depending on the manufacturing process; 'dia' is the nominal shaft diameter in mm.

The minimum film thickness ratio (h_{min}/c_r) is a significant term when predicting the power loss, oil flow and maximum bearing temperature. Using this ratio, prediction charts in nomograph form have been devised which permit quick and accurate determination of the various parameters with very little calculation.

4.4.3.1 Minimum Oil Film Thickness

The film thickness ratio may be obtained from Fig.13 for known operating conditions. Basically three grids are used to define the problem, and by linking these along the appropriate guide lines, as indicated by the arrows on the chart, a point in the fourth, 'answer' grid is obtained.

This chart is also useful for predicting the minimum film thickness in a bearing which has a different clearance to the minimum value shown in Fig.11, ie a bearing which cannot be considered on the slide charts. The acceptability of any film thickness can be checked on Fig.2.

4.4.3.2 Misalignment

Good alignment between shaft and bearing can be critically important because of the dramatic reduction in oil film thickness which misalignment causes. There are many ways in which misalignment can occur, from poor build of the machine to mechanical distortions due to load or temperature, and each of them produces different conditions within the oil film. However, a general guide to the derating effect of misalignment is given in Fig.14. This shows the reduction in oil film thickness from the perfectly aligned case (Fig.13) for a given misalignment across the length of the bearing (δ). The resulting minimum film thickness, which occurs at one axial end of the bearing, should be checked on Fig.2 for acceptability.

4.4.3.3 Power Loss and Oil Flow

Power loss and oil flow may be determined from Figs.15 and 16. The method of use is shown on each chart and is further illustrated by the example in section 4.6. The flow given is the bearing requirement flow, and should be

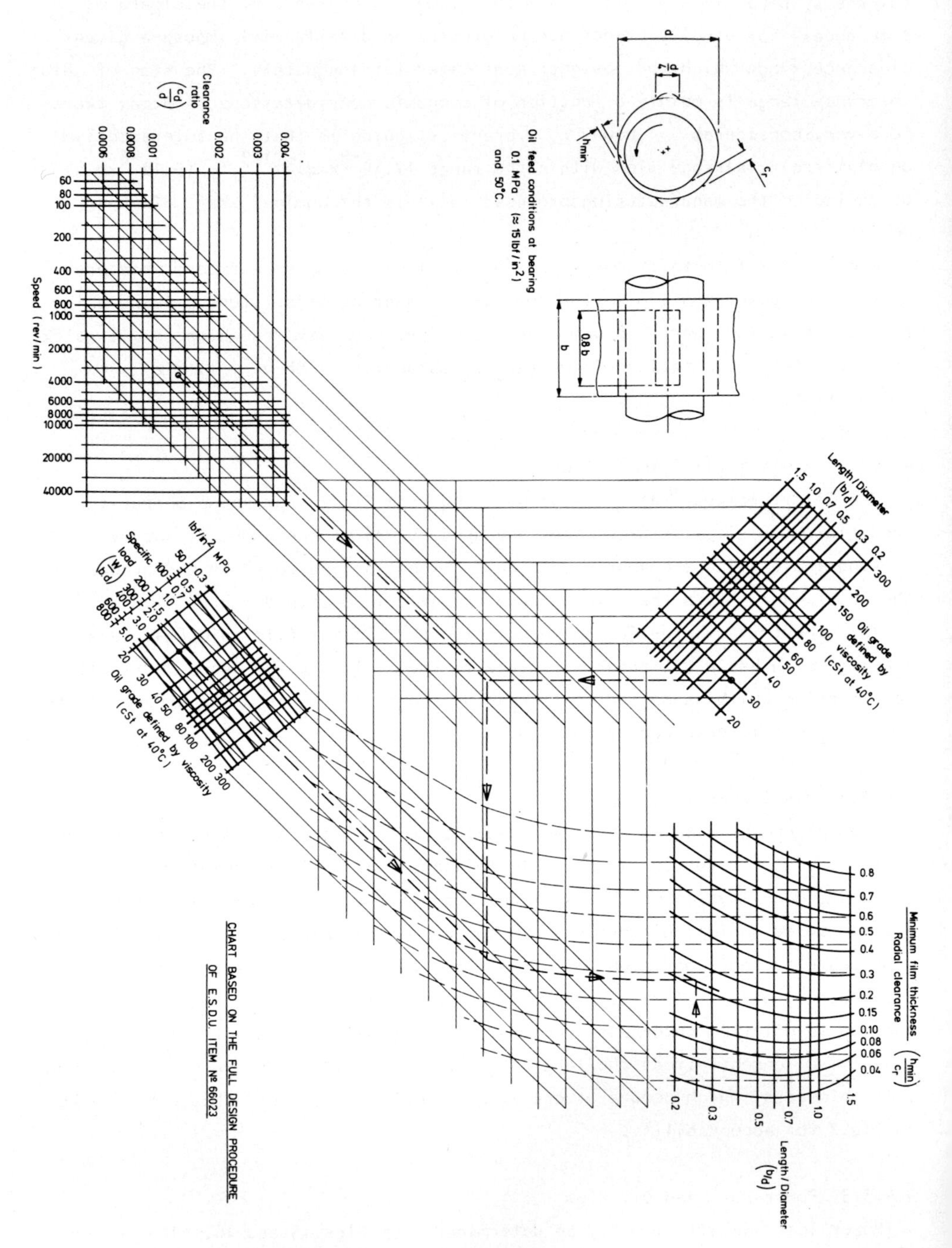

Fig.13 Prediction of minimum oil film thickness

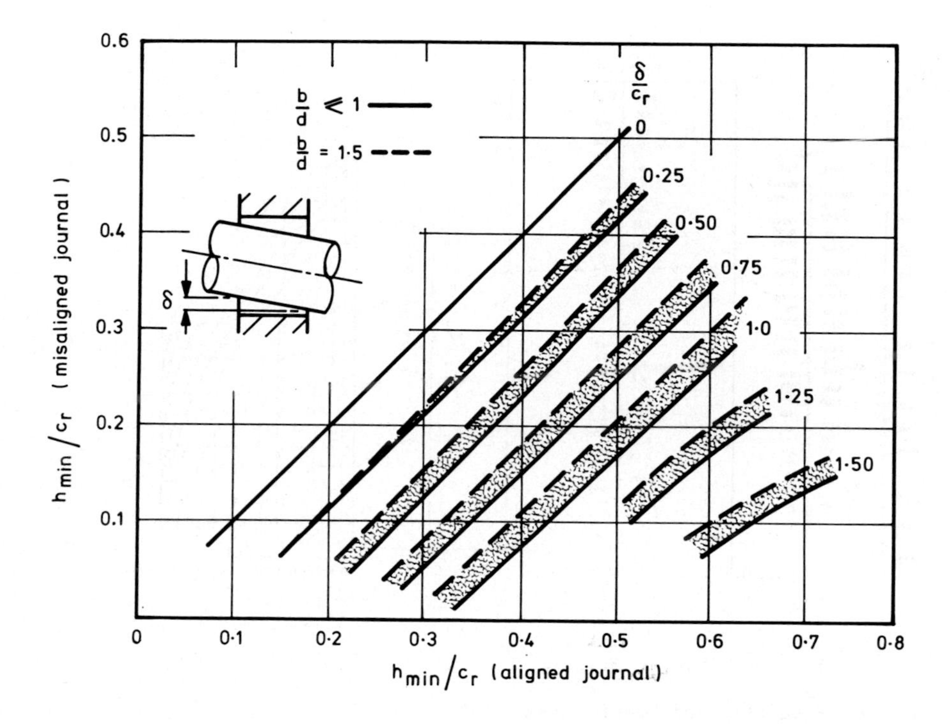

Fig.14 The derating effect of misalignment on film thickness

increased when determining pump capacity to allow for bearing wear, unequal flow to multi-supplied bearings etc. Typically the figures should be increased by 20-25%, and a pump chosen which can adequately supply this amount.

4.4.3.4 Temperatures

The value of the oil outlet temperature may be estimated from the power loss and oil flow values, as read from the charts, using the equation:-

Outlet temperature = inlet temperature + temperature rise

$$= 50 + A\ H/Q\ (°C) \qquad (1)$$

where A = 0.0005 for H kW and Q m^3/s

A = 5 for H hp and Q gal/min.

(A = 6 for hp and US gal/min)

The maximum bearing temperature may be determined from Fig.17; again an example of use is shown on the chart and is further amplified in section 4.6.

It should be noted that this temperature occurs at the surface of the lining material and at some circumferential position which is not well defined; it roughly coincides with the position of the minimum oil film thickness.

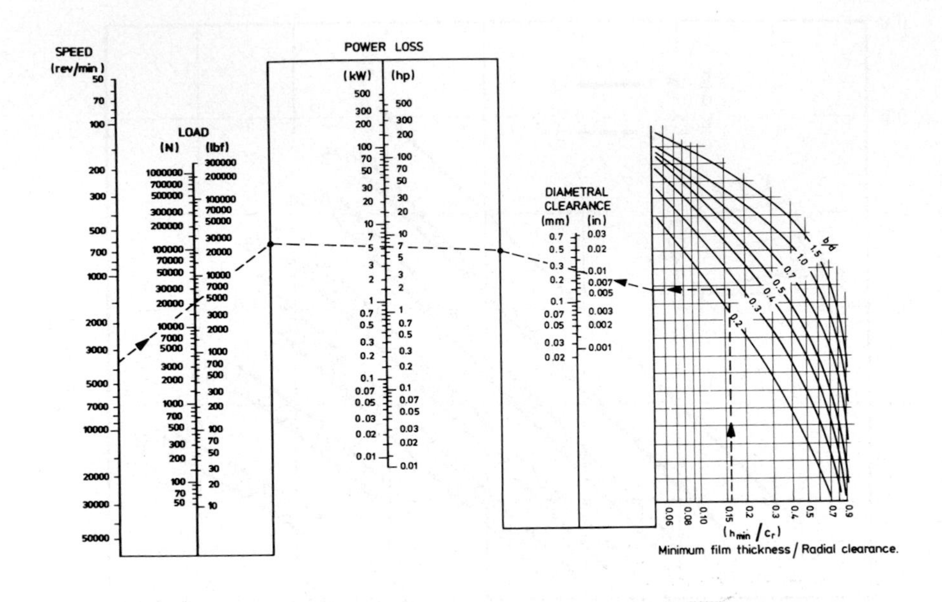

Fig.15 Prediction of bearing power loss

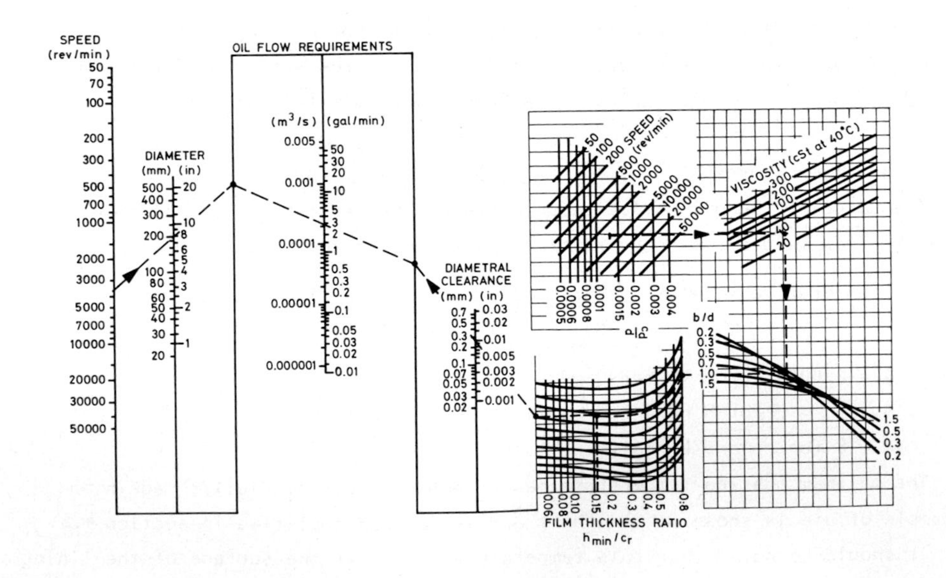

Fig.16 Prediction of bearing oil flow requirement

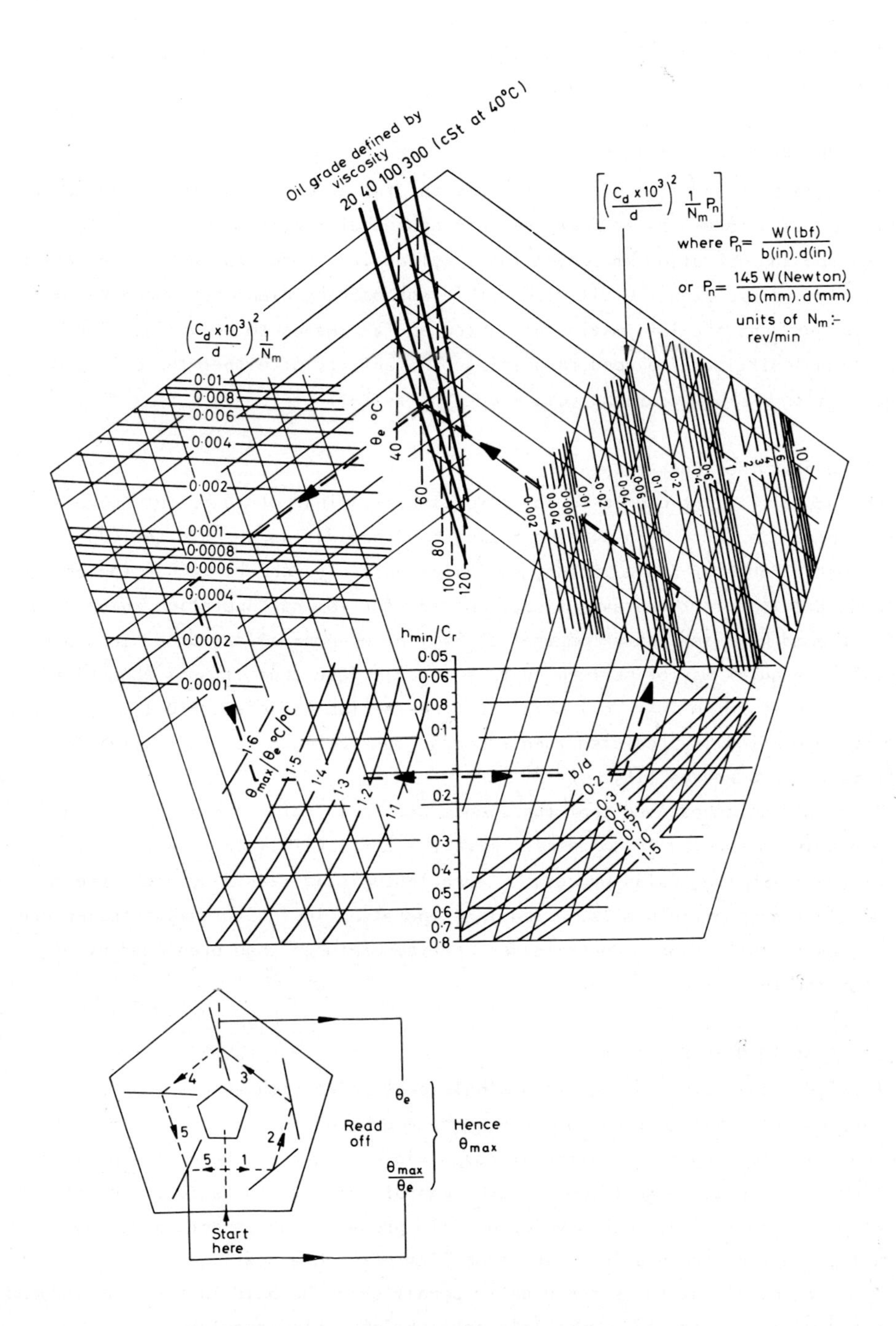

Fig.17 Prediction of maximum bearing temperature

Unless a great deal of thought and care is taken in determining the position of temperature instrumentation then the recorded temperature will be less than the actual maximum value.

4.5 HIGH SPEED APPLICATIONS

The general trend towards larger sizes and faster operating speeds in equipment such as compressors and turbines has caused difficulties with both the design and the operation of conventional cylindrical bore bearings. The design difficulty occurs when the oil flow within the bearing clearance space becomes non-laminar, and the information presented so far ceases to be valid. The operational difficulties concern possible instability of either the oil film (see Section 4.2.1.4) or of the complete rotor system.

4.5.1 Non-Laminar Operation

Colloquially it is common to refer to either laminar or turbulent operation, but in practice there is a wide operating band separating the two regimes in which other forms of fluid motion, eg Taylor vortices, occur. Since the design methods detailed in the previous sections are for laminar operation only it is necessary to have some check on whether laminar or non-laminar conditions apply. Fig.18 is a quick way of performing this check; note that it uses the value of film thickness ratio (h_{min}/c_r) obtained from the laminar chart Fig.13. If non-laminar operation is predicted then recourse must be made to alternative design methods and procedures [15].

It should be emphasised that turbulence per se is not a problem as far as safe bearing operation is concerned, merely a difficulty (albeit a relatively major one) with the design process. Turbulence is increased by high speeds, large clearances or thin oils. Bearings operating in the turbulent regime can have significantly higher power losses and temperatures than predicted by laminar theory.

4.5.2 Profile Bore Bearings

A cylindrical bore bearing has a single converging clearance space (which may in some circumstances be interrupted by a groove) in which oil film pressure is generated to support the external load. In contrast the bore of a profile bearing is arranged so that two or more separate converging regions are present under normal operation, each developing film pressures which act in various directions around the bearing. The stability of such a bearing is better than a cylindrical bore, but there are usually penalties to be paid in terms of reduced load capacity or increased power loss and lubricant flow requirement.

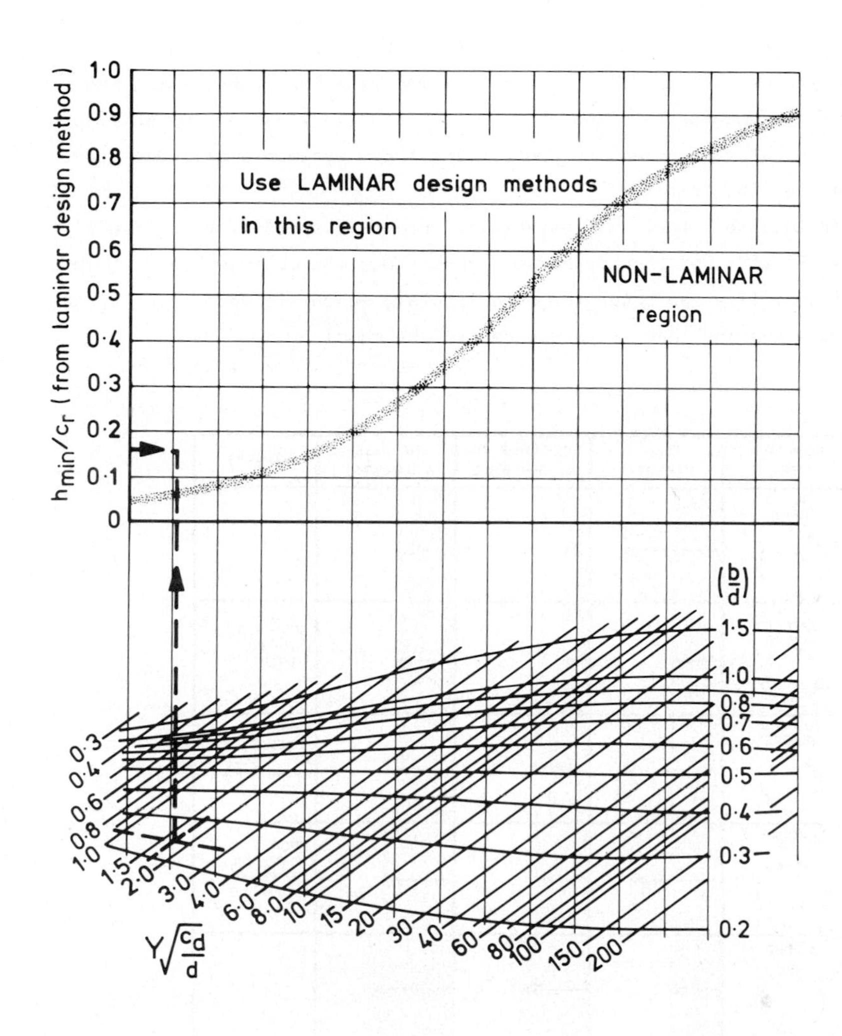

Fig.18 Guidance on the likelihood of laminar or non-laminar operation

The range of possible profiles is almost infinite but the more common forms are shown in Fig.19, together with an indication of their relative merits in terms of various operating parameters. Of necessity these comparisons are approximate only, since minor changes to either of the two clearances involved in profile bore bearings (the so called 'shake clearance' which is a measure of possible shaft movement and the hydrodynamic clearance which is the difference between radii of individual lobes and the shaft) can significantly alter performance. Experience has shown that as a general rule the 'shake clearance', as a ratio (ie c_d/d), should be no tighter than 0.001 mm/mm. This then avoids possible problems with loss of clearance on start-up due to differential thermal expansion of the shaft relative to the bearing and housing.

Information on stiffness and damping is important when considering the dynamics of the machine as a whole, with stiffness influencing the critical speeds of the rotor, and damping controlling the vibration amplitudes when running through the criticals.

In order to give a feel for instability threshold speeds for realistic clearances, Fig.20, has been prepared. This shows the minimum speed at which oil film instability can occur for a horizontal, weight loaded, rotor system (ie where the bearing load is purely due to the mass).

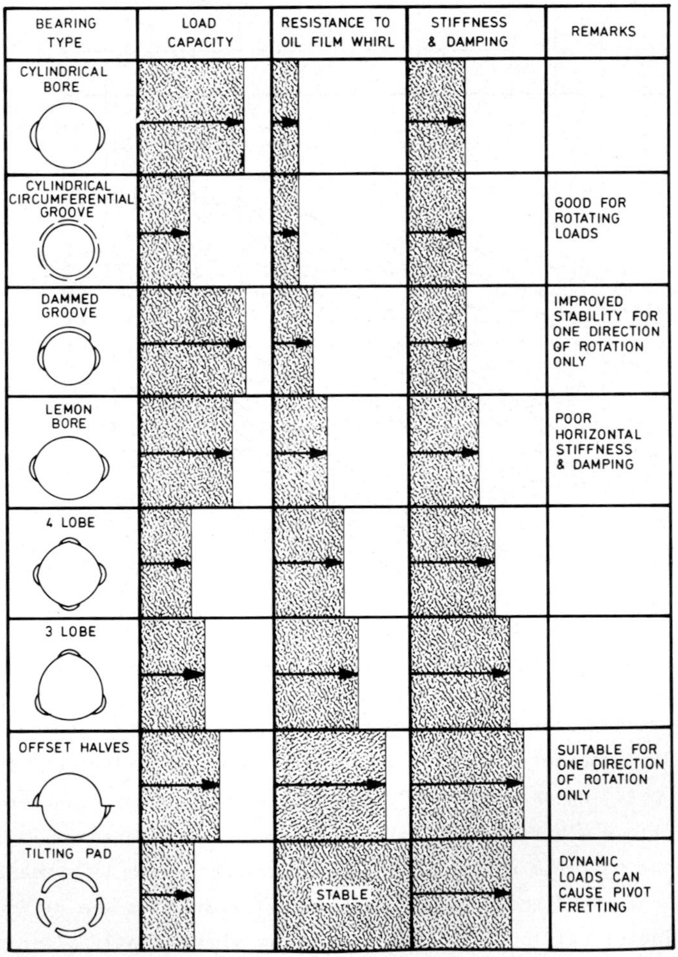

BEARING TYPE	LOAD CAPACITY	RESISTANCE TO OIL FILM WHIRL	STIFFNESS & DAMPING	REMARKS
CYLINDRICAL BORE				
CYLINDRICAL CIRCUMFERENTIAL GROOVE				GOOD FOR ROTATING LOADS
DAMMED GROOVE				IMPROVED STABILITY FOR ONE DIRECTION OF ROTATION ONLY
LEMON BORE				POOR HORIZONTAL STIFFNESS & DAMPING
4 LOBE				
3 LOBE				
OFFSET HALVES				SUITABLE FOR ONE DIRECTION OF ROTATION ONLY
TILTING PAD		STABLE		DYNAMIC LOADS CAN CAUSE PIVOT FRETTING

Fig.19 Comparison of static and dynamic characteristics of commonly used journal bearing types

Comprehensive and detailed procedures for the design of profile bore bearings have not been published, and currently the designer must either resort to a fundamental theoretical study or seek advice, usually on a specific case basis, from specialists [16].

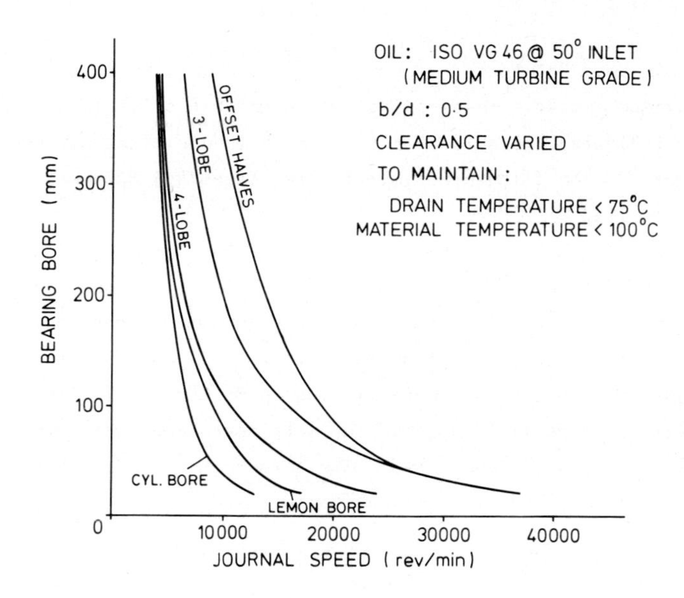

Fig.20 Oil film whirl instability of various types of journal bearings - Limiting speeds for horizontal rotors

4.6 EXAMPLE OF USE OF DESIGN AIDS

4.6.1 Problem

A gearbox bearing 200 mm diameter and 50 mm long has to carry a load of 20 kN at 3600 rev/min. The bearing has two axial grooves extending 0.8 of the bearing length, situated at ± 90° to the load line. The oil is within the ISO VG 32 specification (actual typical viscosity values are given as 30 cSt at 40°C and 5.2 cSt at 100°C, the oil density being 880 kg/m^3). The oil feed temperature is 50°C and the oil feed pressure 1 bar (0.1 MPa).

Check that this bearing will operate safely and estimate oil film thickness, power loss, oil flow requirement, maximum bearing temperature and oil outlet temperature.

4.6.2 Procedure

This particular example has been incorporated on the design aids as dashed lines.

4.6.2.1 Diametral Clearance

From Fig.11, the minimum diametral clearance, for a 200 mm diameter bearing operating at 3600 rev/min, is 0.26 mm.

4.6.2.2 Maximum Specific Load Rating

From Fig.12, the maximum specific load rating (based on thin oil film limit), is obtained by following the guide lines, through the relevant oil grade, length, speed and diameter. The allowable limit is seen to be 4.3 MPa, well above the actual specific load of 2 MPa.

4.6.2.3 Region of Safe Operation

Use Fig.10(a), transparent copy, with Fig.10(b), backing sheet. Mark point on grid, Fig.10(a), corresponding to oil grade (30 cSt at 40°C) and diameter (200 mm). Place transparency on backing sheet, lining up marked point with datum point (see Fig.21(a)). The limiting line for safe operation, with b/d equal to 0.25, can then easily be interpolated. The temperature limit transparency, Fig.10(b), is positioned in an identical manner (see Fig.21(b)).

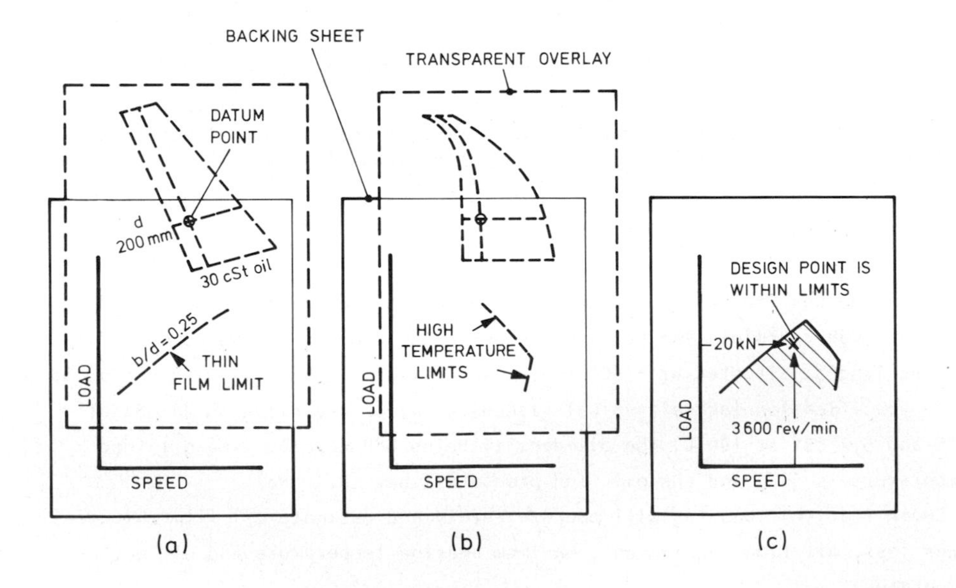

Fig.21 Example use of slide charts

These two combined give guidance to a region of safe operation on a load, speed framework as shown in Fig.21(c). The actual operating conditions are well within this limiting region, indicating a reliable design.

4.6.2.4 Prediction of Minimum Oil Film Thickness

Fig.13 predicts minimum oil film thickness and is very easy to use, although looking fairly formidable. First mark points on the following three grids.

1) Clearance ratio and speed (0.0013 and 3600 rev/min)
2) length to diameter ratio and oil (0.25 and 30 cSt at 40°C)
3) specific load and oil (2 MPa and 30 cSt at 40°C)

Join these points as shown, following the guide lines, and continue to the answer grid in the top right hand corner; at b/d equal to 0.25 the film thickness ratio (h_{min}/c_r) is read off directly as 0.16.

Minimum oil film thickness h_{min} = 0.16 x 0.13 = 0.021 mm

4.2.6.5 Check for Laminar Operation

Fig.18 gives guidance on whether laminar, as presented here, or non-laminar design methods [15] should be used.

Calculate turbulence variable $$Y = \frac{d^2N^2}{P} \rho \left(\frac{d}{c_d}\right)$$

In consistent units d = 0.2m, N = 60 rev/s, ρ = 880 kg/m^3, P = 2.10^6 N/m^2 and c_d/d = 0.0013 - from which Y = 48.74.

In Fig.18 mark point where $Y = \sqrt{\frac{c_d}{d}}$ = 1.76 and b/d = 0.25 (on lower grid).

Project vertically upwards as indicated by dashed line to film thickness ratio of 0.16. The resulting point appears in the laminar region, therefore the design aids presented here are suitable.

4.6.2.6 Prediction of Power Loss

On Fig.15, using the value of h_{min}/c_r already obtained (0.16), join up the appropriate points on the scales as shown.

Power loss = 5.3 kW

4.6.2.7 Prediction of Oil Flow

On Fig.16, starting at the speed-clearance ratio grid, join up the appropriate points as shown.

Oil flow requirement = 0.00021 m^3/s (0.21 1/s)

4.6.2.8 Prediction of Maximum Bearing Temperature

From Fig.17, the maximum bearing temperature is obtained by following the numbered steps shown in the sketch in the lower left hand corner.

Calculate $\left(\frac{c_d \times 10^3}{d}\right) \frac{1}{N_m}$ (=0.00047)

Multiply this by the modified specific load P (=0.136)

Starting at h_{min}/c_r equal to 0.16 follow the guide lines as shown, through values of b/d ratio, oil grade, and the two terms calculated above. A value of effective temperature (θ_e), can be obtained at the end of step 3 (=65°C), and a value of θ_{max}/θ_e (=1.55) from step 5 (which completes the pentagon).

Maximum temperature = 101°C

4.6.2.9 Prediction of Oil Outlet Temperature

From equation (1) - Section 4. 4.3.4.

$$\text{Oil outlet temperature} = 50 + 0.0005 \frac{H}{Q}$$

$$= 50 + \frac{0.0005 \times 5.3}{0.00021}$$

$$= 63°C$$

REFERENCES

1 Elwell,R.C. and Booser,E.R., 'Low Speed Limit of Lubrication Part 1, What is a "Too Slow" Bearing?', Machine Design, 15 June, 1972.

2 Martin,F.A., 'Minimum Allowable Oil Film Thickness in Steadily Loaded Journal Bearings', Proc. Lubrication and Wear Convention 1964 (Inst. Mech. Engrs. London), Vol.178, Pt. 3N pp. 161-167.

3 Booser,E.R., Ryman,F.D. and Linkinhoker,C.L., 'Maximum Temperature for Hydrodynamic Bearings under Steady Load', ASLE Trans. 1970, Vol.26, No.7.

4 Tribology Handbook, 1973, Section B1, Butterworths.

5 Lund,J.W., 'Self-excited Stationary Whirl Orbits of a Journal in a Sleeve Bearing', Thesis, 1966, Rensselaer Polytechnic Institute, NY.

6 Pinkus,O. and Sternlicht,B., 'Theory of Hydrodynamic Lubrication', 1961, Chapter 4, McGraw-Hill.

7 Cameron,A. 'Principles of Lubrication', 1966, Chapter 13, Longmans.

8 Lund,J.W. and Thomsen,K.K., 'A Calculation Method and Data for the Dynamic Coefficients of Oil-Lubricated Journal Bearings' from 'Topics in Fluid Film Bearings and Rotor Bearing System Design and Optimization', 1978, ASME.

9 'Calculation Methods for Steadily Loaded Pressure Fed Hydrodynamic Journal Bearings', Engineering Sciences Data Unit Item 66023, Sept. 1966 (Inst. Mech. Engrs., London).

10 'Computer Service for Prediction of Performance of Steadily Loaded Pressure Fed Hydrodynamic Journal Bearings', Engineering Sciences Data Unit, Item 69002, Sept. 1972, Amendment B.

11 Martin,F.A. and Garner,D.R., 'Plain Journal Bearings under Steady Loads Design Guidance for Safe Operation', First European Tribology Congress, 1973 (Inst. Mech. Engrs., London) paper C313/73.

12 Blok,H., Discussion to Conference. Proc. Conf. Lubrication and Wear, 1957 (Inst. Mech. Engrs., London), pp. 745-747.
13 Moes,H. and Bosma,R., 'Design Charts for Optimum Bearing Configuration 1. The Full Journal Bearing' April 1971, ASME Trans. Vol.93, Series F, No.2, pp. 302-306.
14 Brown,T.W.F. and Newman,A.D., 'High-Speed Highly Loaded Bearings and Their Development', Proc. Conference on Lubrication and Wear, 1957 (Inst. Mech. Engrs), pp. 20-27.
15 Garner,D.R., Jones,G.J. and Martin,F.A. 'Turbulent Journal Bearings Design Charts for Performance Prediction', July 1977, ASLE Trans. Vol.20, No.3, pp. 221-232.
16 Garner,D.R., Lee,C.S, and Martin,F.A., 'Stability of Profile Bore Bearings: Influence of Bearing Type Selection', October 1980, Tribology International, pp. 204-210.

5 THE DIAGNOSIS OF PLAIN BEARING FAILURES

R.W. WILSON and E.B. SHONE
Shell Research Ltd., Thornton Research Centre, P.O. Box 1, Chester CH1 3SH

5.1 INTRODUCTION

A few years ago the casual observer might have concluded that both the metallurgy of bearings and the understanding of their failure mechanisms had reached a settled stage in which bearing materials were available to meet almost every requirement and that characteristic failures were well-documented. Thus, most bearing failures could be described quite adequately from a strictly metallurgical or strictly engineering viewpoint. The lubricant was mentioned only in passing; as long as it had the right viscosity, provided adequate cooling and did not corrode the bearing alloy, it gave little cause for concern. Operators frequently blamed lubricants for bearing failures, but bearing manufacturers acknowledged that very few failures could be attributed to inferior lubricants. There was no special pressure on lubrication technologists to become experts on bearing failures; in any case, bearing manufacturers generally provided excellent technical service to their customers. However, engineering developments in recent years have made increasingly severe demands on plain bearings and have given rise to rather more complicated failure mechanisms. Many of these failures can be described as tribological failures; they cannot be ascribed to a particular defect in the design or in the metallurgy of the bearing or to a shortcoming in the lubricant - they are the consequence of the interaction of a number of factors.

Before it is possible to talk about plain bearing failures it is essential to have an understanding of the materials used, how they are made and the requirements that must be met. There are a number of very good publications describing the construction and properties of plain bearings and we intend to briefly summarise these.

5.2 PROPERTIES REQUIRED OF BEARING MATERIALS

It is evident that a bearing must withstand a variety of imposed conditions. No bearing is equally good with respect to all requirements, and the selection of the most suitable bearing material for a particular set of circumstances requires a careful evaluation of the most critical operating factors. Some of the factors that have to be considered are outlined in Sections 5.2.1 - 5.2.8.

5.2.1 Fatigue Resistance

This is the most important property in those applications where the load varies. However, fatigue failure in bearings is frequently associated with some other factor, such as corrosion, which reduces the strength of the bearing, or wear, which reduces the load-carrying area. The fatigue strength of a fundamentally weak bearing alloy can be increased by making the bearing alloy thin and bonding it firmly to a bronze or steel backing.

5.2.2 Compressive Strength

This is the steady load that the bearing alloy can support without extruding. There is little correlation between compressive strength and fatigue resistance when a single class of bearing material is considered.

5.2.3 Conformability

This is the ability to compensate for misalignment that occurs as a consequence of bad design or manufacture or that may develop in service.

5.2.4 Embeddability

This is the ability to tolerate and absorb foreign particles, thereby avoiding scoring or wear.

5.2.5 Strength at Elevated Temperatures

High sliding speeds and heavy loads can generate considerable heat, even when a bearing is operating hydrodynamically. One of the main functions of the lubricant and the bearing alloy is to conduct heat away from the sliding surfaces; even so, bearings are often required to operate at elevated temperatures, and a lack of high-temperature strength may result in extrusion of the bearing alloy and/or fatigue failure.

5.2.6 Compatibility

All bearing assemblies experience some metal-to-metal contact at some stage in their lives; the resistance of the bearing metal/journal combination to seizure is therefore important. However some of the harder bearing alloys do not function satisfactorily against unhardened steel journals.

5.2.7 Corrosion Resistance

Bearings may be exposed to weak organic acids formed as a result of the oxidation of lubricating oils in service. They may also be subject to corrosion by weak organic acids and strong mineral acids derived from fuel combustion products. Sometimes the operating environment is corrosive - for example, on many chemical plants and marine installations.

5.2.8 Cost

It is essential that bearing alloys are cheap and easy to manufacture.

5.3 TYPE, CONSTRUCTION AND CHARACTERISTICS OF PLAIN BEARING MATERIALS

Some of the more widely used bearing alloys are listed in Table 5.1; the more commonly used overlays are described in Table 5.2.

Sections 5.3.1 - 5.3.7 outline the methods by which bearings are usually manufactured from these alloys and describe the characteristics of the various bearing materials. The characteristics of the various bearing materials are summarised in Table 5.3 and the general properties briefly discussed in Section 5.3.8.

5.3.1 White Metals (Babbits)

These are tin-based or lead-based alloys, the original tin-based white metal being invented by Isaac Babbitt in 1839. This alloy is widely used today for the manufacture of both thick, large, individually manufactured bearings and thin-walled bearings made from continuous strip.

The coated strip is cut into sections of a suitable size which are press-formed to shape. Newer manufacturing techniques permit the use of bearing metal combinations and controlled microstructures that could not otherwise be produced.

Another innovation is the three-component (or tri-metal) bearing, in which one of the harder bearing materials such as bronze, already bonded to a steel backing, is covered with a thin (20-200 μm), electrodeposited or cast overlay of white metal.

The tin-based alloy consists of a tin-rich matrix, with some antimony and copper in solid solution; dispersed in the matrix are cuboids of SbSn and needles of Cu_6Sn_5. This microstructure led to the erroneous belief that bearing alloys must consist of hard crystals dispersed in a soft matrix. In fact, the intermetallic compounds in white metals appear to serve no purpose other than to strengthen the alloys.

White metals are outstandingly good bearing alloys in many respects, their main defect being lack of load-carrying capacity, particularly at elevated temperatures. The corrosion resistance of lead-based white metals is inferior to that of tin-based white metals; nevertheless, lead-based alloys are quite widely used, particularly in the U.S.A.

TABLE 5.1

Some lubricated bearing materials

Bearing material	Major alloying elements	Remainder	Special features
1. Lead-based white metal (lead-Babbitt)	8-16% antimony 5-11% tin	Lead	
2. Tin-based white metal (tin Babbitt)	7-14% antimony 3- 9% copper	Tin	
3. Sintered copper-lead	20-50% lead	Copper	Generally used with 90% lead-10% tin or 96% lead-4% indium electrodeposited overlay, 25 µm thick; lead bronzes can also have thin-Babbitt overlays
4. Cast copper-lead	20-30% lead	Copper	
5. Lead-bronze	20-30% lead 3- 5% tin	Copper	
6. Aluminium-low tin	6% tin, 1.5% silicon	Aluminium	Hardened journals preferred
7. Aluminium-high tin	20% tin, 1.0% copper	Aluminium	
8. Aluminium-silicon (overlay plated)	11% silicon, 1% copper	Aluminium	Still under development
9. Aluminium-Babbitt	10% lead, 2% tin	Aluminium	
10. Phosphor-bronze (cast)	8-12% tin, 0.2-1.0% phosphorous	Copper	
11. Silicon-bronze	1.5-4% silicon	Copper	Primarily bushing material
12. Lead-bronzes (generally steel-backed)	5-10% tin 8-12% lead	Copper	
13. Silver	-	-	Electroplated, 250-500µm thick, often used with lead-indium overlay
14. Porous bronze	8-12% tin		Self-lubricating, impregnated with oil
15. Bronze-graphite composites	5-40% graphite	Tin, bronze	Can be used dry
16. Laminated resins	Fibre-reinforced phenolics or epoxies		Usually water-lubricated
17. PTFE and filler in metal matrix	PTFE and lead	Metal	Good under conditions of marginal lubrication
18. PTFE matrix and metal fillers	Bronze and graphite	PTFE	
19. Metal-backed thermoplastics	Nylon or polyacetal		

TABLE 5.2

Commonly used overlays

Overlay type	Major alloying elements	Special features
A lead-tin, electroplated	8-12% tin	Simultaneously precision-plated, about 0.001 inch thick (25 μm)
B Lead-tin-copper, electroplated	8-12% tin 1% copper	Simultaneously precision-plated, about 0.001 inch thick (25 μm)
C Lead-indium, electroplated	4-8% indium	Precision-plated, first lead then indium, then alloyed by diffusion; about 0.001 inch thick (25 μm)
D Lead-Babbitt, cast Tin-Babbitt, cast	Antimony-tin Antimony-copper	0.005-0.015 inch thick (125-375 μm thick)
E Satco alloy, cast	Lead, with 2% tin and 0.5% calcium	

TABLE 5.3

Characteristics of bearing materials

Material (see Table 1)	Relative load-carrying capacity lb/in²	Relative load-carrying capacity MN/m²	Embeddability (tolerance for dirt)	Seizure resistance	Maximum operating temp. °C	Tolerance for misalignment	Corrosion resistance Organic acids	Corrosion resistance Mineral acids	Special features
1	2000	14	Excellent	Very good	130	Very good	Moderate	Fair	
2	2000	14	Excellent	Very good	130	Very good	Excellent	Very good	
3	3000-5000	20-35	Moderate without overlay; good with overlay	Moderate without overlay; good with overlay	150	Good	Fair	Poor	Resistance to organic acids greatly improved by overlay (materials 3-5)
4	4000-6000	28-41			160	Moderate	Poor	Poor	
5	5000-8000	35-55			170	Fair	Fair	Poor	
6	8000	55	Poor	Fair	180	Poor	Good	Fair	
7	5000	35	Poor	Fair	170	Fair	Good	Fair	
8	8000	55	Poor	Moderate	180	Fair	Good	Moderate	
9	5000	35	Poor ?	Moderate	160	Fair	Poor ?	Fair ?	
10	8000+	55+	Poor	Moderate	220	Fair	Fair	Fair	Excellent for high loads at low speeds
11	8000+	55+	Poor	Fair	220	Poor	Moderate	Fair	
12	6000	41	Moderate	Moderate	180	Moderate	Fair	Moderate	
13	8000+	55+	Poor	Poor	180	Poor	Good except for sulphur	Moderate	
14	2000-4000	14-28	Moderate	Good	130	Fair	Poor	Fair	
15	Very variable			Good	Withstands occasional very high temp.	Fair	Fair	Fair	
16	High		Moderate	Good	100-300 depending on resin	Moderate	Good	Good	Some plastics are damaged by the solvent action of lubricants (materials 16-18)
17	Comparatively poor; depends on operating conditions (materials 17-19)		Fair	Very Good	250	Good	Good	Good	
18			Fair	Very Good	250	Good	Good	Good	
19			Fair	Good	100-350 depending on plastic	Good	Good	Very Good	

5.3.2 Copper-lead Alloys

These can be regarded as the first of the modern alloys. They are blanked from continuously coated steel strip, the bearing alloy being applied either by casting or by sintering. Lead and copper are immiscible in both the solid and the liquid state, and, in order to avoid complete segregation when casting, the moving steel strip is chilled immediately after casting by a water spray on the underside. This results in a pronounced dendritic structure, with crystals of copper and lead normal to the steel backing and the bearing surface.

In the sintering process the copper-lead powder is compacted after application and then sintered. This method of manufacture gives good control of lead distribution and results in a homogeneous and equiaxial structure.

Both types of copper-lead bearing are said to rely on a thin, extruded surface film of lead for satisfactory performance. In many ways they are similar, but the different methods of manufacture give rise to important differences in performance in certain respects. In the cast alloy the long copper dendrites enhance the load-carrying capacity and facilitate heat flow away from the bearing surface. It is, however, difficult to obtain a satisfactory lead distribution. On the other hand, it is much easier to ensure a uniform lead distribution by sintering techniques, and the sintered alloy is therefore less susceptible to corrosion.

Copper-lead bearings have greater strength and better high-temperature performance than white metals, but in most other respects they are inferior. In particular, they are susceptible to corrosion, cannot tolerate as much dirt and generally require hardened journals. In order to overcome these disadvantages, copper-lead bearings are generally overlay-plated, a thin ($\sim$ 25 μm) electrodeposit of a 90% lead - 10% tin or a 95% lead - 5% indium alloy being applied to the bearing surface. In the case of lead-tin, the alloy is co-plated; in the case of lead-indium, the lead is plated first and then the indium, which is finally diffused into the lead. It is often stated that this overlay is merely a running-in aid. This is not the case; the overlay is expected to last the life of the bearing. It provides a seizure-resistant surface, allows soft shafts to be used, increases the ability of the bearing to absorb dirt and combats corrosion of the pure lead in the underlying copper-lead. Since it is so thin it derives considerable support from the underlying bearing alloy, and the fatigue strength of the composite is hardly impaired.

5.3.3 Lead Bronzes

The high-temperature performance and load-carrying capacity of copper-lead alloys can be improved by tin additions. The tin dissolves completely in the copper, thereby strengthening the bearing; the lead remains unalloyed and subject to corrosion. Like copper-lead bearings, lead-bronze bearings can be cast or sintered, and show the same disadvantages. For this reason they are usually overlay-plated. Larger-size bearings may have a thin layer (100-200 μm) of tin-based white metal cast on them; such bearings are sometimes called micro-Babbitt or tri-metal bearings.

This micro-Babbitt layer should always be separated from the lead-bronze by a diffusion barrier, otherwise the bronze will alloy with the tin-rich Babbitt to form a hard, intermetallic phase which can damage the shaft.

Lead-bronzes are often used for little-end bushes in pistons for turbo-charger bearings.

5.3.4 Aluminium Alloys

Two alloys are in widespread use, one containing about 6% tin and the other 20% tin. The 6% tin alloy may be used in massive form or bonded to steel, but, unless it is overlay-plated, it requires hardened journals if wear is to be kept within acceptable limits. The 20% tin alloy is a more recent development and its method of manufacture provides a good example of the advanced technology applied to bearing manufacture.

High-tin aluminium alloys, as cast, have very poor mechanical properties, because the tin forms a continuous network enclosing the primary aluminium crystals. By cold-working and low-temperature heat treatment this continuous phase can be broken up to produce an interlocking network structure, the so-called reticular structure which has greatly improved mechanical properties. However, the alloy is still in strip form and must be bonded to a thin, steel strip backing. This is achieved by a continuous pressure-welding operation carried out between rollers, the bonding between the two strips being promoted by a very thin sheet of pure aluminium.

This type of bearing has a higher load-carrying capacity than copper-lead alloys and yet can be used in conjunction with soft journals. Its tolerance for dirt is not good and it is occasionally overlay-plated.

The latest development in aluminium bearings is the aluminium-Babbitt alloy. At least three suppliers are known to be active in this field, each using a distinctly different manufacturing technique. In one case the method of manufacture is not unlike that used for the 20% tin-80% aluminium bearings. In another, a continuous sintering process is used. The Babbitt alloy is about 90% lead-10% tin, and since tin is only sparingly soluble in aluminium, the tin remains in solution in the lead. This means that the lead-rich phase (the Babbitt) should remain corrosion-resistant - in contrast to the situation with copper-lead and lead-bronze alloys.

5.3.5 Phosphor and Silicon Bronzes

These alloys are cast as either individual bearings or tubes from which bearings can be machined. In recent years, centrifugal and semi-continuous casting procedures have been used. Materials of this type are used mainly for bushes, particularly little-end bushes, where heavy loads and high temperatures are encountered. The presence of phosphorus at about the 0.5% level has a marked effect on resistance to pounding, i.e. the alloy has outstanding resistance to wear when subject to heavy loads at low sliding speeds.

5.3.6 Silver

Silver bearings, sometimes plain, sometimes with a thin overlay of lead-indium, have been adopted by the aircraft industry and are used on one well-known make of diesel engine. The silver is electrodeposited on a steel backing and is about 0.5 mm thick. Electrodeposited silver is much harder than cast silver, and silver bearings are unequalled with respect to load-carrying capacity and fatigue resistance. Unfortunately, they are prone to seizure and are very sensitive to the nature of the lubricant and to certain lubricating-oil additives.

5.3.7 Porous and Self-Lubricating Bearings

Most porous bearings consist of sintered bronze or iron powders with interconnecting pores. These pores may take up 10-30% of the total volume and, in operation, lubricating oil is stored in them and is subsequently fed to the bearing surface. Any oil escaping from the loaded zone is reabsorbed by capillary action. A typical bronze contains 90% copper and 10% tin; 1-4% graphite may be added to the mix to enhance the self-lubricating properties. High porosity and high lubricating-oil content are required for high-speed, light-load applications, whereas a low-porosity material with a high graphite content is better for oscillatory and reciprocating movement, where it is difficult to establish an oil film.

5.3.8 Discussion on Metallic Bearing Materials

No bearing material combines all the desirable properties. However, the important properties of bearings can be grouped in two main categories. One includes surface characteristics, such as wear resistance, journal compatability, conformability and embeddability, the other includes mechanical properties, such a fatigue strength and load-carrying capacity. Since these two categories are divergent, to have optimum surface characteristics strength must be sacrificed and vice versa. However, by using modern manufacturing methods, it is generally possible to arrive at a good compromise solution as, for example, in the case of tri-metal bearings.

There is no clear understanding of the way in which bearing alloys function, and general theories regarding their behaviour can almost always be discredited by reference to some particular bearing. In the formulation of bearing alloys, practice has always been ahead of theory. This is not to say that bearing design and manufacture is a backward industry; in fact, the manufacturing techniques described prove the opposite, and bearing manufacturers can provide satisfactory solutions to almost any bearing problem. Nevertheless, many bearings still fail prematurely in service and there is a great need to tell designers about the bearing materials already available or under development.

5.4 BEARING FAILURES

Sections 5.1 - 5.3 have provided the reader with a brief introduction to the subject, and it is now possible to describe and discuss some of the likely causes of bearing failures, giving particular emphasis to the identification of factors that give rise to failures.

5.4.1 Metallurgical Defects in New Bearings

Modern plain bearings can be very complex constructions and much may go wrong during their manufacture. Nevertheless, very few defective bearings enter service, since the major bearing manufacturers maintain a high standard of quality control. Defects are more likely to occur on individually manufactured bearings, e.g. large, white-metal bearings, than on mass-produced bearings of the copper-lead or aluminium-tin varieties. Defects that may be encountered are described in Sections 5.4.1.1 - 5.4.1.5.

5.4.1.1 Bad Bonding

Bonding is still a major problem on large, white-metal bearings, although poor bonds are extremely rare on other types of bearing. Bearing shells for white-metal bearings must be tinned before the white metal is cast in place, and the temperature of both the shell and the white metal must be carefully controlled. A metallurgical bond between the bearing and the shell strengthens the white metal and facilitates the flow of heat away from the bearing surface. In the past, some "white metallers" have argued that if they machined dovetail grooves in the shell, these held the white metal in place and there was no need to insist on good bonding. Experience proves that this is an extremely short-sighted policy; dovetail grooves are no substitute for good bonding and, in addition, the sharp edges at the shoulders of the grooves act as localised stress raisers and can initiate cracks (Fig.1).

A general view of a copper-lead bearing which failed due to bad bonding, the copper-lead bearing alloy having separated cleanly from the steel shell, is shown in Fig.2.

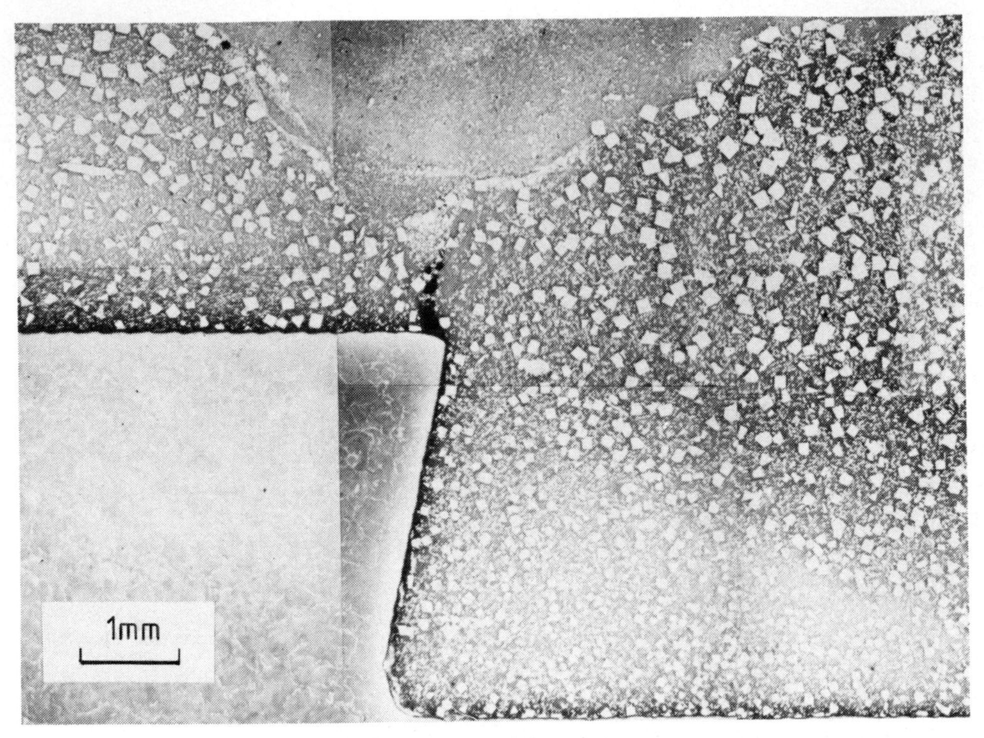

Fig.1 Dovetail in white-metal thrust pad bearing.

Fig.2 Bad bonding: copper-lead bearing.

Bad bonding on new bearings can be detected readily using ultrasonic or electrical resistance measuring techniques, and there are a number of commercially available test instruments. On a failed bearing, bad bonding can be distinguished from fatigue failure because with the former, the bearing metal detaches cleanly from the shell (Fig.3).

5.4.1.2 Gas Cavities

In large, white-metal bearings, too low a casting temperature or insufficient preheating of the shell can give rise to gas cavities near the shell surface. Such cavities weaken the bearing metal and adversely affect heat flow; sometimes, the hydrostatic pressure above the cavities may cause their collapse (Figs. 4 and 5).

Cast copper-lead alloys sometimes exhibit casting porosity, which in this instance is generally assumed to be due to hydrogen evolution from the copper. This porosity due to out-gassing must be distinguished from corrosion of the lead phase by oil-oxidation products. Casting porosity is sub-surface, whereas with corrosion, the lead in the surface layers is preferentially dissolved (Fig.6).

5.4.1.3 Oversize Cuboids

This problem is specific to white-metal bearings. In larger bearings, slow cooling through the solidification range can give rise to oversize (> 0.1 mm) tin-antimony cuboids (Fig.7). This intermetallic compound is brittle and the large cuboids can crack and break-up in service, causing scoring of the journals and damage elsewhere in the oil system (Fig.8). In general, if cuboids are readily visible to the naked eye they are too big.

5.4.1.4 Excessive Lead Content in Tin-Based White Metals

The lead content of tin-based white metals should be below 0.5% to prevent the formation of a low-melting, lead-tin eutectic in the grain boundaries. The eutectic weakens the alloy and makes it more susceptible to wiping. For similar reasons, it is bad practice to mix lead-based and tin-based white-metal half-bearings, as is sometimes done to save initial costs, tin being used on the loaded half and lead on the unloaded half. Carry-over of lead from one half-bearing to the other can result in the formation of the low-melting-point eutectic.

5.4.1.5 Uneven Lead Distribution in Copper-Lead and Lead-Bronze Alloys

Copper-lead and lead-bronze bearing alloys can be manufactured by melting and casting or by sintering techniques. Lead distribution is more readily controlled by powder-metallurgy techniques, so uneven distribution is mainly

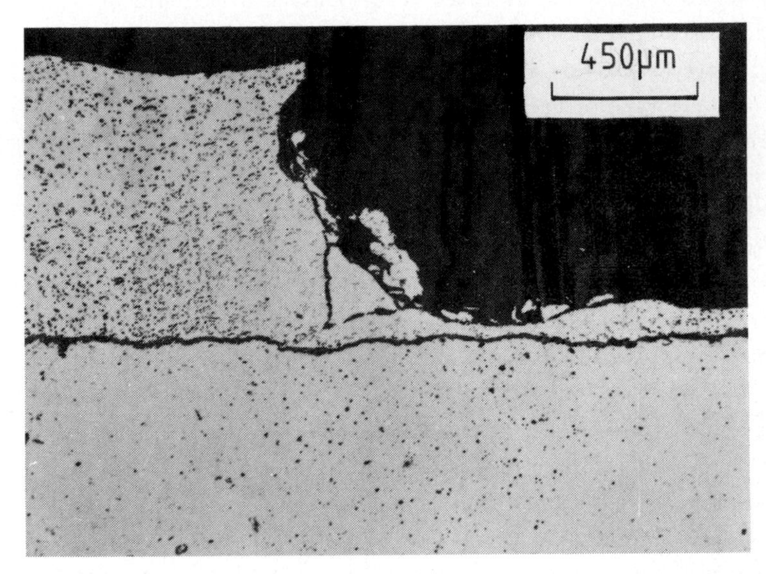

(a) Fatigue

(b) Bad bonding

Fig.3 Micro-sections.

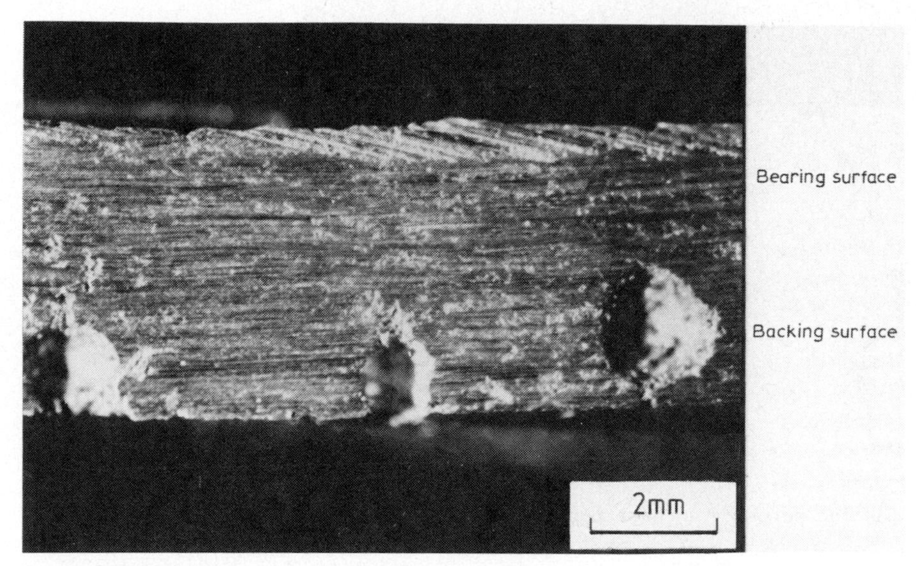

Fig.4 Cross-section of White-metals Showing Gas Cavities

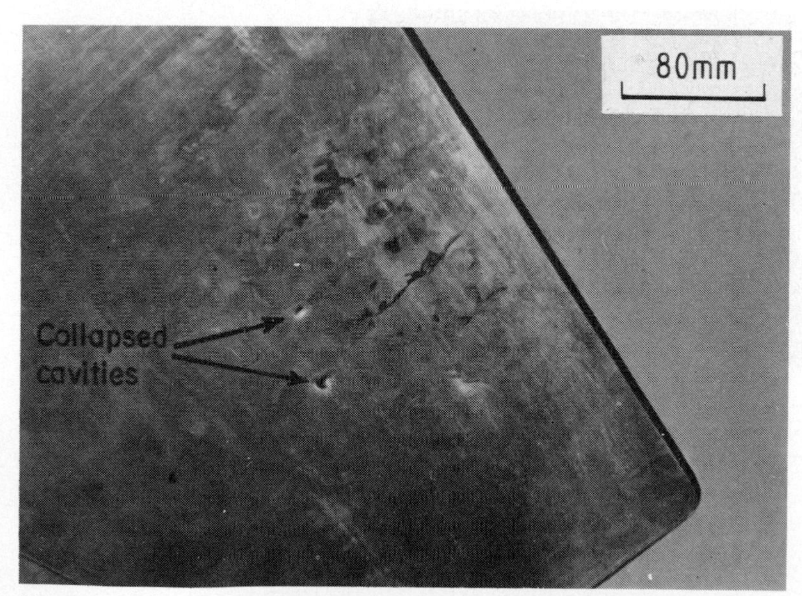

Fig.5 Collapsed surface of white-metal thrust pad bearing.

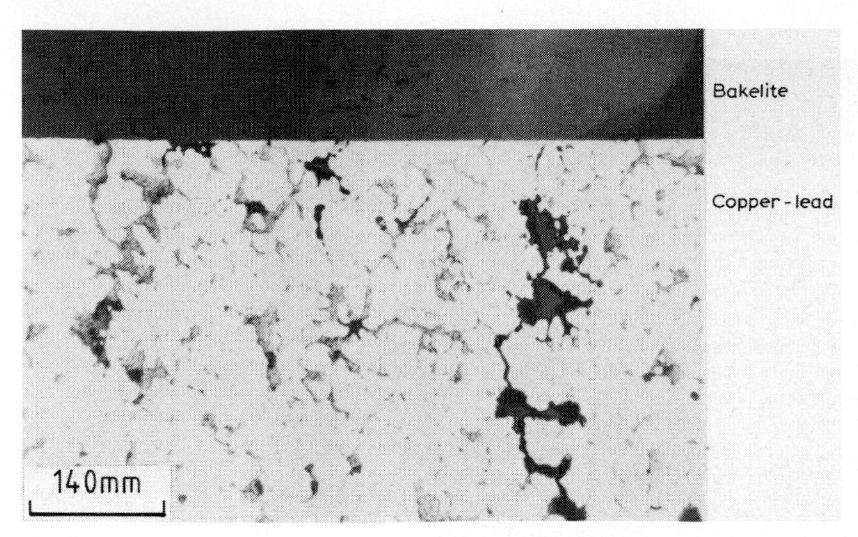

Fig.6 Casting porosity in cast copper-lead big-end bearing.

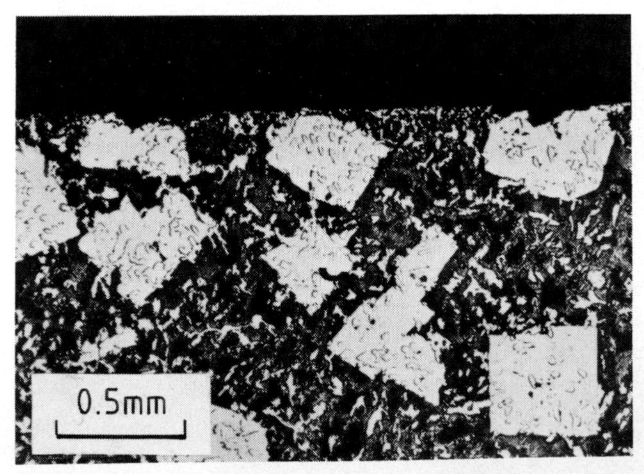

Fig.7 Oversize tin-antimony cuboids in white-metal thrust pad from steam turbine

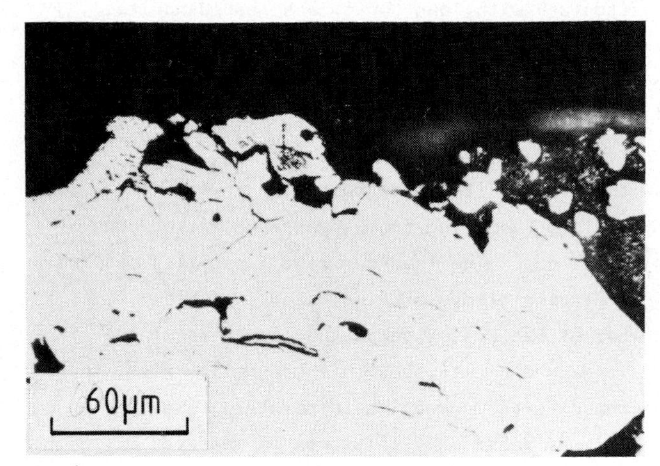

Fig.8 Single cuboids showing cracks.

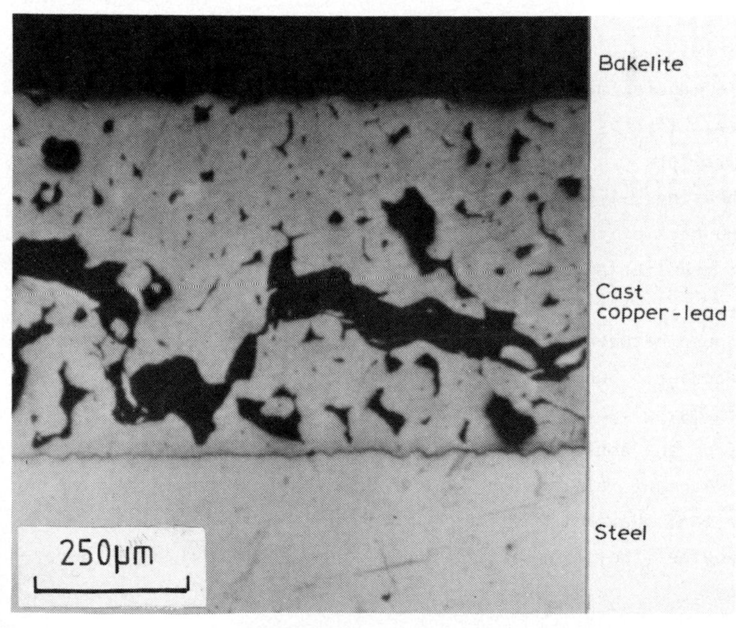

Fig.9 Bad lead distribution.

encountered on cast bearings. Copper and lead are immiscible and good lead distribution in cast alloys depends on special manufacturing techniques, in particular on rapid cooling of the alloy after casting. This rapid cooling results in a highly oriented structure with long copper and lead dendrites normal to the surface of the bearing shell. If manufacturing controls are relaxed, the lead can be unevenly distributed; for example, it may be present in large slugs, normal or parallel to the surface (Fig.9).

5.4.2 Failures Associated with Design, Fitting or Operating Environment

The final manifestation of failure resulting from adverse operating conditions of this type is generally fatigue. The load-carrying capacities and temperature limitations of typical bearing alloys are indicated in Table 5.2. However, these depend on a number of factors, such as the thickness of the bearing alloy and the support it gets from its substrate or shell. Also, the strength of bearing alloys decreases with increasing temperature, and bearing surface temperatures are generally at least 20°C in excess of measured oil temperatures.

The design of bearings, particularly the location of oil feeds and oil grooves, is outside the scope of this paper. Nevertheless, it should be emphasised that load-carrying capacity is rarely enhanced by a proliferation of oil grooves.

5.4.2.1 Bad Fitting

Modern bearings are manufactured to very close tolerances and should be assembled with great care. If they are not accurately aligned, the loading will be uneven and premature failure due to fatigue may occur. If the bearing does not fit properly in its housing, slight movement may occur, which can result in severe fretting on the back of the bearing. Loose fitting and/or fretting can adversely affect heat transfer and can give rise to excessive bearing-surface temperature (Fig.10).

Dirt particles trapped between a bearing and its housing constitute another cause of poor heat transfer. Such particles can also cause high spots on the bearing surface, which may give rise to localised wiping and fatigue (Fig.11).

It will be clear from the above that examination of the back of a failed bearing can provide important information regarding its mode of failure. The comment has been made that the best way to evaluate "experts" on bearing fatigues is to note whether they examine the backs of bearings with as much care as the sliding surface.

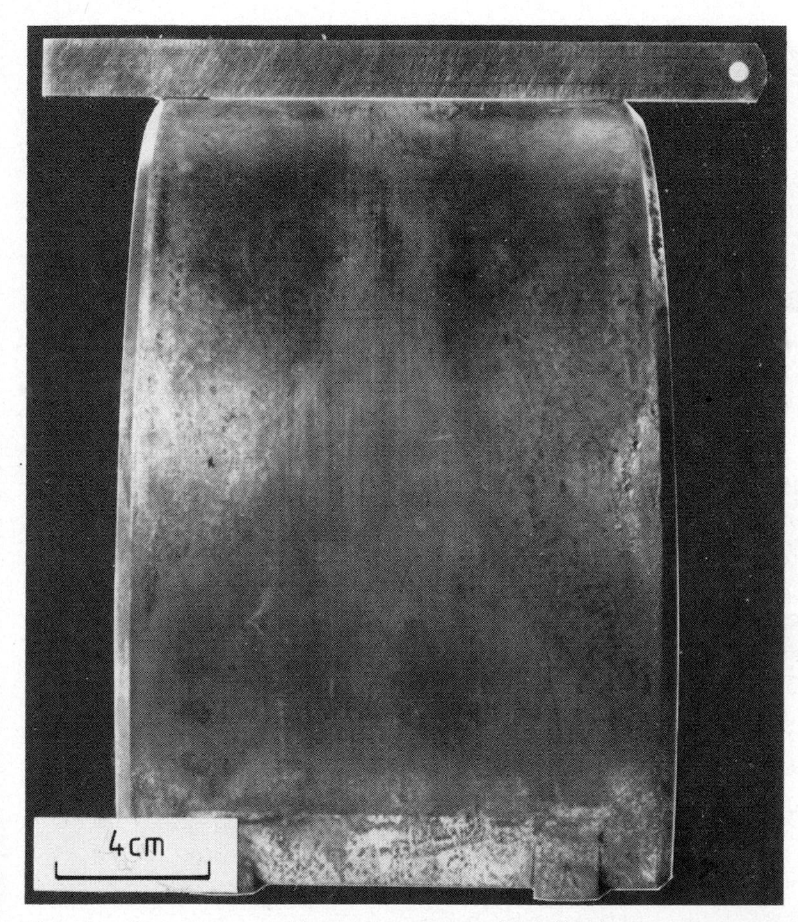

Fig.10 Fretting and distortion due to bad fitting.

5.4.2.2 Extraneous Particles

This is probably the most common cause of bearing failure. The more expert we become at detecting and identifying foreign material, the more obvious it is that cleanliness on assembly and good lubricant filtration in service are essential for long, trouble-free service lives. All too frequently, engineering components are assembled without adequate cleaning in dirty environments, with the result that machining swarf, moulding sand and other kinds of debric circulate with the lubricating oil. Large particles are generally removed by the oil filters, but medium-sized particles may embed in the bearings and very small particles continue to circulate with the oil. Foreign particles and wear debris can also accumulate during service, and very long oil-change periods, now much in favour, aggravate the situation.

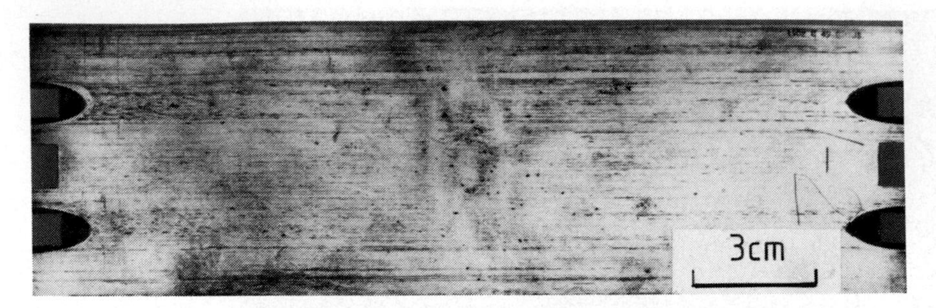

(a) Back of bearing, showing mark due to trapped particle.

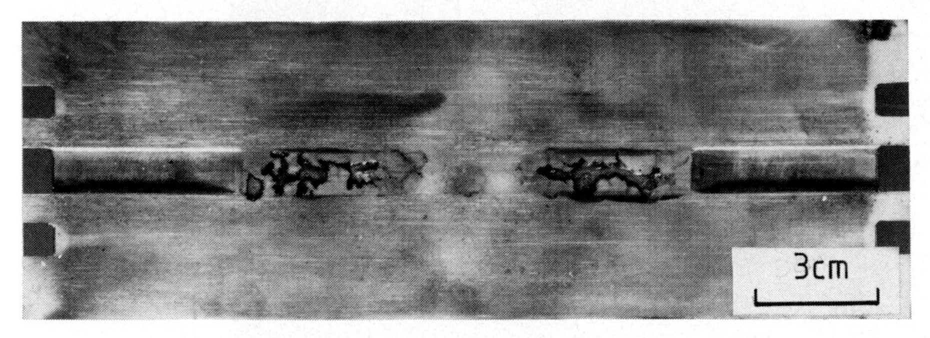

(b) Localised fatigue on high spot above particle.

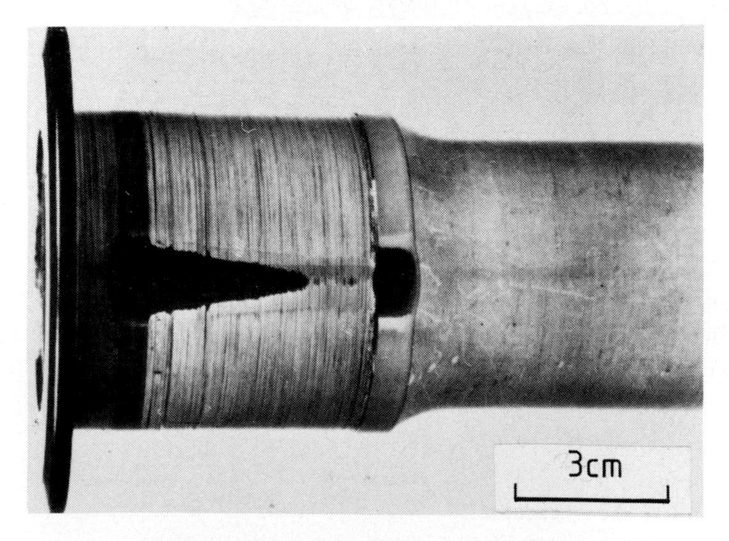

Fig.12 Shaft wear due to embedded particles in bearing.

Bearing alloys, particularly white metals, have a considerable capacity for abrasive materials. When this capacity is exceeded, the bearing surface can act as a lap, abrading away the journal surface, thereby increasing clearances and leading to fatigue failure. This lapping action by embedded particles is the prime cause of journal wear (Fig.12).

In most instances, the bearing surface will be in relatively good condition (unless complete failure has occurred), although the overall thickness of the bearing can actually increase owing to the amount of embedded material.

Particles embedded in bearing surfaces displace metal and therefore are often surrounded by a raised, burnished halo (Fig.13).

If the composition of embedded particles can be determined, their origin may be established and the appropriate steps taken to eliminate them.

A variety of identification techniques are now available. For example, the particles may be extracted chemically or mechanically for identification by X-ray diffraction or X-ray spectrometry. They can be examined in situ using an electron-probe. Alternatively, chemical extraction techniques have much to recommend them; the amount of debris generally obtained is enough to horrify most plant operators (Fig.14).

Particles in electrodeposited lead overlays can be extracted with an acetic acid/hydrogen peroxide mixture, which dissolves the overlay; those in aluminium -tin bearings can be extracted with caustic soda. Both mixtures can loosen particles in tin-based white metals.

Embedded ferrous particles (the largest single group) can be quickly identified by "iron printing" (Fig.15).

This involves soaking an unglazed paper in a 5%w solution of potassium ferri-cyanide to which a few drops of hydrochloric acid and wetting agent have been added. The surplus liquid is drained off the paper, which is then placed in contact with the degreased surface of the bearing for about 30 seconds. The paper is then removed, and blue spots indicate the presence of ferrous particles. The particles that remain in the bearing surface are also stained blue. Similar techniques are available for the identification of particles of most common non-ferrous metals.

Since ferrous particles constitute the most common type of abrasive wear particle, the use of magnetic filters, in addition to ordinary filters, has much to recommend it.

Fine particles, smaller than normal bearing clearances, can circulate with the lubricant and erode the bearing surface. Hard particles erode deep, well-defined channels, while soft particles give rise to more general erosion (Fig.16), particularly on soft electrodeposited overlays.

Erosion by fine particles is most prevalent on high-speed bearings and may be associated with cavitation erosion.

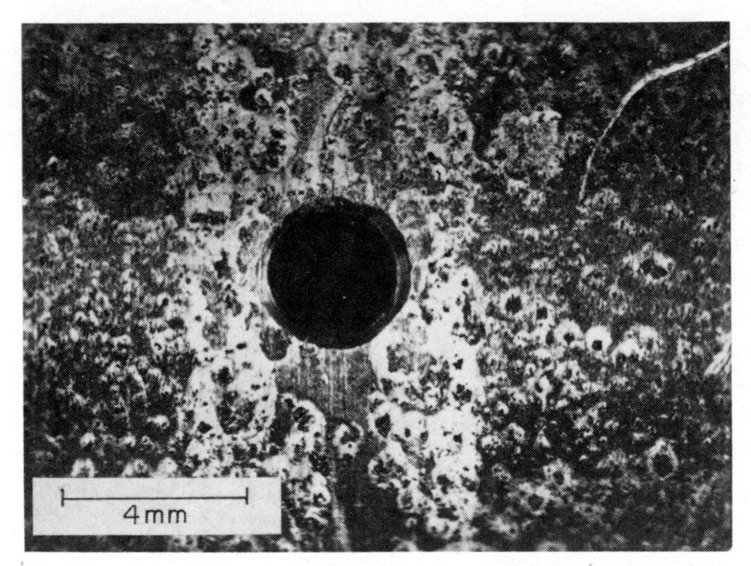

Fig.13 Particles embedded in bearing surface.

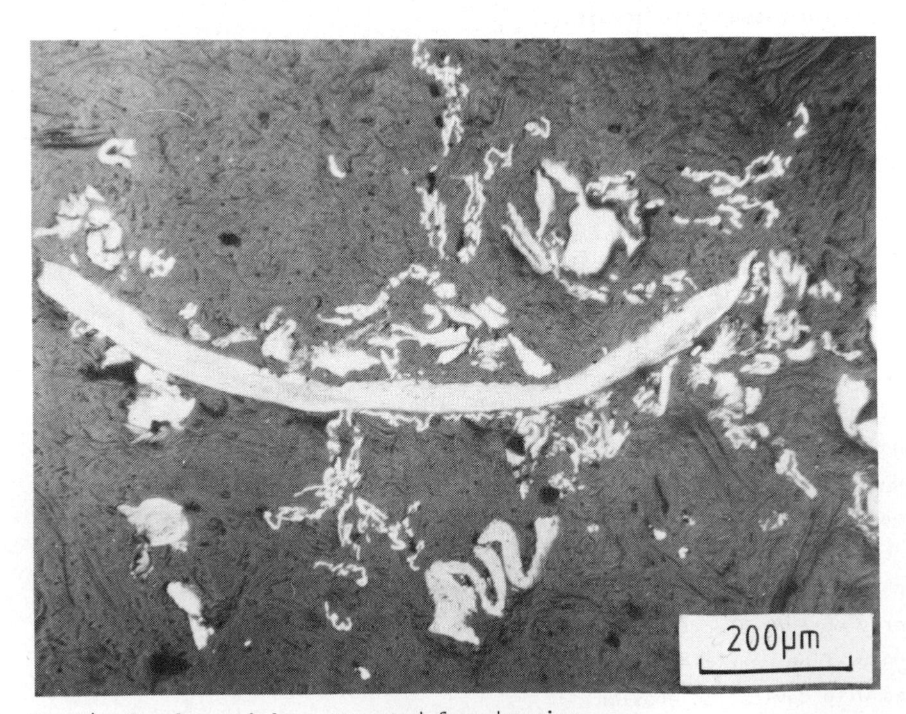

Fig.14 Steel particles extracted from bearing.

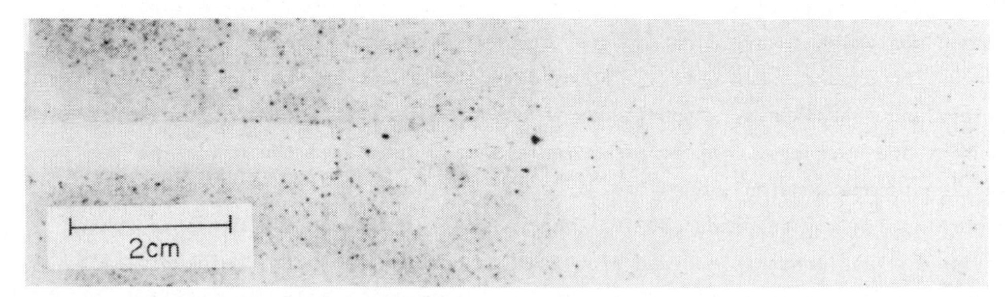

Fig.15 Iron print.

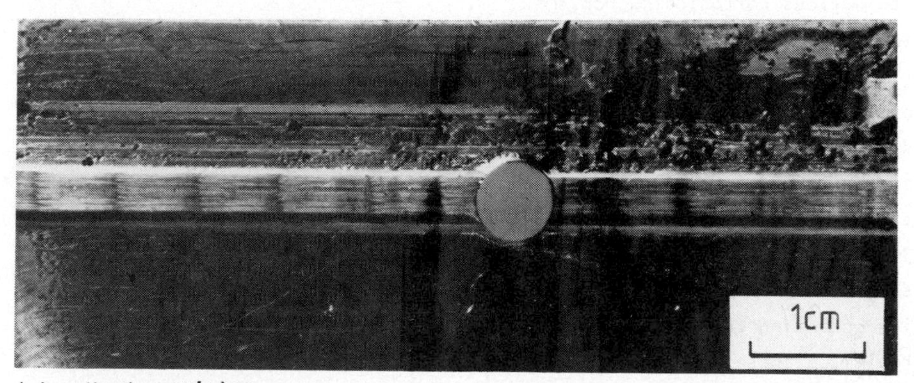

(a) Hard particles

(b) Soft particles

Fig.16 Erosion damage.

5.4.2.3 Corrosion

Lubricants deteriorate in service in two ways - they become contaminated and they undergo physical and chemical changes due to oxidation. In engines, the common contaminants are airborne dust and wear products, unburnt fuel, fuel combustion products and water. The oxidation products are mainly acidic materials and asphaltenes. Asphaltenes in association with fuel contaminants form sludges and lacquers. The acidic materials resulting from the oxidation of lubricants are generally weak organic acids, although in extreme cases strong mineral acids may be produced. However, almost all strong mineral acid contamination in lubricants is derived from fuel combustion products, sulphuric acid being a major contaminant in diesel engines and halogen acids in gasoline engines.

Since bearings are of complex construction, the way in which their structure and composition affects their corrosion resistance will be discussed with reference to various corrosion mechanisms.

It should be emphasised that bearing failures that are primarily due to any form of corrosion are comparatively rare and that bearing failures caused by inherent deficiencies on the part of the lubricant are extremely rare. Nevertheless, certain specific forms of corrosion, such as tin oxide formation on tin-based white-metal bearings and sulphur corrosion of phosphor-bronze alloys, have attracted considerable attention. Also, as will be apparent in Section 5.4.2.3 (corrosion by weak organic acids) and 5.4.2.4 (cavitation and erosion), increasingly severe operating conditions, such as longer-oil-change periods, very high operating temperatures and increased speeds, can give rise to particular problems. These are problems in which corrosion plays a part, but in which other factors are also operative.

(i) Corrosion by weak organic acids

Weak organic acids arise either from prolonged exposure of the lubricant at elevated temperatures or by contamination of the lubricant with partially burnt combustion products. These acids attack lead far more readily than other metals, and can dissolve the lead phase in copper-lead and lead-bronze bearings (Fig.17).

Some engine lubricating oil specifications include tests to determine the corrosivity of lubricating oils towards copper-lead bearings. The loss of lead is assessed by weighing the bearing shells before and after test. In service failures, where the weights of new bearings are not available, lead corrosion can be detected by metallurgical sectioning (Fig.18).

Copper-lead and lead-bronze bearings may be manufactured by sintering or casting. In sintered alloys it is possible to ensure that the lead

Fig.17 Corroded copper-lead bearing

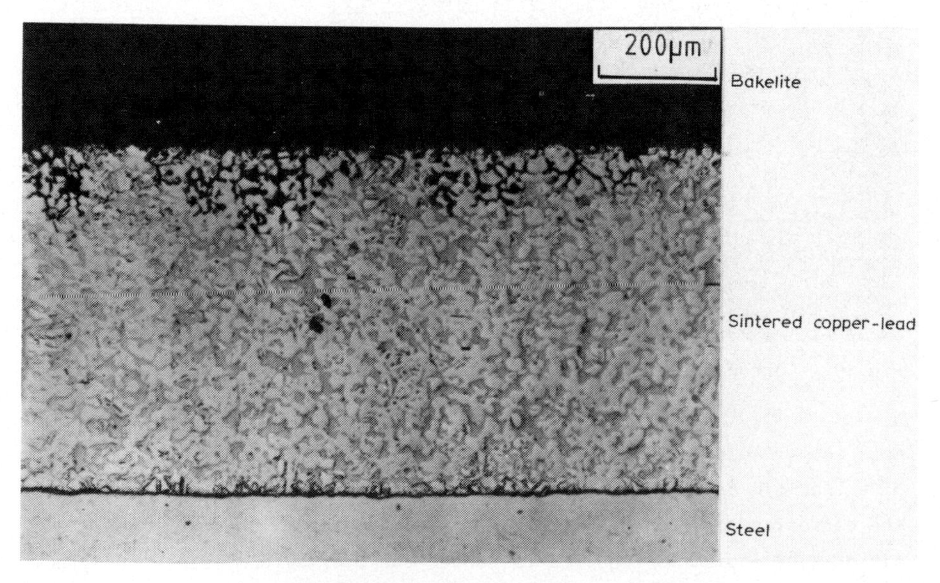

Fig.18 Corroded sintered copper-lead bearing

phase is well distributed and discontinuous, so the corrosion will be restricted to the surface layers. On the other hand, with cast alloys, satisfactory lead distribution depends on highly specialised manufacturing techniques and rapid cooling of the alloy after casting. This rapid cooling gives rise to a highly oriented structure, with long copper and lead dendrites normal to the surface of the bearing shell. The long lead dendrites provide an easy path for the penetration of a corrosive lubricant. In these circumstances it is possible for almost all of the lead phase to be leached out of a cast copper-lead alloy (Fig.19).

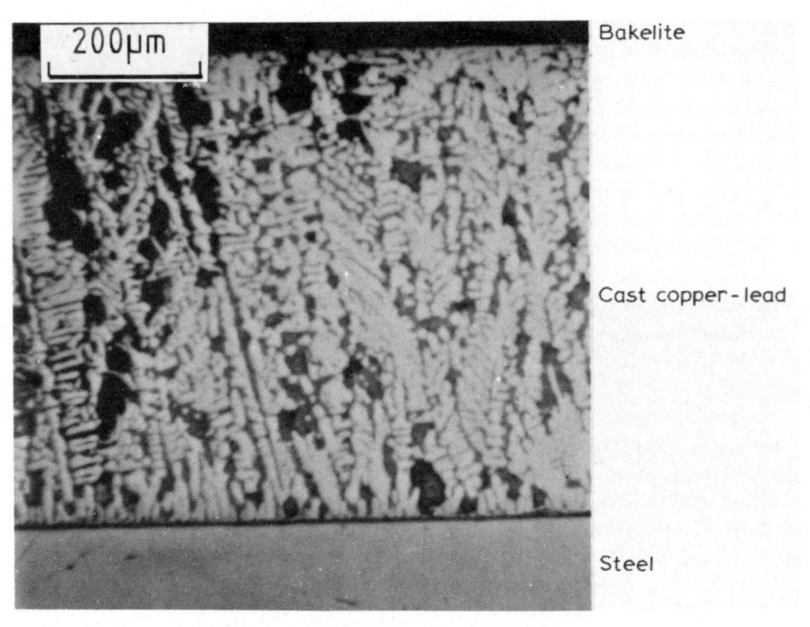

Fig.19 Corroded cast copper-lead bearing.

Most copper-lead bearings now have a precision electrodeposited overlay of a lead-tin or lead-indium alloy. The indium or tin additions improve the corrosion resistance of the lead. We have established that about 3% tin or 4.8% indium is required to render the overlay completely resistant to corrosion (Fig.20).

Bearing manufacturers generally provide overlays with about 5% indium or 10% tin. However, at the high temperatures that bearings can experience in service, the indium or tin diffuses quite rapidly and migrates into

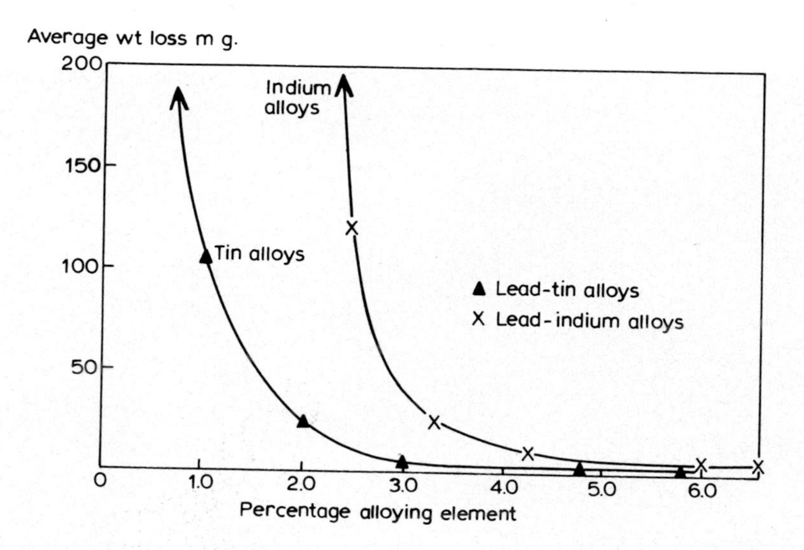

Fig.20 Corrosion of lead-tin and lead-indium alloys.

the underlying copper-lead. This diffusion can readily be observed both on bearings removed from service and on bearings tested in the laboratory. In the case of tin, the tin alloys with the copper to form copper-tin intermetallic compounds. In the case of indium, enrichment occurs adjacent to the copper, but we have no evidence of compound formation. Similar effects can be observed on silver bearings with lead-rich electrodeposited overlays.

This migration of indium and tin from the surface layers of the overlay leaves them susceptible to corrosion (Figs.21-23). It is said that one way to prevent this diffusion is to interpose a dam between the overlay and the underlying copper-lead (Fig.24).

Very thin nickel, iron or brass layers have been used as dams. However, many bearings users are unwilling to face the increased cost associated with the use of dams and, in any case, their effectiveness is being questioned.

There is another way in which the indium in lead-indium overlays may be depleted. If the indium content falls to below about 3%, internal oxidation of the indium can occur in the grain boundaries (Fig.25). The indium oxide formed in this way embrittles the alloy and renders it very susceptible to fatigue failure as well as to corrosive attack (Fig.26).

Fig.21 Corrosion of bearing overlay.

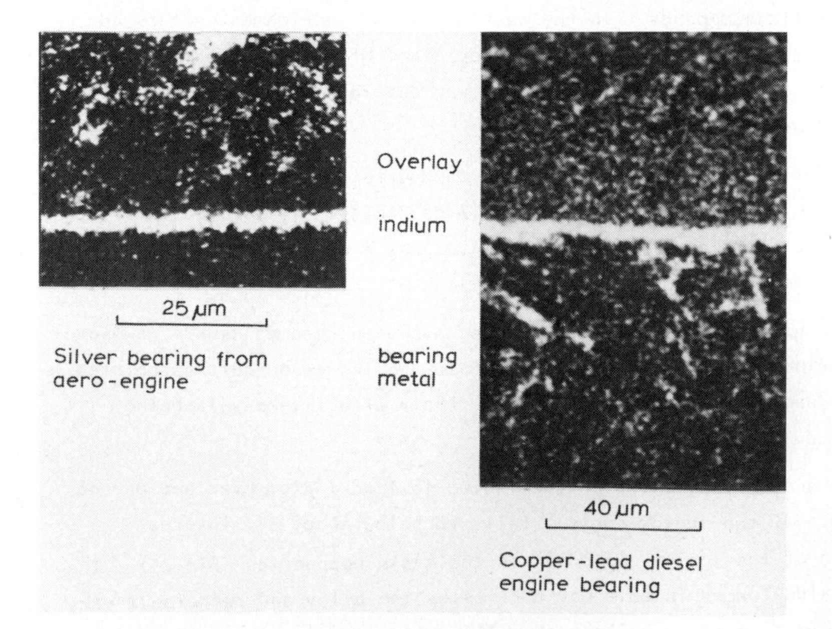

Fig.22 Electron probe micrographs showing indium distribution.

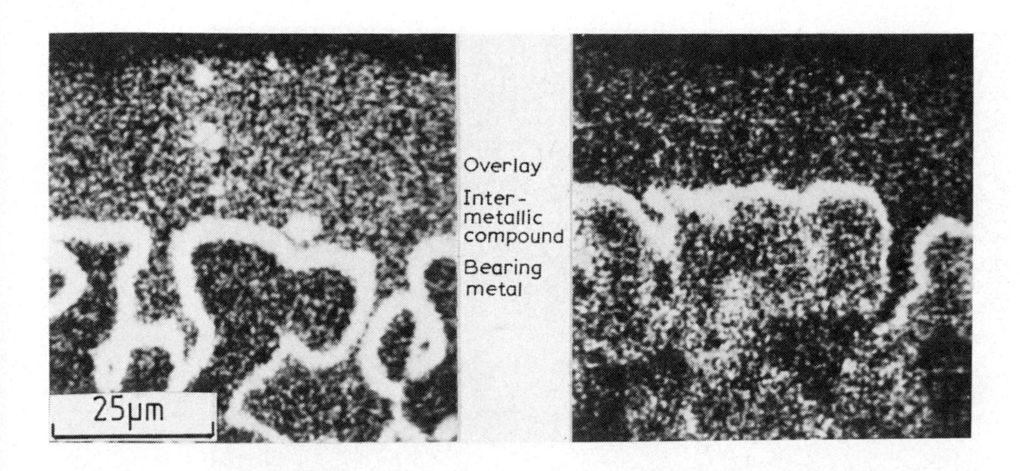

After 13,000 hrs. After 200 hrs. at full load.

Fig.23 Intermetallic compound formation in overlay bearings (copper-lead).

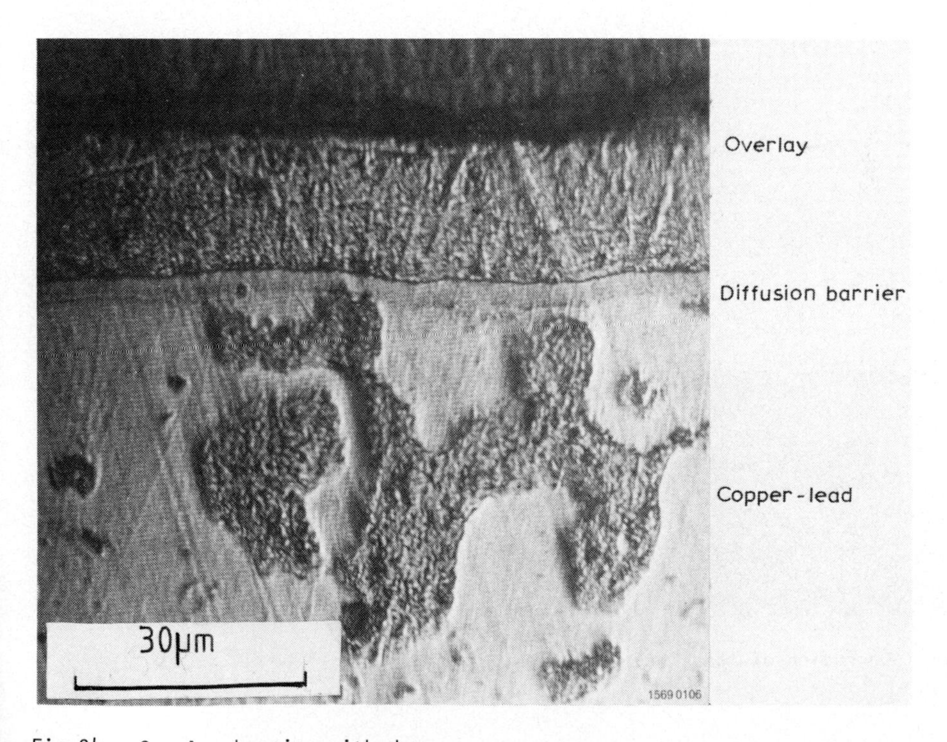

Fig.24 Overlay bearing with dam.

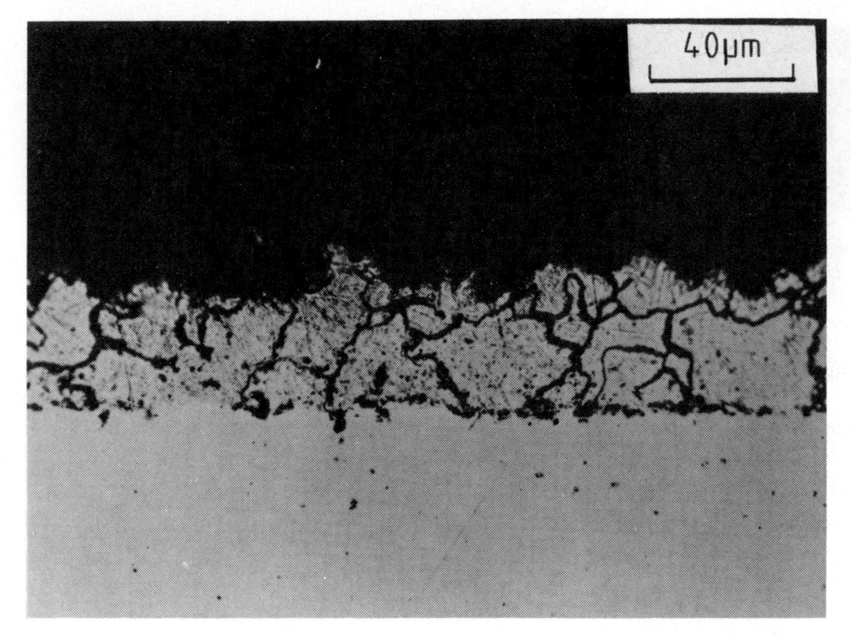

Fig.25 Internal oxidation of lead indium overlay.

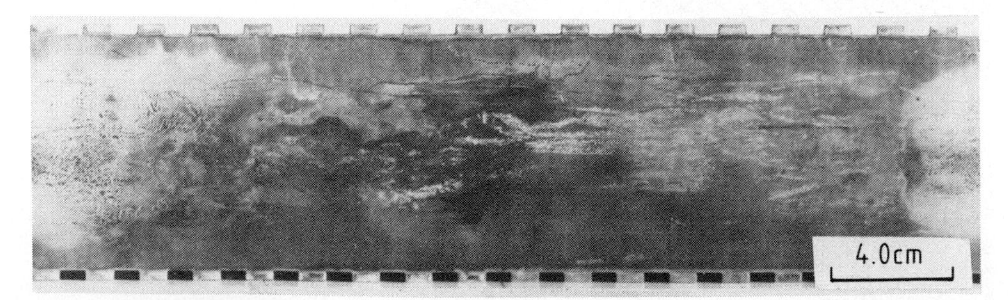

Fig.26 Corrosion of lead indium overlay.

Another consequence of the diffusion of tin from lead-rich overlays is the roughening of the overlay surface (Fig.27). Although copper has a much higher affinity for tin than has lead, tin diffuses much more rapidly in lead than in copper. This means that where the lead phase in copper-lead alloy is continuous with the overlay, the tin can penetrate deeply into the alloy in the lead phase (Fig.28) and cause sinking of the overlay. This movement of tin away from the overlay can give the overlay a roughened appearance, which may be mistaken for corrosion.

(ii) Corrosion by strong mineral acids

The main source of strong mineral acids in lubricants is contamination by fuel combustion products. Diesel fuels, particularly marine diesel fuels, contain significant quantities of sulphur; this sulphur is the source of sulphuric acid, which may find its way into the lubricant. Gasolines, on the other hand, contain very little sulphur, but do contain chlorine and bromide compounds which are added to scavenge the lead antiknock compounds. Thus, lead halide compounds may accumulate in the lubricating oil in gasoline engines and, in certain circumstances, halogen acids can form.

The strong acids generally attack bare steel surfaces rather than bearing alloys. This usually results in journals being roughened by corrosion, and bearings then fail either because of this or as a result of damage by corrosion products (rust).

In the presence of aluminium and moisture, lead halides can deliquesce, giving rise to halogen acids, which corrode aluminium. This corrosion reaction was first observed on aluminium-alloy pistons and can readily be duplicated in the laboratory. Fortunately, examples of this type of attack on aluminium-tin bearings in engines are extremely rare, although bearings are susceptible to halide corrosion after removal from engines.

Halogen attack has also been observed on lead overlay bearings operated at very high engine temperatures, when mixed bromide/chloride layers can form on the overlay surface and cause blackening (Figs. 29 and 30). Electron-probe studies of the reaction layer show that it contains indium at the same concentration as in the bearing, indicating that the lead halides have been formed in situ as a result of corrosion and have not been derived directly from the fuel.

(iii) Sulphur corrosion

This is a general description which is applied to most forms of corrosion encountered on silver-rich or copper-rich bearing alloys.

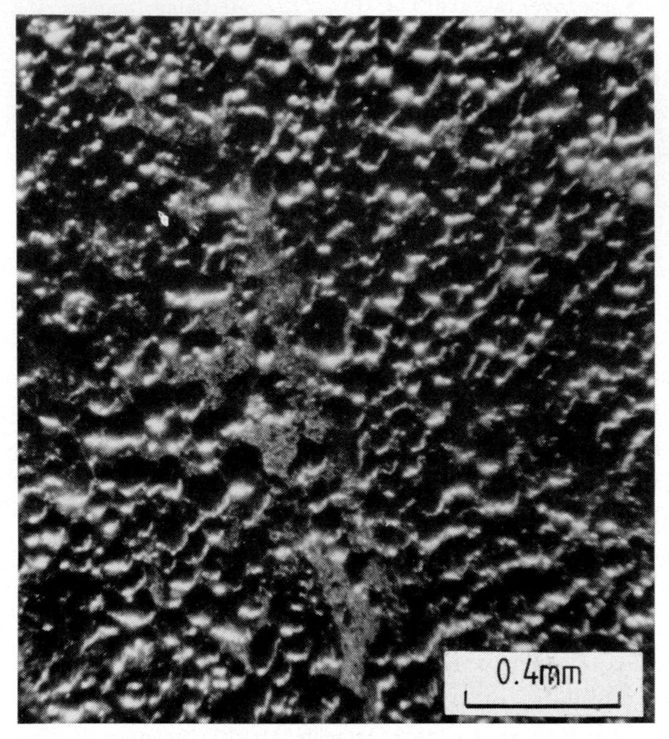

Fig.27 Roughening of overlay due to diffusion.

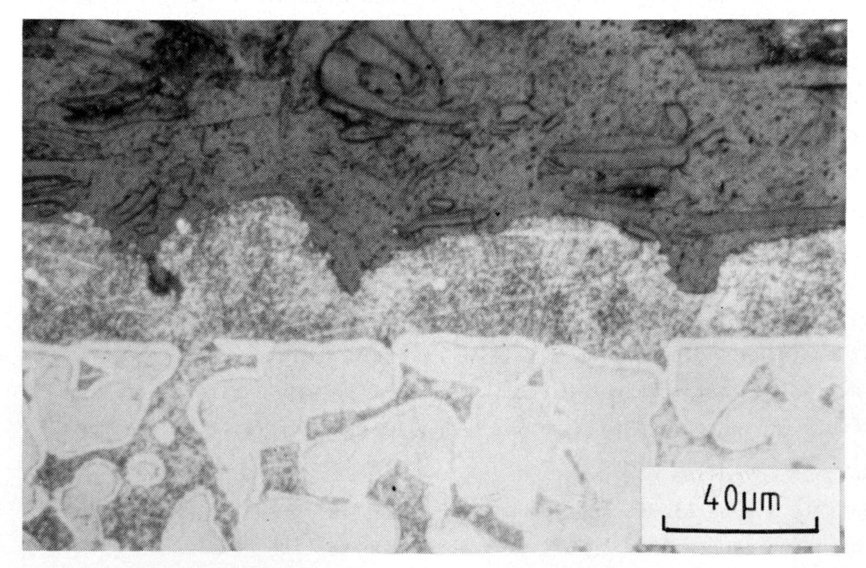

Fig.28 'Sinking' of overlay.

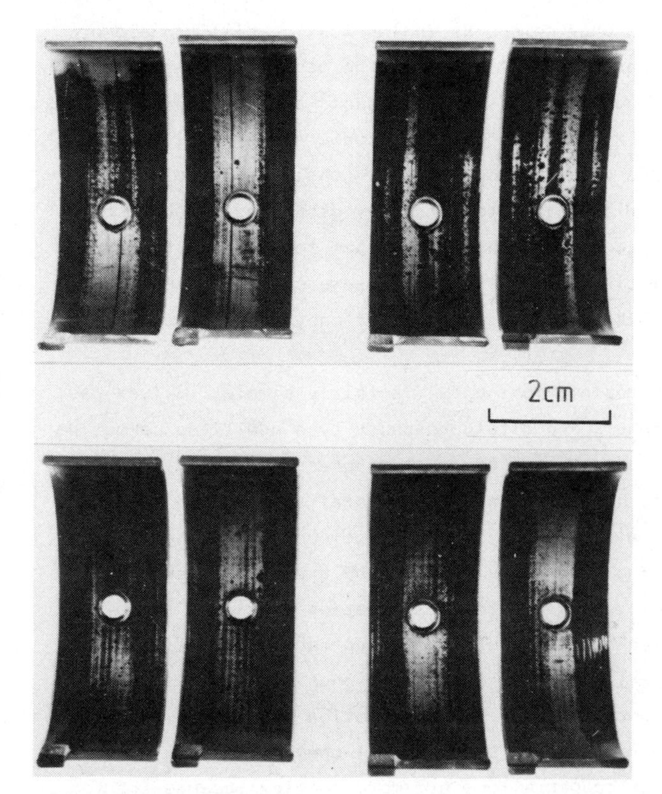

Fig.29 Blackened big-end bearings.

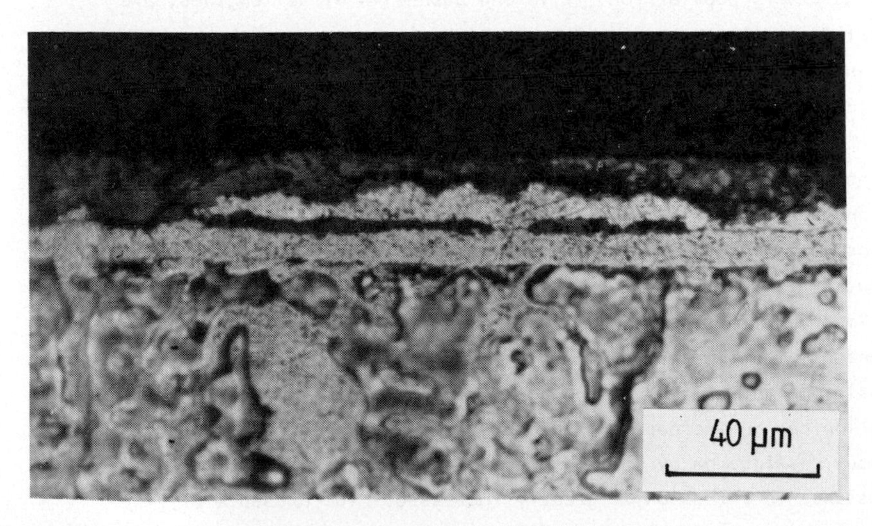

Fig.30 Lead halide layer.

There is no doubt that sulphur compounds in lubricating oils can promote the corrosion of these particular alloys. On the other hand, both naturally occurring sulphur compounds in lubricants and sulphur-containing additives (e.g. zinc dialkyldithiophosphates) confer beneficial properties on lubricants. Thus, the dithiophosphates show considerable antioxidant and anti-wear effects. With regard to naturally occurring sulphur compounds, modern refining techniques can remove them completely, but experience shows that this is most unwise, since some of these compounds play a large part in inhibiting the corrosion of many metals, particularly lead.

On silver bearings the problem, in theory, is fairly simple. Silver has a high affinity for sulphur, and dithiophosphate-type additives cannot be used. However, a small amount of e.p. (extreme-pressure) activity on the part of the lubricant is necessary, and the natural sulphur can sometimes meet this requirement. Alternatively, a chlorine-containing additive may be used. A silver bearing after service would be expected to show a certain amount of discoloration due to the formation of silver sulphide or silver chloride, which have a beneficial effect on the frictional characteristics. However, too much reactive sulphur can cause severe corrosion. It can be concluded that the formulation of oils to lubricate silver bearings requires the expertise of an oil chemist.

The corrosion problems of copper alloys are more complex because the alloys themselves are complex. On simple copper-lead alloys the copper phase may be attacked by sulphur, but this is a comparatively rare occurrence, the problem of sulphur corrosion being much more acute on phosphor-bronze alloys (Fig.31). This is because phosphor-bronze alloys, which are very popular for little-end bushes in diesel engines, are expected to operate at considerably higher temperatures than copper-lead bearings. In any case, there are very few bare copper-lead bearings in use today.

There is no general agreement about the corrosion mechanism. Some engine manufacturers and users hold dithiophosphate additives entirely responsible, but this opinion cannot be substantiated, for severe corrosion can occur when oils containing only natural sulphur compounds are used.

Two important factors influencing the severity of corrosion are the amount of alloying element in solution in the copper-rich phase and the porosity of the alloy. In phosphor bronzes, if the amount of tin in solution can be increased by special casting techniques, such as continuous casting, or by solution treatment after manufacture, the resistance to corrosion is greatly increased. The presence of zinc and/or silicon as alloying elements in copper alloys also increases the resistance of these alloys to sulphur corrosion. However, it is difficult to make sound silicon-bronze castings.

If the alloy is porous, the lubricant is drawn into the pores, where it stagnates and, if operating temperatures are high, can become very corrosive (Fig.32). The particular temperature at which corrosion becomes severe depends on the type of dithiophosphate used; very active varieties can start to corrode at about 130°C, whereas other varieties may be comparatively stable up to 180°C. Natural sulphur compounds give little trouble below 170°C. Most phosphor-bronze castings are microporous, and the greater the porosity the greater the risk of corrosion. Cases are known where only certain bushes corroded in a particular engine, and metallurgical examination showed that the bushes which corroded were porous. Continuous-casting techniques give sounder alloys than other casting techniques and, in addition, a greater proportion of tin stays in solution, thereby improving the intrinsic corrosion resistance. However, if the problem is to be controlled, the only satisfactory solution is to use an alloy that is resistant to sulphur corrosion. Laboratory tests followed by extensive field experience extending over ten years have shown that alloys of the gun-metal type, i.e. copper-tin alloys with 2-4% zinc, are completely immune from sulphur corrosion. These gun-metal alloys are rather softer than the traditional phosphor bronzes, and nickel may also be added to compensate for the reduced hardness.

The importance of microstructure in this type of corrosion was recently demonstrated by some phosphor-bronze bushes which corroded in regularly spaced bands, despite the fact that they were free from porosity and had been operated on a lubricant containing only a small amount of a very stable dithiophosphate. Metallurgical examination showed that they had been manufactured by a semi-continuous casting process, which gave rise to marked segregation in the alloy, making it very susceptible to corrosion in certain areas (Fig.33).

Recently some cases of what is claimed to be sulphur corrosion have been encountered on tin-based white-metal bearing alloys. It appears that very active sulphur compounds can selectively attack the copper-rich, copper-tin intermetallic compounds in the white metal, and that the resulting corrosion products, rich in Cu_2S, can spread over the bearing surface. The darkening caused by these corrosion products should not be confused with the darkening due to tin oxide formation.

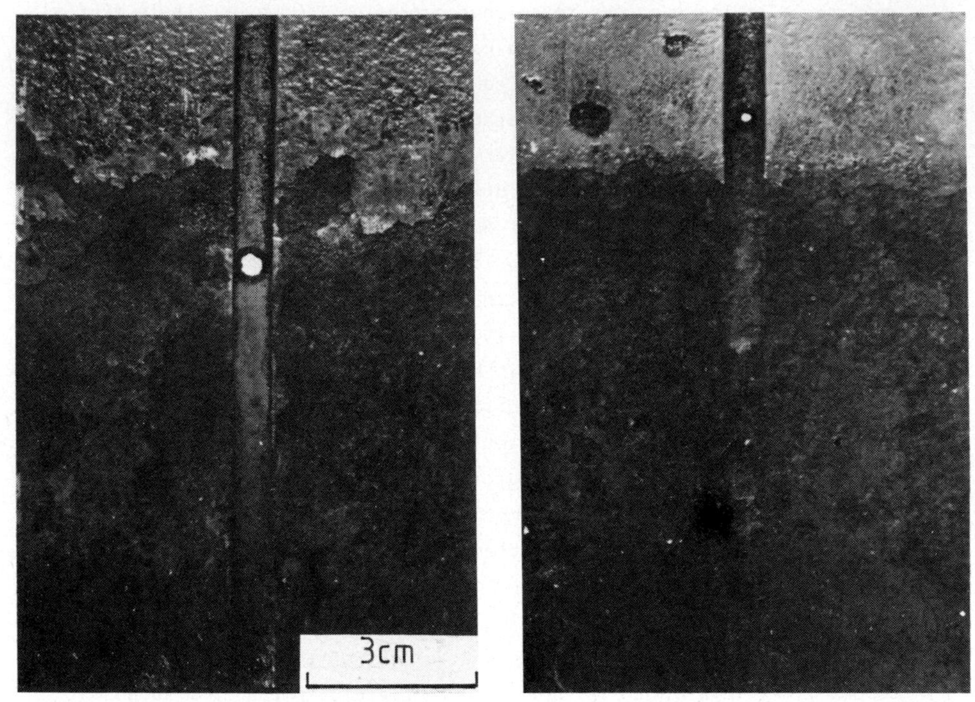

Fig.31 Corrosion of phosphor bronze.

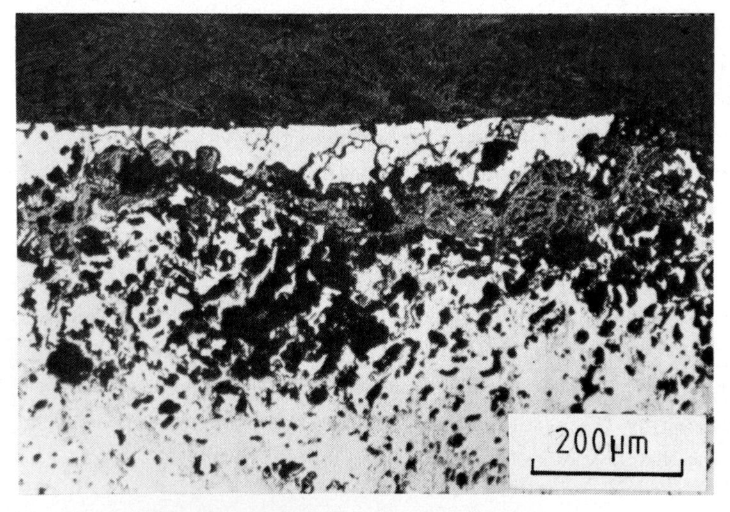

Fig.32 Subsurface attack on phosphor-bronze.

(iv) Corrosion of tin-based white metals: tin oxide formation

In recent years the formation of hard, black layers on the surface of tin-based white-metal bearings in ships' turbines has attracted a lot of attention. Yet this is by no means a new problem; it has been encountered in the main diesel engines of ships for many years. Some turbine builders and ship owners have blamed the corrosion on the increasing use of additives in lubricating oils, but the fact that diesel engine bearings operating on base oils have suffered from this corrosion for many years throws some doubt on their assumptions.

The characteristic features of the corrosion are as follows. A hard, black layer forms on the surface of the bearings, sometimes only on the working surfaces but generally on all the exposed surfaces (Fig.34). The presence of this layer is extremely harmful for two reasons: firstly, it is very hard and may damage the adjacent steel surfaces, and secondly, its formation decreases normal bearing clearances. Metallurgical sections from failed bearings show that the hard layer is formed from the tin-rich matrix of the bearing alloy (Fig.35). The copper-tin and antimony-tin intermetallic compounds are completely unattacked. Microhardness tests show that the hardness of the black matrix is between 200 and 600 DPN, and usually between 400 and 500 DPN. The hardness of the untransformed matrix is about 25 DPN.

X-ray diffraction examination shows that the black layer consists mainly of cassiterite, SnO_2, usually with a little stannous oxide, SnO, associated with it. On thrust pads, where bearing clearances are large, the black layer can grow to a considerable thickness, but eventually it disintegrates and the hard debris circulates with the lubricant.

Some years ago, Bryce and Roehner discussed this problem in detail, but failed to arrive at a satisfactory explanation. However, it has been established that the corrosion occurs only when aqueous electrolytes are present in the lubricant, which suggests that the tin dioxide is formed by an electrochemical mechanism. Some investigators claim to have reproduced the corrosion in the laboratory by making a piece of white metal the anode in sea-water but, although this produces a black layer, the layer is soft and amorphous. We have found that the natural current between pieces of white metal (tin-based to B.S.S. 3332/2) and copper, partially immersed in full-strength or diluted synthetic sea-water under oil at 60°C, will cause the white metal to corrode, forming thick, hard, adherent stannic oxide.

The black layers formed in this way have the crystallographic and metallographic characteristics of the layers found in engines. With this simple set-up it has been possible to investigate the corrosion mechanism

Fig.33 Effect of micro-structure on corrosion.

Fig.34 Tin oxide on thrust bearing

and to study the influence of various oil additives on the corrosion rate. Some additives reduced or prevented tin dioxide formation in the test cells; unfortunately, almost all the additives that behaved in this way would have an adverse effect on other components in engines. For example, one additive that completely prevented corrosion of the tin was corrosive to copper alloys.

The mechanism of tin oxide formation on bearings is still not fully understood. However, it has been definitely established that corrosion occurs only when electrolytes are present in the lubricant and when there is a restricted supply of oxygen. The fact that bearings on bronze supports appear to experience more trouble than bearings on steel supports suggests that this is primarily galvanic corrosion. This is not a problem that can be easily overcome by changes in lubricating oil composition or changes in alloy composition. The best remedy is to keep electrolytes out of the system.

(v) Corrosion of copper-lead alloys by water present in the oil

This form of corrosion in uncommon and indicates that considerable quantities of water are present in the lubricating oil system. Analysis may indicate that the oil is in excellent condition and that the water content is not abnormal. However, the water content of a sample taken for analysis can depend very much on when and where it was taken. If substantial amounts of water are present it could be exemplified by the need to change filters frequently - as they swell and block when in contact with water.

The damage associated with the presence of water in the oil takes the form of removal of the lead-based overlay material and severe localised corrosion of the lead phase in the copper-lead bearing alloy (Figs. 36 and 37). The overlay is even removed from the unloaded shoulders of the bearing, and the intensive localised nature of the attack is indicative of galvanic corrosion, the less-noble (anodic) lead being preferentially attacked.

5.4.2.4 Cavitation and Erosion

Cavitation and erosion are increasingly a cause of failure on bearings, owing to increasingly high and variable loads and speeds. It is important to define these terms. Two types of cavities can form in lubricants, vaporous cavities and gaseous cavities. Gaseous cavities, which are formed by outgassing of the lubricant, can form and collapse only slowly and cannot cause any mechanical damage directly. However, their presence reduces the load-carrying capacity of the lubricant film, thereby promoting other kinds of failure, such as fatigue.

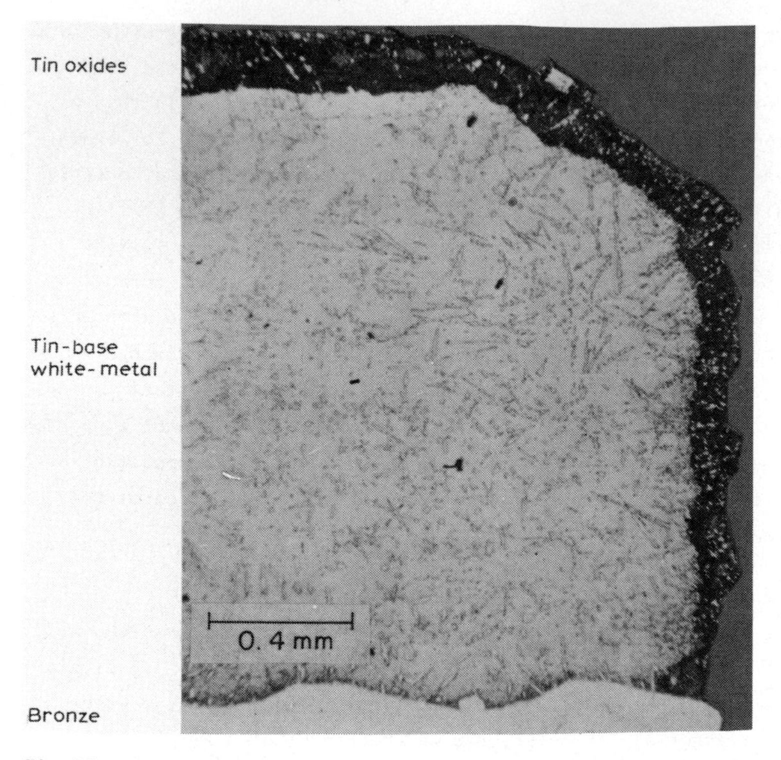

Fig.35 General oxidation of white-metal.

Fig.36 Damage associated with the presence of water in oil.

Vaporous cavities, which contain vapour of the liquid concerned, and little or no gas, can form and collapse extremely rapidly, and the very high pressure associated with their collapse can cause mechanical (impact) damage on metal surfaces. Both types of cavity can be generated by the pressure fluctuations associated with the flow of lubricant through a bearing and the fluctuating loads imposed on it. Vaporous cavitation can also be caused by the vibration of a metal surface in contact with a lubricant, as in an ultrasonic cleaner. The damage associated with both types of cavitation may be aggravated by the presence of fine particles in the lubricant; it is in such circumstances that it may be justified to talk of erosion damage and cavitation erosion. However, it should be emphasised that vaporous cavitation can cause severe damage even in the absence of solid particles.

Unlike other types of damage, vaporous cavitation damage is generally encountered on the unloaded areas of bearings, where oil-film pressures are low, and this provides a useful means of identification (Fig.38). Microsections of damaged areas show signs of local work-hardening and fatigue cracking. When the damage is due solely to cavitation, the texture in the damaged areas is rough (Fig.39); when particles are present (cavitation erosion), the damaged surfaces are smooth (Fig.40).

Vaporous cavitation can remove protective films, such as oxides, from metals and initiate corrosion. In addition, the very high local pressures and temperatures associated with the final stage of cavity collapse can induce chemical reactions which would not normally take place. There is some evidence that certain oil additives are unstable under cavitating conditions and that the decomposition products can be corrosive. In such circumstances, bearing surfaces can be subjected to the combined effects of cavitation and corrosion.

Work carried out at Thornton Research Centre indicates that there is very little one can do to a commercial lubricant to eliminate the effects of cavitation, and cavitation must be regarded primarily as a design problem.

5.4.2.5 Electrical Discharge Damage

On electrical machinery, and occasionally on other types of machinery, potential differences can be built up and electrical discharges may occur across the bearing surfaces.

Each discharge gives rise to a small pit, and a large number of discharges can eventually cause damage of the type illustrated in Fig.41. This type of damage is characterised by the fact that the pitting occurs on both bearing and journal surfaces.

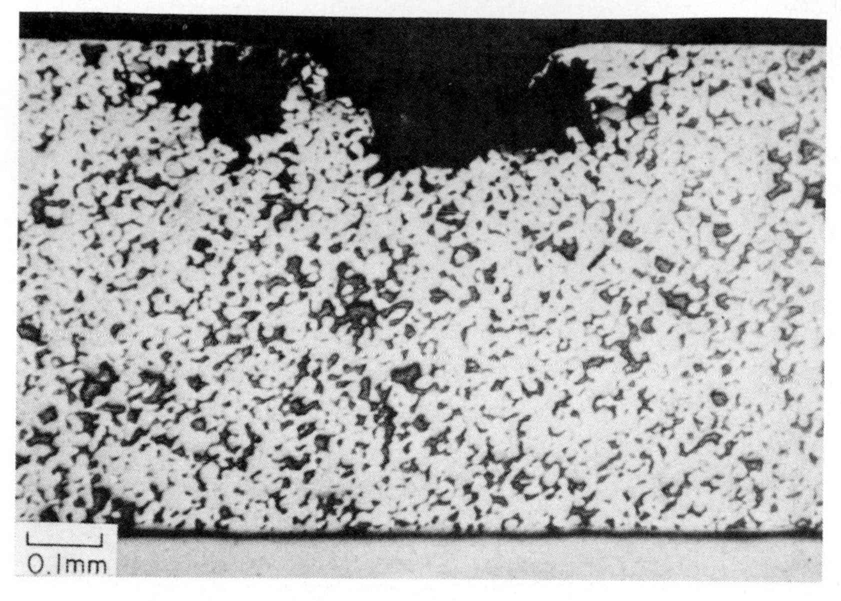

Fig.37 Section showing the intensive localised nature of the attack.

Fig.38 Cavitation damage.

Fig.39 Cavitation of white metal.

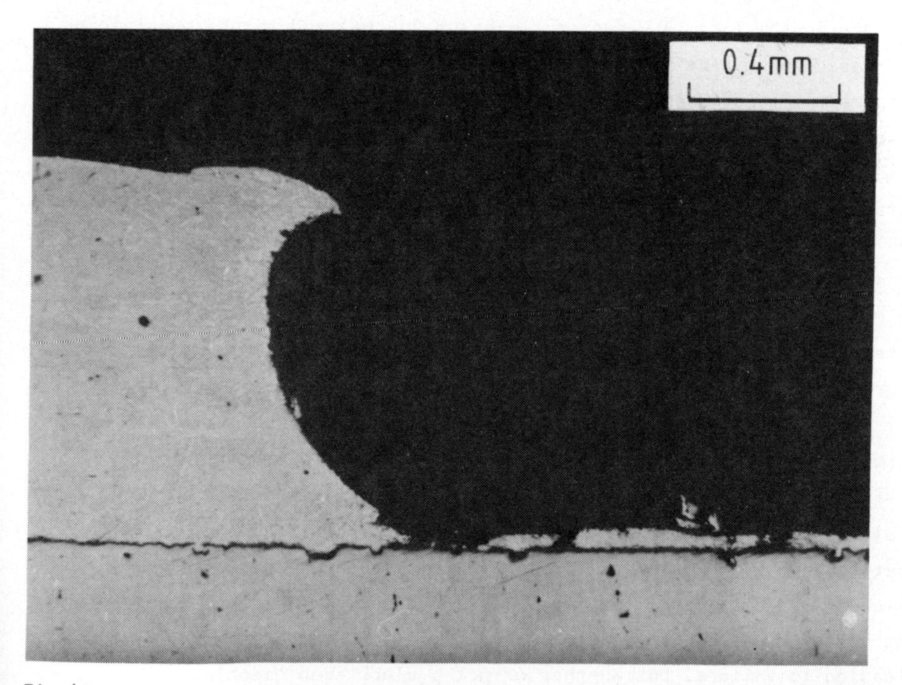

Fig.40 Cavitation erosion of white metal.

Fig.41 Electrical pitting in lead base babbitt bearing.

5.4.2.6 Wire-wool Failures

This is the name given to a catastrophic type of failure that has been encountered during the last 10 years on tin-based journal and thrust bearings on both land-based and marine turbine installations. A hard, black scab (Fig.42), which in some cases has been shown to be an amorphous mixture of iron and tin containing dispersed iron carbides and possibly nitrides, forms on the surfaces of the tin-based white-metal bearings and proceeds to machine-away the adjacent steel surfaces. Steel slivers are usually found in and around the bearing, sometimes in sufficient quantity to give the appearance of wire-wool. The failures almost always occur within a few hours of start-up, either from new or after an overhaul, and are characterised by a very high wear rate (sometimes centimetres of steel may be lost) without high coefficients of friction or much temperature rise. The failures have been attributed both to the use of chromium-steel rotor shafts and to e.p. oils. There is no clear understanding of the sequence of events leading to failure, and in particular of the roles played by metal surfaces, the base oil and the additives present in the base oil. All that can be said is that this is indeed a tribological failure. However, various investigators have established that the triggering agent responsible for wire-wool failures is a dirt particle, partially embedded in a white-metal bearing. It is not certain what causes particular particles to trigger the sequence of events leading to failure, but whether or not a black scab "machine tool" develops depends very much on the nature of the lubricant and the composition of the rotor steel. Oil additives (e.g. sulphur-containing e.p. additives) that can prevent black scab formation with chromium-containing-steel rotors may actually promote scab formation when used with other rotor steels such as mild steel and ½% molybdenum steel. With these latter steels it has been claimed that chlorine-containing e.p. additives also increase the susceptibility to failure, but this is not supported by our investigations at Thornton Research Centre. What is more probable is that some lubricants are more likely to allow failure when their specific anti-rust/anti-wear agent becomes depleted.

The best way to avoid wire-wool failures is to avoid dirt. The system from the outset must be as clean as possible. The oil should be continuously filtered and the oil pressure in the thrust housing under running conditions should always be at least 7 lbf/in^2 (48 kN/m^2).

Now that black scab failures have been widely publicised, it appears that they may have been encountered previously but passed unrecognised. For example, severe wear of nitrided stainless-steel shaft journals on power-recovery turbines of aircraft piston engines has been observed from time to time and these appear to have suffered severe machining-type wear. It is particularly significant that the bearings in contact with these failed journals are in generally quite good condition, showing much less damage than would be expected from the state of the journals. This is also a characteristic of black scab failures. However, the aircraft bearings were silver with lead-indium overlays. If these failures are of the black scab type, then black scab is not peculiar to white-metal bearings.

5.4.2.7 Fatigue

A plain bearing may fail by fatigue when it has achieved its designed life expectancy; however, if failure occurs prematurely, this will be because either an incorrect bearing material has been used or the bearing has been incorrectly fitted.

In fatigue failures the cracks start at the bearing surface, propagate normal to the surface until they approach the shell, then turn through 90° and extend parallel to the bond between the bearing metal and the shell leaving a thin layer of bearing metal attached to the shell (Fig.43).

Bad bonding and fatigue are superficially similar. However, with bad bonding the bearing metal separates cleanly from the shell, whereas with fatigue the cracks start at the bearing surface, propagate normal to the bearing surface until they approach the shell and then extend parallel to the bond between the bearing metal and the shell, always leaving a layer of bearing metal attached to the shell. For a comparison of the two types of failure see Fig.44.

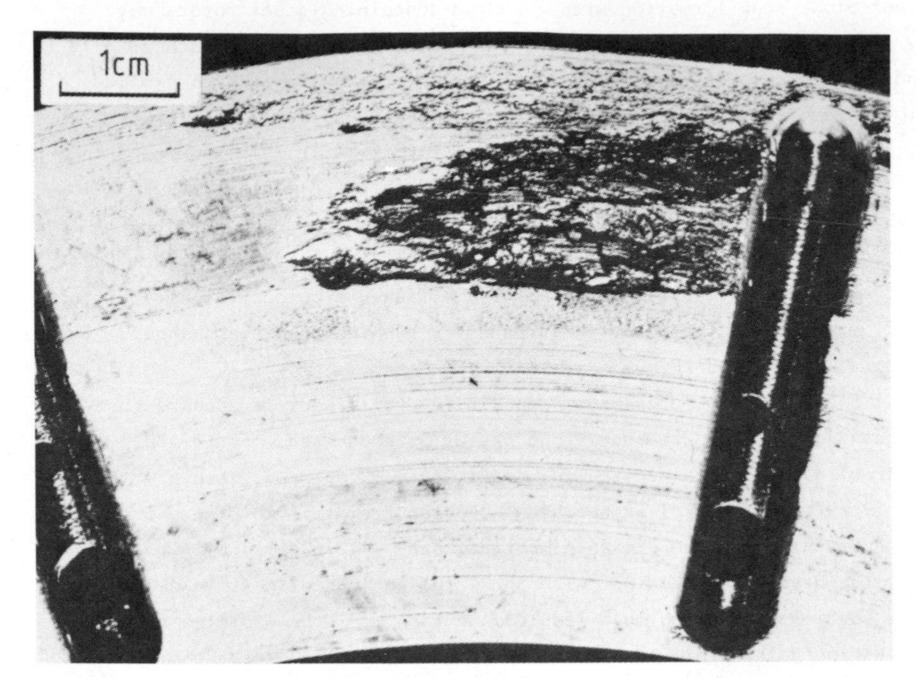

Fig.42 Black scab formation on a thrust bearing.

Fig.43 Fatigue of lead bronze bearing.

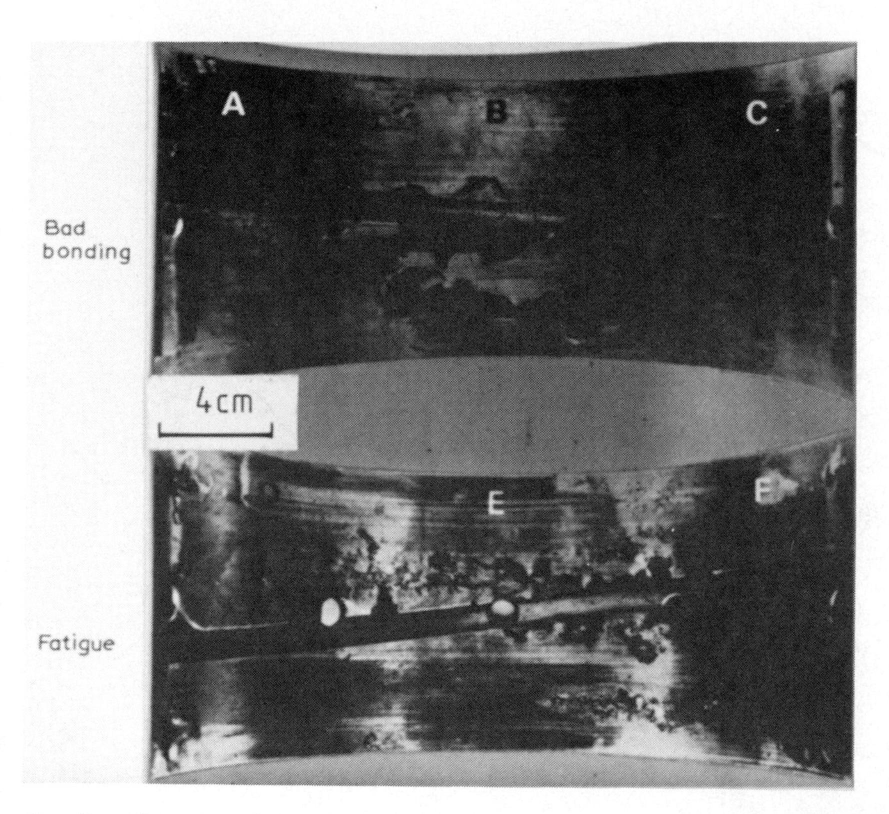

Fig.44 White metal bearings.

5.4.2.8 Thermal Cycling Damage

This is only a problem on tin-rich bearing alloys. The thermal expansion of tin crystals is anisotropic, i.e. the linear coefficient of thermal expansion is different along the three principal crystallographic axes. Consequently, tin-based white-metal bearings exposed to considerable thermal cycling can experience grain-boundary distortion and cracking (Fig.45). Bad bonding facilitates this type of failure.

5.4.2.9 Alloying in Service

This can be a problem on electroplated lead-alloy overlay bearings and on micro-Babbitt bearings. In both cases the tin in the surface layer migrates towards and alloys with the copper-rich phase or silver phase in the underlay. Hard, intermetallic compounds are formed which, if they are exposed, can score the journal surfaces. This problem is most acute on micro-Babbitt bearings, where complete alloying of the 200-μm thick, tin-rich surface has been observed in service (Fig.46). Obviously, the alloying reaction is dependent on both time and temperature; experiments show that it can proceed very rapidly at 150°C (Fig.47).

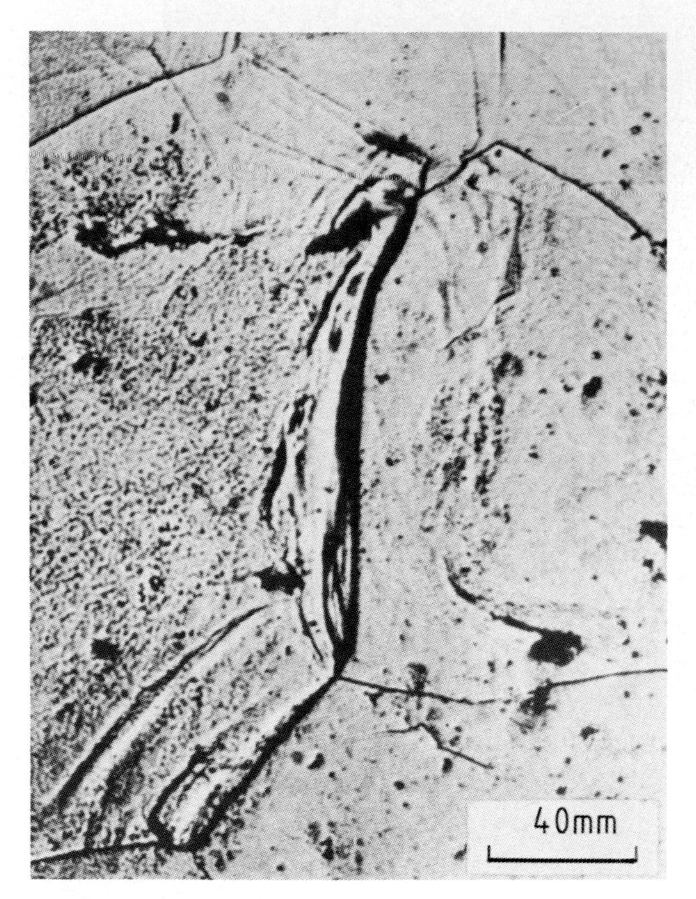

Fig.45 Distortions in tin due to thermal cycling.

5.4.2.10 Inadequate Viscosity and Lack of Lubricant

Should the supply of lubricant to a bearing be interrupted, even for a matter of seconds, catastrophic failure can occur. The damage usually takes the form of complete seizure and extensive melting of the bearing alloy. If the flow of a lubricant is liable to interruption, some degree of temporary protection can be provided by the use of special surface treatments on the steel surfaces, by using plastic or plastic-impregnated bearings and by the use of graphite or molybdenum disulphide if the latter is correctly applied.

Some failures due to oil starvation arise because of lack of lubricant on start-up, and on many items of heavy machinery it is essential to have a means of circulating the oil prior to start-up to provide lubrication and cooling.

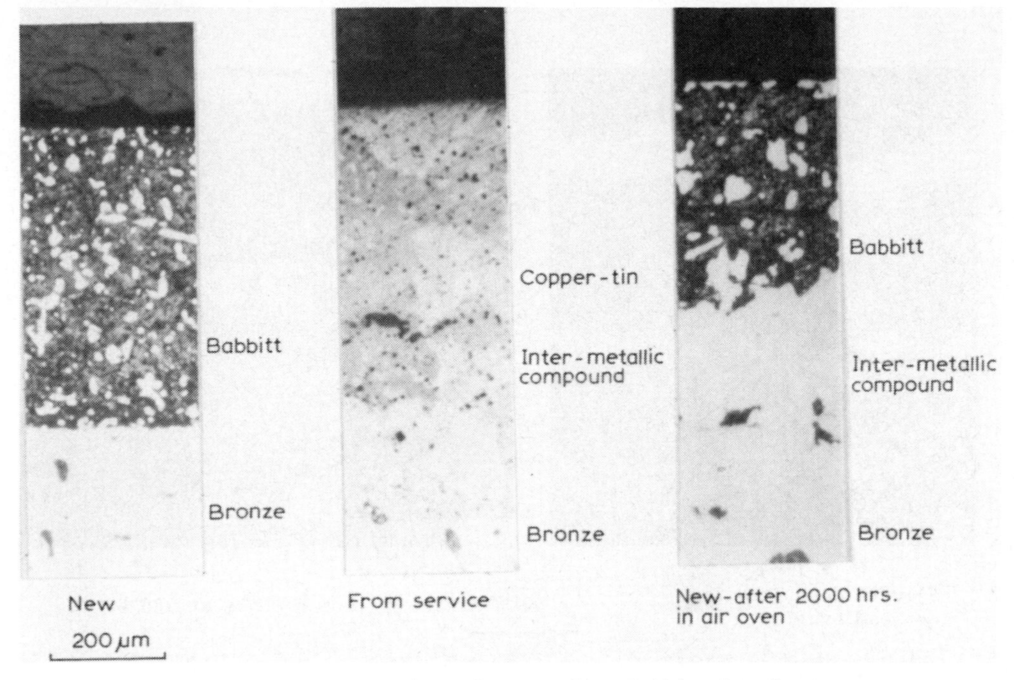

Fig.46 Intermetallic compound formation on micro-Babbitt bearing.

Dilution of the lubricant by fuel can occur in both gasoline and diesel engines. It is most frequently encountered on lightly loaded, cool-running engines, especially those with poor combustion characteristics. The presence of fuel in the lubricant lowers its viscosity, thereby reducing its load-carrying capacity. If the reduction in viscosity is marginal, premature failure by fatigue may occur; however, if the reduction is severe, then sudden, catastrophic failure may take place.

5.5 CONCLUSIONS

In the preceding pages we have attempted to describe the metallurgical features of some of the bearing failures that we have investigated in recent years. Bearing failure mechanisms are becoming increasingly complex and it is no longer always possible to give reliable, on-the-spot diagnosis. Metallographic examination of sections from failed bearings will often provide the necessary information to establish the cause of the trouble. For the more difficult and complex cases there is a wide variety of investigational techniques that can be applied to the problem. To assist others in the diagnosis of plain bearing failures we have compiled Table 5.4, in which the features of various failures are described. Whilst this is not a substitute for experience, we believe that, coupled with the accompanying bibliography, it may be an aid to establishing the cause of failures.

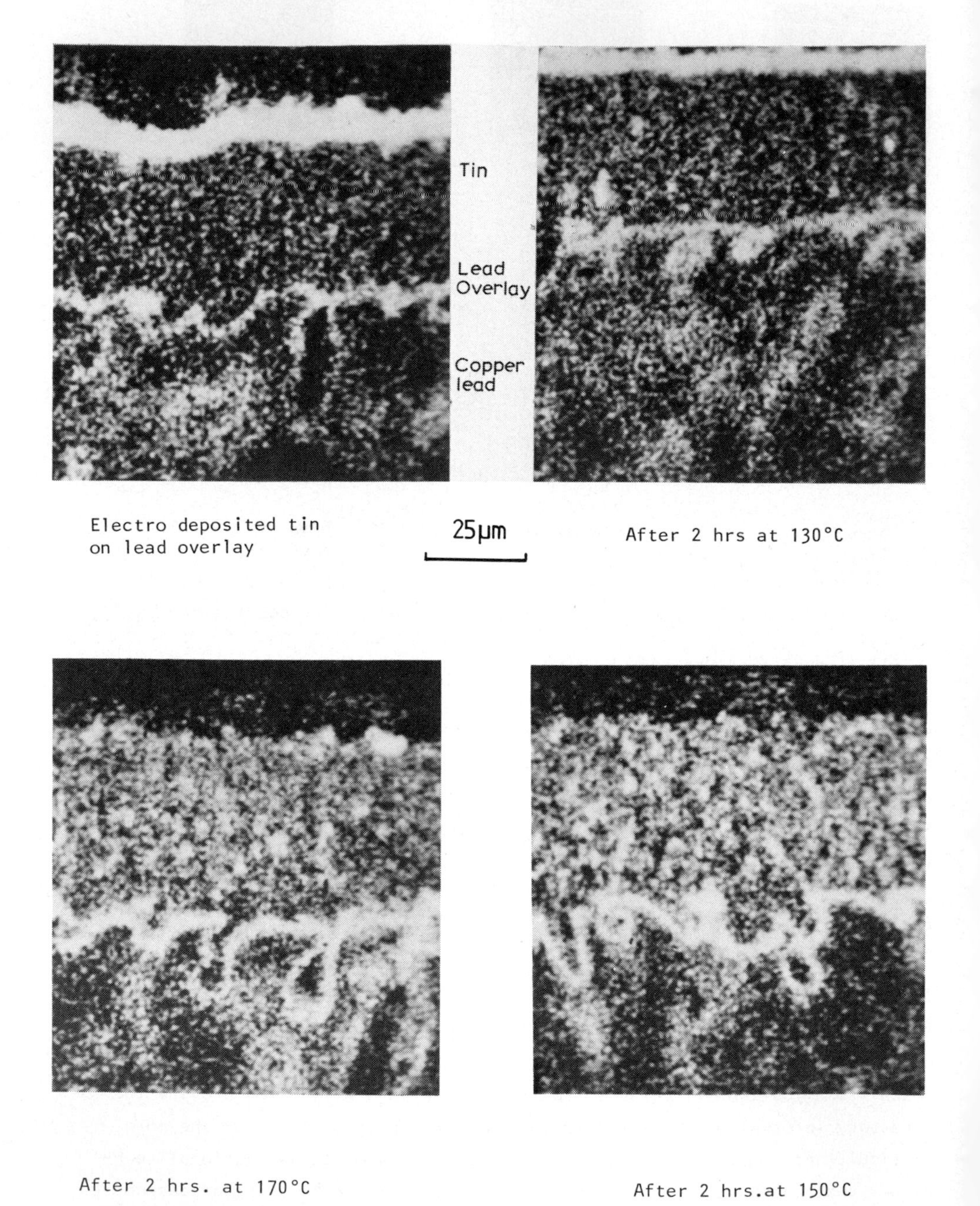

Fig.47 Electron-probe micrograph showing rate of diffusion of tin.

TABLE 5.4

Cause of failure	Typical features
Extraneous matter circulating in lubricant	Soft particles, e.g. carbonaceous matter, can erode white metals and overlays. Hard particles, e.g. metallic wear debris cuts well-defined channels.
Extraneous matter embedded in bearing	Burnished spots round embedded particles. Wear of journal - can lead to fatigue.
Fatigue	Cracks, initiated at bearing surface, propagate normal to surface; then, near backing, turn and run parallel with bearing surface.
Bad bonding	Bearing alloy lifts clearly away from backing - no evidence of alloying. Common on large, white-metal bearings.
Thermal cycling	Can cause roughening of tin-based white-metal bearing surfaces
Surface pitting of tin-based white metals	Cracking of oversize tin-antimony cuboids
Hard, black surface layers all over tin-based white-metal bearings	Tin-rich matrix of alloy transformed to hard tin oxides. Antimony-tin and copper-tin phases unchanged.
Sudden catastrophic wear of a single turbine journal or thrust bearing shortly after start-up	Black scab formation in white-metal bearing, which "machines" shaft, producing wire-wool.
Electrical pitting	Fine pits on both bearing surface and journal.
Cavitation in lubricant film	Localised metal removal, generally in unloaded areas of bearing.
Corrosion by weak organic acids	Surface roughening and filling. In copper-lead and lead-bronze, pure lead phase is leached out and surface may disintegrate. Lead overlays deficient in tin or indium may be corroded.
Corrosion by strong acids	Takes a variety of forms; bearing metal or journal may be attached. See text.
"Sulphur" attack	Discoloration and corrosion of copper and silver-rich alloys. Can be very severe in phosphor bronzes.
Corrosion by water in the oil	Takes the form of removal of the overlay and localised intensive attack of the lead phase of copper-lead bearing alloys.
Excessive operating temperatures	Wiping of surface layers. Fatigue failure.
Inadequate viscosity of lubricant	If marginal, may give rise to premature failure by fatigue. If severe, for example due to excessive fuel dilution, can give sudden catastrophic failure.
Lack of lubricant	Complete seizure. Extensive melting of bearing alloy.
Bad fitting	Fretting of backs of bearing. Lack of thermal contact with housing may cause bearing to bow.
Misalignment	Uneven contact and wear across bearing surface.
Manufacturing defects, e.g. poor lead distribution in copper-lead alloys	Requires expert metallurgical examination.
Surface hardening of tin Babbitt overlay	Cracking of overlay, wear of journal.
Internal oxidation of lead-indium overlays	Corrosion and fatigue of overlay.

REFERENCES

1 Sleeve Bearing Materials, 1949, ASM Cleveland, Ohio.
2 Metals Handbook, 1961, 8th edition, Vol.1, p.843-863, ASM Cleveland, Ohio.
3 Lubrication, 1953, 8, (3), p.29, "The Modern Bus and Truck - Fuels and Lubricants".
Lubrication, 1953, 8, (5), p.57, "Some Problems Associated with Lubrication of Large Engines".
Lubrication, 1953, 8,I (11), p.45, "Automotive Engine Bearings".
Lubrication, 1955, 10, (4), p.37, "Industrial Bearing Lubrication".
Lubrication, 1957, 12, (8), p.85, "Abrasives and Wear".
Lubrication, 1958, 13, (9), p.245, "Petroleum Laboratory Investigations".
Lubrication, 1963, 49, (6), p.81, "Diesel Power Plants".
Lubrication, 1964, 50, (7), p.77, "Plain Bearing Failures".
4 Forrester, P.G., Modern Materials, 1964, 4, p.173, Academic Press, New York and London, "Materials for Plain Bearings".
5 Hunter,M.S., Churchill, J.R. and Mear, R.B., "Electro-graphic Methods of Surface Analysis", Metal Progress 1942, 42, p.1070.
6 Crooks, C.S. and Eastham, D.R. "Plating for Bearing Applications". Trans. Inst. of Metal Finishing 1982 Vol.60.
7 Rafique, S.O., Inst. Mech. Eng. Lubrication and Wear, Second Convention 1964, p.180. "Failures of Plain Bearings and their Causes".
8 Love,P.P., Forrester,P.G. and Burke,A.E., Inst. Mech. Eng. Auto. Div. Proc., 1953-54, 2, p.29. "Function of Materials in Bearing Operation".
9 Lubricants and Lubrication (editor Braithwate, E.R.) Elsevier, London, 1967.
Morris, J.A., Ch.7. p.310, "Metallic Bearing Materials".
Pratt, G.C., Ch.8, p.377, "Plastic-based Bearings".
10 Pratt,G.C. and Perkins,C.A., "Silicon Aluminium Bearings for High-speed Diesels", Diesel and Gas Turbine Worldwide Vol. XIII, No.10, p.76-78.
11 Davis,T.A., "Plain bearing wear in IC Engines", Automotive Engineer, Aug./Sept. 1981.
12 Principles of Lubrication (editor A. Cameron) Longmans Green and Co., London, 1966.
Holligan,P.T., Ch.25, p.511. "Plain Bearings - Bearing Materials and Diagnosis of Bearing Failures".
13 Engineering, 1967, 20, p.260. "Bond Strength of White Metalling".
14 Rose, A. Trans. Inst. Mar. Eng. 1967, 79, p.233, "Marine Bearings".
15 Wilson,R.W. and Shone,E.B., Joint Course on Tribology, Institution of Metallurgists, London, 1968, Paper 4, "Metallurgical Studies of Bearing Failures".
16 Wilson,R.W. and Shone,E.B., Anti-Corrosion Methods and Materials, 1970, 17, p.9. "The Corrosion of Lead Overlay Bearings".
17 Quayle, J.P., Copper, 1969, 3, (5), p.12. "The Science of Tribology - Part 3".
18 Bryce,J.B. and Roehner,T.G., Trans. Inst. Mar. Engs., 1961, 73, p.377, "The Corrosion of Tin-Base Babbitt Bearings in Marine Steam Turbines".
19 Lloyd,K.A. and Wilson,R.W., Inst. Mech. Engs., Tribology Convention 1969, Paper 10, p.76. "Formation of Tin Oxides on White Metal Bearings".
20 Hiley, R.W., "Corrosion of Tin Base Babbitt Bearings to form Tin Oxides", Trans Inst. Mar. Eng., 1979, 91, (2) p. 52-66.
21 Dawson, P.H. and Fidler, F., Inst. Mech. Engs. (Lubrication and Wear Fifth Convention), 1967, 181I, p. 207, "Wire-wool Type Failures; The effect of Steel Composition, Structure and Hardness".
22 Dowson, D., Godet, M. and Taylor, C.M., "Cavitation and Related Phenomena in Lubrication", University of Leeds, Yorks, England.
Leeds-Lyon Symposium on Tribology, 1st, Prog. Pap and Discuss., University of Leeds, Yorks, England, Sept. 1974, Publ. by Mech. Eng. Publ. Ltd., New York, NY, 1974, 248. Publ. for Institute of Tribology, Leeds University, Yorks, England.

23 James, R.D., "Cavitation Damage in Plain Bearings". Tribology Inst. Feb. 1978, 11, (1) p. 22-23.
24 Garner, D.R., James, R.D. and Warriner, J.F., "Cavitation Erosion in Engine Bearings - Theory and Practice", 13th CIMAC Conf. Vienna 1979.
25 Forrester, P.G., "Bearing Materials", Metallurgical Reviews, 1960, 5, p. 507.
26 Forrester, P.G., "Electrodeposition in Plain Bearing Manufacture", Trans. Inst. Met. Finishing, 1961, 38, p.52.
27 Booser, E.R., "Plain Bearing Materials", Machine Design, 1970, 42, p. 14.
28 Standard Handbook of Lubrication Engineering, ASLE, McGraw-Hill, 1968. Chapter 18, "Sliding Bearings".
29 Tribology Handbook (editor M.J. Neale), Butterworths, London 1973.

6 ROLLING ELEMENT BEARINGS

D.G. HJERTZEN and R.A. JARVIS, SKF (U.K.)LTD.

6.1 INTRODUCTION

The present development of rolling bearings is characterised by numerous apparently small internal improvements in the bearings and to material quality rather than the introduction of radical new designs. There is considerable technical research into rolling bearing technology and important progress is being made in manufacturing and inspection methods, improved lubricants and lubrication equipment, etc.

It is possibly surprising, but true, that although rolling bearings are extensively used, there are many misconceptions and the methods of selecting the most suitable bearings are not always fully understood or applied. It has been known for some unfortunate experience with a certain bearing type to result in loss of confidence; consequently, the characteristics of that particular bearing are often not fully utilised.

6.2 BEARING SELECTION

The major function of bearings is to transmit loads between machine parts in relative motion, but additionally there may be special performance or environment requirements affecting choice of bearings.

It is essential for the designer to select from the wide range of bearings available the bearing design suitable to carry the loads involved under the various operating conditions and to satisfy the requirements of rotational speed, temperature variations, bearing housing misalignment, and rigidity etc. It is also important to ensure that correct fits are used between the bearing inner ring to shaft and outer ring to housing. The choice of fit depends essentially on the working conditions, but there are other considerations such as bearing type and size, bearing internal clearance and method of assembly, etc.

6.3 BEARING TYPES

6.3.1 Single Row Deep Groove Ball Bearing (Fig.1)

The depth of the ball tracks coupled with a relatively large ball size and high degree of conformity (ratio of track radius to ball radius) gives this bearing considerable axial-carrying capacity in addition to radial capacity, even at high speeds. These bearings normally have a cylindrical bore and are mounted direct onto the shaft. For location purposes, a groove can be provided in the outer ring for a snap ring which can be used for axial location.

These bearings can be fitted with shields or seals. The shields are intended as grease retainers and to keep out a certain amount of foreign matter, but the seals are actually rubbing seals and designed for dusty environments, etc. Bearings with two shields or seals are initially charged with the correct quantity of grease and, consequently, do not require relubrication. These bearings are frequently called lubricated-for-life bearings.

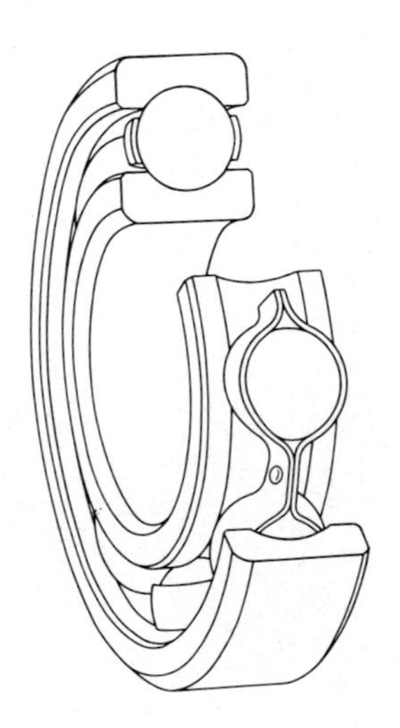

Fig.1 Single Row Deep Groove Ball Bearing.

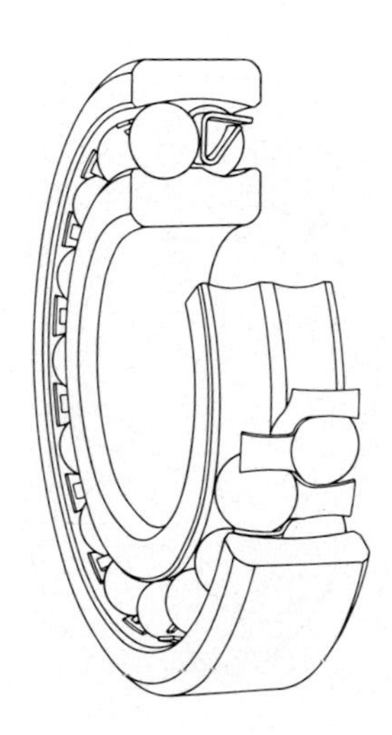

Fig.2 Self-Aligning Ball Bearing.

6.3.2 Self-aligning Ball Bearings (Fig.2)

These have two rows of balls each in its own groove on the inner ring, but in a common sphered track in the outer ring. The inner ring and balls form a unit which can align freely about the bearing centre. This feature is an advantage in cases where it is difficult to obtain accurate parallelism between the shaft and housing bore, or where there is a deflection of the shaft. Due to the sphered outer ring track, the bearing does not have high load-carrying capacity and the axial-carrying capacity is limited.

The degree of mis-alignment of the shaft or housing is limited by the balls contacting the edges of the outer ring and permissible mis-alignment normally varies from 2° - 3°.

These bearings are manufactured with cylindrical or taper bore and the latter is usually mounted on a split sleeve.

6.3.3 Angular Contact Ball Bearing (Fig.3)

The direction of load through the balls is at an angle to the bearing axis which makes these bearings particularly suitable for carrying combined radial and axial loads. A radial load imposed on a single row angular contact ball bearing gives rise to an induced axial load which must be counteracted; therefore these bearings need to be arranged so that they can be adjusted against a second bearing.

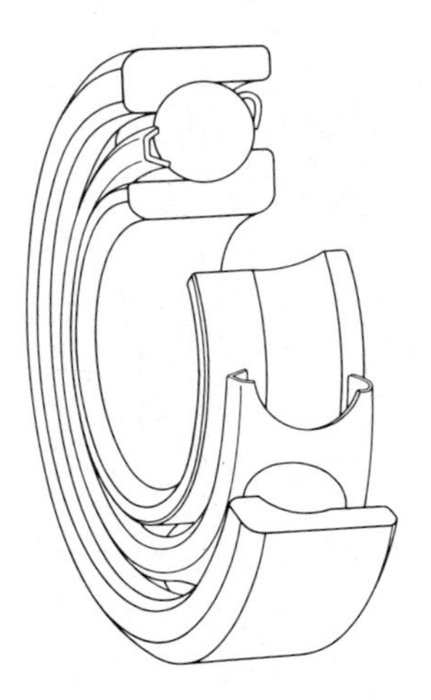

Fig.3 Angular Contact Ball Bearing.

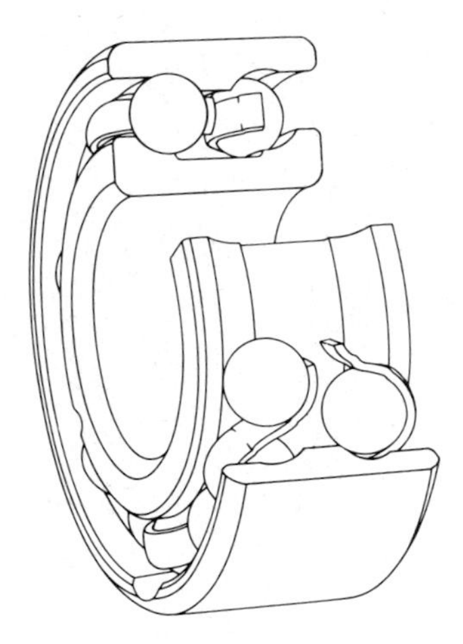

Fig.4 Double Row Angular Contact Ball Bearing.

These bearings are frequently mounted in pairs in face-to-face, back-to-back or tandem formation, and in order to achieve this the side faces of the bearing rings are ground to achieve the correct residual internal clearance.

The contact angle can vary between 15° - 25° - 30° - 40°.

6.3.4 Double Row Angular Contact Ball Bearing (Fig.4)

This bearing has similar characteristics to two single row angular contact ball bearings mounted back-to-back; consequently, the lines of pressure or contact through the balls are directed outwards, thereby giving increased rigidity. These bearings have very little axial clearance, giving close axial location of the shaft and eliminating the necessity for axial adjustment. These bearings can be supplied to give a pre-load condition when mounted so that even when subjected to axial load, the axial displacement of the shaft is very small.

6.3.5 Cylindrical Roller Bearings (Fig.5)

The rollers in these bearings are guided between integral flanges on the outer or inner ring, thereby allowing the rings to move axially relative to each other, which is an advantage when the shaft expansion is greater than the housing expansion in an axial plane. The flanged ring and rollers are held together by a cage to form an assembly which can be removed from the other ring. This separable feature is often utilised to ease assembly problems. The bearing has high radial load-carrying capacity with accurate guiding of the rollers, resulting in a close approach to true rolling. Consequently, the low friction permits high speeds.

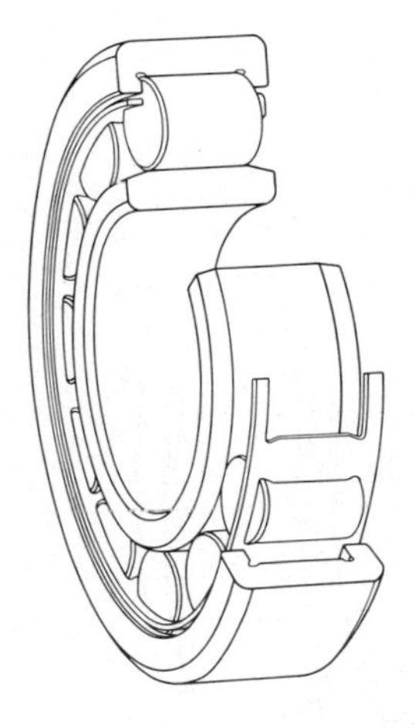

Fig.5 Cylindrical Roller Bearing.

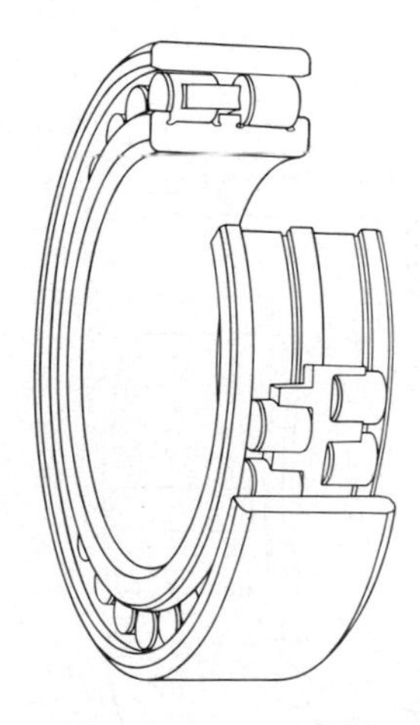

Fig.6 Double Row Cylindrical Roller Bearing.

Various designs are made with different flange arrangements and additional loose flanges can be incorporated to axially locate the rollers in both directions. The roller ends and ring flanges are capable of sustaining relatively high axial loads.

6.3.6 Double Row Cylindrical Roller Bearings (Fig.6)

These bearings are of separable design which allows a flanged ring with roller and cage assembly and flangeless ring to be fitted independently, thus facilitating mounting and dismounting.

Low cross-sectional height and high load-carrying capacity give rigid bearing arrangements for the accommodation of heavy radial loads. They are, therefore, mainly used for machine tool and rolling mill applications. The inner ring normally has a tapered bore to enable the ring to be driven up a tapered journal to achieve a given radial clearance or even pre-load. They can be supplied in special or ultra-precision execution.

6.3.7 Needle Roller Bearings (Fig.7)

These are similar to cylindrical roller bearings and have high load-carrying capacity. They are made in various designs and are particularly suitable for applications in confined spaces. These bearings are used in gudgeon pins and universal joints, but they are not recommended where there is likely to be misalignment or shaft bending or any condition where the rollers can be subjected to tilting forces which can cause a clutch action.

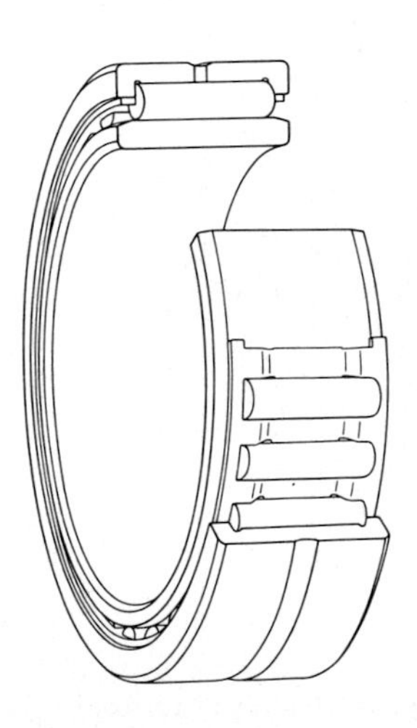

Fig.7 Needle Roller Bearing.

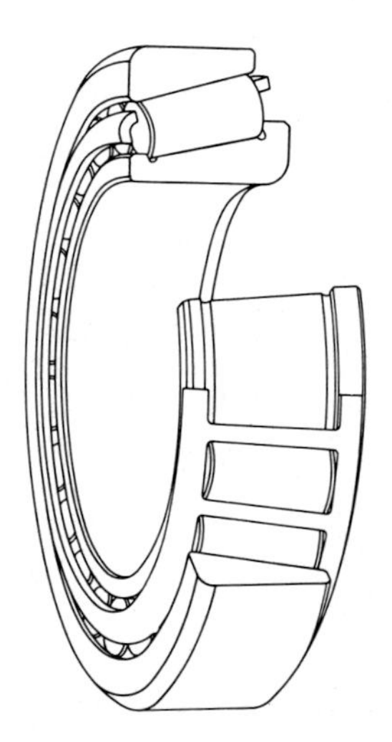

Fig.8 Taper Roller Bearing.

The diameters of needle rollers are small, usually 1.5 to 5 mm and the roller length is approximately 2.5 times the diameter.

6.3.8 Taper Roller Bearings (Fig.8)

The rolling elements in these bearings are truncated cones and the track of both the inner ring (cone) and the outer ring (cup) are tapered. The tapers of the rollers and the inner ring have a common apex on the bearing axis and the greater the inclination of the rollers, the greater the axial carrying capacity. As with single row angular contact ball bearings, taper roller bearings must always be mounted in pairs or adjusted towards another bearing capable of dealing with axial forces acting in the opposite direction. The taper bearing can also be supplied as a paired unit with a common inner or outer ring. This bearing is used extensively in the Automobile Industry.

6.3.9 Double Row Spherical Roller Bearings (Fig.9)

Due to the size, shape, and number of rollers, together with the accuracy with which the rollers are guided, this bearing has excellent load-carrying capacity in both a radial and axial direction. Since the bearing is self-aligning, angular misalignment between shaft and housing has no detrimental effect and full capacity is always available. Accurate roller guidance in the normal 'C' design is by means of a loose relatively narrow guide ring of thick radial section between the rows of rollers and by the cage.

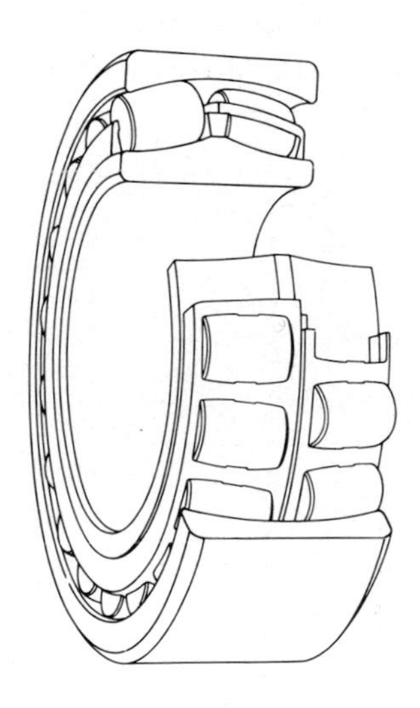

Fig.9 Double Row Spherical Roller Bearing

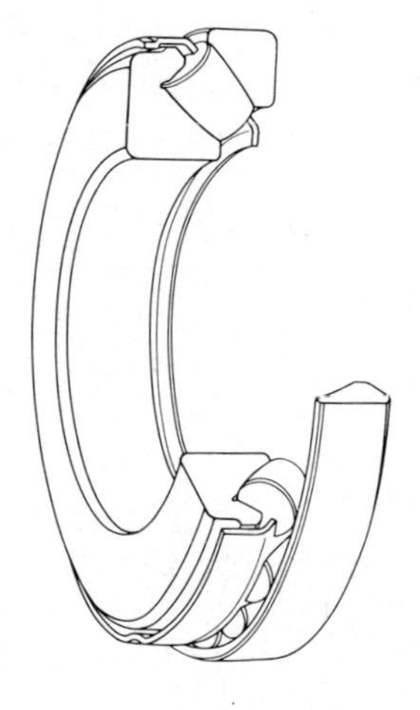

Fig.10 Spherical Roller Thrust Bearing

Spherical roller bearings are manufactured with cylindrical or tapered bores. They can be mounted on tapered sleeves and are used in Railway and Heavy Engineering. These bearings are regarded as the heavy-weight of the standard bearing types and are used extensively in plummer blocks.

6.3.10 Spherical Roller Thrust Bearings (Fig.10)

Unlike most other types of thrust bearing, the line of action of the load at the contacts between the tracks and the rollers forms an angle with the bearing axis which makes these bearings suitable for carrying a radial load up to 55% of the simultaneously acting axial load. The sphered track of the housing washer provides self-aligning properties which permit a certain angular displacement of the shaft relative to the housing, due to mounting errors or shaft deflection.

In order to prevent unacceptable sliding at the roller-to-track contacts under the influence of centrifugal force and gyratory moments, it is necessary to apply a certain minimum axial load to the bearing.

The shaft washer, rollers and cage form an assembly which is separable from the housing washer, allowing the washers to be mounted independently. These bearings are suitable for heavy loads and for relatively high speeds.

6.3.11 Ball Thrust Bearings (Fig.11)

These consist of a row of balls, retained in position by a cage and two washers known as the shaft and housing washer, each with a shallow ball track groove. The shaft washer has a smaller bore than the other washer and is located by the shaft. The housing washer has a larger outside diameter than the shaft washer for location in a housing.

These bearings are only capable of carrying axial load in one direction.

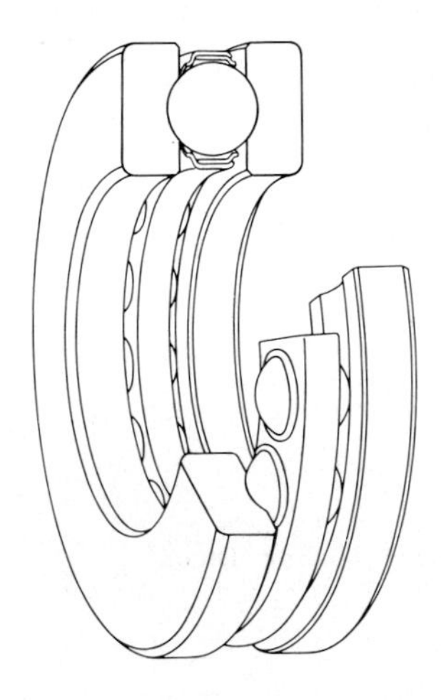

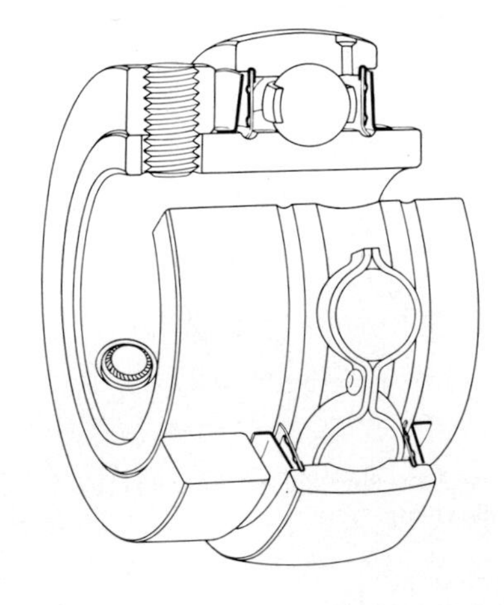

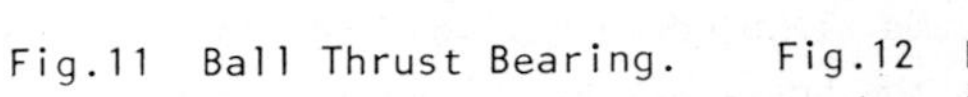

Fig.11 Ball Thrust Bearing.

Fig.12 Bearing with Spherical Outside Surface and Extended Inner Ring Width.

6.3.12 Bearings with Spherical Outside Surface and Extended Inner Ring Width (Fig.12)

These bearings are essentially a variation of a deep groove ball bearing and are normally used in conjunction with a range of cast iron or pressed steel housings. They have a sphered outer surface and this, when fitted into a sphered seating in the housing, allows the unit to accommodate any initial mis-alignment of the shaft which may occur during mounting. These bearings normally have an extended inner ring with some form of locking device which is used to lock the bearing and, hence, the unit to the shaft. Although the bearing is made to the same degree of accuracy as deep groove ball bearings, the method of locking the bearing to the shaft does not give the same centering accuracy as when bearings are mounted with an interference fit on the shaft. These bearings are used extensively in machinery where the rotational accuracy require-ments are not too stringent, such as agricultural machinery and conveyors, etc. where they offer a simple and economical solution.

6.4. FATIGUE LIFE AND LOAD-CARRYING CAPACITY

The concept of a rolling bearing is that load between the stationary and rotating machine components is transferred through the bearing by means of interposed rolling elements. Pure rolling seldom occurs and the net effect is usually a combination of rolling and sliding. The exact proportions are a function of the bearing type, but 90-99% of the load is related to rolling motion and 1-10% to sliding.

If a bearing is properly handled, correctly mounted, lubricated, and protec-ted, all causes of failure are eliminated except one - fatigue of the material. The life of a rolling bearing as defined by ISO (International Standards Organ-isation) is the number of revolutions (or number of operating hours at a given constant speed) which the bearing is capable of enduring before fatigue occurs on one of its rings or rolling elements. Repeated tests have verified that when a group of apparently identical bearings are run under the same conditions of load and speed they have different lives. A clear definition of the term "life" is, therefore, essential for the calculation of bearing size. Dynamic load ratings given in bearing manufacturers catalogues are based on the life that 90% of a sufficiently large group of apparently identical bearings can be expected to attain or exceed. This is called the basic rating life (or the nominal life) and agrees with the ISO definition.

A typical life dispersion curve is shown in Fig.13, and it can be seen that half the bearings achieve an average life five times greater than the nominal life on which the calculations are based.

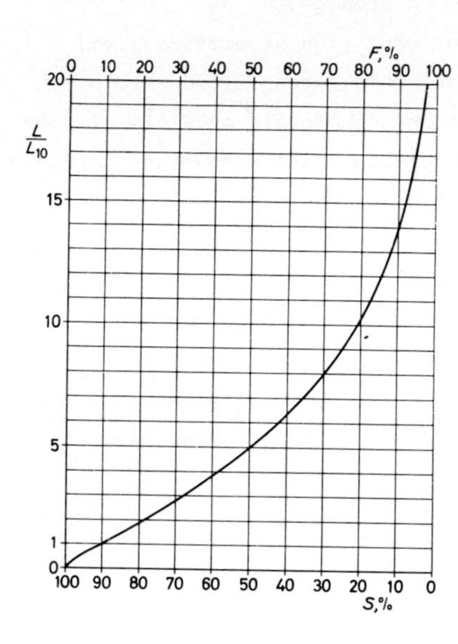

The classic curve of bearing life shows the life distribution for identical bearings run under identical conditions of load and speed. S is the percentage of still serviceable bearings, F the percentage showing signs of fatigue, and L, the life of an individual bearing, given in terms of L_{10} as explained in the text

Fig.13 A Typical Bearing Life Dispersion Curve.

6.4.1 The Hertzian Contact

Whilst the Hertz theory relating to the contact between solid bodies is still of interest in connection with calculating contact pressures and deformations in rolling bearings, the greatest interest is in the shear stresses beneath the contact surface as these are considered to be the cause of both plastic deformations and fatigue cracks in the contact zone.

Figure 14 shows the stress distribution in a Hertzian contact and from a fatigue aspect the shear stresses at the edge of the contact zone are the most dangerous ones. They are parallel to the contact surface and are situated beneath the surface. As these shear stresses change direction with the passage of the rolling body, material fatigue will occur if any weak point such as a slag inclusion comes within the subsurface zone where these shear stresses are considerable. It has been proved that even tiny slag inclusions can constitute weak points of this kind and that under the influence of the alternating shear stresses these inclusions result in micro cracks which subsequently increase in size, find their way up to the surface, and lead to surface fatigue failure (see Fig.15).

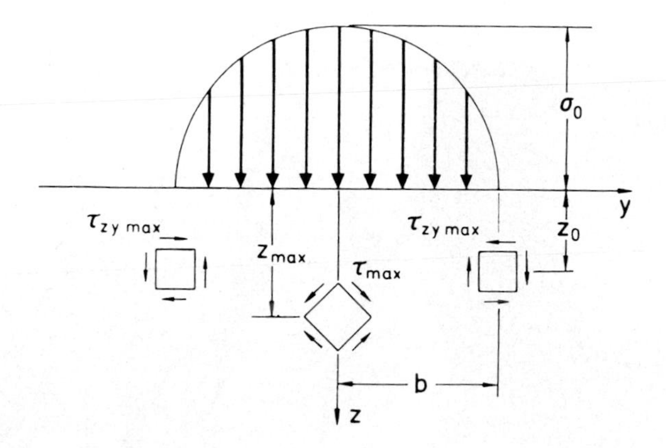

Fig.14 Shear Stresses in the Contact Zone.

τ_{max} = greatest shear stress in the contact zone and occurs at a depth z_{max} below the contact surface forming an angle of 45° with the plane of the contact surface. For line contact of infinite length $\tau_{max} = 0.30\,\sigma_o$ and $z_{max} = 0.786$ b where 'b' is half the width of the contact zone.

$\tau_{zy\,max}$ = greatest shear stress parallel to the plane of the contact surface and occurs at a depth "z_o" beneath the contact surface. For line contact of infinite length $\tau_{zy\,max} = 0.25\,\sigma_o$ and $z_o = 0.50$ b.

6.4.2 Relationship between Load and Life

The relationship between bearing load and life shown in Figure 16 has been obtained by testing bearings under loads of different magnitude. Three test series were run under loads P1, P2 and P3. The lives obtained are indicated by the dots on the horizontal lines. The lives obtained by 90% of the bearings in each test series, i.e. the nominal bearing life, has been indicated by the numerals 1, 2 and 3. It will be seen in the graph that these three points lie on an approximately straight line. This means that in view of the fact that the scales of both axes are logarithmic, the life can be expressed as a power of the load, hence:

$$L = \left(\frac{C}{P}\right)^{p}$$

where L = nominal bearing life in millions of revolutions;

P = equivalent bearing load in Newtons

C = basic dynamic load rating of the bearing in Newtons

p = 3 for ball bearings and 10/3 for roller bearings.

Fig.15 Normal Fatigue Failure

The basic dynamic load rating of the bearing is defined as the load that gives a nominal bearing life of one million revolutions. On the graph, the load at the point at the intersection of the ordinate for the life one million revolutions with the line representing the relationship between load and life is the basic dynamic load rating "C". This is the load rating given in SKF catalogues and is used for the calculation of the bearing life.

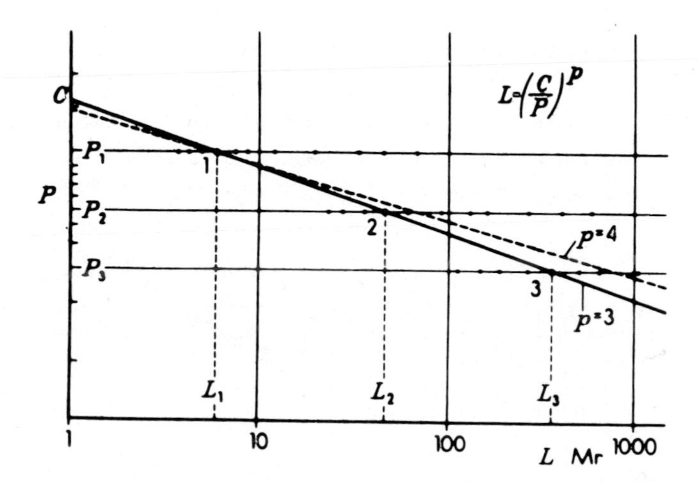

Fig.16 Relationship between Load and Life.

6.4.3 Further Development of the Life Equation

With the advancement of technology, greater reliability is required from rolling bearings and of the calculation methods used. It is necessary to consider factors not included in the basic life equation even though the results given by the basic life equation are satisfactory in the majority of cases.

ISO has suggested that the life equation should be as follows:-

$$L = a_1 \times a_2 \times a_3 \left(\frac{C}{P}\right)^p$$

a_1 = reliability factor, which enables the bearing life to be calculated for any given probability of fatigue (a_1 = 1 for 90% probability).

a_2 = material factor (a_2 = 1 for a rolling bearing of good quality steel with normal hardness and structures.

a_3 = lubrication factor (a_3 = 1 for normal lubrication).

Although the new life equation is a development of the old equation, a multiplicative combination of the three modifiers a_1, a_2, and a_3 does not always result in an improvement. The negative effect of inadequate lubrication, for instance, is not compensated for by using say a vacuum remelted steel bearing material.

6.5 BOUNDARY DIMENSIONS AND INTERNAL CONTROLS

Every standard bearing with metric boundary dimensions belongs to a "dimensional series" which forms part of general plans prepared by the International Organisation for Standardisation. For any given bore there are a series of different outside diameters and within each diameter series there are bearings of different widths, as shown in Fig.17. Each standard bearing belonging to a dimension series is designated by a two-digit number. The last two digits in a four or five digit number are 1/5 of the bore when this is in millimetres; i.e. bearing 6004 has a bore of 20 mm.

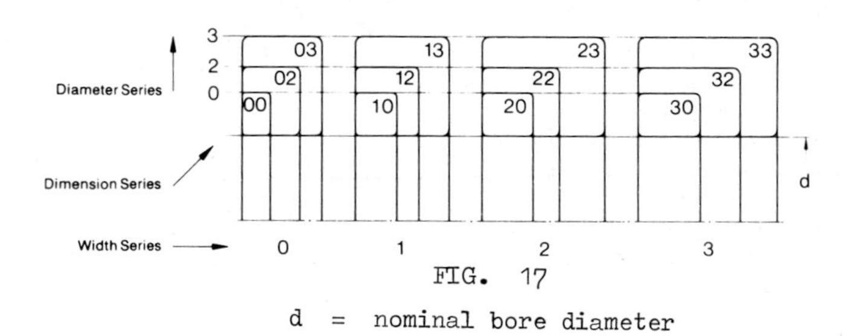

FIG. 17

d = nominal bore diameter

Supplementary symbols fall into various groups relating to internal design features such as cage, material, sealing, internal radial clearance and precision etc.

The radial clearance for single row deep groove ball bearings has been agreed both internationally and by British Standards Organisations. A general summary is as follows:-

SKF Designation	I.S.O. & B.S. Group	Also referred to as
C2	2	0
Normal	Normal	00
C3	3	000
C4	4	0000

SKF also manufacture certain bearings with smaller radial clearances (C1) and larger radial clearances (C5).

6.6 USAGE

C1 - used on machine tool spindles where minimum movement and maximum rigidity are required.

C2 - suitable for fractional horse power motors for domestic appliances particularly where silent running is required.

Normal clearance - used for normal applications where there are no temperature, speed, or interference fit problems.

C3 - often used on large electric motors, particularly where the inner ring temperature exceeds that of the outer ring, or when the rings are mounted with heavy interference fits.

C4 - used frequently on applications such as large traction motors for diesel electric and electric locomotives where there are temperature differentials between the inner and outer ring and the interference fits are much greater than those used for C3 bearings.

C5 - this control is often used for bearings in furnace trucks where there are large temperature differentials between the inner and outer rings coupled with heavy interference fits.

6.7 SPEED LIMITS

Due to the many factors combining to determine the maximum speed limits for rolling bearings such as bearing type, size, radial clearance, cooling conditions, method of lubrication, load, degree of precision and environment etc., it is difficult to give precise limits and all attempts to give limiting speeds must be to provide an approximate general guide.

For comparison purposes between speed and bearing size, rolling bearing speeds are usually expressed in terms of "ndm" and $n\sqrt{DH}$,

where $ndm = A \times f_1 \times f_2$ for radial bearings

and $n\sqrt{DH} = A \times f_1 \times f_2$ for thrust bearings

n = speed, rpm

dm = mean diameter of bearing, $0.5(d+D)$mm

d = bearing bore, mm

D = bearing outside diameter, mm

H = height of single thrust bearing, mm

f_1 = correction factor for bearing size (Fig.18)

f_2 = correction factor for bearing load (Fig.19)

A = constant, which is a function of the lubricant and lubrication method (Fig.20).

Although the speed limit formula is based on practical experience, extensive research into more accurate determination of the factors A, f_1 and f_2 has resulted in limiting speeds in close approximation to the conditions found in practice.

Approximate speed limits for a load giving the bearing a minimum life L_h = 100,000 hours (f_2 = 1) are shown on Figs.21 and 22. If the load is greater

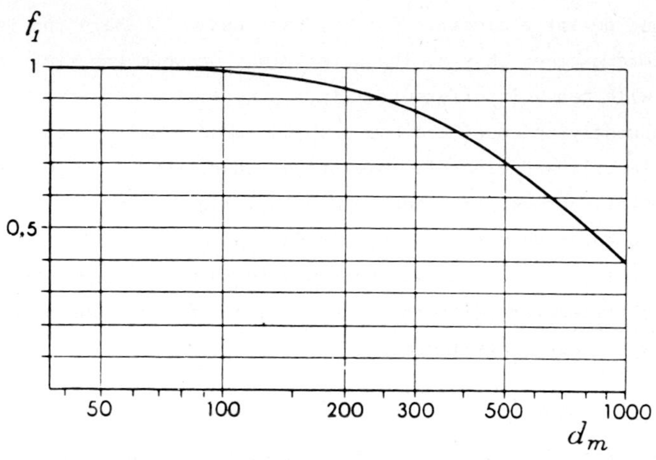

Correction factor f_1 for bearing size (d_m=mean diameter of bearing, mm)

Fig.18 Correction Factor for Bearing Size.

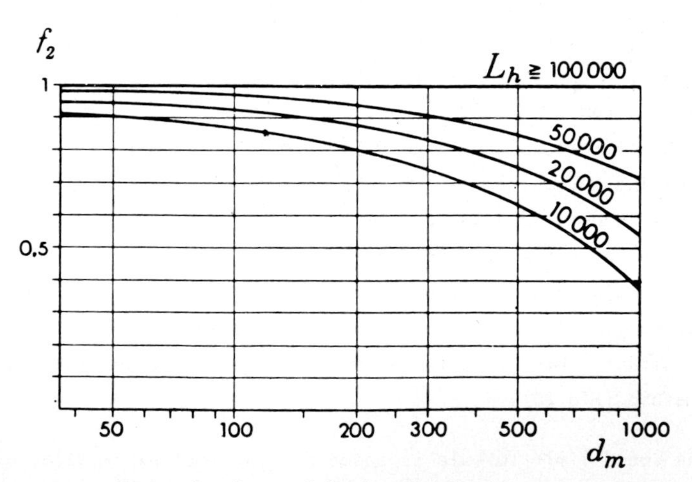

Correction factor f_2 for bearing load, expressed as the resulting life L_h in hours (d_m=mean diameter of bearing, mm)

Fig.19 Correction Factor for Bearing Load

Bearing type	Factor A		Remarks
Radial bearings: $n\, d_m = f_1 f_2 A$			
Deep groove ball bearings	Normally	500 000	Pressed-steel cages
	Maximum	1 000 000	Solid brass cages
Self-aligning ball bearings	Normally	500 000	Pressed-steel cages
	Maximum	800 000	Solid brass cages
Single row angular contact ball bearings	Normally	500 000	Pressed-steel cages
	Maximum	800 000	Solid brass cages
Double row angular contact ball bearings	Normally	200 000	
	Maximum	400 000	
Cylindrical roller bearings	Normally	400 000	Pressed-steel cages
	Maximum	800 000	Solid brass cages
Spherical roller radial bearings	Normally	200 000	
	Maximum	400 000	
Taper roller bearings	Normally	200 000	In the case of a predominant thrust load 20-40% lower limit values apply, depending on the working conditions
	Maximum	400 000	
Thrust bearings: $n \sqrt{DH} = f_1 f_2 A$			
Ball thrust bearings	Normally	100 000	
	Maximum	200 000	
Spherical roller thrust bearings	Normally	200 000	Good, natural cooling generally sufficient.
	Maximum	300 000	Effective cooling necessary.

At high speeds the bearing slackness must usually be greater than normal

Thrust bearings operating at high speeds must carry a certain minimum load $F_{a\,min}$, as shown in the graph in Fig.

Fig. 20

the limit values are reduced by modifying the factor f_2. Figures 21 and 22 show two speed limits (normal and maximum) for each bearing type. Bearings operating up to normal limits can be fitted with standard cages and, as a general rule, grease lubrication may be used. It should be noted that Ball Thrust and Spherical Roller Thrust bearings shown in Fig.22 should be mounted free from slackness and carry a certain minimum axial load, otherwise the tracks may be damaged (due to smearing) as a result of the gyratory forces acting on the rolling elements.

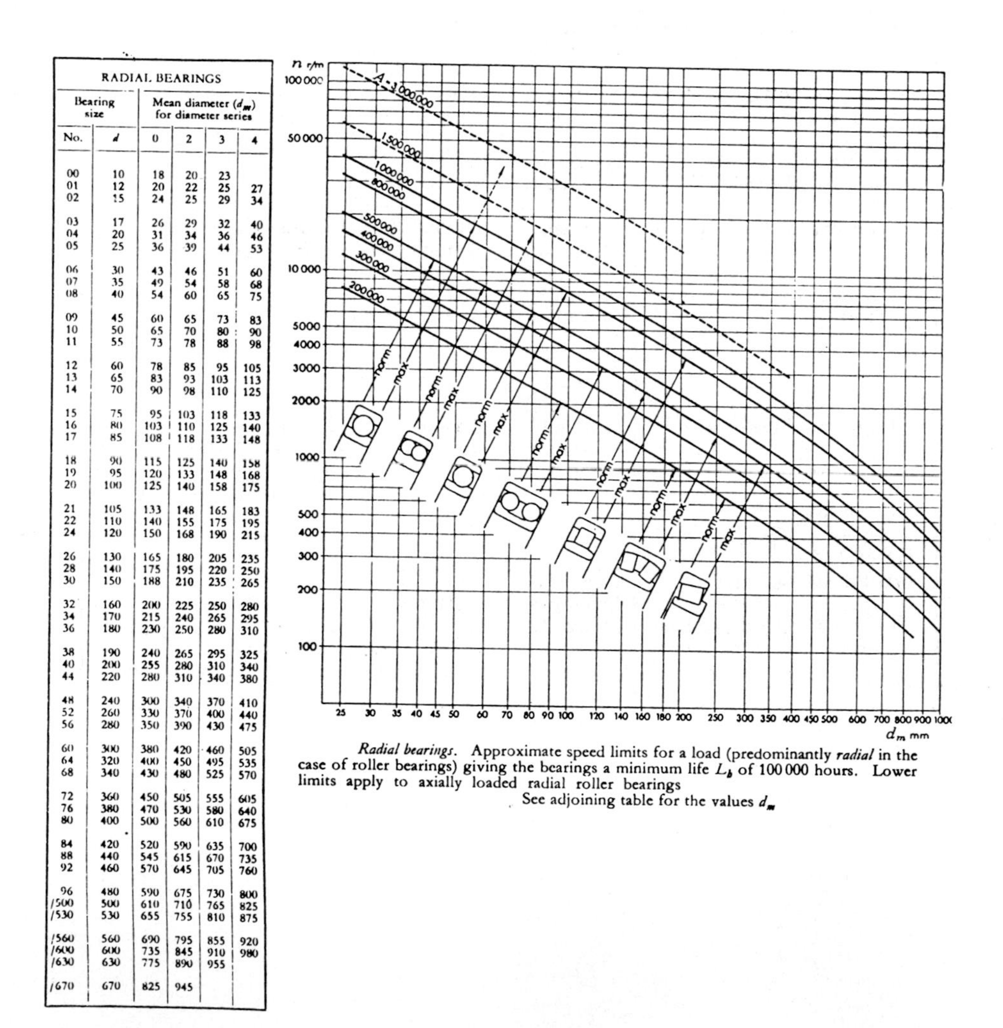

RADIAL BEARINGS

Bearing size		Mean diameter (d_m) for diameter series			
No.	d	0	2	3	4
00	10	18	20	23	
01	12	20	22	25	27
02	15	24	25	29	34
03	17	26	29	32	40
04	20	31	34	36	46
05	25	36	39	44	53
06	30	43	46	51	60
07	35	49	54	58	68
08	40	54	60	65	75
09	45	60	65	73	83
10	50	65	70	80	90
11	55	73	78	88	98
12	60	78	85	95	105
13	65	83	93	103	113
14	70	90	98	110	125
15	75	95	103	118	133
16	80	103	110	125	140
17	85	108	118	133	148
18	90	115	125	140	158
19	95	120	133	148	168
20	100	125	140	158	175
21	105	133	148	165	183
22	110	140	155	175	195
24	120	150	168	190	215
26	130	165	180	205	235
28	140	175	195	220	250
30	150	188	210	235	265
32	160	200	225	250	280
34	170	215	240	265	295
36	180	230	250	280	310
38	190	240	265	295	325
40	200	255	280	310	340
44	220	280	310	340	380
48	240	300	340	370	410
52	260	330	370	400	440
56	280	350	390	430	475
60	300	380	420	460	505
64	320	400	450	495	535
68	340	430	480	525	570
72	360	450	505	555	605
76	380	470	530	580	640
80	400	500	560	610	675
84	420	520	590	635	700
88	440	545	615	670	735
92	460	570	645	705	760
96	480	590	675	730	800
/500	500	610	710	765	825
/530	530	655	755	810	875
/560	560	690	795	855	920
/600	600	735	845	910	980
/630	630	775	890	955	
/670	670	825	945		

Radial bearings. Approximate speed limits for a load (predominantly *radial* in the case of roller bearings) giving the bearings a minimum life L_h of 100 000 hours. Lower limits apply to axially loaded radial roller bearings

See adjoining table for the values d_m

Fig.21 Approximate speed limits. Radial Bearings.

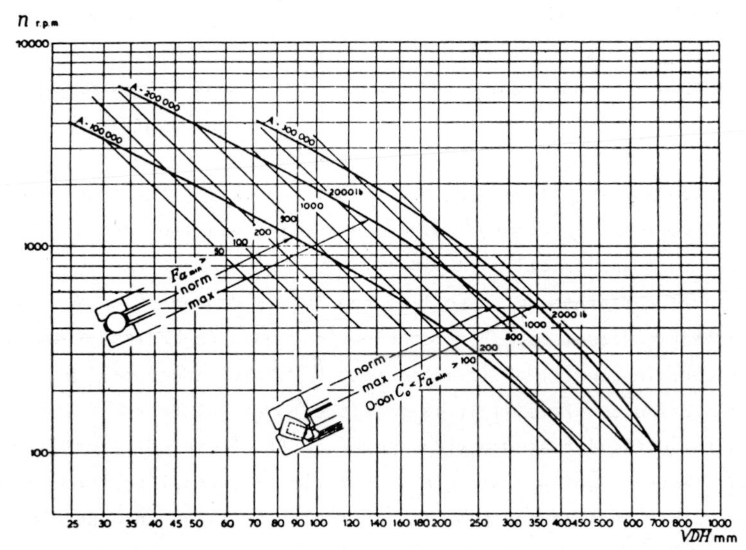

Thrust bearings. Approximate speed limits for a load giving the bearings a minimum life L_h of 100000 hours. See adjoining table for the values $\sqrt{DH}$. It should be noted that the value 0.001 C_o (C_o=static carrying capacity of bearing) is often greater than the numerical value of $F_{a\,min}$ obtained from the graph. The former value is then taken. It should also be noted that the $F_{a\,min}$ lines have different values for ball and roller bearings

BALL THRUST BEARINGS					
Bearing size		$\sqrt{DH}$ for series			
No.	d	511	512	513	514
00	10	15	17		
01	12	15	18		
02	15	16	20		
03	17	16	20		
04	20	19	25		
05	25	22	27	31	38
06	30	23	29	36	44
07	35	25	33	40	51
08	40	28	36	45	57
09	45	30	38	49	63
10	50	31	41	54	69
11	55	35	47	61	76
12	60	38	50	62	81
13	65	40	52	64	89
14	70	41	53	71	95
15	75	44	55	77	102
16	80	45	57	79	107
17	85	46	62	86	114
18	90	51	69	88	121
20	100	58	76	97	134
22	110	60	78	109	148
24	120	62	81	121	160
26	130	71	92	130	172
28	140	75	96	139	177
30	150	77	104	141	190
32	160	79	107	153	204
34	170	86	115	156	214
36	180	88	118	169	224
38	190	94	129	183	
40	200	96	132	193	
44	220	100	137		
48	240	116	163		
52	260	120	169		
56	280	136	174		
60	300	153	200		
64	320	159	204		
68	340	164	210		
72	360	169	235		

SPHERICAL ROLLER THRUST BEARINGS				
Bearing size		$\sqrt{DH}$ for series		
No.	d	292	293	294
12	60			74
13	65			79
14	70			85
15	75			90
16	80			96
17	85			102
18	90			107
20	100			118
21	105			
22	110		95	130
24	120		106	140
26	130		114	152
28	140		120	154
30	150		122	164
32	160		135	174
34	170		137	187
36	180		148	198
38	190		158	209
40	200		170	221
44	220		175	226
48	240	143	180	232
52	260	147	200	252
56	280	151	204	275
60	300	175	229	280
64	320	179	234	300
68	340	183	251	325
72	360	206	261	330
76	380	210	282	342
80	400	214	286	362
84	420	235	302	368
88	440	239	314	400
92	460	243	326	406
96	480	259	331	436
/500	500	262	335	441
/530	530	278	358	
/560	560	294	386	
/600	600	312	402	
/630	630	335	425	

Fig.22 Approximate speed limits. Thrust bearings.

In the top part of Fig.21 the speed curves with A = 1500000 and A = 3000000 have been drawn in and the broken line shows the extreme values which have been achieved with bearing types in a few known cases. Such speeds, however, do require experience in mounting and the greatest care must be exercised.

6.8 FRICTION

The extensive use of rolling bearings is due to rolling friction being less than sliding friction. Frictional losses in rolling bearings are usually very low, hence the term antifriction bearings.

The frictional resistance of a rolling bearing is dependent on several factors such as bearing load, speed of rotation, and the properties of the lubricant. Under certain conditions (bearing load $P \cong 0.1C$, adequate lubrication and normal operating conditions) it is possible to calculate the frictional resistance with sufficient accuracy using the coefficients of friction given in Table 6.1. Where rubbing seals are used their frictional resistance, which can be greater than that in the bearing, must also be taken into account. The friction torque "M" of a bearing is obtained from the equation:

$$M = \mu \times F \times \frac{d}{2} \text{ (Nmm)}$$

where μ = coefficient of friction for the bearing (see Table 6.1)
F = bearing load, N
d = bearing bore, mm.

TABLE 6.1 Coefficient of friction

Bearing type	μ
Self aligning Ball Bearings	.001
Cylindrical Roller Bearings	.0011
Thrust Ball Bearings	.0013
Deep Groove Ball Bearings	.0015
Spherical Roller Bearings	.0018
Taper Roller Bearings	.0018
Spherical Roller Thrust Bearings	.0018
Angular Contact Ball Bearings	
Single Row	.002
Double Row	.0024
Needle Roller Bearings	.0025
Cylindrical Roller Thrust Bearings	.004
Needle Roller Thrust Bearings	.004

Higher coefficient of friction values than those given in Table 6.1 are obtained with new bearings and this applies particularly to roller bearings which have not been run in. Higher values will also be achieved when starting and using excessive quantities of lubricant.

The friction loss is:

$$W_F = \frac{9.8}{10^3} \times M \times n \text{ (Watts)}$$

where M = friction moment, Nmm

n = speed, rpm.

6.9 LUBRICATION

Rolling bearings must be lubricated to prevent metallic contact between the rolling elements, tracks and cage and also to protect the bearing from corrosion and wear. The most favourable running temperature for a rolling bearing is achieved when the minimum of lubricant necessary to ensure reliable lubrication is used.

Lubricating properties deteriorate due to ageing and mechanical working and all lubricants become contaminated in service and must, therefore be replenished or changed periodically.

Rolling bearings may be lubricated with grease or oil, or in special cases with a solid lubricant. When considering lubrication for bearings the choice is between oil and grease and various aspects need to be considered. Grease lubrication has certain advantages which are:-

1. Costs involved in mounting are lower than with oil.
2. Less maintenance is required and it is not necessary to incorporate piping or pumping equipment.
3. Constant oil level devices not required.
4. Easier to contain grease in housing than oil.
5. Cleaner to use grease as there is no splashing as with oil.
6. Cheaper to seal for grease than for oil.
7. Grease assists in sealing against the entry of moisture and other impurities.

Tests have shown that only small amounts of grease adhere to the surfaces of the bearing. Reservoirs of grease form on the cage and against the side faces of the bearing. The bulk of the grease collects outside the bearing and in the grease cavities of the housing and, as a result, this is usually inactive. It can be argued that this reserve of grease helps to maintain an oil bleed to the bearings, but experience suggests that although this reserve may be in reasonable quantities in the cavities, it is still possible for the bearing to fail due to insufficient lubricant. The basic rule for a normal bearing arrangement is that the bearing should be well packed with grease with the housing no more than half full. If the space round the bearing is excessively filled with grease then churning of the grease in the bearing can occur which could lead to a rapid breakdown of the grease structure due to overheating. In such cases the grease softens and the oil in the grease tends to bleed from the soaps. The stiffness or hardness of a grease is called consistency and is usually quoted in terms of the National Lubricating Grease Institute (NLGI) scale and Consistencies 2 or 3 are completely satisfactory in normal applications for ball and roller bearing lubrication.

A '3' consistency grease would be used in an application such as an axlebox or traction motor where there is considerable vibration and a risk of the grease slumping.

6.9.1 Greases

Lubricating greases are thickened mineral oils or synthetic fluids. Metal soaps are mainly used as the thickening agent. The consistency of the grease depends mostly on the type and quantity of the thickening agent used. When selecting a grease, its consistency, temperature range and rust-inhibiting properties are the most important factors to be considered.

6.9.1.1 Temperature Range

(i) Sodium Base Greases. These greases may be used at temperatures between -30 to +80°C, although some special versions may be used up to +120°C Sodium base greases are water-soluble, i.e. they absorb water to a certain extent and form a rust-inhibiting emulsion without their lubricating properties being impaired. These greases will protect the bearings sufficiently against rust providing that water cannot enter the bearing arrangement. Where water can enter, such greases are easily washed out of the bearing housing.

(ii) Calcium Base Greases. Most calcium based greases are stabilised with 1 to 3% water. With increased temperature the water evaporates and separation of the grease into mineral oil and soap occurs. The upper temperature limit for these greases is therefore approximately +60°C. Some heat-stable calcium base greases are available which permit operating temperatures up to +120°C.

(iii) Lithium Base Greases. These greases are generally suitable for use at temperatures between -30 to +110°C, but a few greases of this type are suitable for working temperatures up to +150°C.

Lithium and calcium base greases are virtually insoluble in water and do not therefore give protection against corrosion. Such greases should therefore never be used unless they contain a rust-inhibitor.

For heavily-loaded rolling bearings, e.g. rolling-mill bearings, greases containing EP additives are used since these increase the load-carrying ability of the lubricant film. Such greases are also generally recommended for the lubrication of medium and large sized roller bearings. The rust inhibiting properties of calcium and lithium base greases containing EP additives (mainly lead compounds) are good. These greases adhere well to the bearing surfaces as well as being insoluble in water. They are, therefore, particularly suitable for applications where water can penetrate the bearing arrangement, e.g. paper-making machines or rolling-mills.

Greases containing inorganic thickeners instead of metal soaps, e.g. clay or silica, may be used for short periods at higher temperatures than lithium base greases. Synthetic greases, e.g. those made from diester or silicone fluids, may be used at both higher and lower temperatures than greases made from mineral oils.

Grease re-lubrication intervals as recommended by SKF are given in graphs in Fig.23; these intervals are conservative and are known to give a wide safety margin.

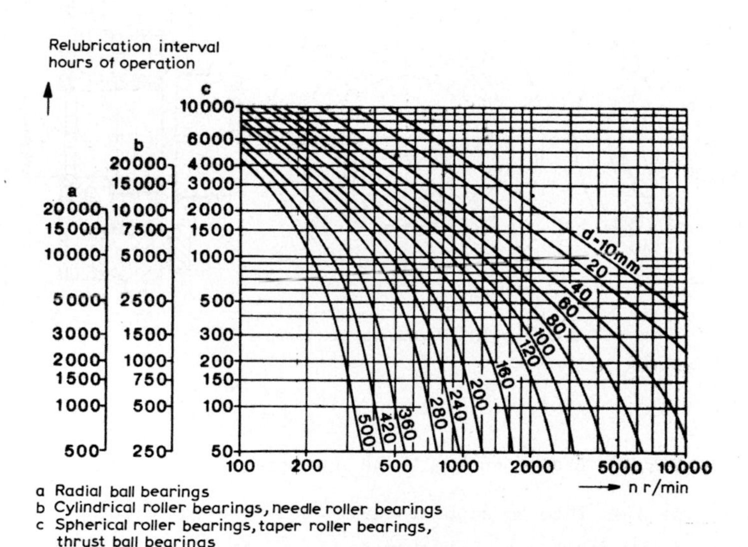

Fig.23 Re-lubrication Intervals

To prevent the possibility of mixing incompatible greases with its inherent problems, it is advisable to ensure that only greases having the same thickener and with a similar base oil are used when re-lubricating.

There is often doubt as to the quantity of grease to be used in a bearing and whilst it is difficult to be exact, the following guide can be used:

$$G = 0.005 \times D \times B$$

where G = grease quantity, grams

D = bearing outside diameter, mm

B = bearing width, mm.

For high-speed bearings necessitating frequent re-lubrication intervals it is essential to avoid over-filling the housings, since too much grease causes the grease to churn, resulting in an excessive rise in temperature. Churning can lead to a breakdown in the lubricating properties of the grease with a further rise in temperature and the bearings operating in a pre-load condition. This problem can be avoided by utilising a grease escape valve arrangement, as shown in Fig.24. The valve consists of a disc which rotates with the shaft and, in conjunction with a housing end cover, forms a narrow radial gap. Excess grease

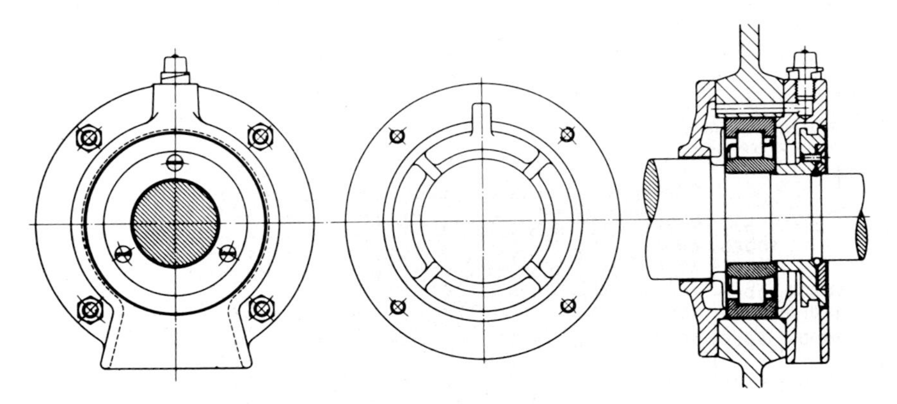

Grease valve for end-frame type electric motor

Fig.24 Grease Escape Valve Arrangement

is thrown out by the disc into an angular cavity and ejected through an opening on the underside of the end cover. This system of escape valve is extremely useful from a maintenance aspect as examination of the ejected grease can give valuable evidence regarding the condition of the bearing, i.e. if the bearing is fitted with a brass cage and this is beginning to wear the grease will become discoloured, and this can be easily seen and recognized. Figure 25 shows the temperature effect of operating bearings at high speeds with and without grease valves. Experiments in the machine tool industry where bearings are run with only a light smear of grease on the tracks indicated that initial lubrication could be as little as 1 cm^3 per 50 mm mean bearing diameter. The use of such small quantities in 90mm bore cylindrical roller bearings in jig borers pre-loaded 0.0025 mm and operating at a speed of 2500 rpm has reduced the temperature rise to a slow as 8^oC. Although long-running periods have also been achieved using this technique, the method is extremely delicate and necessitates special training for the fitters.

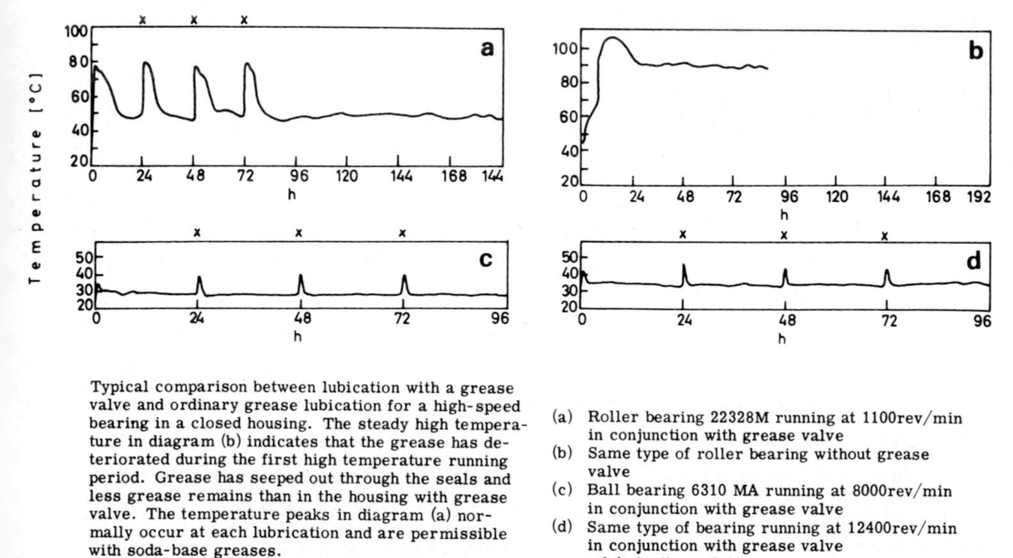

Typical comparison between lubication with a grease valve and ordinary grease lubication for a high-speed bearing in a closed housing. The steady high temperature in diagram (b) indicates that the grease has deteriorated during the first high temperature running period. Grease has seeped out through the seals and less grease remains than in the housing with grease valve. The temperature peaks in diagram (a) normally occur at each lubrication and are permissible with soda-base greases.

(a) Roller bearing 22328M running at 1100rev/min in conjunction with grease valve
(b) Same type of roller bearing without grease valve
(c) Ball bearing 6310 MA running at 8000rev/min in conjunction with grease valve
(d) Same type of bearing running at 12400rev/min in conjunction with grease valve
x = relubrication point

Fig.25 Effect on Temperature of Incorporating Grease Valve.

6.9.2 Oil Lubrication

Oil has several advantages compared to grease such as ease of draining and replenishing when necessary and particularly when the relubrication interval for grease is very short. Oil lubrication is generally used when high speeds or operating temperatures prohibit the use of grease and is useful when it is necessary to dissipate frictional or applied heat from the bearing. The selection of a lubricating oil is easier than the choice of a grease. Oils are more uniform in their characteristics and if resistant to oxidation, gumming, and evaporation can be selected on the basis of a suitable viscosity (Fig.26).

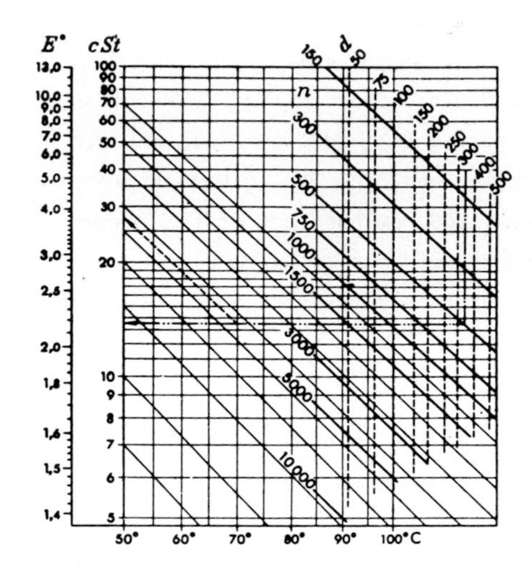

Graph for selection of oil. d = bore of bearing, mm., n = speed, r.p.m. Example: spherical roller thrust bearing 29468: d = 340 mm, n = 500 r.p.m. Use an oil which has a *minimum* viscosity of 13.5 cSt at working temperature. When the working temperature is known, the approximate viscosity of the oil required at 50° C. can be obtained with the aid of the thin, oblique lines. At a working temperature of 70° C. the oil in the example should have a viscosity of approx. 27 cSt at 50° C. Select an oil from those available whose viscosity is the *nearest above* this value, for example, Gargoyle DTE Oil Medium (27.3 cSt at 50° C.)

Figure 26. Selection of Oil.

Due to the fact that heat-generated in rolling bearings increases with viscosity, it is necessary to select a thin oil for high-speed operation, otherwise the bearing temperature would be too high. For very slow speeds, applications such as spherical roller thrust bearings in cranes, an extremely thick oil (minimum viscosity 400 c St at 50°C) is used to ensure a sufficiently strong oil film. For normal ambient temperatures and working conditions, an oil with a viscosity of 12-22c St at 50°C is suitable. At moderate speeds, no special oil is required for bearings in gear-boxes since they can be lubricated by the gear oil provided the bearings are adequately protected against wear particles from the gears entering the bearings. If this cannot be prevented, then the bearings must be separately lubricated, usually with oil. Grease can be used, but the seals must have the ability to prevent the gear oil flushing the grease out of the bearing. It must be mentioned that, as with grease, an excessive quantity of oil can cause churning and considerable heat; therefore, for normal reservoir systems it is essential for the maximum oil level to be no higher than the centre of the lowest rolling element. If the oil level exceeds this there could be a temperature rise due to churning. Figure 27 shows the effect on temperature rise and friction torque of increasing the quantity of oil. When the quantity of oil reaches a minimum level, i.e. the dotted zone, metal-to-metal contact occurs, resulting in rapid temperature rise and possible bearing seizure. Bearing friction torque is also a function of oil quantity and it can be seen that the torque increases with the quantity of oil. As this could represent considerable power loss it is essential to ensure that the oil level does not exceed a level compatible with adequate lubrication. Oil lubrication can be by circulation, drip feed, wick feed, or oil mist. In a circulatory system the outlet must be greater than the inlet to prevent the possibility of excessive oil in the bearing. For high-speed applications such as grinding spindles, oil mist lubrication is often used. In this system a mist of oil and air is transported through pipes to the bearings. Condensing nipples immediately before each bearing position cause the oil to be supplied to the bearing in droplet form. The small quantities of oil can be accurately regulated, consequently the lubrication friction is negligible. Figure 28 shows a typical oil mist unit.

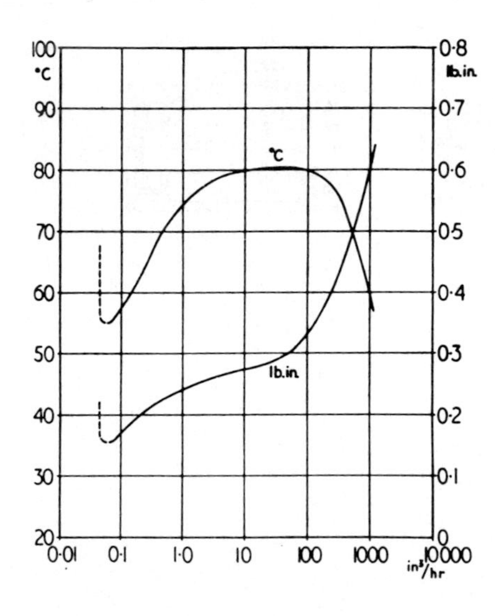

The dependence of bearing friction on quantity of oil (in in^3/hr)
Lower curve: friction torque, lb. in.
Upper curve: outer ring temperature, °C

Fig.27 Effect on Bearing Friction of Varying Quantity of Oil

6.10 SEALS

Selecting the correct bearings for a particular application necessitates more than ascertaining correct type and size. If bearings are to function satisfactorily, they must be properly lubricated and protected from the operating environment by means of correctly designed seals.

Seals are normally intended to prevent foreign material entering the bearing and in certain cases to prevent the ingress of moisture and corrosive media. The friction developed by a rubbing seal must be considered relative to the power input. The seal or side plate must also retain the lubricant in the bearing or housing. Rotational speed or rubbing seals must also be considered to ensure excessive heat is not developed.

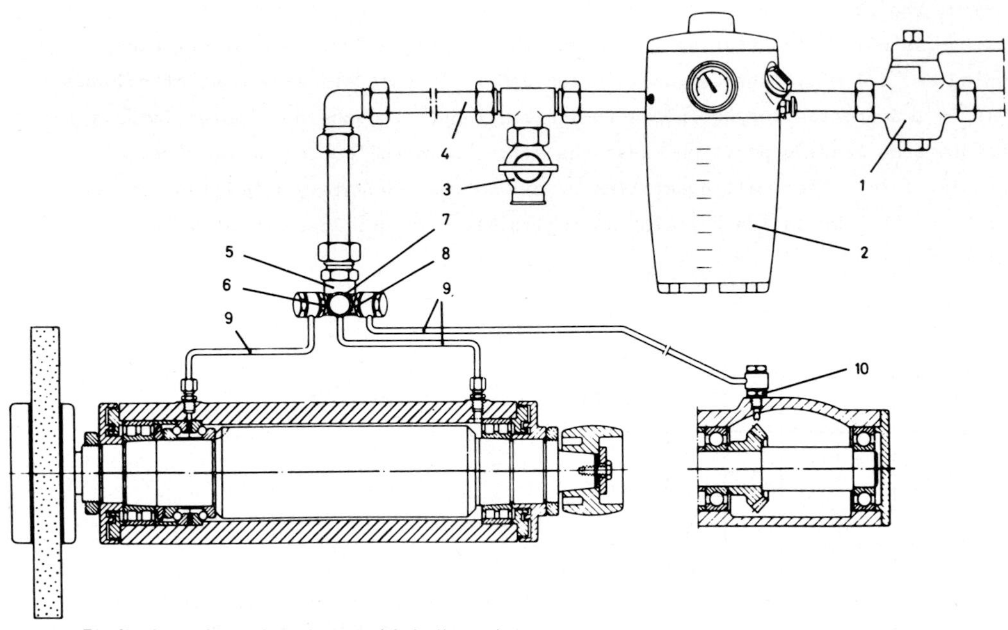

Pipeline layout for condensed oil mist lubricating system

1 Compressed air cut-off valve	5 Distribution box
2 Oil mist lubricator	6-8 Condensing nipples
3 Pressure control valve	9 Branch pipelines
4 Main pipeline	10 Spray nozzle

Fig.28 Oil Mist Lubrication System

Simple gap seals are efficient using grease lubrication, the purpose of the gap seals being to keep out relatively small amounts of foreign material and to retain the grease without excessive temperature problems. A gap seal should be long axially and as small as possible radially. For normal applications the gap should be 0.1 to 0.3 mm. Gap seals can be used with oil lubrication, but it is advisable to incorporate a groove in the shaft or, alternatively, fit a ring, both arrangements causing the oil to flow back into the housing. A further alternative would be to machine a small groove in the shaft adjacent to the gap seal (which has a drain hole) thereby encouraging the oil to flow back into the housing.

Labyrinth seals are extensively used with grease lubrication in dirty and wet conditions. The labyrinth consists of a number of radially separated tongues with a small radial clearance and can be considered as elaborate grooved gap seals. The grooves are often filled with grease to prevent the ingress of dirt and a typical labyrinth system for a traction motor bearing is shown in Fig.29.

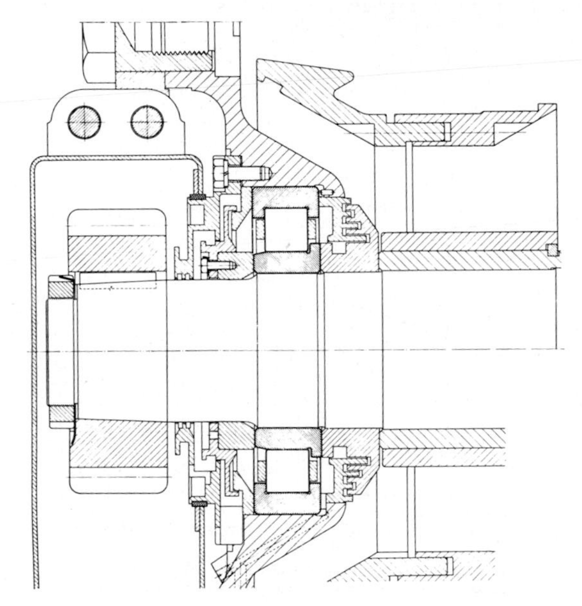
Bearing scheme for traction motor armature

Fig.29 Bearing Scheme for Traction Motor Armature

6.11 NOISE

Noise originates from an oscillating body which produces a moving longitudinal wave consisting of variation in pressure. In order to reduce noise irritation to a minimum there is a constant demand for silent running bearings particularly in domestic and office equipment such as vacuum cleaners, floor polishers, circulatory pump motors for central heating and fans etc.

A clean high quality rolling bearing runs with an even purring sound and depending on the bearing arrangement, the noise may be so low as to be inaudible or it can be unpleasant and disturbing. Bearing noise is a function of the level of vibration in the bearing and the quieter the bearing the lower the level of vibration. Vibrations in a bearing depend on many factors such as surface finish, speed, load and accuracy of geometric form etc.

Clearance in a bearing is a further contributory factor to noise and it is necessary to select bearings and fits to achieve zero clearance in the bearing under operating conditions. Fig.30 shows the effect on noise of different bearing clearances. Freedom from clearance in mounted ball bearings can also be achieved by adjusting the bearing against each other by means of disc springs as shown in Fig.31. This method ensures practically zero clearance under all working conditions. The total spring pressure should be approximately 5 Newtons per mm of shaft diameter i.e. 100 Newton pressure for a 20 mm shaft.

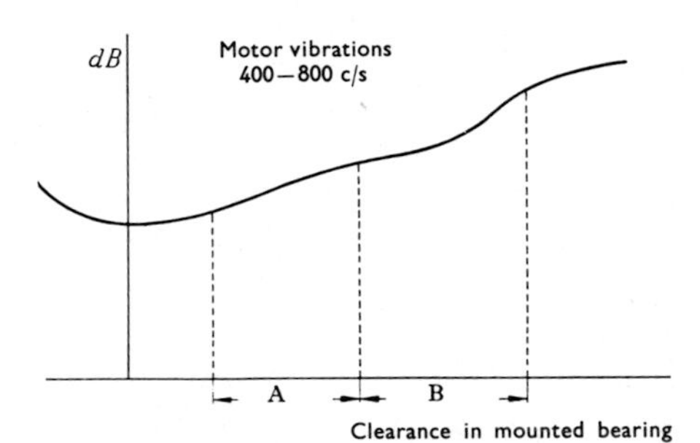

The dependence of motor noise on bearing clearance. Measurement of vibration of electric motor over the frequency range 400—800 c/s with progressive variation of bearing clearance

A=clearance range obtained with C2 bearings and normal fits
B=clearance range obtained with normal bearings and normal fits

Fig.30 Effect of Noise on Different Bearing Clearances

For ball and cylindrical roller bearings operating at high speeds (ndm 500000 and 400000 respectively) the temperature differential between the inner and outer rings is greater than at normal speeds and the effect of this must be taken into account. Any reduction in clearance due to temperature differential must be compensated for by using bearings with greater initial radial clearance.

Whilst vibration levels in bearings can be reduced, it is equally important to ensure that the other components in the machine are also manufactured to similar accuracy to ensure they are not the cause of vibration, otherwise the quiet running properties of the bearing will not be utilised. From a noise aspect the ovality and taper of the shaft and housing seatings should be accurate and lie within half the tolerance range for grades IT5 and IT6 for shaft and housing respectively. Bearing alignment must also be considered, as misalignment can also be a source of noise. Figure 32 shows the high and corresponding diametrically opposed low spots on a lobed type of out-of-roundness, and Fig.33 shows the effect of alignment errors on noise intensity. There are

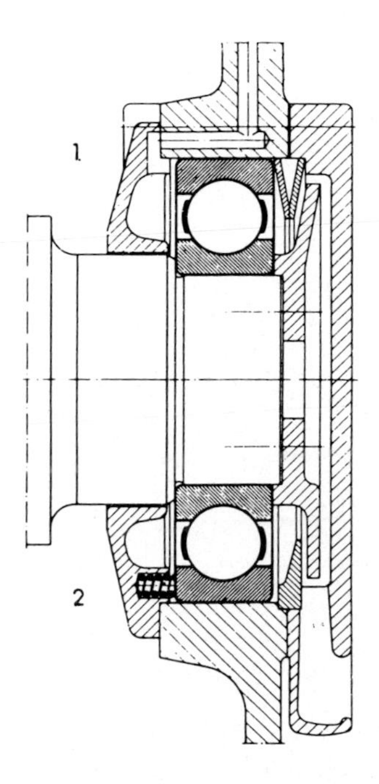

Fig.31 Spring preloading of deep groove ball bearings. 1 with spring washers, 2 with helical springs.

Fig.32 Lobing. For every high spot there is a corresponding low spot diametrically opposite. There are always an odd number of lobes.

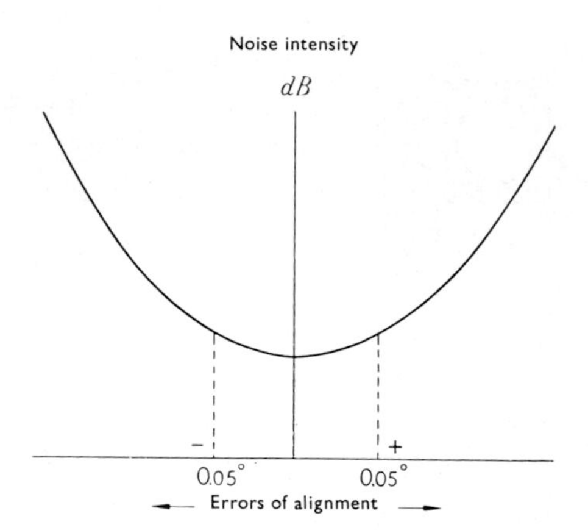

Fig.33 Effect of Alignment Errors on Noise

many applications where noise must be tolerated in order to achieve a satisfactory bearing life. In electric traction motors geared to the driving axles of railway vehicles, the operating conditions dictate heavy fits which in turn necessitate bearings with large initial radial clearance. The increased radial clearance results in combined roller and cage drop as they come out of the load zone which can cause increased bearing noise, although it has no effect on the life of the bearing and cannot be heard when the motor is fitted under the locomotive. A high pitched whine can occasionally occur at certain speed ranges with cylindrical roller bearings, despite adequate lubrication. This has been attributed to resonance caused by certain frequencies of the rolling elements coinciding with the natural frequency of the bearing end frame and, in certain cases, whine has been significantly reduced by modifying the mass distribution of the housing.

6.12 ANTICIPATING BEARING DAMAGE

The running performance of rolling bearings from a noise aspect can be checked with reasonable accuracy using a wooden listening stick and listening through the handle to the transmitted noise. A wooden stick is particularly useful in that it transmits noise relating to the condition of the bearing and cuts out most of the extraneous noise from other machine components which can cause problems for the more sensitive stethoscope. With experience, an operator becomes tuned to certain noises and can ascertain that bearing examination should be carried out at the next shutdown period. Naturally, if say a consistent knocking develops with the normal noise level, then an immediate inspection of the bearing should be made. Damaged tracks caused by careless mounting produce pronounced low pitch noise and inadequate radial clearance produces metallic noise.

6.13 DETECTION OF BEARING DAMAGE BY SHOCK PULSE MEASUREMENT

When a rolling bearing suffers fatigue failure, flaking occurs in the rolling element surfaces or in the tracks. A bearing component coming into contact with the damaged zone causes a mechanical shock, causing transient vibrations which are transmitted to the bearing housing. These vibrations generate electric voltages in an accelerometer and their amplitude is determined by a shock pulse meter. The condition of the bearing is monitored, taking readings at suitable intervals. Incipient damage can, therefore be detected at an early stage and bearing replacement planned accordingly.

Earphones can be used to listen to the rhythm of the shocks and, provided the damage is not too complex, the rhythm will suggest which component of the bearing is damaged.

Temperature is a further method for gauging bearing condition, and bearing failures are sometimes preceded by a drop in temperature followed by a rapid increase in temperature, usually caused by metal-to-metal contact with subsequent fatigue failure or even seizure. The old method of checkina the bearing temperature by feeling the housing is not satisfactory, and in applications where breakdowns cause maintenance problems it is usual to use thermal-couples positioned as close to the bearing outer ring as possible. When using the temperature method of checking bearing condition it must be noted that there will always be a temperature rise with new or freshly-greased bearings. This only applies during initial running and once the new lubricant has distributed itself, the temperature will start to fall.

6.14 FITS (SHAFT AND HOUSING)

Tolerances for the bore and outside diameter of metric rolling bearings are internationally standardised and the required fits are achieved by selecting shaft and housing tolerances using the ISO tolerance system (incorporated in BS 4500: Part 1:1969). Only a small selection of the ISO tolerance zones need to be considered for rolling bearing, and Fig.34 shows these relative to the bearing bore (a) and bearing outside diameter (b).

The shaft tolerance is indicated by a small letter and a number and the housing bore by a capital letter and a number. A typical shaft and housing bore tolerance combination would be written j6 - J7, the values for each being obtained from tolerance tables.

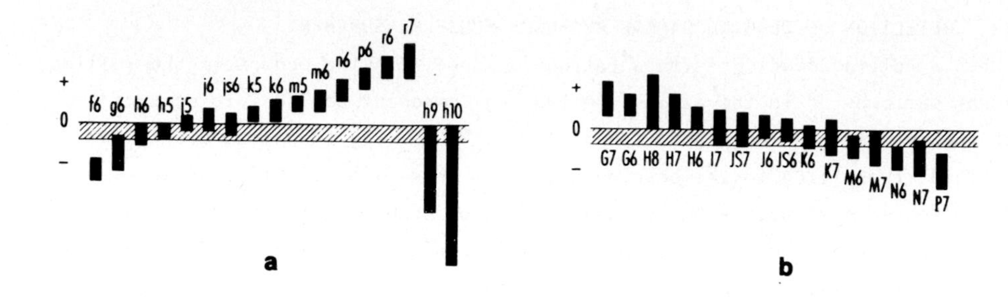

Fig.34 Shaft and Housing Fits.

The various symbol gradings are as follows:-

Shafts	
Clearance	Bearings always have clearance fit if the shaft tolerance is "f".
Transition	Bearings can be clearance or interference fit depending on the actual dimensions, if the shaft tolerance is within the range "g - j".
Interference	Bearings will always have an interference fit if the shaft tolerance range is within "k - r".

Housings	
Clearance	Bearings will always have clearance if the housing tolerance range is "G - H".
Transition	Bearings will have a clearance or an interference fit depending on the actual dimensions if the housing tolerance is within the range "J - N".
Interference	Bearings will have interference fit if the housing tolerance is "P".

The most important factors to consider when selecting fits are;-

1. Conditions of rotation
2. Magnitude of the load
3. Temperature conditions

6.15 CONDITIONS OF ROTATION

When a bearing ring is to rotate relative to the direction of the radial load, it must be mounted with sufficient interference on the shaft or in the housing to prevent "roll" or "creep". If creep occurs between heavily-loaded dry surfaces, the surfaces undergo rapid wear. It is therefore essential that the fit used is tight enough to ensure that no clearance exists and none can develop due to the action of the load. If the load is always directed towards the same point on the ring, however, no creep occurs and a clearance fit is generally permissible.

Various loading conditions can be classified as follows;-

(i) Rotating inner ring load. The shaft rotates relative to the direction of the load so that all points on the inner ring track are subjected to load during one revolution.

Example - Shaft with belt drive.

(ii) Stationary inner ring load. The shaft is stationary relative to the direction of loading so that the load is always towards the same sector of the inner ring.

Example - Automobile front hub.

(iii) Stationary outer ring load. The bearing housing remains stationary relative to the direction of loading so that the load is always directed towards the same sector of the outer ring.

Example - Shaft with belt drive.

(iv) Rotating outer ring load. The bearing housing rotates relative to the direction of loading so that all points on the outer ring track come under load during one revolution.

Example - Automobile front hub.

In many applications operating conditions cannot be related to any of these simple loading cases and variable external loads or out-of-balance forces influence the direction of loading, an appropriate fit being classified as "direction of loading indeterminate". In this instance, bearing ring "creep" can only be prevented by using an interference fit for both rings and in such cases bearings with increased radial clearance are usually necessary.

6.16 INFLUENCE OF LOAD AND TEMPERATURE

The load compresses the inner ring in a radial direction which stretches the ring in a circumferential direction and compresses the shaft, thereby loosening the fit. Similarly, as the bearing inner ring track is warmer during operation than the bearing bore, this also has the effect of loosening the fit of the bearing inner ring on its journal.

In the majority of applications it is unnecessary to calculate the required interference since experience is a good guide in selecting suitable fits for different operating conditions and only in particularly difficult or unusual conditions is it necessary to resort to special calculations.

6.17 BEARING APPLICATIONS

When designing bearing arrangements there are certain basic rules which should be followed. Wherever possible only one bearing should be used for location purposes. This means that the bearing outer ring should be held axially in its housing with the inner ring located on the shaft in a similar manner. All other bearings on the same shaft should be axially free, either on the shaft or in the housing, as shown on Fig.35. If this basic rule is ignored and two axially located bearings are used, any shaft expansion occurring due to generated or external heat could cause severed locking (preloading) across the bearings, resulting in premature bearing failure. In certain applications using angular contact ball or taper roller bearings it is not possible to use only one bearing for location and the bearings must be adjusted by end covers.

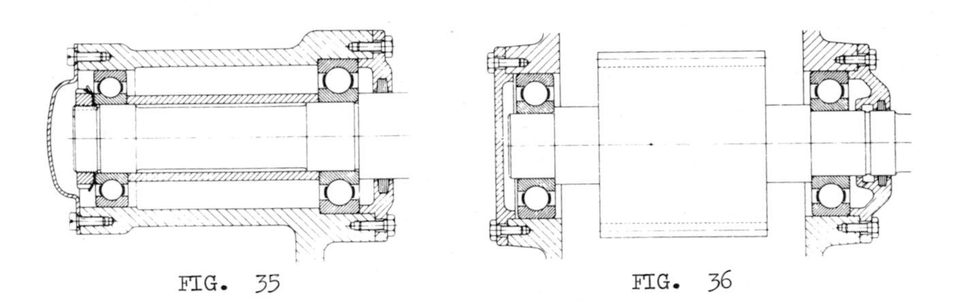

FIG. 35 FIG. 36

With these bearings great care must be taken to ensure that the bearings can accommodate axial variations and in some applications a small gap is left between the bearing outer ring and the abutment cover, as shown in Fig.36.

The axial expansion problem can be resolved by incorporating a cylindrical roller bearing at the non-located position, this arrangement being used extensively on electric motors where a ball bearing is used at the commutator end and a roller bearing at the drive end. In addition to resolving the thermal expansion problem such an arrangement has an added advantage in that a cylindrical roller bearing has significantly higher load-carrying capacity (compared to a dimensionally equivalent ball bearing) which makes it particularly suitable for reacting heavy drive forces.

6.18 BEARING CARE

Cleanliness is of paramount importance when handling and mounting bearings, but in spite of the fact that bearings are precision made, one has only to look around the average workshop to see open bearings left on benches or in dusty environments. It should be noted that if foreign matter is allowed to enter a bearing the rolling elements in passing over it during service can cause indentations leading to fatigue failure in the rolling elements or tracks which can shorten the life of the bearing considerably. As a general rule, in order to ensure the bearings remain free from impurities, they should not be removed from their original packing until they are required for installation. Before packing bearings are ultra-sonically washed and coated with a bearing preservative which mixes readily with most lubricants, apart from certain clay or synthetic-based greases, in which case it is advisable to contact the bearing manufacturers.

Although initial washing of bearings by the user is not recommended due to the possibility of the washing fluid not completely evaporating, if washing is necessary then the bearings should be washed with clean white spirit after which the bearings must be thoroughly dried before adding lubricant.

6.19 BEARING MOUNTING

If the bearing inner ring is to be a tight fit on the shaft, it can be driven onto the shaft journal by means of a tubular drift which should bear evenly against the face of the inner ring, as shown in Fig.37. If the bearing outer ring is to be a tight fit in its housing then the ring should be driven into the housing by applying force to the outer ring.

Hollow drift for bearings with an interference fit on the shaft

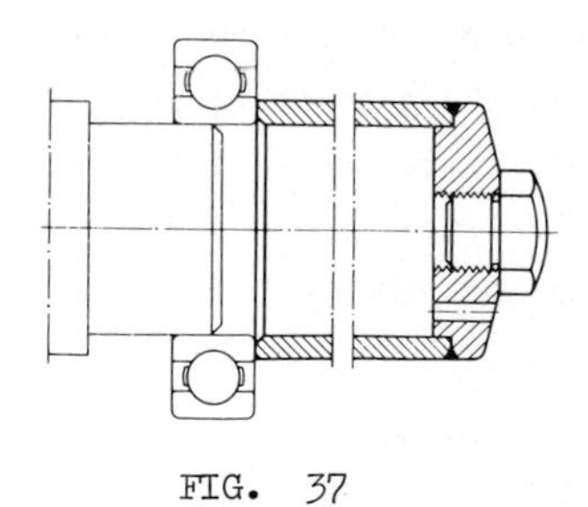

FIG. 37

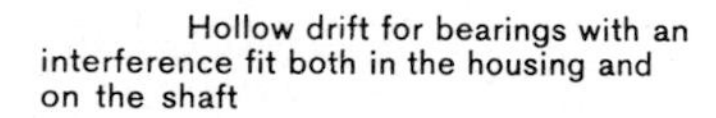

Hollow drift for bearings with an interference fit both in the housing and on the shaft

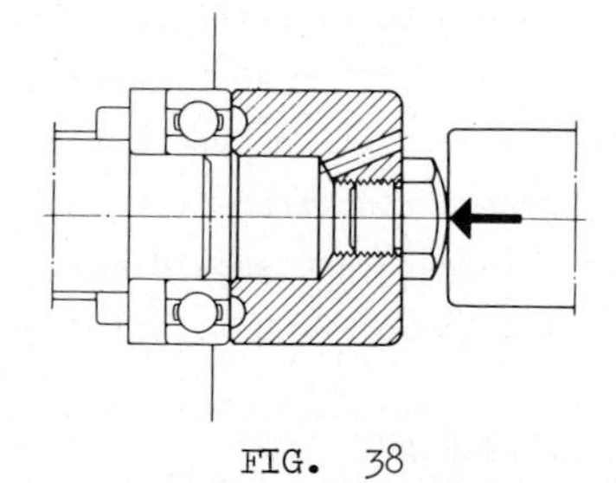

FIG. 38

Pressure must never be applied to the inner or outer bearing ring in order to mount the other ring which has a tight fit, otherwise there is a danger of the tracks being indented and the bearing would probably fail in service after a short time. If both the inner and outer rings are a tight fit, then a tool of the type shown in Fig.38 must be used which contacts both side faces of the bearing rings.

Care must be taken to ensure that the bearing ring being pressed on is correctly aligned, particularly in the case of small rings. Excessive misalignment puts severe stress on the cage, which can result in premature bearing failure.

Cylindrical roller bearings sometimes give assembly problems due to the rollers being scored as they are fed over the tracks. This problem can be resolved by using a special mounting sleeve as shown on Fig.39.

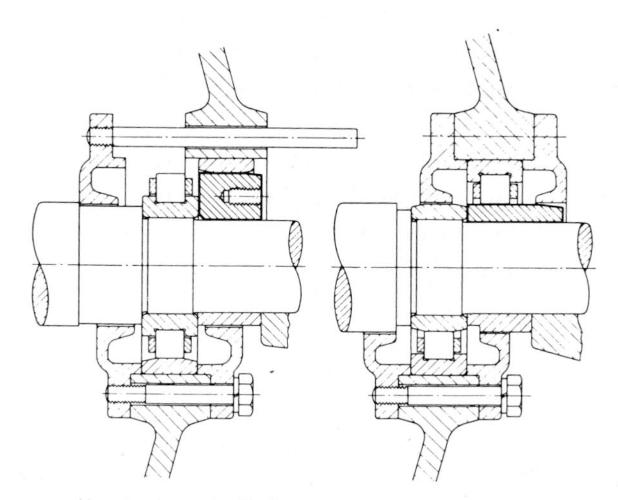

Mounting sleeves simplify the mounting of cylindrical roller bearings and prevent damage to the tracks

FIG. 39

For large rolling bearings with cylindrical bores where the inner ring is a tight fit, mounting can be simplified by heating the complete bearing in an oil bath at a temperature of 70° - 80°C above ambient. If the bearing needs to be transported some distance it can be heated to a higher temperature to compensate for cooling down during transit. Bearings can be heated up to 120°C in this way, but this temperature should not be exceeded as, above 120°C, there is a danger of reducing the bearing material hardness.

6.20 DISMOUNTING BEARINGS

If an interference fit has been used to locate a bearing on a shaft it is essential to use a suitably designed withdrawal tool to remove the bearing. The basic principles for mounting a bearing apply also to dismounting. The dismounting force must be applied to the bearing ring having the interference fit and not to the loose ring and through to the other ring by means of the rolling elements. Claw-type pullers are often used to remove ball bearings, but designers must take this into consideration during the design stage by incorporating suitable grooves in the abutment shoulder adjacent to the bearing face to enable the bearing inner ring to be gripped by the puller claws. Naturally, the puller force must be applied to the inner ring and not transmitted through the rolling elements. If the inner ring is inaccessible it is necessary to pull on the outer ring, but bearing damage can be avoided by locking the puller centre screw with a spanner and rotating the claws. The outer ring will then rotate during the withdrawl process, distributing the load over the tracks and thereby preventing the possibility of indentations. The bearing is then pulled off sufficiently to enable the inner ring to be gripped with the puller (Fig.40).

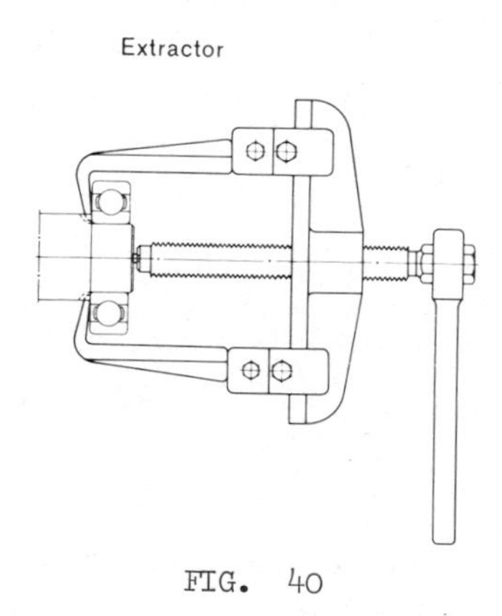

FIG. 40

6.21 MOUNTING AND DISMOUNTING BY OIL INJECTION

In order to overcome many of the mounting and dismounting problems SKF, in 1940, introduced an oil-injection technique which is now widely used in many branches of engineering. Oil under high pressure is injected between the bearing inner ring and shaft journal during mounting or dismounting. An oil film is formed which both separates and lubricates the contact surfaces. The oil separates the contact surfaces except for a narrow zone at each end of the ring. The surface pressure is greater in these zones due to the influence of the shaft material beyond the end of the ring and the zones act as an oil lock which retains the oil between the contact surfaces. When the oil pressure is released the oil is forced automatically back through the supply ducts, thereby restoring the original friction.

The advantage of using oil-injection is that the force required to move the component is significantly reduced and that by eliminating direct contact and the resulting friction between the contact surfaces, the possibility of damage to the surfaces occurring during the mounting or dismounting process is minimised.

A further advantage is that components can be dismounted or adjusted without the risk of the fit deteriorating.

If the bearing ring is mounted on a tapered journal with a self-releasing taper, the ring will be ejected with some force when the oil is introduced and some for of axial restraint such as a nut will be necessary. If, however, a locking taper is used then an external force additional to the oil injection force will be required. Figures 41, 42 and 43 show typical arrangements.

Position of distribution groove in a tapered and a cylindrical seating for a rolling bearing

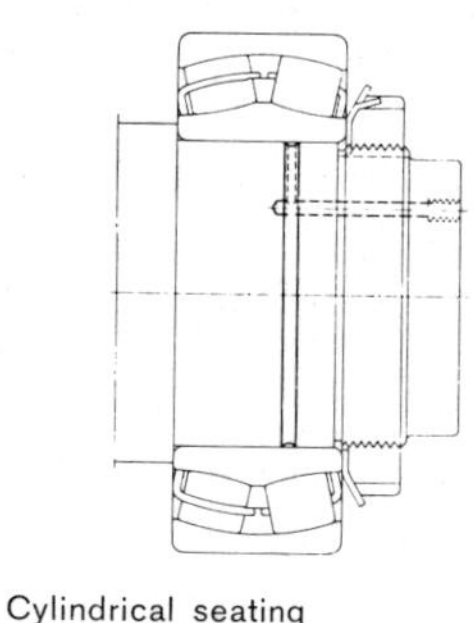

Cylindrical seating

FIG. 41

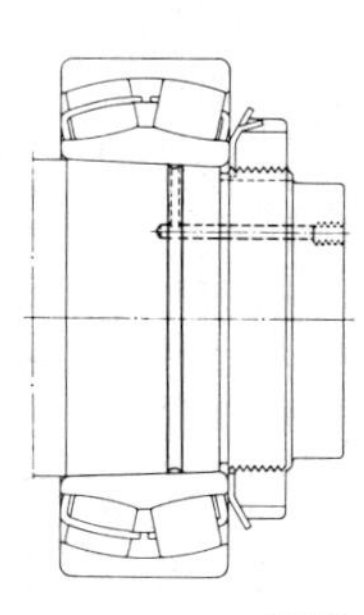

Tapered seating FIG. 42

The first two show a bearing on a parallel shaft and a bearing on a tapered shaft. The third shows oil injection being applied to a gear wheel mounted on a parallel shaft. It should be noted that, with a parallel shaft, once the bearing ring is past the oil entry hole there is no longer any oil pressure and the ring could lock. It has been found, however, that if the ring is withdrawn rapidly then it will travel over the remaining area without too much difficulty. Large rolling bearings with a taper bore and mounted on adaptor or withdrawal sleeves can be easily mounted or dismounted using hydraulic nuts designed by SKF. The hydraulic nut consists of an internally threaded steel ring with a groove in one face into which is fitted an annular piston sealed with O-rings. When oil is pumped into the annular space behind the piston it is forced outwards, thereby forcing the bearing on or pulling the bearing off the sleeve or shaft. Figure 44 shows a hydraulic nut being used to mount a spherical roller bearing on an adaptor sleeve and Fig.45 shows an arrangement for dismounting a bearing on an adaptor sleeve. An arrangement for dismounting a bearing from a withdrawal sleeve is shown on Fig.46.

Cylindrical mating surface having two distribution grooves

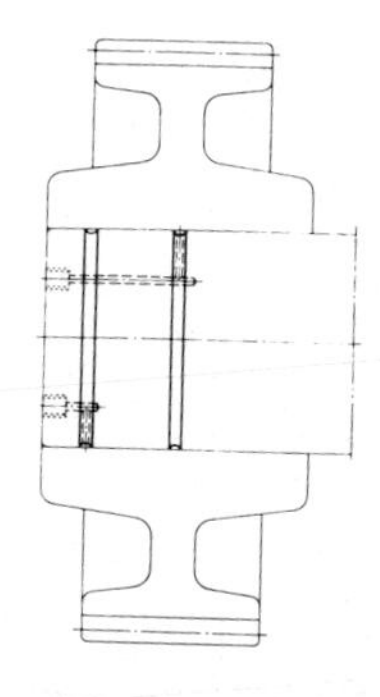

FIG. 43

HMV nut for driving up a bearing on an adapter sleeve

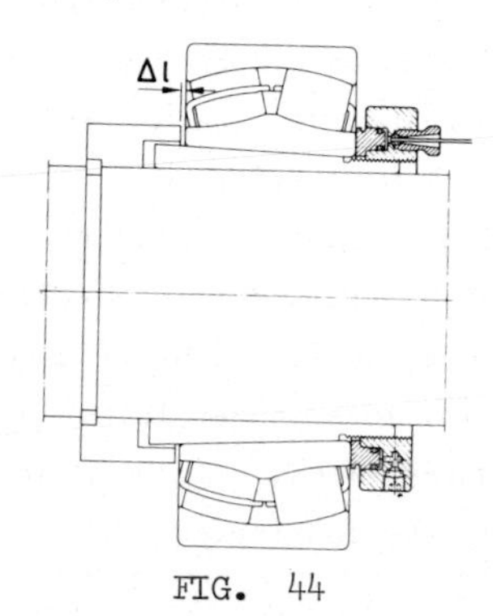

FIG. 44

HMV nut and stop rings in position to press loose an adapter sleeve

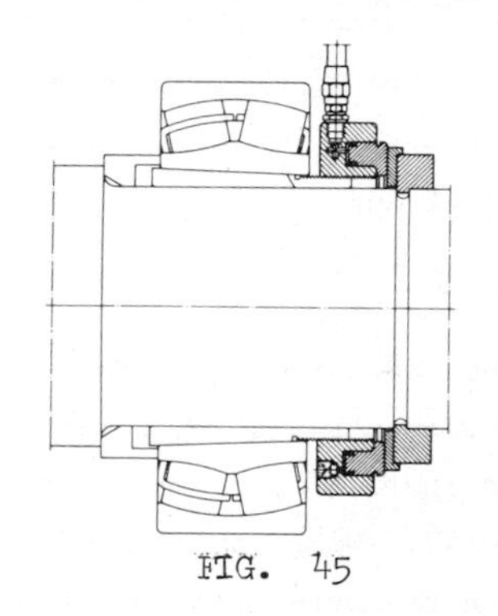

FIG. 45

HMV nut used to free a withdrawal sleeve

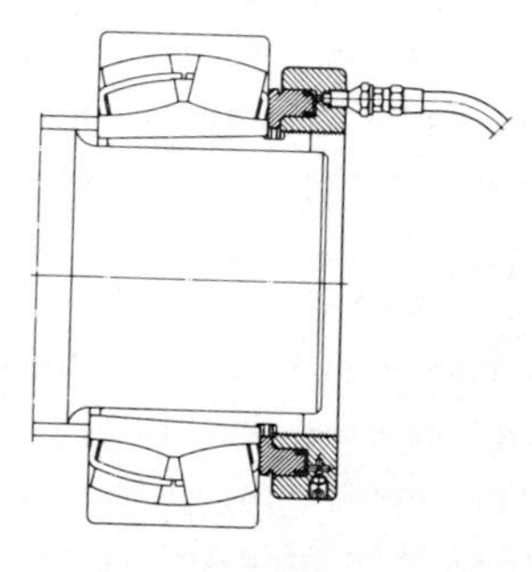

FIG. 46

6.22 CLEANING OF BEARINGS

As previously mentioned, it is not necessary to clean bearings taken direct from their packages and any attempt to do so could be detrimental from both a cleanliness and future lubrication aspect. There are occasions, however, when bearings need to be dismounted, cleaned, and inspected in spite of the fact that unnecessary dismounting may cause deterioration of the shaft and housing fits as well as possible damage to the tracks.

After dismounting for cleaning purposes all used grease should first be carefully removed. Small bearings should be immersed in white spirit or other cleaning fluid and swirled round, the residual grease and dirt being removed using a good quality brush. Care must be taken to ensure that none of the bristles are trapped between the cage and rolling elements.

After this preliminary washing the bearings should be washed in one or two additional baths of white spirit. A warm (100°C maximum) light mineral oil should then be flushed through the bearing as it is slowly rotated. Under no circumstances should the bearing be rotated until the oil passes through it, otherwise any foreign particles in the bearing will be rolled into the tracks causing indentations. If the bearings are not intended for immediate mounting they should be lubricated with a good quality oil containing a water-repellant additive.

6.23 RECOGNITION OF BEARING FAILURES

Obviously premature bearing failures occur, but in the majority of cases if the symptoms are recognised the cause of the failures can be eliminated. If rolling bearings are properly mounted, sealed, lubricated and maintained, they will run until fatigue failure (flaking) occurs on the bearing inner or outer ring or rolling elements. As already mentioned, bearing life based on fatigue failure can be predicted as a function of the life dispersion curve, and if a bearing fails well short of its normal L_{10} life it is important to ascertain the cause of the failure before fitting new bearings.

6.23.1 Wear

If a bearing housing is inadequately sealed, abrasive particles can enter the bearing causing wear in the tracks, rolling elements and cage, and in certain bearings wear in the guide flanges. The worn surfaces are dull in appearance and an example of worn tracks is shown in Fig.47. Wear can also occur between the inner ring bore and its journal or between the outer ring outside diameter and housing bore due to creep. This is usually the result of excessive clearance and is particularly serious because not only the bearing but also the shaft or housing could be damaged and require replacing. Creep between mating surfaces causes fretting corrosion and wear, and particles of rust from the corroded areas may also enter the bearing causing wear in the tracks and rolling elements. Wear can also be caused by inadequate lubrication, the worn surfaces having a highly planished finish. This type of wear develops into fatigue failure.

Fig.47 Worn Tracks due to Abrasive Particles.

6.23.2 Incorrect Mounting

Faulty mounting often results in the bearing being subjected to high loads which can cause material fatigue in the bearing rings or rolling elements well before the nominal calculated life of the bearing has been reached. Indentations with the same pitch as the rolling elements can be caused by the mounting force being applied through the rolling elements. During service an overload occurs each time a rolling element passes over an indentation, and after a relatively short time small fragments of bearing material break away, the condition being generally referred to as "flaking". Figure 48 shows the inner ring of a deep groove ball bearing with advanced areas of flaking on the right-hand side of the track.

Fig.48 Flaking caused by Faulty Mounting.

Figure 49 shows the inner and outer ring of a ball bearing with a filling slot. The flaked wear on the right-hand side of the inner ring and on the left-hand side of the outer ring show that failure has been caused by a heavy thrust load due to cross-location. Fatigue failure of this type begins at a point below the surface of the track or rolling elements. Small cracks develop, which gradually work up to the surface where, under repeated overload at such weak points, eventually cause fragments to break away, these fragments being rolled into other parts of the track thereby initiating further areas of weakness.

Fig.49 Failure caused by a Heavy Thrust Load due to Cross Location.

6.23.3 Cage Failures

With an adequately lubricated bearing operating in normal conditions, the cage is the most lightly-loaded component. If the lubrication is inadequate, wear will occur where the cage makes contact with the rings and rolling elements and eventually the cage may fracture. Highly planished surfaces are usually associated with inadequate cage lubrication. Excessive misalignment of the inner and outer bearing ring relative to each other is another cause of cage failure and severe misalignment can result in the cage, or even the bearing rings, cracking. Wear can also be caused by foreign particles entering the bearing, the particles jamming or wedging between the cage and rolling elements. Figure 50 shows a spherical roller bearing case with worn roller prongs and wear in the bottom of the roller pockets. Figure 51 shows a more advanced case where the prongs have actually sheared.

Fig.50 Spherical Roller Bearing Cage with Worn Roller Prongs and Wear in the Bottom of the Roller Pockets.

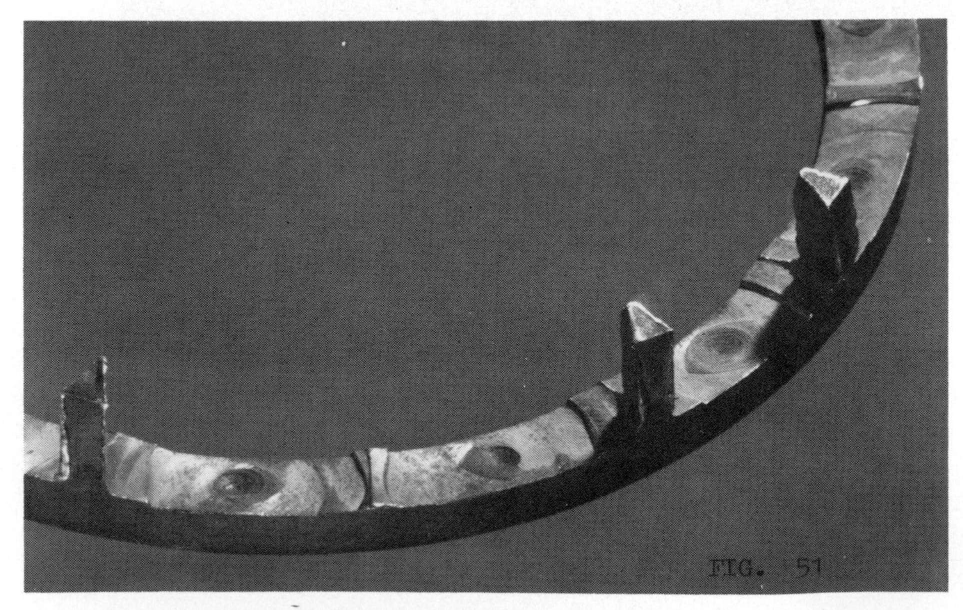

Fig.51 More Advanced Case than Fig.50, with Prongs Sheared.

6.23.4 Smearing

Smearing is a transference of material from one body to another when two inadequately lubricated surfaces slide against each other. Local stress concentrations are induced by smearing, and these produce cracks in the surface layers and subsequent flaking. Figure 52 shows a smeared cylindrical roller which has been caused by rapid acceleration of the roller in the loaded zone coupled with inadequate or incorrect lubrication. If cylindrical rollers are subjected to severe thrust forces combined with inadequate or unsuitable lubrication, smearing develops at the ends of the roller and on the guide flanges

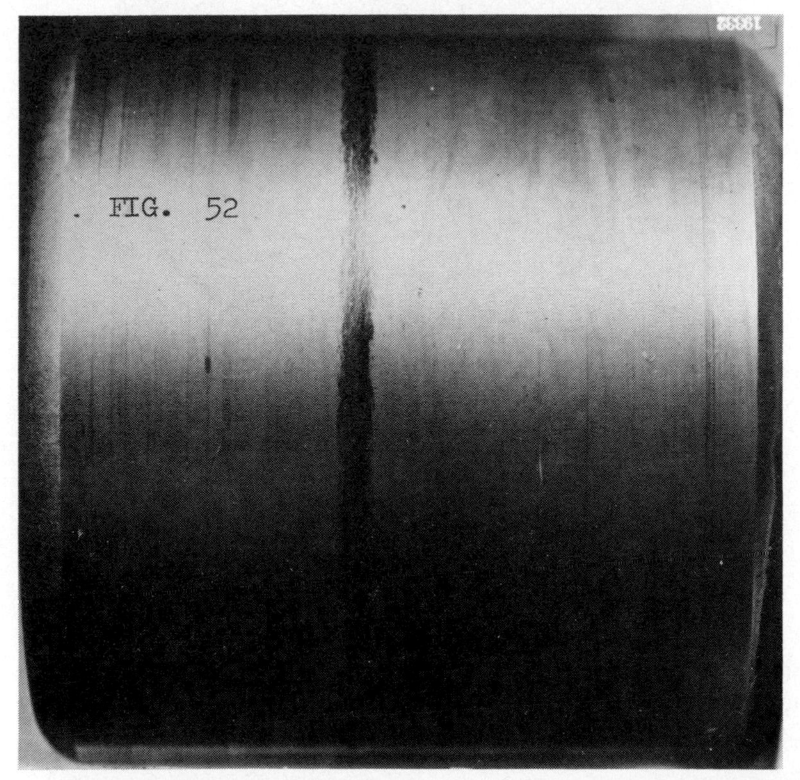

Fig.52 Smeared Cylindrical Roller caused by Rapid Acceleration of the Roller in the Loaded Zone coupled with Inadequate or Incorrect Lubrication.

of bearing rings, and Fig.53 shows smearing on the end of a roller. Smearing can also occur on lightly-loaded high-speed ball thrust bearings where the gyratory moment may force the balls to slide tangentially to the direction of

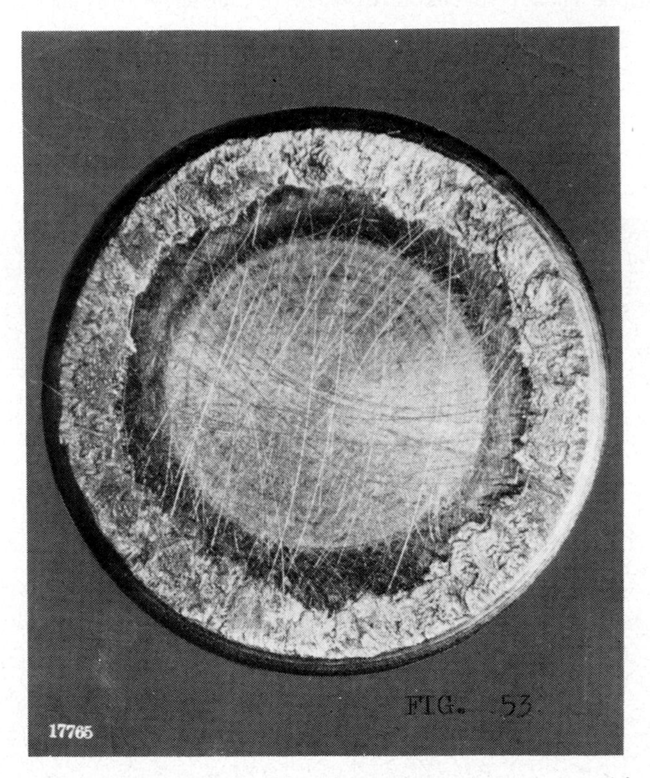

Fig.53 Smearing Caused by Severe Thrust Forces combined with Inadequate or Unsuitable Lubrication.

rolling, causing spiral-shaped smearing. Lightly-loaded high-speed ball thrust bearing must therefore always have a small pre-load or spring load when mounted as given on Fig.22.

6.23.5 Vibrations

A phenomenon known as 'False Brinelling' can occur in rolling bearings fitted to stationary machines, due to vibrations from adjacent machinery. Bearings in ships ancillary equipment subject to vibrations from the ship's machinery are particularly prone to this damage and this can apply to bearings in machines being transported by sea.

Fig.54 Vibration Damage.

Fig.55 Vibration Damage.

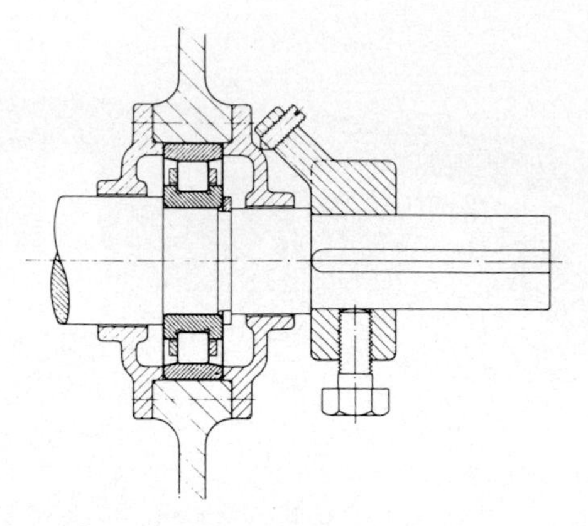

Rotor clamp for preventing damage due to vibration during transport

FIG. 56

Fig.57 Corrosion at the Contact Surfaces due to Water.

A characteristic feature of vibration damage is that the damaged areas are spaced at the same pitch as the rolling elements. The vibrations induce small movements at the contact surfaces between the rolling elements and the rings, and the resulting wear causes microscopic particles of material to break away. These particles oxidise and, on mixing with the lubricant, cause a lapping action, and hence an increase in the oxidation rate. The higher the frequency of the vibrational movement at the contact surfaces, the more rapid the damage, each rolling element gradually forming a cavity.

Cylindrical roller and needle roller bearings are more prone to this type of damage than ball bearings and spherical roller bearings, and one method of overcoming the problem is to use spring-loaded ball bearings. The pre-load should be approximately equal to 20 Newtons per millimeter of shaft diameter. Before pre-loading, however, it is necessary to check that the combination of external forces and additional pre-load does not overload the bearing.

Typical examples of vibration damage to bearings during transportation due to relative movement between the shaft and housing, whether by sea, rail, or road, can be resolved by driving a wooden wedge between the shaft or coupling and a robust part of the crate. Alternatively, the shaft can be locked relative to the housing by means of a clamp, as shown in Fig.56.

6.23.6 Rust and Other Types of Corrosion

Rolling bearings will rust if water or moisture is present and the bearings are not protected by a lubricant containing a rust-inhibitor. Pitting develops in the corroded areas, releasing small particles of rust, and if there are such areas in the tracks the rust mixes with grease and acts as a lapping agent.

Fig.57 shows corrosion on a spherical roller bearing and in this case the grease was unable to protect the bearing from water. Electrolytic action due to the water has resulted in corrosion at the contact surfaces between the rollers and rings.

6.23.7 Passage of Electric Current Through Bearings

Electric current passing through a rolling bearing causes damage to the tracks or rolling elements, which can result in premature failure of the bearing. Alternating and direct current have a similar effect and where there is a risk of electrical leakage, rolling bearings must be protected against the passage of current. If a continuous current passes through a rotating bearing a dark coloured film is produced on the tracks and rolling elements which gradually develop into a washboard formation as shown in Fig.58.

The balls in ball bearings subjected to electric current do not usually develop the washboard surface, but become uniformly dark-coloured over the whole surface. This is due to the balls spinning when the bearing rotates. The

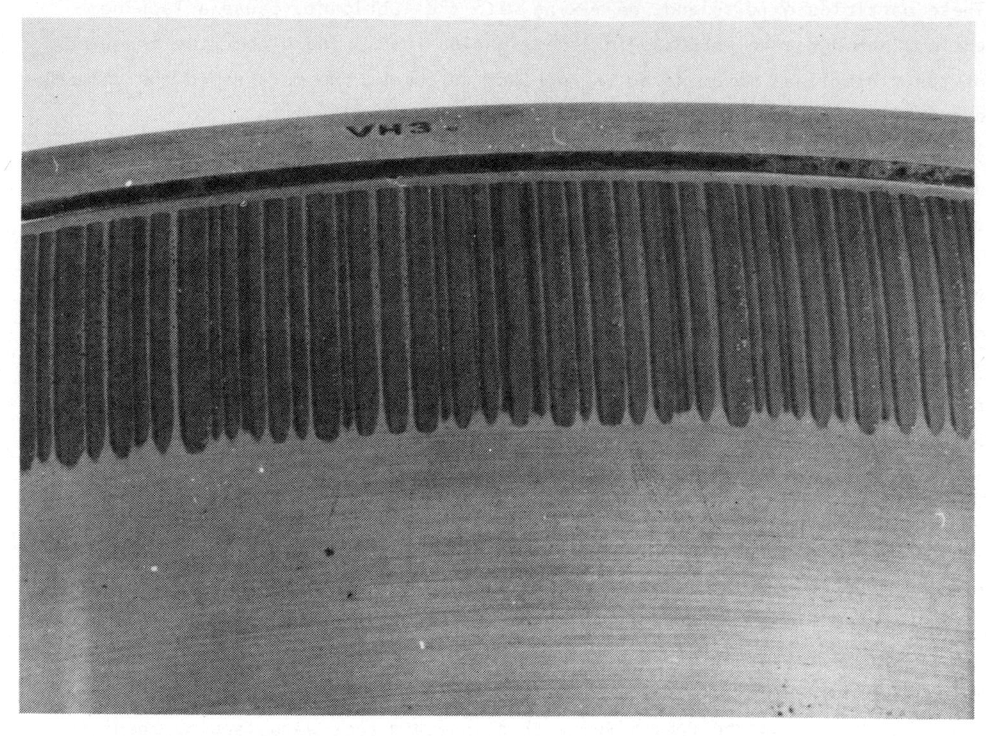

Fig.58 Washboard Effect Caused by Continuous Passage of Electric Current.

rings have fluting similar to that on rollers, but the bands are much narrower.

On applications such as traction motors, craters and burns occur instead of the washboard formation and a typical burn is shown on Fig.59. Generally pitting from electrical leakage does not mean rapid failure and on railway applications, it is known that pitted bearings have run for very long periods before requiring replacing. The main problem is removing the debris released during the formation of the craters, and provided this is carried out and fresh lubricant added, preferably by means of a grease escape valve, the bearings should function successfully for a further period.

There are many problems relating to ball and roller bearings, but providing the basic principles governing bearing selection and a knowledge of lubrication is developed, then the calculated nominal bearing life can be achieved.

Fig.59 Typical Burn Caused by Electric Current Leakage in A Traction Motor.

7 PRACTICAL GEAR TRIBOLOGY

T.I. FOWLE, Consultant, Tenterden

7.1 INTRODUCTION

The identification of the causes of the various forms of distress appearing on gear teeth is seldom an easy matter because of their great variety and because few engineers have the opportunity to see even a minority of them at first hand. The troubleshooting charts given at the end of this chapter are intended to simplify the identification of possible causes and the selection of appropriate remedies from the observed symptoms. Systematic consideration of the various possibilities should at least narrow down the number and suggest tests which might be applied to confirm the final identification. The following notes are given to amplify and explain the reasons for the most important effects.

7.2 ALIGNMENT

There are two aspects of gear alignment: external and internal. Errors in external alignment, that is alignment with the connected machines, place overloads on the bearings and couplings, risking failure or at least noise which could be erroneously attributed to the gears themselves. Errors in internal alignment cause uneven distribution of the load along the gear teet with consequent risk of damage and noisy running and are of particular concern in this chapter.

With parallel-shaft gears there are basically two errors of internal alignment: the shafts may not be parallel, and they may not be in the same plane. Both these errors may, of course, be present together. The various combinations produce patterns of contact, or of damage such as pitting or scuffing, as shown in Figure 1. Only in a few cases are gears provided with means for adjusting the parallelism of the two shafts, and indeed, with modern machining it is unlikely that the shafts will not be parallel. However, if the gearcase is not evenly supported on its feet by the foundations, the shafts will not be in the same plane.

Checking that the two shafts are in the same plane is best carried out by removing the top of the gear casing and, if necessary, the top halves of bearings or bearing keeps. A straight edge with a spacer block to allow for any difference in diameter supporting a precision spirit level can then be placed across both ends of the shafts in turn to check that the ends are in the same plane. An accuracy of 1 per 60 000 is satisfactory. The thickness of the shims required can then be easily determined.

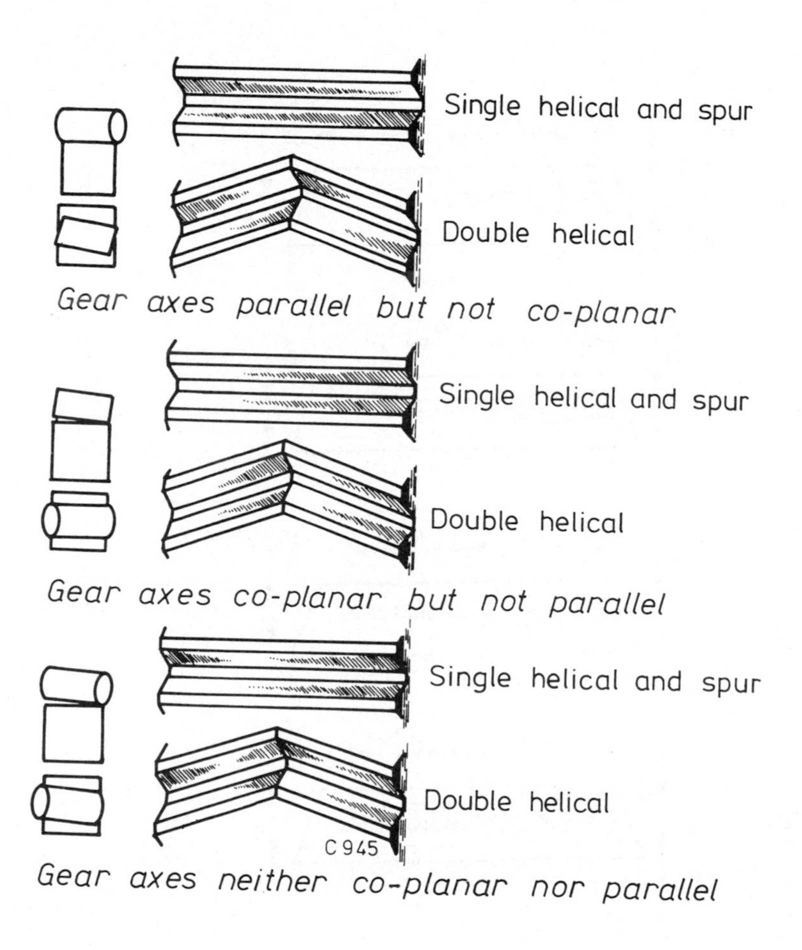

Fig.1 Internal misalignment patterns on parallel shaft gears

Internal alignment can also be checked by determining the extent of tooth contact with marking blue. After ensuring that the journals cannot lift out of their bearings and that the journals are lubricated, a thin coating of marking blue is applied in a thin axial band on one of the gears. The gears are then turned so that the marking is transferred onto the other gear. A record may then be taken by means of a Sellotape impression. The degree of contact required depends on the conditions of service as indicated in Figure 2(a) to (d) which is based on BS 1807 for turbine gears and similar drives. 'Split Marking' as shown in Figure 2(e), which is due to the hob not having been concentric with its mandrel, is undesirable because it prevents the most favourable part of the tooth profile, i.e. that with the least sliding, from supporting the load and transfers it instead to less favourable parts.

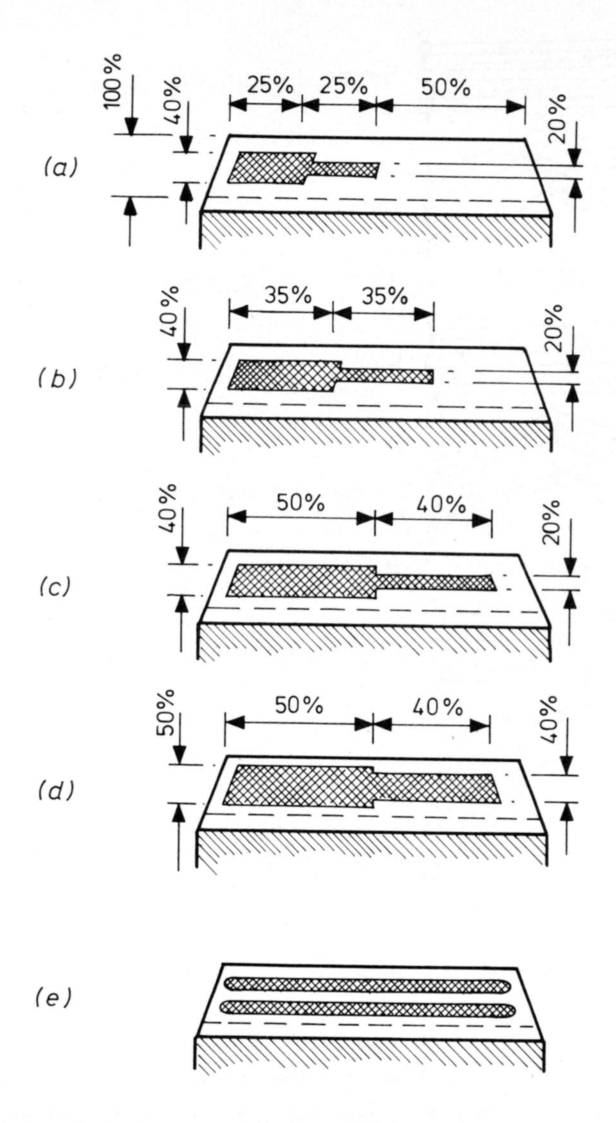

Fig.2 Contact area

(a), (b) and (c) represent the minimum contact areas required by BS 1807 for turbine gears and similar drives, classes B, A2 and A1 respectively. The specification does not stipulate the distribution of the contact areas, only their depth and total length. Classes A2 and A1 are for gears with pitch line speeds over 50 m/s, A2 being suggested for 50-100 m/s and A1 for 50-150 m/s. (d) represents the requirements of some authorities for precision gears. (e) Split marking.

With non-parallel shaft gears, such as bevel gears and worm gears, allowance has to be made for the inevitable distortion under load, as shown in Figures 3 and 4. In all cases contact should not extend to the ends of the teeth as loads

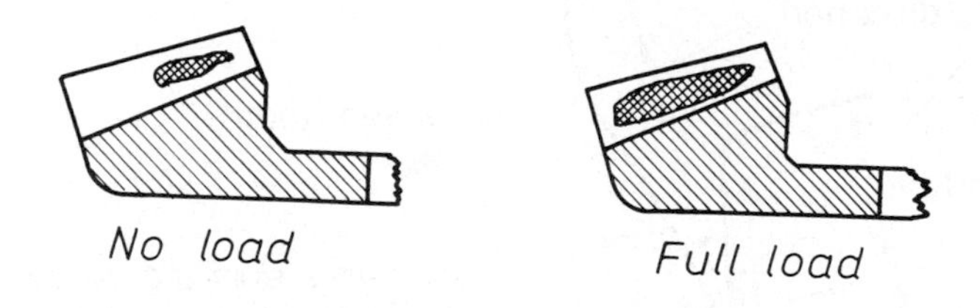

Fig.3 Contact marks on bevel gears

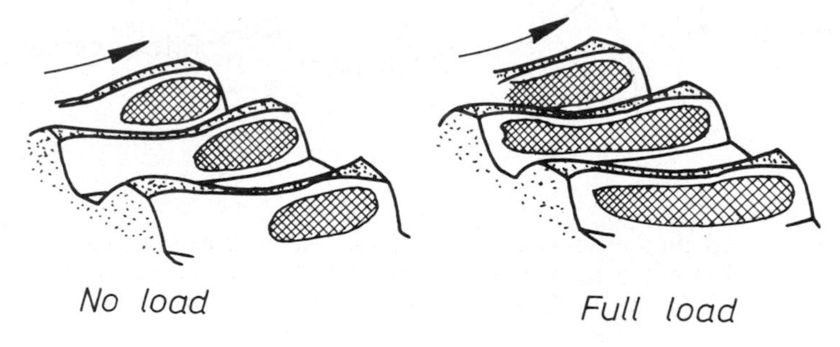

Fig.4 Contact marks on worm gears

there would be highly concentrated. In the case of worm gears there must also be clearance where the worm enters the contact, even at the highest loads, so that the oil on its surface is not scraped off by the edge of the tooth. If this clearance is not available the friction and wear are greatly increased and the transmitted power limited.

7.3 TOOTH ACTION

An understanding of tooth action helps to explain many aspects of gear tribology. In spur, helical and bevel gear tooth action is as represented in Fig.5. The point of contact moves continuously over both teeth and in this sense they roll over one another. The transient nature of the contact enables very heavy pressures to be carried. At the pitch point both tooth surfaces are moving in the same direction at the same speed and so momentarily roll over one another without sliding. At all other points the surfaces are moving at different speeds so that there is sliding as well as rolling.

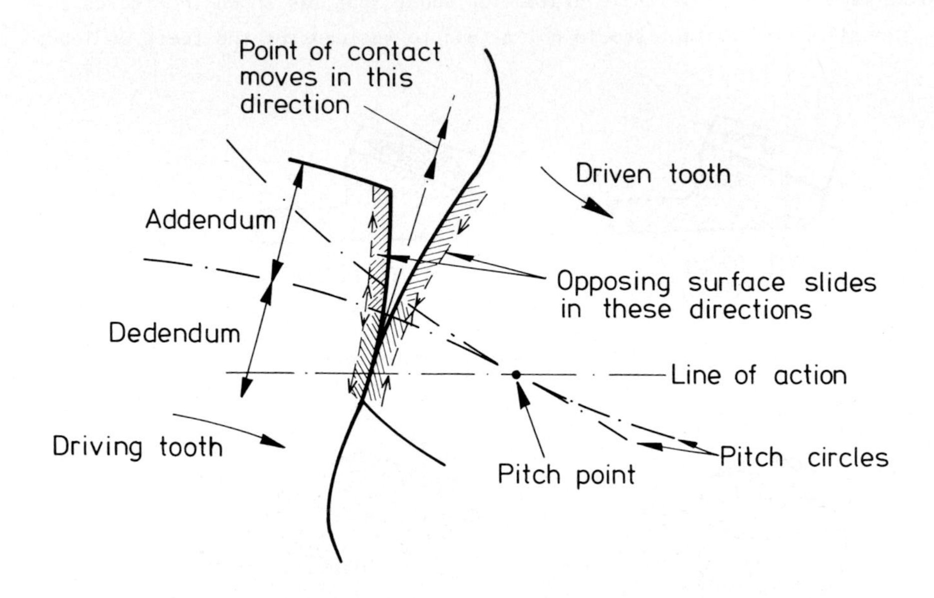

Fig.5 Contact condition in Spur, Helical and Bevel teeth. The hatching indicates the direction of distortion of the surface layers.

On the surface of the driving gear the sliding is always away from the pitch line, while on the driven tooth it is always towards the pitch line. When lubrication is inadequate the sliding shear forces thus tend to draw material away from the pitch line of the driver and to pile it up at the pitch line of the driven tooth. This produces the condition known as grooving and ridging (refer symptom 11 of the Trouble-shooting chart). Another effect of this system of sliding is that on the dedendum surface of both gears any cracks are pulled open in advance of the contact. The cracks being also inclined in the most favourable direction, the oil readily enters the crack. Further movement of the contact then seals the mouth of the crack and compresses the oil in it to extend the crack still further. In contrast, any cracks on the addendum surfaces slope away from the on-coming contact which, furthermore, tends to push the sides of the crack together in advance of contact so that oil is not encouraged to enter the crack. The result is that fatigue pitting tends to occur almost exclusively on the dedendum surface of gear teeth, both driver and driven.

An important feature of spur and helical gears is that there is no sliding along the line of contact, which is also virtually the case with bevel gears. In hypoid gears and worm gears, however, there is a considerable component of sliding along the line of contact which makes for greater difficulty in lubrication, because any asperity on one surface is in contact with the other for a greater time and distance. In worm gears particularly, this sliding considerably increases friction and temperature rise. Much of the friction, moreover, merely serves to distort the wheel towards critical conditions of contact (see symptom 20).

7.4 TOOTH SURFACE DISTRESS

Of the three main forms of surface distress: pitting, abrasive wear and scuffing or adhesive wear, the first tends to occur at the lower end of the speed range while scuffing tends to occur at the higher end as indicated in Figure 6. Scuffing may follow pitting, but pitting does not occur where there is scuffing or where there is abrasive wear. The ideal condition of full-film or elastohydrodynamic lubrication (EHL) occurs at low loads and high speeds.

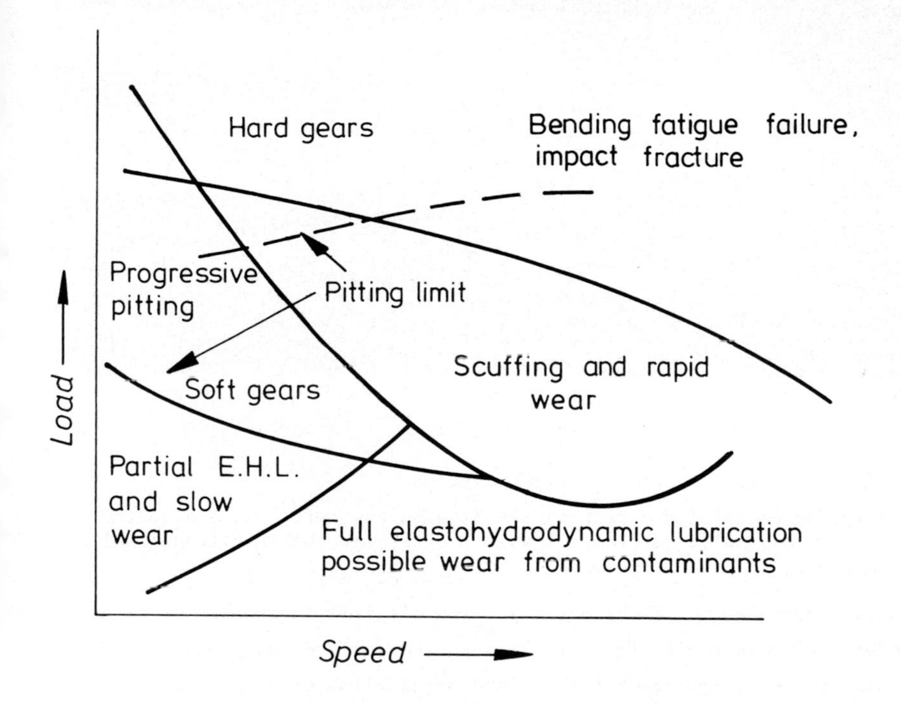

Fig.6 Zones of gear tooth distress.

The different slopes of limits for pitting in hard and in soft gears illustrate separately the two effects of speed on film thickness and tooth errors.

7.4.1 Pitting

About the most common form of surface distress in soft or through-hardened gears is pitting, which is a type of fatigue failure. As indicated in Figure 6, pitting is rather unusual in surface-hardened gearing. Various sub-types and causes are listed against sympton 2, and an example is shown in Figure 7. In this Figure the typical oyster-shell shaped craters produced by the oil prising the fragments loose can be seen.

Fig.7 Surface fatigue pitting.

Note characteristic oyster-shell shape on the left. Note also the effects of 'split marking'. (Photo courtesy of Shell International Co.Ltd.)

As well as the strength of the material, the oil film thickness, or rather the ratio of theoretical oil film thickness to the surface roughness of the harder tooth,is a most important factor, and thick films or fine finishes or both are desirable to reduce the incidence of pitting. Shock-loading is also to be avoided as much as possible as it considerably increases the surface stresses. This is indicated by the negative slope of the pitting limit for soft gears in Figure 6.

Pitting may be either of the initial or the progressive type. With the former the removal of asperities and prominent areas by pitting increases the actual area of contact and reduces the stresses so that it arrests itself. With overloaded surfaces, however, pitting continually reduces the area of actual contact and promotes further pitting. But if one gear is able to resist pitting and can maintain its shape, pitting will progress less rapidly on the other (see sympton 2(f)).

The size of the pits is generally related to the size of the gears, but there may be wide variations due, perhaps, to surface asperities and other details. In particular, where the oil film thickness is relatively very thin the surface traction can be very high and produce very large shallow pits (see 2(j)).

The effect of extreme pressure (EP) additive oils is varied. In laboratory tests using very accurate gears or discs, EP oils are generally found to accelerate pitting, but where practical gears having surface undulations or similar inaccuracies are concerned, EP oils are often able to delay or suppress the problem. Apparently, at least some EP agents can prevent the crests of such undulations from work-hardening so the contact zones can deform to increase their area and thus reduce stresses and temperatures.

7.4.2 Scuffing

Severe adhesive wear takes several different forms in gear teeth according to conditions and is given even more different names. For example, scuffing, scoring, galling and plucking. The nomenclature is even further confused by the fact that what is called scuffing in the UK is known as scoring in the USA. In laboratory test rigs the various forms are not too difficult to recognise and are reasonably well defined, as for example scuffing and scoring in IP 166 for the IAE Gear Rig, but in non-standardised tests distinctions are sometimes made between high-speed and low-speed scuffing (less than about 4 m/s pitch line speed according to some authorities) and between self-propagating and self-healing forms in both cases.

In practical gears appearances may vary to an even greater extent, but are differences in degree, not of kind. In comparison with laboratory test gears it should be borne in mind that in the latter a large part of the power is used to overcome tooth friction in the test gears so that when lubrication fails and scuffing occurs, there is a considerable increase in the power required. There is thus an immediate fall in speed which further increases friction and noise, as well as producing smoke and sometimes sparks. Under these conditions scuffing often does not last for more than a fraction of a minute and, in the FZG rig, for example, never more than 15 minutes. In practice, however, the extra power absorbed when gear teeth scuff is only a very small fraction of the power being transmitted so that the onset of the damage is usually unnoticed and may continue for many hours. Under such conditions the very severe type known as 'galling' arises.

Except possibly at low speeds, scuffing appears to be an essentially thermal phenomenon due to tooth sliding friction. The speed of sliding depends mainly on

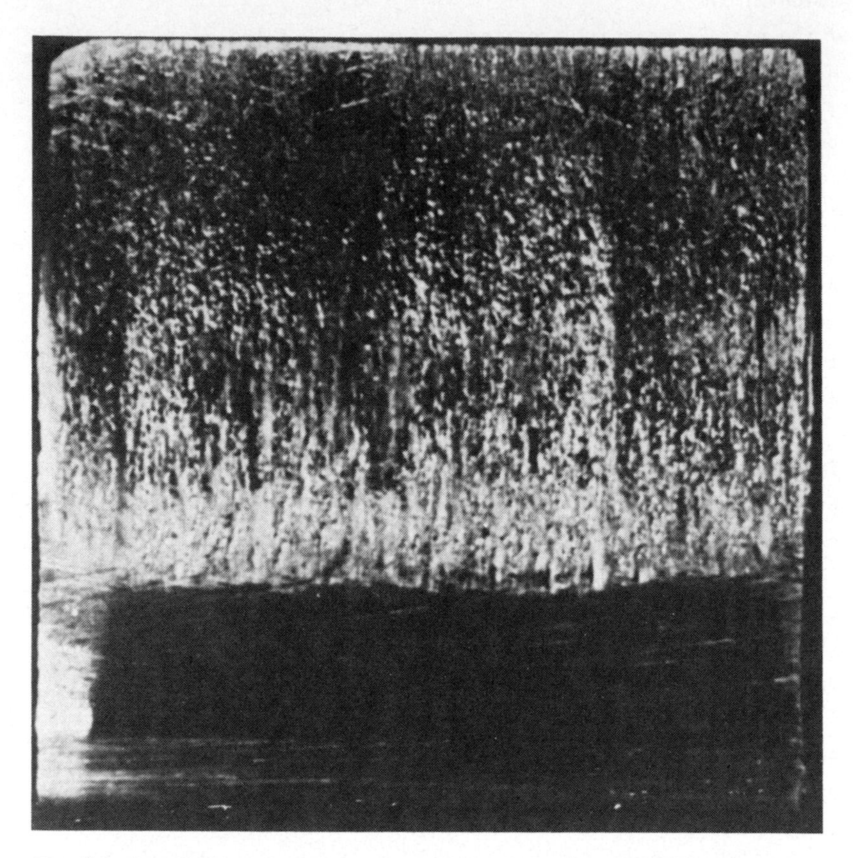

Fig.8 Scuffed pinion tooth from FZG test. (Photo courtesy of Shell International Co. Ltd.)

the peripheral speed of the gears and the size of the teeth. And since large teeth can be more heavily loaded than small teeth, tooth size tends tohave a preponderant influence on the incidence of scuffing, as indicated in Table 7.1 below, which summarises general experience, also in section 3(f).

Table 7.1 Danger of scuffing in spur, helical and bevel gears.

Module pitch (mm)	Danger of scuffing
1.25	None
2.5	Only at very high speeds with thin oil.
5	At moderate speeds even with medium oil.
10	At low speeds even with heavy oil.

As indicated in the above Table, and also in symptom 3(d), the higher the viscosity grade of the oil the greater the protection it affords against scuffing. If the grade is too viscous, however, there could be difficulties in starting up from cold and the power loss and temperature rise in high-speed bearings might be excessive. Compromises have to be made, therefore, and lighter grades have to be used for high-speed gears which, in any case, are more accurately made and have smaller teeth and correspondingly smaller loads. Viscosity grade recommendations for spur, helical and bevel gears are, therefore, often given in terms of speed only, as in Table 7.2 below, which is for gears operating at ambient temperatures between about 10°C and 25°C.

Table 7.2

Oil viscosity grades for spur, helical and bevel gears

Pitch line speed		Range of ISO Viscosity Grades cSt at 40°C
m/s	ft/min	
0.5	100	460 - 1000
1.3	250	320 - 680
2.5	500	220 - 460
5	1000	150 - 320
12.5	2500	100 - 220
25	5000	68 - 150
50	10000	46 - 100

Higher viscosity grades may be needed where the ambient temperature exceeds 25°C, where the gears are subject to shock loads, or where both gears are made of through-hardened nickel-chrome steels. Lower viscosity grades may be used where the ambient temperature is below 10°C, or where the teeth have been given a scuff-resistant coating to assist running-in.

For double-reduction gearing the low viscosity end of the range for the low speed train should be taken, and for multiple trains the mean for the two slowest trains.

Worm gears need to be treated rather more generously because tooth friction is of much greater importance. To minimise tooth friction HVI oils are preferred to other types and viscosity grades are higher than for other types of gearing, as indicated in Table 7.3 below.

For designers, the risk of scuffing can be assessed by the use of Blok's Critical Contact Temperature theory or the Niemann and Seitzinger bulk tooth temperature criterion [1] and a decision made on whether to recommend the use of straight or EP oils. For field use simpler guidance is needed and consideration of Table 7.1 is recommended. EP oils should, furthermore, be considered for parallel-shaft gears where the tooth contact markings do not reach the extent required by the speed as indicated in Figure 2, where the gears are subject to dynamic overloads, and where the gears step up the speed.

Table 7.3
ISO oil viscosity grades required for enclosed worm gears

Centre distance (inches)	(mm)	Output or wormwheel r.p.m. 50 and under	800	150 and over
2.5	64	HVI 1000	HVI 460	HVI 320
4	100	460	320	320
10	250	320	220	220
20	500	320	220	220

EP oils should generally be used for hypoid gears because of their high component of sliding along the line of contact, their large teeth and their liability to considerable dynamic overloads in automotive service. They should not, however, be used for worm gears unless the oil temperature is consistently below about 60°C because of the risk of excessive corrosive wear of the bronze. Since scuffing is a thermal phenomenon, overheating can be a cause, and its courses and the appropriate counter-measures detailed under symptom 23 should be considered.

7.4.3 Abrasive Wear

There are two kinds of abrasive wear. One, as in symptom 4, where a rough, hard surface rubs against a softer one, which is known as 'two-body' abrasion; the other, as in symptom 5, where abrasive dirt acts between two rubbing surfaces, which is known as 'three-body' abrasion.

The first kind occurs where a rough surface-hardened pinion runs against a soft steel or plastic wheel, and also where a rough surface-hardened worm runs against a bronze wheel. In the latter case the associated high friction may so distort the gears that contact is brought onto the inlet edge of the wheel teeth and friction further increases to the extent that power transmission may be limited (see symptom 20). The remedy is to stone or lap the harder member to a smoother finish, e.g. 0.5 to 0.2 microns Ra.

Typical contaminants causing three-body abrasion are sand and millscale. Filtration down to 20 microns (nominal) is normally the best practicable solution, though smaller particles can still cause abrasion. The maximum amount of contaminant tolerable in the oil depends on its hardness relative to that of the gears concerned. For example, in steel mill practice a typical limit for millscale is 0.3% w by DIN 51 592 , i.e. retained on a 0.45 micron millipore filter). The corresponding limit for sand would be 0.1% w. Greases are particularly liable to permit abrasion since they tend to keep the abrasive contaminants and the wear products in the vicinity of the mesh.

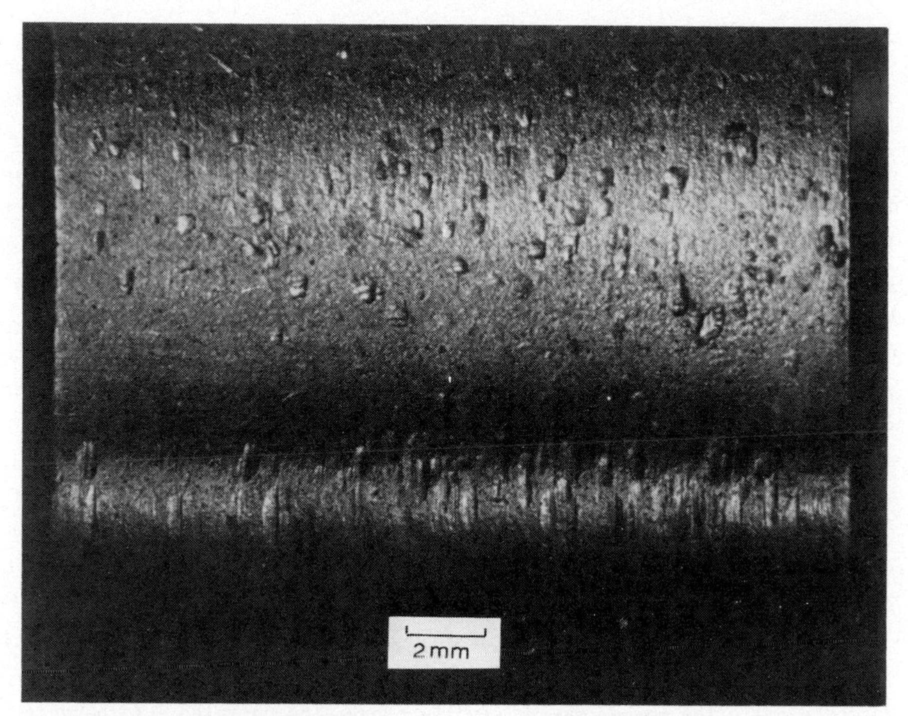

Fig.9 Three-body abrasion of a gear tooth by sand in the oil. Note the short length of the scars. (Photo courtesy of Shell International Co.Ltd.)

7.4.4 Other forms of gear wear

The other forms of gear wear listed under symptoms 6 to 19 are relatively rare and space does not permit special discussion here.

7.5 PROBLEMS IN LUBRICATION SYSTEMS

The main problems arising in supplying and controlling the flow of oil to and from the tooth mesh are: overheating, windage (in very high speed gears), excessive foaming (in bath lubricated systems) and excessive aeration (in large circulation systems).

One frequent source of overheating in bath systems is when one gear dips too deeply into the bath (see symptom 23(a), (b) and (c)). This can sometimes be avoided by surrounding the lower part of the dipping gear by a special trough. When the gear is at rest the trough fills with oil, but during operation the excess is thrown out and the meshing teeth receive sufficient oil for lubrication and cooling from holes in the bottom of the trough.

Above about 15 m/s pitch line speed even this special form of bath lubrication tends to cause excessive power loss and temperature rise so that spray lubrication from circulation systems is normally adopted. Pump pressures are usually in the range 0.3 to 1.5 bar. Pressures below 0.3 bar may be satisfactory, but whenever the gauges read below 1 bar a visual check should be made to ensure that the oil is reaching the parts required. (Upper limits to oil pressure may be dictated by the safety limits for pumps, coolers, filters, etc.).

At speeds above about 50 m/s windage effects become noticeable and at around 100 m/s and over they need to be specially catered for in the design by arranging for plenty of space between the gears and the housing, by arranging completely separate or even dual drain lines with shields, otherwise the casing may become choked with oil.

The location of the oil sprayers needs special consideration where pitch line speeds are above 50 m/s. Above that speed the sprayers should not be directed straight into the mesh, but a little in advance. In this way all the oil serves to cool the gears, the excess above that required for lubrication is thrown off before the gears mesh. This arrangement is especially required for 'down-going' gears. Above approximately 75 m/s, about 80-90% of the flow should be directed onto the outgoing side for cooling, with the remainder being sprayed slightly in advance of mesh for lubrication.

Excessive foaming and aeration of the oil is due either to a deterioration of the properties of the oil from contamination or to excessive ingress of air into the oil. The former may be readily checked by carrying out the appropriate tests on the oil, e.g. IP 146 and IP 313, the latter by reference to symptom 24(a), (b) and (c).

REFERENCES

1 Fowle, T.I., Lubn.Engg., 1976, 32 No.1, 17.

GEAR PROBLEMS : CAUSES AND REMEDIES

Symptom	Possible Causes	Remedies
1. Broken teeth	(a) Fatigue (i) Load unevenly distributed. (ii) Sharp corner in tooth root. (iii) Notches in tooth root caused by improper filing or grinding. (iv) Overload. (v) Torsional vibrations. (vi) Bruises on teeth, e.g. caused by dropping. (vii) Coarse martensitic structure in hardened steel. (viii) Blow-holes in roots of cast teeth. (ix) Surface fatigue pits adjacent to root of tooth. (b) Fracture (i) Repeated heavy loads. (ii) Hard foreign objects jammed in gear mesh.	Temporary corrections may be made by cutting out broken teeth and cracked sections until only sound teeth are left, and running at proportionally reduced load; by inserting pegs and building up new teeth on them by welding, followed by reshaping. Apart from eliminating overloads the only permanent correction is, however, to procure gears without the faults listed, to have them correctly lined-up and run with any shock loads limited by shock absorbing couplings.
2. Pitting		
(a) Small widely scattered pits on working surfaces.	(a) Fatigue at surface asperities during initial running.	(a) None: the pitting will cease to spread and will be of no consequence.
(b) Pits concentrated at one of tooth or helix.	(b) Concentration of load due to slight misalignment.	(b) Check alignment of shafts and casing for distortion. The pitting may cease to spread.
(c) Concentration of pits in regular bands along tooth width.	(c) Concentration of load on surface undulations.	(c) The pitting may cease to spread, but lapping or stoning advisable in severe cases.

Symptom	Possible Causes	Remedies
2. Pitting (contd.)		
(d) Pits concentrated along pitch line.	(d) (i) Faulty profile. (ii) Excessive wear or scuffing has produced a ridge along the pitch line which becomes overloaded.	(d) (i) As above. (ii) Use oil of higher viscosity or of higher EP activity.
(e) Pits concentrated on dedendum surfaces of one gear.	(e) (i) Insufficient hardness of softer gear or excessive free-ferrite in micro-structure relative to loading. (ii) Excessive amounts of hard non-metallic inclusions in the metal. (iii) Initial surface finish too rough and oil viscosity too low.	(e) The pitting will cease to spread if opposing gear unaffected. An increase in oil viscosity could be beneficial
	(iv) Overload, especially by shock, torsional oscillations and high starting torque.	Fit shock absorbing coupling.
(f) Pits on dedendum surfaces of both gears, often with distinct step at the pitch line.	(f) Insufficient hardness of both gears relative to the loading.	(f) Reduce loading insert shock absorbing type of coupling between source of shock and gear. If possible, increase operating viscosity of oil. Change to EP oil, mainly to prevent subsequent scuffing of remaining contact areas. If possible, lap or stone teeth to improve surface and rub in dry MoS_2 powder.
(g) Pitting in case-hardened gears.	(g) Teeth too soft either from accidental decarburisation or inadequate quenching.	(g) Reduce load.
(h) Micro-pitting or 'frosting' in surface hardened gears (closely spaced pits smaller than 0.1 mm across).	(h) (i) Nitrided gears: thin surface layer of brittle super-rich nitrides (white layer).	(h) (i) A 25 micron layer or less is generally harmless but removal of thicker layers by lapping or grinding should be considered.
	(ii) Inadequate oil film.	(ii) Increase operating viscosity of the oil.

Symptom	Possible Causes	Remedies
2. Pitting (contd.)		
(i) Very large oyster-shaped pits extending over practically the whole of the active flank of case-hardened teeth.	(i) High tooth friction due to direct contact between the teeth; theoretical oil film thickness less than 0.5 micron.	(i) Increase operating viscosity of the oil. NB. Pitting of soft and through-hardened steel gears is so common that it cannot be counted as a failure. But, if possible, it is prudent to correct as detailed above. In some cases, a fresh start can be made by turning round one or both gears and lapping them together as appropriate. In severe cases where noise is excessive or reduced areas of contact cannot be prevented from scuffing, or there is a danger of surface pitting initiating tooth breakage, it may be necessary to replace gears.
3. Scuffing	(a) Tooth loading too high.	(a) Reduce loading
	(b) Insufficient lubrication	(b)
	(i) Oil-bath level too low.	(i) Raise oil level so that gear dips 1-3 tooth heights when running.
	(ii) Oil sprayer nozzles choked.	(ii) Check oil nozzles, clear as necessary, filter or change oil.
	(iii) Pump suction strainer choked.	(iii) Check suction strainer, clean as necessary, filter or change oil.
	(iv) Bearings rob oil from sprayer nozzles.	(iv) Restrict oil supply to bearings.
	(v) Windage in high speed gears deflects oil spray.	(v) Increase oil pressure. Reposition sprayer nozzles.
	(vi) Lubricant channels at low starting temperatures.	(vi) Use lubricant with lower channel point.
	(vii) Clearance between gears and casing too small: oil cannot flow back to bath.	(vii) If possible use lower viscosity oil. Increase clearance.
	(c) Operating temperatures excessive.	(c) Reduce temperatures (see Symptom 23)
	(d) Oil viscosity too low (particularly soft or through-hardened gears).	(d) Use higher viscosity grade; improve cooling.

Symptom	Possible Cause	Remedies
3. Scuffing (contd.)	(e) Lubricant has insufficient EP activity (particularly hardened gears).	(e) Use more active EP oil.
	(f) Teeth have excessive addendum height for the speed.	(f) Check design. Use more active EP oil.
	(g) Teeth do not have sufficient contact because of undulations, split markings or misalignment.	(g) Reduce undulations by stoning, lapping or shaving. Check alignment. Check that casing has not distorted from uneven settling of foundations. Check that bearings are not worn.
	(h) Teeth are not adequately relieved.	(h) Apply tip-relief by shaving or stoning.
	(i) Full-load applied before gears have been adequately run-in.	(i) Use active EP oil to prevent further scuffing while continued running makes surfaces smoother.
	(j) Both gears have high nickel-chromium content and are not case-hardened.	(j) Use active EP oil or higher viscosity grade.
		NB. Provided scuffing has not roughened the surfaces unduly, especially with spiral bevel and helical gears, once the basic cause has been eliminated further running, preferably with an EP oil, will correct the problem.
4. Wear of softer member only of gear pair.	Surface finish of harder member too coarse.	Stone or lap teeth to a finer finish.
5. Wear of harder member is greater.	Abrasive dirt in lubricant becoming embedded in softer member.	Change oil or pass it through a fine filter. Fit filters on air vents.
6. Wear at low speed.	Lubricant film too thin.	Use higher viscosity lubricant.
7. Wear at high speed.	Excessive friction caused by overload, overspeed, loss of backlash, or faulty lubrication.	Reduce overloads or overspeed, improve oil flow and distribution.

Symptom	Possible Causes	Remedies
8. Wear of worm wheel teeth.	Material combination may be unsuitable. (See also No.20).	If possible use case-hardened steel worm and centrifugally cast phosphor bronze.
9. Teeth tips rounded and dedendum surfaces gouged.	Interference : gears not properly matched or centre distance too small.	Check design. If possible extend centre distance.
10. Plastic flow of tooth surfaces with pronounced fin at tips.	(a) Combination of soft material and repeated shock loads leading to separation of teeth with re-contact insufficiently damped by oil film (peening).	(a) Reduce shock loads, use higher viscosity oil, reduce backlash, change to harder gear material.
	(b) Very heavy steady loads and soft materials (rolling).	(b) Reduce loads, increase gear surface hardness.
11. Groove along pitch line of driving teeth and ridge along pitch line of driven teeth.	When associated with scuffing of the rest of the teeth it may be due to complete failure of the lubricant supply. When not, oil film may be too thin.	Check oil supply, e.g. that oil bath level is correct when gears are running, that oil supply pipes, filters and sprayer nozzles are not choked.
12. Grooves along pitch line of both driving and driven teeth.	Erosion by spark discharge.	Establish source of stray electric currents and lead to earth by (a) earthing brushes of generous size, and (b) stopping other possible paths through the gear mesh by insulating pads under pedestals and insulating bushes for holding-down bolts.
13. Rippling.	Excessive surface friction at low speeds.	Use hypoid type full EP oil.
14. Bulk plastic deformation of teeth especially at middle of tooth width.	Very severe overheating due to failure of oil supply.	Repair teeth or renew gears. Eliminate cause of oil supply failure. Install alarms so that unit can be stopped quickly in event of oil supply failure.

System	Possible Causes	Remedies
15. Indentations.	Hard particles in system, often swarf, occasionally from EP oil carbonised on highly rated heaters in system.	Thoroughly clean system, check heater surfaces, clean and reduce surface temperature as necessary.
16. Cracks in surface of hardened gears, often in net-like pattern.	Overheating during grinding, incorrect heat treatments, or both.	Check with manufacturer.
17. Longitudinal cracking and flaking in case-hardened gears.	Case is too thin and core too soft so that surface has collapsed under load.	Check with manufacturer.
18. Red-brown spots on surfaces of case-hardened gears.	Attack by corrosive substances such as salts from hardening process.	Clean spots with emery stick. Clean, flush and refill lubrication system.
19. Red-brown contact marks on teeth and red-b-own powder in nominally stationary gears.	Fretting due to vibration while under stationary load.	Arrange for the gears to be flushed with oil and slowly rotated.
20. Worm gear fails to transmit full torque.	(a) Insufficient allowance for distortion under load has brought contact onto entry side of wheel teeth.	(a) Adjust position of wheel so that even under full load contact is not on entry side of wheel teeth and oil can be drawn into contact.
	(b) Excessive tooth friction, due either to rough worm surface or to unsuitable combination of gear materials, causing excessive distortion.	(b) Improve surface finish of worm, reduce oil temperature or use higher viscosity grade. Best material combination is case-hardened steel worm and centrifugally cast phosphor bronze wheel.

Symptom	Possible Causes	Remedies
21. Vibration	(a) Defective bearings or couplings.	(a) Check bearings and coupling and replace as necessary.
	(b) Shafts misaligned.	(b) Check that casing is not distorted, realign shafts.
	(c) Check and rebalance as necessary.	(c) Check and rebalance as necessary.
22. Unusual noise.	(a) Defective rolling element bearing.	Check these parts and replace as necessary.
	(b) Defective oil pump.	
	(c) Defective coupling.	
	(d) Tooth surfaces excessively pitted or roughened.	(d) Stone or lap teeth and rub in MoS_2 powder.
	(e) Continuous tooth double helical gears running 'down-going' with apex trailing, squirting oil out of mesh.	(e) Reduce excessive oil supply to mesh by reducing immersion in bath or reducing flow to sprayers and directing spray in advance of mesh.
23. Overheating.	(a) Oil level too high in bath.	(a) Adjust level when gear running to dip 1-3 tooth heights.
	(b) Oil viscosity too high.	(b) Change to lower viscosity grade.
	(c) Speed too high for bath lubrication.	(c) Change to spray lubrication system.
	(d) Too much oil sprayed too close to ingcing mesh of high-speed gears.	(d) Direct oil spray further in advance of mesh; restrict amount of oil to sprayers.
	(e) Inadequate drainage from housing.	(e) Improve drainage or restrict amount of oil to sprayers.
	(f) Clogged cooler.	(f) Check oil and water sides and clean as necessary.
	(g) Cooler inadequate.	(g) Change cooler for larger size or switch to cooler water supply.
	(h) Heat radiated from surroundings.	(h) Interpose radiation shields.

Symptom	Possible Causes	Remedies
23. Overheating (contd.)	(i) Inadequate air flow over gearbox.	(i) Increase ventilation of surrounding air space.
	(j) Dirt accumulations on casing.	(j) Clean dirt away.
	(k) (Worm gears), unsuitable lubricant.	(k) Change to HVI mineral oil or, preferably, polyglycol type synthetic oil.
	(l) Excessive power loss in plain bearings.	(l) Check bearing design.
24. Excessive foaming and aeration.	(a) Gear dips too deeply into oil bath.	(a) Adjust level when gear running to dip 1-3 tooth heights.
	(b) Air leaks on suction side of circulating system.	(b) Remake suction-side joints including pump gland.
	(c) Oil cascades down vertical return pipes into reservoir.	(c) Rearrange return lines to allow smooth flow into tank below oil level.
	(d) Oil contaminated by grease, jointing compound, another, and incompatible oil etc.	(d) Renew oil charge.

8 MATERIALS FOR TRIBOLOGICAL APPLICATIONS

D. SCOTT, Consultant, Editor of Wear

8.1 INTRODUCTION

Engineering design is the creation of instructions for making an article to satisfy a specific requirement. From a tribological point of view, the materials of construction and the lubricant are important factors in such specifications. For tribological applications the important properties of materials are those properties which must be taken into account in designing a component to withstand the mechanical and thermal stresses to which it will be exposed and the effects of the environment in which it has to function [1,2]. There is a continuous demand for materials of improved properties and with better strength to weight ratios. Mechanisms operating under arduous conditions of high speed, heavy load or extremes of environment require materials of high strength. If subjected to relative motion they may require materials of great hardness, wear and corrosion resistance and structural and dimensional stability [3]. Newer materials [4] may meet stringent design requirements beyond the capabilities of the more commonly used materials but availability and cost make conventional materials more attractive commercially and encourage innovation.

The designer has a vast range of materials from which to select. One reference book [5] gives 35000 proprietary material compositions. The designer however, besides searching for improved materials must often seek the cheapest material to satisfy his requirements and sometimes the more readily available indiginous materials. A determining factor in the extensive use of many materials is the amenability of the material to manipulation and the extent to which the designer can control and vary properties such as strength, hardness and ductility within the range of specific engineering requirements. The choice of materials is often restricted by the manufacturing facilities at his disposal.

8.2 TYPES OF MATERIALS

Materials may be conveniently divided into four principal types:- ferrous, non-ferrous, non-metallic and composite materials. The abundance of iron and its alloys comprising the bulk of metals made, their favourable economics and diverse properties make ferrous materials the desirable choice for tribological applications. Modern cast irons and steels find extensive use in tribo-engineering as alloying and heat-treatment enables them to be tailored to specific applications. The principal methods of strengthening steel include work-hardening, decreasing the grain size, solid solution and dispersion strengthening. In currently used steels, the martensite transformation produces the best combination of strength and ductility but as hardness increases, ductility decreases and at the highest strength levels produced by conventional heat treatment procedures, ductility is diminished to levels considered at present unacceptable for most engineering applications [6]. If a thermomechanical treatment is used whereby austenite is strain hardened before transformation to martensite unusual ductility, fatigue and impact properties are obtained. Ausforming may thus allow increased strengths above the present usable limits without sacrifice of ductility [7].

With materials generally, high hardness is usually associated with a high melting point. Powder metallurgy has widened the field of available hard metallic materials by making possible metal combinations unobtainable by conventional melting and casting techniques. Hard sintered carbides may be used to advantage in many applications requiring a high degree of wear resistance. However, such materials are usually expensive to manufacture and difficult to form and surface treatments and coatings on ordinary materials may be used to increase strength and improve wear resistance.

For use at elevated temperatures metals must form a dense, tough, impervious oxide layer which resists cracking under load and prevents attack of the metal by hostile environments. The established non-ferrous metal alloy systems in current use are based on nickel rich and cobalt rich alloys. In both cases the necessary resistance to oxidation and corrosion is conferred by the introduction of chromium. The nickel based alloys are stiffened principally by the addition of titanium and aluminium. In cobalt alloys stiffening is effected by complex carbides of molybdenum, niobium and tantalum.

High speed tool steels and similar special steels are also used for elevated temperature service. Corrosion resistance requires careful material selection and the use of stainless steels, non-ferrous metals such as aluminium, nickel, chromium, titanium and their alloys or non-metallic materials such as plastics or elastomers. To resist severe abrasive wear, cemented carbides, cermets or even diamond may be required.

As conventional materials have been improved by orthodox methods almost to the limit of their potential mechanical properties, new types of material are being developed. Composites which combine materials of dissimilar mechanical and physical properties, can have properties superior to one or both of their constituents. There are two principal types. In one, a matrix may be reinforced with fibres or particles to improve its properties. In the other the

the matrix is essentially a glue to hold together fibres or particles which have desirable properties but which by themselves cannot be used as engineering materials. Certain difficulties require to be surmounted before composites achieve their full potential. Whiskers and fibres are expensive and have problems with stress concentration at their ends which can influence crack initiation. Conventional methods are not suitable for the manufacture of components from composites nor for the formation and joining of fibre reinforcing materials. By using reinforcements of oxide and non-metallic whiskers which approach the theoretical strength, very high ultimate strengths in composites are possible. Glass, carbon, silicon nitride and alumina are attractive non-metals. Besides replacing metals, ceramics may be used as coatings to complement desirable metal characteristics by adding refractory properties, insulation, and erosion, wear, oxidation and corrosion resistance.

The strengthening of metals for use at high temperatures can be achieved by dispersing non-metallic particles in them to maintain useful properties to within 50-100°C of the melting point of the matrix metal. Only small amounts of the dispersoid are required and nickel alloys with thorium, TD nickel, are commercially available. Other newer materials for arduous conditions include synthetic diamond and sapphire, new graphites and materials such as the carbides, borides and nitrides of certain metals which approach the hardness of diamond.

8.3 MATERIALS FOR SPECIFIC APPLICATIONS

Adequate material properties for design are usually ensured by indirect means mainly by the designer specifying chemical analysis, heat-treatment and mechanical properties although such specified properties may not be directly representative in service. For instance, the most important material property may be resistance to abrasion or resistance to scuffing and seizure, or to rolling contact fatigue or to lubricant attack or corrosion. A property such as dimensional stability may completely determine the service life. As the ultimate assessment of a material is performance in practice, full scale testing and service simulation testing are usually resorted to as a means of material selection.

Tribo-engineering depends in many instances upon bearings, components which allow relative motion between members of a mechanism whilst transferring load. Bearings may take many forms but the most widely used types are plain bearings, gears and rolling bearings.

8.3.1 Plain Bearings

In plain bearings the load is transmitted between moving parts by sliding contact and the criterion of satisfactory bearing performance is minimum wear of the components together with freedom from seizure and freedom from mechanical failure by deformation or fatigue. To carry a hard steel shaft, usually specified for its mechanical properties, a bearing material must be comparatively soft to avoid wear of the harder material yet strong enough to withstand heavy loads without distortion and without suffering fatigue. Soft bearing materials also allow abrasive particles to become embedded and thus reduce abrasive wear. As a low hardness is usually associated with a low melting point, high spots of soft bearings are removed by sliding contact without damage to the mating surface and without the risk of seizure. However, low hardness is usually associated with low fatigue strength and, as stress levels are raised, the demand is for harder bearing materials to improve the load carrying capacity but with the minimum loss of friction and wear properties. As a general rule it is advisable to use the softest bearing material possible.

White metal, a widely used plain bearing material is based on tin or lead or their intermediate alloys. A typical tin based alloy contains 7-10% Sb and 3-5% Cu, the principal constituents being SbSn, Cu_6Sn_5 and a ternary peritectic complex, Fig.1. Hardness and mechanical properties are little affected by composition [8]. At 100°C, the hardness ranges from 11-16 HV and the fatigue strength for 10^7 cycles from 1.6-1.9 MN/m^2.

Lead based alloys containing Sb and Sn and Cu in the form of intermetallic SbSn and Cu_6Sn_5 may be cheaper than similar tin based alloys but are slightly inferior regarding wear and fatigue properties. Intermediate alloys of high lead and tin content are widely used but appear to have no advantages over the other white metals. The success of white metals is generally regarded as being due to the correct compromise between softness to avoid wear and strength to resist fatigue.

Copper-based alloys, stronger bearing materials than the white metals at operating temperatures, range from the phosphor bronzes (10% Sn, 0.5% P) through the leaded bronzes to the copper lead alloys of up to 50% Pb, Fig.2. The wear properties of leaded bronze are better than those of white metal. The hardness of copper-lead varies according to composition from 30-70 H.V. Journal wear increases with increase in hardness but fatigue strength increases roughly in the same proportion as journal wear. Increased journal hardness can help to minimize wear. A disadvantage of copper-lead alloys is their susceptibility to lubricant corrosion of the lead phase.

A compromise between alloys soft enough to avoid wear, those hard enough to resist fatigue and those able to resist corrosion has evolved by the use of overlay bearings in which a strong metal, such as copper-based metal, has a soft metal overlay. For economic reasons, the copper-lead may be used as an interlay between a steel base and the overlay, Fig.3. To avoid fatigue under the applied loads, the overlay is usually thinner than 5 μm. The overlay plated

Fig.1 (x75) Structure of gravity cast tin based white metal.

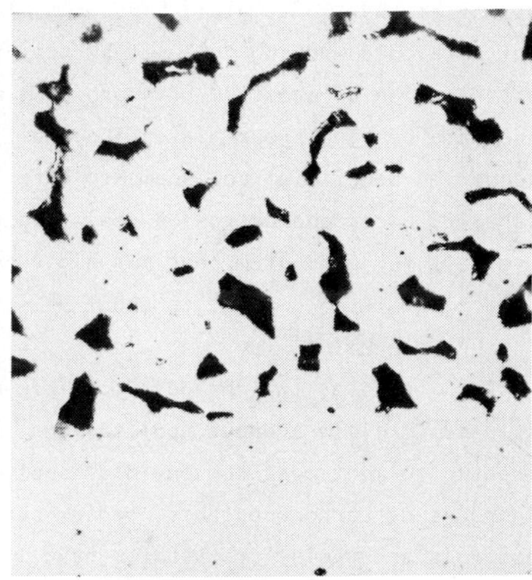

Fig.2 (x75) Structure of copper-lead on steel base.

copper-lead bearing is widely used for high duty engine bearings but the continuous search is for superior readily available materials.

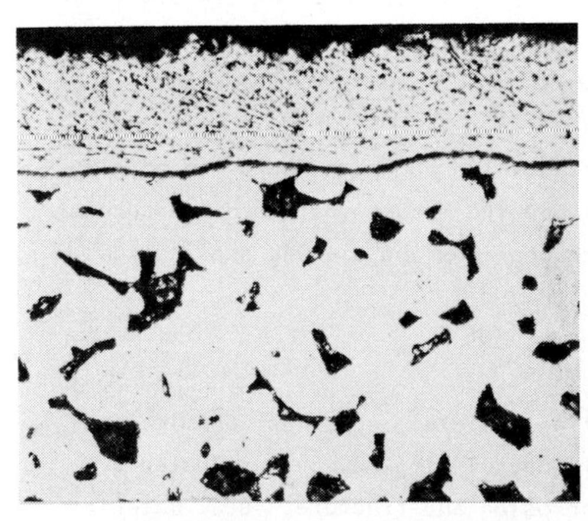

Fig.3 (x75) Soft overlay on copper lead bearing.

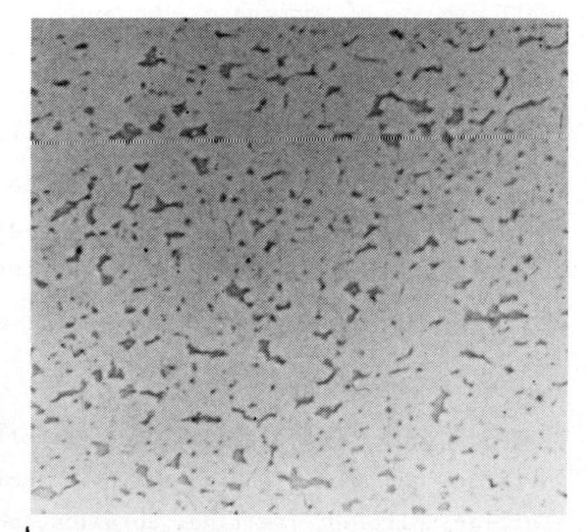

Fig.4 (x120) Structure of aluminium 20% tin bearing.

Aluminium, a comparatively cheap material in abundant supply, has met with some success by the conventional approach of using hard metal compounds in the aluminium matrix to produce a structure analogous to that of white metal. The use of another soft metal, tin, with aluminium has produced good results. By suitable cold working followed by heat-treatment and recrystallization the distribution of about 20% tin produces a reticular structure, Fig.4, with adequate bearing properties. The addition of a small amount of hardener such as copper is beneficial and seems to offer the best combination of load carrying capacity, wear and corrosion resisting properties currently available. Suitable overlays for aluminium bearings aid running-in, minimise journal wear and allow reduction of tin content to about 9%. Similar aluminium-lead bearings effect further economies.

For journals, the cheapest possible material is usually chosen. Mild steel is used for less arduous applications, and plain carbon steels can be heat-treated to meet most conventional applications, the properties improving with increase of carbon content. Medium-carbon steels used for smaller sizes of journals and engine crankshafts have low hardenability. For more massive parts, which are required in the hardened and tempered condition, low-alloy steels are needed to enable satisfactory properties to be obtained. Depending upon the specific properties required, manganese, nickel, chromium, molybdenum and vanadium, used separately or in various combinations, provide a wide range of materials for highly stressed transmission components and the more stringent applications. Surface-hardening techniques, such as carburizing and nitriding, are beneficial for providing an extremely hard, wear-resistant surface usually on specially manufactured low-carbon manganese or tough alloy steels containing small amounts of manganese, chromium, nickel and molybdenum. Nickel steels are particularly suitable for case hardening as such steels provide a strong, tough, wear-resistant case with a ductile core whilst the hardest nitrided cases are obtained with aluminium-containing steels.

Compatability of metals in sliding contact is a dominant factor in bearing performance and both the bearing material and the harder shaft require careful selection. Materials for high temperature bearings and sliding applications have been surveyed [9].

8.3.2 Gears

Gears in service are subjected to rolling, sliding, abrasive, chemical, vibratory and shock-loading action. Their useful life may be terminated by scuffing, pitting, fretting, abrasion, corrosion and fracture. Gear materials must be chosen to resist these phenomena. The most extensively used gear steels are the carbon-manganese steels; manganese contributes markedly to strength and hardness but its effect depends upon the carbon content. It also enhances hardenability, and fine-grained manganese steels attain unusual toughness and strength. For more stringent gear applications, alloy steels, heat-treated to provide the optimum properties, are used. Nickel provides solid-solution strengthening and increases toughness and resistance to impact, particularly at low temperatures, lessens distortion in quenching, improves corrosion resistance and allows more latitude in heat treatment. Chromium increases hardenability and has a strong tendency to form stable carbides which hamper grain growth and provide fine-grained, tough steels. Vanadium forms stable carbides that do not readily go into solution and which are not prone to agglomeration by tempering. It inhibits grain growth, thus imparting strength and toughness to heat-treated steels. Molybdenum and vanadium are generally used in combination with other alloying elements. Lead may be added to gear steels to attain faster machining rates, increased production and longer tool life.

Surface hardening to reduce wear is extensively applied to gear steels without sacrificing desirable core properties. Carbon and alloy steels can be liquid, gas or pack carburized. Nitriding is usually applied to special aluminium-containing steels. Flame and induction hardening methods are also used. Other surface treatments such as Sulphinuz, phosphating and soft nitriding which reduce friction and aid lubrication, can be beneficial.

8.3.3 Rolling Bearings

Although ball and roller bearings are basically rolling elements, in operating mechanisms they are also subjected to some wear by sliding and to chemical attack by lubricant and environment. Their useful life is usually limited by surface disintegration, pits being formed by a fatigue process dependent upon the properties of the material, the nature of the lubricant and the environment [10,11,12]. The principal qualities of ball-bearing materials are dimensional stability, high hardness to resist wear, high elastic limit to avoid plastic deformation under load, and good fatigue resistance to contend with high alternating stresses. A high-carbon steel satisfies these requirements if a carbide-forming element is incorporated to increase hardness, give hardenability and allow oil quenching to minimize distortion during heat-treatment. EN 31, 535A99 or SEA 52100 (1.0% C, 1.5% Cr) through hardening steel is used for conventional bearings, Fig.5. Vacuum degassed, vacuum remelted and electro-slag refined material of improved mechanical properties may be used for increased rolling contact fatigue resistance. For convenience, in the manufacture of the larger sizes of roller bearings case-hardening steels containing chromium, nickel and molybdenum according to the degree of hardenability, shock resistance and core hardness required are used. For use in a corrosive environment martensitic stainless steels are used with some loss of fatigue resistance.

For use at elevated temperatures, conventional rolling bearing steels are not satisfactory owing to loss of hardness and fatigue resistance and high speed tool steels with high tempering temperatures are used, Fig.6. High speed tool steels containing principally tungsten, molybdenum and vanadium are also less prone to deleterious lubricant effects than EN 31 steel [11]. Material combination and material lubricant combination are important to ensure adequate rolling contact fatigue life [13,14,15].

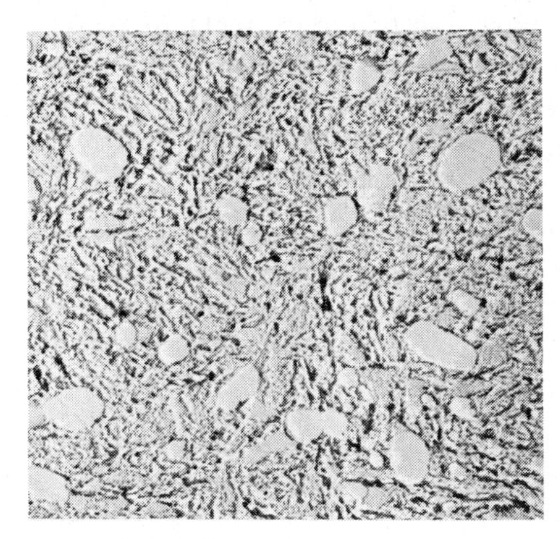

Fig.5 (x4000) Structure of EN 31 ball bearing steel.

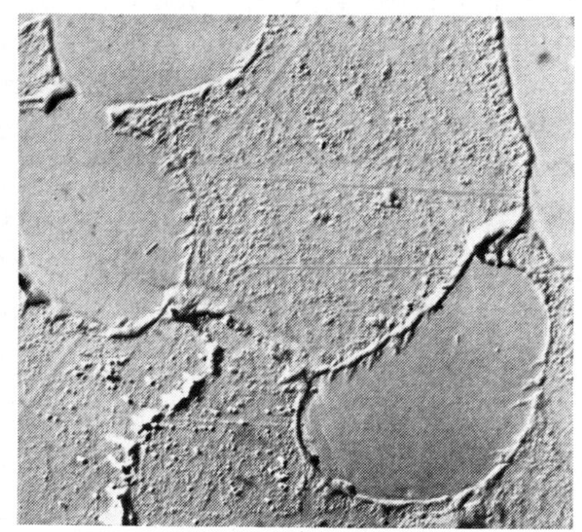

Fig.6 (x6,500) Structure of 18%W high speed tool steel.

Under conditions of unlubricated rolling contact, failure occurs not by the usual fatigue mechanism but by excessive wear limiting useful life, owing to vibration and rough, noisy running. Various superalloys with base composition of chromium, molybdenum and cobalt and containing significant amounts of nickel, tungsten, vanadium and other alloying elements, also cermets and ceramics, are potentially suitable and have been used under arduous test conditions. Under certain test conditions [13], tungsten carbide was the best of the materials tried, giving the lowest wear rate and being relatively unaffected by temperature. The best results were obtained with the smallest carbide size and the lowest percentage of matrix material, Fig.7. Hot pressed silicon nitride was also suitable.

Fine grained

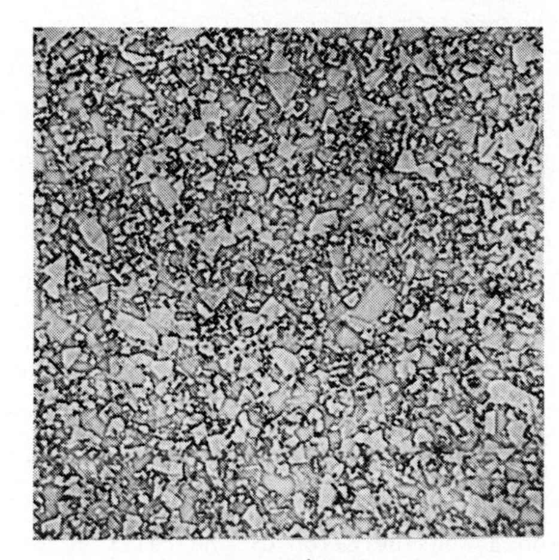

Coarse grained

Fig.7 (x500) Structure of 6% Co tungsten carbide.

8.3.4 Wear Resistant Materials

Wear resistant materials require the correct combination of hardness to resist abrasion and ductility to contend with shock loads and cyclic stressing. As these are conflicting requirements, suitable material selection involves compromise. Other factors such as the types of wear, the nature of any abrasive, the operating temperature and the environment affect the choice of material.

For conditions of high stress and impact, the toughness and work hardening properties of austenitic manganese steel are usually required. For lower stress sliding conditions where toughness is less important, depending upon the cost factor, hardened steels, alloy cast irons, hard facing materials, concrete or ceramics may be used. Where maximum wear resistance is required and cost is immaterial, cemented carbides may be used. Corrosion resistance requires the use of stainless metals, suitable rubbers or plastics. Suitably reinforced polymer material may be used where a low coefficient of friction as well as wear resistance is required. High chromium alloys of iron and steel offer the best wear resistance to elevated temperature problems of cracking, spalling and thermal shock.

The wear resistance of a metal varies with different abrasives and the effective hardness of an abrasive has been defined as the maximum value of hardness of a metal that can be abraded by it. This is of importance in material selection where a known abrasive is to be encountered in practice. It is also of importance in abrasive selection for material removal. Next to diamond, the hardest abrasive available, silicon carbide is the most important abrasive for lapping and grinding and for grinding wheels. Diamond can be synthesized from carbon and in a similar manner, cubic crystalline boron nitride of similar hardness can be produced. Its extreme hardness and durability make this material of economic significance even though the present initial cost is high.

8.3.5 Tools

The evolution of modern production methods has been largely dependent on the development of tool steels capable of operating efficiently under increasing arduous conditions. High hardness is a requisite of almost every tool steel but the optimum hardness depends upon the application and, as hardness and toughness vary inversely, maximum hardness can only be used in the absence of shock loading. Hardness at elevated temperature is sometimes essential as well as resistance to abrasion and distortion. The choice of steel is usually based on a consideration of the relative importance of the properties required. As no single steel possesses all desirable properties, a compromise is generally necessary. Economic aspects can rarely be ignored.

The simplest tool steels are plain carbon (0.6 to 1.4% C) types, which by water quenching, develop a hard case. Low alloy varieties contain up to 0.25% V for carbide stabilization and to retard grain growth, and up to 0.5% Cr to increase hardenability and case depth. Oil-hardening manganese steels (1.0 to 2.0% Mn) provide high hardness with minimum distortion. Chromium may be added to help hardenability whilst additions of tungsten and vanadium improve wear resistance and control grain growth. For the more arduous applications the high-carbon (2.0% C), high-chromium (14% Cr) steels are used to resist wear and corrosion and to maintain high dimensional stability. They maintain a good cutting edge for cold blanking tools.

If shock resistance is required, steels of lower carbon content (0.4 to 0.5% C) are usually required and alloying elements such as silicon, tungsten, vanadium and chromium may be added to improve abrasion resistance. For hot-working, high-speed tool or hot die steels which retain high hardness and a good cutting edge at elevated temperatures are needed. The most widely used type is the 18-4-1 (tungsten, chromium, vanadium) type and hot hardness may be increased by raising the vanadium content or by addition of cobalt. It is possible to partly replace tungsten with molybdenum as is the case with, for example, M.2, 6.5.4.2 (W, Mo, Cr, V) steel. Tungsten, chromium and molybdenum chromium hot die steels may be regarded as lean type high-speed tool steels in which improved toughness can be developed if the carbon content is reduced to below 0.4% C. Nickel, chromium, molybdenum steels generally used as structural steels, may by suitable heat treatment be used as tool steels. They do not possess the high hardness or abrasion resistance of high-speed tool steels but advantage can be taken of their superior toughness.

8.3.6 Cutting Tools

From an economic point of view, one of the most important factors in machining is the rate at which cutting tools wear [16]. Generally the hardest, most wear-resistant, tool material which resists brittle fracture should be used.

High-speed steels have good shock resistance and can be readily shaped by forging and machining, so are the preferred cutting tool materials for a large range of applications. Surface treatments which increase hardness and minimize adhesion to the workpiece are beneficial. Cast cobalt-chromium-tungsten alloys, the Stellites are suitable for applications involving high temperatures, where cooling is impracticable but where impact is absent. Sintered carbide tools may be successfully used for most metal-cutting operations and their exceptional performance results from high hardness and compressive strengths. The straight tungsten carbides have the highest hardness and are used for general applications, but grades containing titanium and tantalum carbide are more resistant to cratering and used generally for machining steel. The use of thin CVD coatings of TiC improves tool life considerably.

The introduction of new materials which are difficult to machine and the focus on high productivity, has accentuated the need for harder cutting tools. Ceramic and cermet tools are now in use. The heat-resisting properties of ceramic tools enable them to be used at speeds unattainable by carbide tools with consequently improved stock removal rates. Their high rigidity prevents chattering and allows smoother cuts and superior work finish. Their high hardness and wear-resistant properties allow close maintenance of dimensions and enable the machining of high hardness materials. Being non-metallic, ceramics do not weld to the material being cut and their extreme refractoriness obviates the need for coolants. Diamond tools are used for special purpose cutting where the high cost can be justified; tool life, however, is superior to that of carbide.

8.3.7 Piston Rings

Apart from speed, temperature and load, the principal factors which influence wear of piston rings are corrosion, frequency of use and dirt. Whilst high temperatures may cause wear by adversely affecting lubrication, low temperatures cause cylinder and ring wear due to corrosion by the condensate of combustion products, this is particularly so with infrequently used engines. Airborne dust can be a serious problem.

Grey cast iron is probably the most widely used material for compression and oil rings but harder materials, such as carbon steel or even En31 ball bearing steel, find use. For greater hardness and wear resistance, chromium plating in sufficient thickness (0.005 inch) can be used on rings and gives reduced wear rates for both cylinder and rings. Chromium plating does not run satisfactorily against itself, so only one mating surface is plated. Although occasionally used for lighter service on passenger car engines, chromium plating is generally used on compression rings on heavy duty engines; it is widely used in diesel and natural gas engines. As chromium plating reduces fatigue strength it is usual to plate materials of high fatigue strength such as high carbon steels. Molybdenum coatings of hardness over 1000 HV and a porosity of 15-25% may now be favoured in preference to chromium plating. Sintered iron piston rings are finding increasing use. Carbon-graphite, used for many years as a piston ring material for unlubricated compressors, is a satisfactory material but it has little inherent flexibility, needs support and depends for its low wear properties on the presence of condensed water [17]. Non-metallic piston rings, particularly fabric reinforced PF resins, have been used where there are doubts about lubrication so that, in the event of lubrication failure, the soft rings will not damage the cylinder wall.

Dry running PTFE piston rings are successful in medium pressure oxygen compressors as they possess low frictional characteristics, dimensional stability at operating temperatures, good wear resistance, high strength and load-carrying ability and chemical inertness. However, for non-lubricated high-pressure applications, dimensional stability can only be retained if the PTFE is reinforced. Glass-fibre reinforcement appears to be potentially more suitable than steel reinforcement. Other fillers commonly used are carbon, powdered metals, MoS_2, ceramics and carbon fibres. The concept of avoidance of damage to the cylinder wall and increasing experience with PTFE rings on unlubricated compressors have led to the idea of replacing metallic rings on lubricated compressors with PTFE rings to allow reduced lubricant feed rates. A state of the art review of their use has been presented [18]. A literature survey of material and metallurgical aspects of piston ring scuffing has been carried out [19].

8.3.8 Cams and Tappets

In modern high-speed, high-output, small automtoive units distress of cams and tappets can be a major problem due to accelerated normal wear, pitting, scuffing and burnishing. Hardenable grey cast iron is the most widely used camshaft material. Water-quenched high-carbon or oil-quenched alloy steels of high carbon content are used in the through-hardened condition. Carburizing steels may also be used with selective flame or induction hardening of surface areas. For automotive engine cams, chromium and molybdenum containing irons are generally used with individual cams surface hardened. Tappet materials are usually through-hardened high carbon, chromium or molybdenum types of carburized low-alloy steels. The most common tappet materials in automotive applications are grey hardenable cast iron containing chromium, molybdenum and nickel or chilled cast iron. Oxide coatings on hardened steel tappet faces improve frictional qualities and accelerate wearing-in. Surface finishes are important to ensure adequate life; stress-relieving can be beneficial. Salt-bath nitriding treatment can be beneficial to cam followers.

8.3.9 Friction Materials

Technological progress in the aeronautical and automobile industries makes demands on friction materials increasingly more severe. High quality brake materials must have high coefficients of friction, stability at all operating temperatures, good wear resistance and strength, high thermal conductivity and corrosion stability. For arduous applications where surface temperatures are high, organic friction material is no longer suitable. At operating temperatures above 350°C, wear of organic material becomes extremely severe. Copper-base facings can operate up to 1000°C but new materials are under development for aeronautical applications where rubbing surfaces may well reach temperatures well above this. Sintered materials offer a wide range of frictional characteristics by the disposal of carefully graded non-metallic particles in a metallic matrix; the composition can be adjusted to minimise wear, seizure and noise level. Cermet and carbon friction materials are being successfully used in many applications.

8.3.10 Plastic Bearings

Since the first appearance on the market in the 1930's of bearings based on thermosetting resins, there has been a continuous increase in the utilization of polymers and polymer based composites in the bearing field, [20,21]. Nylon was the first of the thermoplastic materials used followed by PTFE and more recently the polyacetals. Plastics offer a number of advantages over metals. Their physical and mechanical properties can be varied over a wide range by suitable choice of polymer type, filler and reinforcement; some are cheap and easy to shape. Many polymers are resistant to chemical attack and exhibit low coefficients of friction during unlubricated sliding. Their wear rates sliding against smooth metal counterfaces are low and they do not normally exhibit scuffing or seizure. Lubrication by fluids can often be dispensed with, but when lubricated, polymers allow elastohydrodynamic lubrication more readily than metals. They have disadvantages compared with metals, especially regarding ultimate strengths, elastic moduli, creep resistance and coefficients of expansion. Polymers may also readily absorb fluids resulting in dimensional instability. Their low thermal conductivities can cause problems with dissipation of frictional heat. They also have temperature limitations regarding softening, melting and thermal degradation.

The prime virtues of thermo-setting resin based bearings, cloth or fibre reinforced, is their high strength and excellent performance under conditions of water lubrication. A typical use is roll neck bearings of steel rolling mills where heat removal by water is essential. On a smaller scale, reinforced resin bearings are used in pumps for water circulation and with oil and grease lubrication in automotive applications.

Nylon and polyacetal bushes provide good dry bearing lives if the PV factor is kept low. These materials sintered to provide controlled porosity are used for oil impregnated bearings at much greater PV values with low wear. Thermoplastic bearings find extensive use in fractional HP electric motors, automotive applications and washing machines. Polyethylene and high density polyethylene are used in prosthesis human joints.

PTFE based bearings are now used almost anywhere where cleanliness, heat resistance, freedom from seizure and lubricant elimination is required. Thin films of PTFE find effective use in many applications, particularly to avoid fretting. Depending upon the application and the properties required, PTFE may be filled with glass, asbestos, carbon fibres, MoS_2, graphite, bronze, nickel and iron oxide. A principal application of PTFE is for gas lubricated bearings to contend with stopping and starting.

8.4 SURFACE TREATMENTS AND COATINGS

Materials for tribological applications must fulfil two important functions. They must have structural properties for load carrying and surface characteristics to allow relative motion with low friction and the minimum amount of wear damage. Surface treatments and coatings allow design scope to meet these demands. Also, with wear problems it is often difficult to make worthwhile innovations in design, lubrication or materials selection, but many surface treatments and coatings are available which can effect wear resistance and improved frictional properties [22,23]. Hard wear resistant materials are usually expensive to manufacture and difficult to shape, and for economic reasons, wear-resistant surface treatments may be applied to more common, cheaper to produce and more easily shaped materials.

Many surface treatments are available to combat wear, and the choice depends largely upon the type and severity of wear involved. Some treatments are short lived but the more common surface treatments are those which are expected to last the life of the component. Both types of treatment may be combined to effect protection during arduous conditions of running in and long service effectiveness. Treatments expected to last the life of a component involve surface hardening without changing the composition, surface hardening by diffusion treatments and by the application of surface coatings. Carbon and alloy steels of a hardenable composition and cast iron may be surface hardened by flame or induction methods. Shot peening work hardens the surface of metals. An extensively used method of surface hardening to improve wear resistance depends upon diffusing specific elements into the surfaces of metal by such treatments as carburizing, carbo-nitriding, nitriding, chromosing, boronising and siliconising. Care must be taken to ensure a sufficient depth of case of the correct structure and hardness and a satisfactory transition zone, Fig.8.

Electro-deposition of hard metal such as chromiumprovides hard surface coatings and alloy coatings may be used to improve wear resistance. Composite coatings may be produced by the co-deposition of hard particles and electro-deposited metal, Fig.9. Coatings of silicon carbide in nickel, tungsten carbide in nickel and cobalt and silicon nitride in nickel have been used effectively. Diamond containing coatings have been developed for specific applications [24].

Hard surfacing or facing finds wide use in many applications to provide specific wear resistant alloy or ceramic coatings. Almost any metal or alloy which can be cast may be used as a welding rod to apply a coating. With plasma spraying even the most refractory materials can be deposited with good surface bonding.

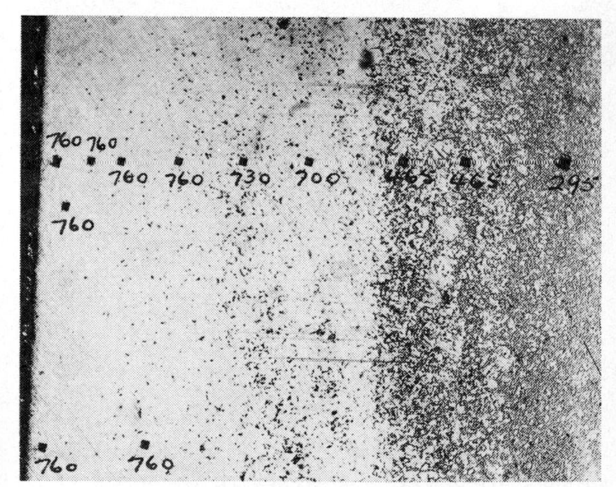

Fig.8 (x75) Micro hardness survey of a case hardened gear tooth with superimposed HV.

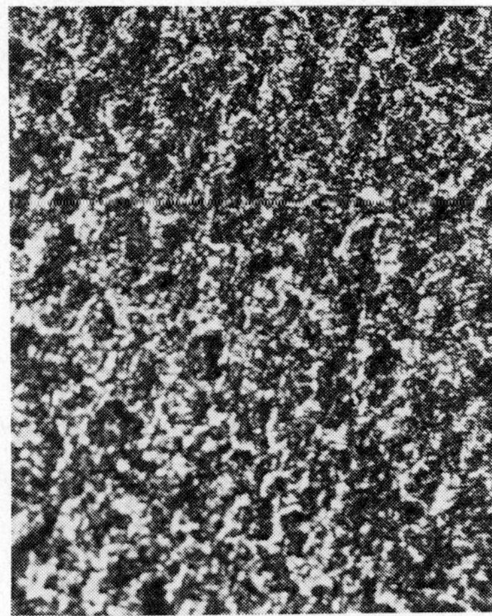

Fig.9 (x150) Co-deposited Fe-4%Ni and chromium carbide.

Some surface treatments, besides effecting wear resistance, may also aid lubrication. Thus Sulphinuz [25] treatment can nitrogen harden a surface with the associated compressive stresses beneficial to fatigue resistance, as well as produce a sulphur-rich surface layer with good lubricant properties under boundary conditions. Soft nitriding carried out in an oxidising bath produces an oxygen-ion rich, low friction surface beneficial in reducing wear and aiding lubrication under arduous conditions. Sulphinuz and soft nitriding treatments may soften hard materials. This can be avoided with the Noskuff process [25] which incorporates a quenching treatment of the Suft BT process [25] which produces a sulphur rich layer, by an electrolytic treatment in a low temperature salt bath. Various types of phosphating treatments provide a thin, porous crystalline lubricating surface film of insoluble phosphate, Fig.10, which can also retain lubricant or provide an effective base for a solid lubricant. Solid lubricant films such as PTFE and MoS_2 can also control wear, reduce friction and aid lubrication.

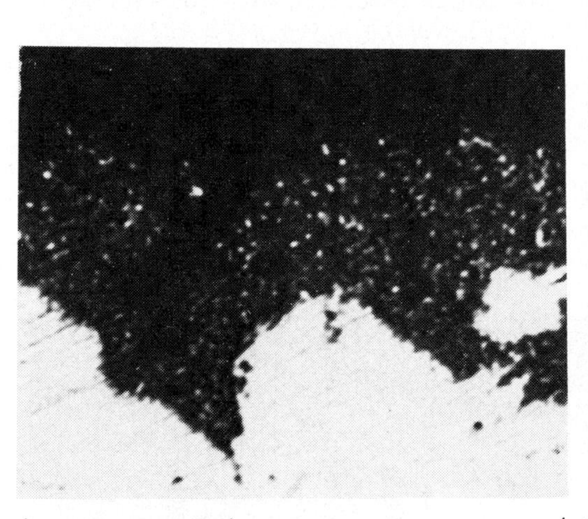

(H=x300, V=x3000) Taper section through a phosphate layer.

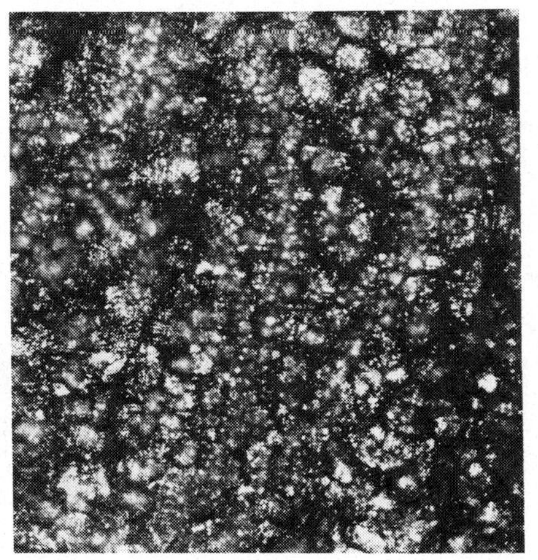

(x75) Phosphated steel surface.

Fig.10 Phosphate coating.

A very hard, strongly adherent, homogeneous thin film usually of hard carbide such as TiC may be applied to metal by chemical vapour deposition (CVD) [26,27]. Refractory coatings such as alumina and silicon nitride may also be chemical vapour deposited.

Physical vapour deposition techniques are a viable approach to the production of surface films of extreme versatility in deposit composition. Virtually any metal, alloy, refractory or intermetallic compound, some polymeric type materials and their mixtures can be deposited [28]. Metal films deposited by ion-plating are strongly bonded to the surface as the film is deposited on a surface cleaned by splutter etching [29,30]. Soft metal lubricant films may also be bonded to a metal surface by ion-plating [31]. Ion-nitriding speeds up the nitriding operation.

8.5 CONCLUSIONS

Mechanisms can only perform satisfactorily if the design and the materials of construction are correctly chosen to contend with the operating conditions and the environment in which they are required. It is only by efficient selection and careful specification that the most effective use of materials can be accomplished to effect material conservation and energy saving in their production.

Economics in materials and manufacturing costs can often be made by judicious use of surface coatings and surface treatments.

REFERENCES

1 Scott,D., Tribology, 1968, 1, 14.
2 Scott,D, In Tribology - Proc. J. Residential Course, 1968, Paper 1, Inst. Metallurgists, London.
3 Scott,D., Proc. Inst. Mech. Engrs., London Int. Conf. on Lubrication and Wear, 1967, 182, (3A), 325.
4 New Engineering Materials, Proc. Inst. Mech. Engrs, London, 1965/66, 180, (3D).
5 Woldman,N.E., Engineering Alloys, 1962, Chapman & Hall, London.
6 Zakay,V.F. and Justusson,W.M., In High Strength Steels - I.S.I. Special Report 76, 1962, 14. Iron & Steel Inst., London.
7 Duckworth,W.E., Leak,D.A. and Phillips,R., In High Strength Steels, I.S.I. Special Report 76, 1962, 22, Iron & Steel Inst., London.
8 Forrester,P.G., Metall Rev., 1960, 5, (20), 507.
9 Amateau,M.F., Nicholson,D.W. and Glaeser,W.A., 1961, O.T.S. PB 171625, Office of Technical Services, Washington D.C.
10 Scott,D., In Fatigue in Rolling Contact, 1963, 103, Inst. Mech. Engrs. London.
11 Scott,D., In Low Alloy Steels, 1968, 203, Iron and Steel Inst. London.
12 Scott,D., In Rolling Contact Fatigue, (Ed. Tourret,R., and Wright,E.P.), 1977, 3, Heyden, London.
13 Scott,D., Wear, 1977, 43, 71.
14 Scott,D., Proc. Inst. Mech. Engrs., 1969, 183, (3L), 9.
15 Scott,D., Rolling Contact Fatigue, In, Wear, (Scott,D., Ed.), Treatise on Materials Science and Technology, 1978, 13, 321, Academic Press, NY.
16 Trent,E.M. Metal Cutting, 1977, Butterworths, London.
17 Summers-Smith,D., Wear, 1966, 9, 425.
18 Summers-Smith,D., Proc. Tribology Conf., 1971, Paper C93/71, Inst. Mech. Engrs.
19 Scott,D., Smith,A.I., Tait,J. and Tremain,G.R., Wear, 1975, 33, 293.
20 Pratt,G.C., Plastic Based Bearings in "Lubrication and Lubricants", (E.R. Braithwaite - Ed.), 1967, 377, Elsevier, Amsterdam.
21 Evans,F.C. and Lancaster,J.K., The Wear of Polymers, in Wear, (Scott,D.), Treatise on Materials Science and Technology, 1978, 13, 85, Academic Press, NY.
22 Scott,D., Wear, 1978, 48, 283.

23 Wilson,R.W., Proc. 1st Euro, Tribology Congr., 1973, 165, Inst. Mech. Engrs., London.
24 Sharp,W.F., Wear, 1975, 32, 315.
25 Gregory,J.C., Tribology, 1970, 3, 73.
26 Gass,H. and Hintermann,H.E., Swiss Patent 452.205, 1968.
27 Hintermann,H.E. and Aubert,F., Proc. 1st Euro Tribology Congr., 1973, 207, Inst. Mech. Engrs., London.
28 Bunshah,R.F. and Juntz,R.S., J. Vac. Sci. Technol., 1972, 9, 1389.
29 Spalvins,T., Przbyszewski,J.S. and Buckley,D.H., NASA Tech. Note TN D - 3707, 1966.
30 Teer,D.G., Tribology, 1975, 7, 245.
31 Sherbiney,M.A. and Halling,J., Wear, 1977, 45, 211.

9 SELECTION OF LUBRICANTS

A.R. LANSDOWN, Director, Swansea Tribology Centre, U.K.

9.1 INTRODUCTION

The variety of available lubricants is enormous. If we simply consider broad, basic types there are probably many hundreds: if we take into account minor differences in composition and the various commercial brands, there are probably tens of thousands. To the non-specialist the problem of proper lubricant selection can therefore seem very confusing.

For many applications the selection of a lubricant is in fact not critical, and a wide variety of lubricants could work quite satisfactorily. For such applications the object of lubricant selection should probably be to ensure the lowest overall life cost for the system as a whole.

For some applications, however, the selection of lubricant may be very critical indeed, and there may be very few or even no lubricants capable of ensuring satisfactory operation.

Because there are thousands of different lubricants available, and because many applications are not critical, there is often a tendency to leave lubricant selection to a very late stage in the design process. The result may be that a design is completed and a machine constructed for which no suitable lubricant exists, and there have been cases where very expensive modification has been necessary to resolve the problem of lubrication.

One important principle is therefore that lubricant requirements should always be considered at an early stage in design.

In order to approach lubricant selection realistically, we should be clear as to what are the objects of lubrication.

The primary object of lubrication is to reduce friction or wear, or usually both friction and wear. This is the factor which defines a material as a lubricant.

There are in addition three secondary functions of a lubricant:-

(i) To act as a coolant. In some systems this will be a vital function of the lubricant because frictional or process heat must be removed and no alternative cooling fluid can be used.

(ii) To remove wear debris or other contaminants, or to prevent other contaminants from entering the system.

(iii) To protect metals against corrosion. There is no good reason why a lubricant should be expected to provide such protection, but because mineral oils are very effective corrosion preventives, many designers have come to expect the same of all lubricants.

The lubricant will often satisfactorily fulfil all these tasks, but where the available choice of lubricants is limited, it may be necessary to choose the lubricant only to meet the friction and wear requirements, and to use other techniques to solve the cooling, contamination and corrosion problems.

9.2 SELECTING THE LUBRICANT TYPE

Lubrication systems as such are outside the scope of this chapter, but the problem of lubricant selection cannot entirely be separated from that of selecting the lubrication system.

The most straightforward way to select both lubricant and lubrication system is probably to start with the simplest technique and to progress from that only as far as is necessary to ensure satisfactory operation. The simplest technique will usually have the lowest initial cost, and will often also be the most reliable.

The simplest system consists of a small quantity of plain mineral oil in place in the lubricated component, without any facility for re-lubrication.

Such a system will cope with a surprisingly wide variety of mechanisms, including watches and clocks and many other precision instruments, door-locks and hinges, sewing-machines, typewriters, bicycles, roller skates, skate-boards, and so on.

It ceases to cope when there is too much load or speed or heat or debris, or when the life required is so long that the oil oxidises, evaporates, or creeps away from the bearing surfaces. It is then necessary to use a more sophisticated oil, a grease, a solid lubricant, or sometimes even a gas lubricant, or to use a more complex lubrication system.

Table 9.1 shows some of the possible choices of alternative systems when the simplest system is no longer adequate. Some of these choices are concerned with the lubrication system rather than the choice of lubricant, but whatever lubrication system is used, a choice of lubricant is needed.

As the demands on a bearing increase, a point is eventually reached where a plain mineral oil is no longer adequate, and it is at this point that the problem of lubricant selection begins.

A convenient approach to lubricant selection will be to consider first the properties which influence the selection of different mineral oils, then the various other oils which can be used, and finally the alternatives to lubricating oils, namely greases, solid lubricants and gases.

It may be useful to refer to Figure 1, which indicates broad limits of speed and load within which different classes of lubricant can be used.

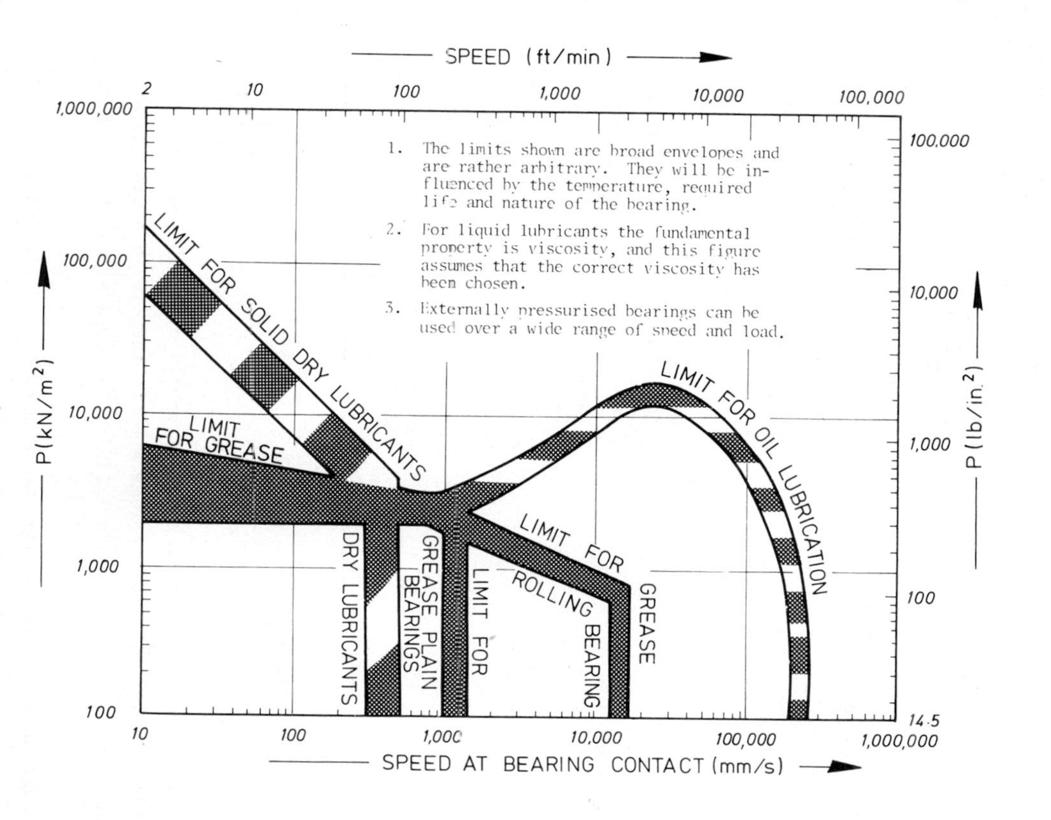

Fig.1 Speed/load limitations for different types of lubricant.

Table 9.1

Possible choices when a small quantity of plain mineral oil in place is no longer adequate.

Problem	Possible choices
Too much load	More viscous oil Grease Extreme pressure oil Extreme pressure grease Externally pressurised lubrication Solid lubricant
Too much speed (which may lead to too much heat)	Greater quantity of oil Less viscous oil Oil circulation system Gas lubrication
Too much heat	Oil with anti-oxidant More viscous oil Greater quantity of oil Oil circulation system Synthetic oil Solid lubricant
Too much debris	Greater quantity of oil Circulation system with filtration
Contamination	Oil circulation system Grease Solid lubricant
Long life required	Oil with anti-oxidant More viscous oil Great quantity of oil Relubrication system Synthetic oil Grease Solid lubricant

9.3 PROPERTIES OF MINERAL OILS

9.3.1 Viscosity

Viscosity is the most important single property of a lubricating oil, since it is the sole property which determines whether the oil in a bearing at a specific load and speed will give full fluid film separation of the bearing surfaces.

Viscosity may be expressed in any one of a number of different units. Probably the most commonly used in English-speaking countries is the centistoke (cSt) which properly describes Kinematic Viscosity, and is directly measured by the various standard Institute of Petroleum, ASTM, British Standard and ISO test methods. The Absolute Viscosity is commonly expressed in terms of the centipoise (cP) and is the viscosity used in the various engineering calculations such as equations (1) to (3).

Each of these units has its SI equivalent, but the introduction of the SI Units has been the subject of considerable dispute. The SI equivalent of the centistoke is the millimetre² per second ($mm^2\ s^{-1}$) and the SI equivalent of the centipoise is the millinewton second per metre² or millipascal second (mNs/m^2).

There are also three other units which are still widely used, although current policy is to discontinue them in favour of kinematic viscosity in centistokes. They are Redwood Seconds, used in Britain, Saybolt Seconds (SUS) in the United States, and degrees Engler, used in Germany and other parts of continental Europe.

There are several equations which relate the oil film thickness to the oil viscosity. For general hydrodynamic lubrication the Reynolds Equation [1] applies,

$$\frac{\partial}{\partial x}\left(h^3 \frac{\partial p}{\partial x}\right) + \frac{\partial}{\partial y}\left(h^3 \frac{\partial p}{\partial y}\right) = 6\mu U \frac{\partial h}{\partial x} + 6\mu h \frac{\partial U}{\partial x} + 12\mu V \tag{1}$$

for elastohydrodynamic lubrication with cylindrical elements, as in roller bearings and many gear configurations, an equation of the Dowson-Higginson type, [2] applies,

$$h_{min}/_R = 2.65 \frac{U^{0.7}\ G^{0.54}}{W^{0.03}} \tag{2}$$

while for elastohydrodynamic lubrication with spherical elements, as in ball-bearings, the Archard-Cowking Equation [3] applies,

$$h_{min}/_R = 1.4 \frac{(UG)^{0.74}}{W^{0.074}} \tag{3}$$

where U is proportional to viscosity.

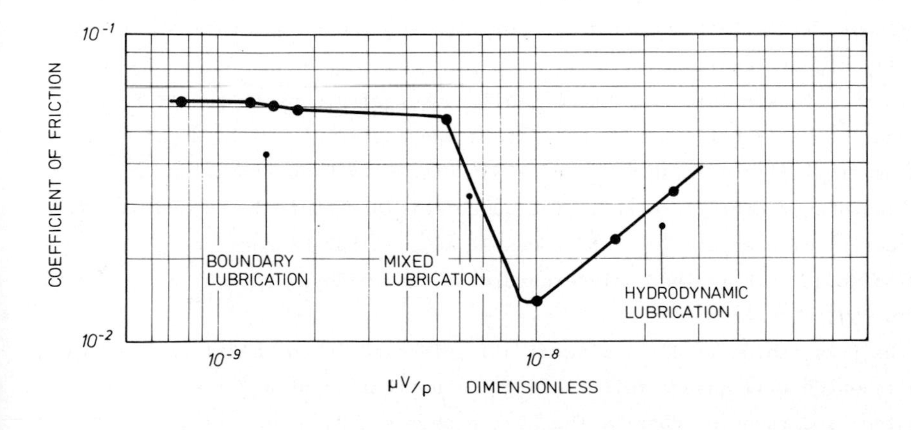

Fig.2 Typical Stribeck curve

Figure 2, sometimes called a Stribeck curve, shows how the viscosity of a lubricant affects the lubrication mode and the coefficient of friction in a bearing.

By the use of the appropriate equation the oil viscosity required for any bearing can be calculated, but if this is done for every bearing in every machine in a factory the result will probably be a list of several hundred viscosities. The number of available mineral oils is so great that it might well be possible to obtain oils with every viscosity required, but there are some very good reasons for not doing so.

(i) If all the bearings in a complex machine can use the same lubricant, then a single oil reservoir and circulation system can be used, and this will generally be cheaper and more reliable.

(ii) A large number of different lubricants causes storage problems.

(iii) The greater the number of lubricants in use in one plant, the greater is the probability of the wrong one being used.

(iv) Most bearings are required to operate over a range of temperature, so the viscosity of the oil will only be at the optimum value for a part of the operating time.

Fortunately the viscosity requirement is often not critical. As long as the viscosity is high enough to ensure the required oil film thickness, there is no harm in going to a slightly higher viscosity. In any case an increase in viscosity is partly self-compensating. The effect of increasing viscosity is to increase viscous friction, which increases the power dissipated in the bearing. The excess power is converted to heat, the heat raises the oil temperature, and the higher temperature reduces the oil viscosity. So the end result of using a more viscous oil is that the power consumption and the temperature stabilise at a slightly higher level.

It is therefore possible to use a small number of viscosity grades to fill a wide range of oil requirements, and in fact many national and international standards (including British Standard 4231 "Viscosity Classification for Industrial Liquid Lubricants") restrict the number of lubricants in such a way that the viscosity of each oil in centistokes at 40°C is 50% higher than that of the preceding grade. This gives only eighteen grades to cover the whole range from 2 cSt to 1500 cSt.

So the first thing to do in mineral oil selection is to calculate the lowest viscosity which will ensure full fluid film lubrication of all the bearings in the system, and then to identify the next higher standard viscosity grade.

The viscosity of an oil is always quoted at a stated temperature, usually 0°F, 70°F, 100°F or 130°F, but increasingly 40°C. These temperatures will rarely be those at which your bearings are intended to operate, so the second important property to consider is the variation of viscosity with temperature.

9.3.2 Viscosity/Temperature Relationship

For all liquids the viscosity decreases as the temperature increases, but the rate of decrease varies considerably from one liquid to another. It follows that if we plot graphs of oil viscosity against temperature, which gives something approaching straight lines, the slope of the lines is different for different oils. Figure 3 shows some typical viscosity/temperature relationships for different oils.

Standard curves are readily available showing the viscosity/temperature relationship for different oils, so that we can convert our required viscosity to the viscosity at a standard reference temperature, provided that we know which line (i.e. which slope) relates to the oil under consideration.

The property which is most widely used to describe the viscosity/temperature behaviour of an oil is the Viscosity Index (V.I). This index relates the change of viscosity with temperature to two arbitrary oils, one based on a Pennsylvania crude oil (V.I. = 0) and one on a Gulf Coast oil (V.I. = 100).

The viscosity index of an unknown oil can be calculated from the measured viscosities at 40°C and 100°C by means of equation

$$V.I. = 100\,(L-U)/(L-H)$$

where U is the viscosity at 40°C of the oil sample in centistokes, L is the viscosity at 40°C of an oil of 0 viscosity index having the same viscosity at 100°C as the oil sample, and H is the viscosity at 40°C of an oil of 100 viscosity index having the same viscosity at 100°C as the oil sample. Tables of values for L and H are contained in the standard Institute of Petroleum and ASTM test methods.

Briefly, the Viscosity Index will be low if the viscosity changes rapidly with temperature, and high if viscosity is less affected by temperature. British Standard D.S.4231 refers to mineral oils having Viscosity Indices of 0, 50 or 95. The highest V.I. for a simple mineral oil is slightly over 100, while certain synthetic oils have V.I. greater than 200.

Generally, a V.I. higher than 100 is only obtained with a mineral oil by the use of a polymeric additive called a Viscosity Index Improver. V.I.Improvers are used in multi-grade engine oils and in hydraulic fluids which are required to operate over a wide temperature range.

Most bearings have to operate over a range of temperatures, and it is the width of this range which determines what V.I. is required. A V.I. of 0 may be acceptable if the temperature range is very narrow, or if a large change in viscosity is acceptable. On the other hand, a V.I. of 160 or more is specified

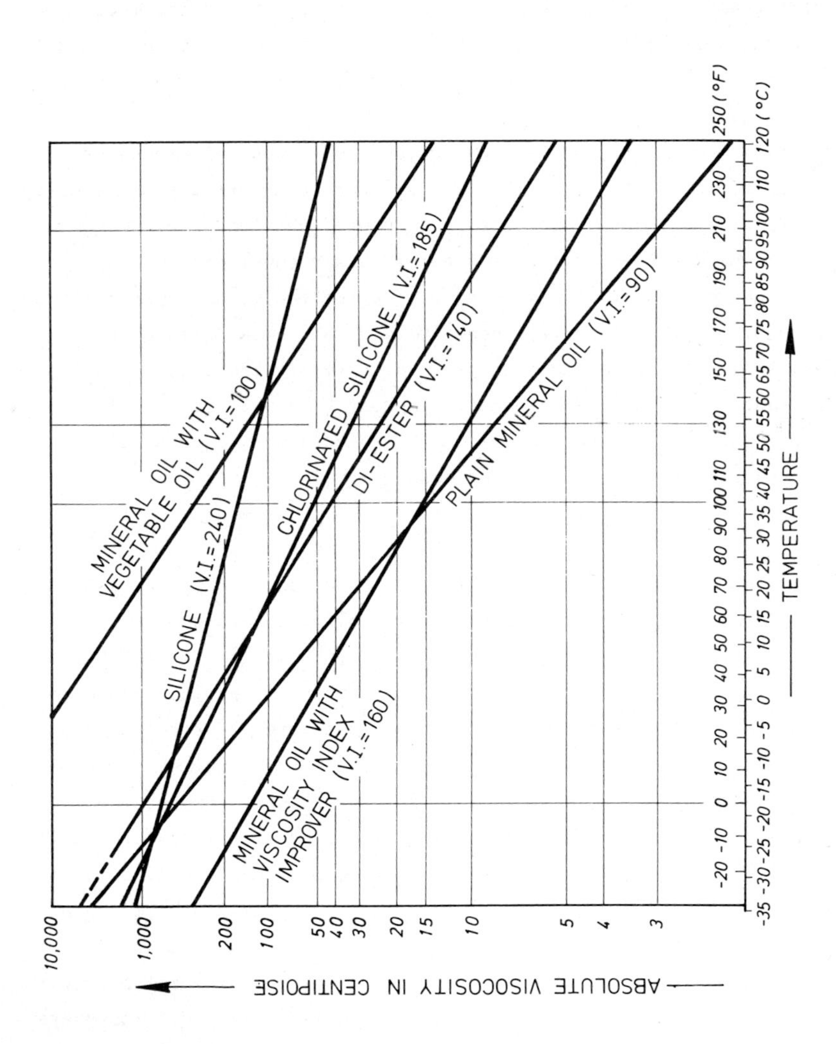

Fig.3 Viscosity/temperature characteristics of various oils.

for an aircraft hydraulic fluid which is required to operate from -40°C to +150°C.

The effectiveness of a V.I.Improver is affected by the shear rate, the rate at which the oil has to pass through confined spaces, and at high shear rate in a high-speed bearing the effective viscosity of a V.I. improved oil may be little or no different from that of the base oil.

Having chosen the required Viscosity Index, the viscosity of the selected oil at the reference temperature can be worked out by means of a graph such as Figure 3 or a chart such as the one in B.S.4231.

9.3.3 Viscosity/Pressure Relationship

Before leaving the subject of viscosity, it is perhaps desirable for completeness to mention the variation of viscosity with pressure. The viscosity of all lubricating oils increases as the pressure is increased. In practice the rate of increase is very low, and in plain hydrodynamic or externally pressurised bearings it can safely be neglected.

In elastohydrodynamic lubrication, however, the pressure generated in the lubricant can be substantial. The increase in lubricant viscosity is correspondingly high, and is a major factor in ensuring successful lubrication of gears and rolling bearings.

It is rare to specify or quote a pressure coefficient of viscosity for a lubricating oil, and this is normally only done for very critical situations such as in certain spacecraft applications.

9.3.4 Anti-Wear, Extreme Pressure and Anti-Friction Properties

In theory, from the point of view of pure physical lubrication, we have completely specified the oil when we have specified the viscosity, the V.I. and perhaps the pressure coefficient of viscosity. In practice this is not always so, and it is often necessary to specify the anti-wear, extreme pressure or anti-friction properties of the oil.

Referring again to Figure 2, the boundary and mixed lubrication regions represent situations in which the bearing surfaces are not completely separated by a film of lubricant, but experience some degree of solid-to-solid contact.

Even in lightly-loaded bearings solid-to-solid contact can occur at low speed when a bearing is starting or stopping, and this is particularly important if a mechanism is intended to operate intermittently. The residual oil film will often give some protection under these conditions, but greater protection can be obtained by the use of an anti-wear additive. This is usually an organic acid such as stearic acid or a natural oil such as rapeseed oil, but synthetic organic phosphorus compounds such as tri-xylyl phosphate are also effective.

In heavily-loaded bearings or intermittent rolling bearings or certain types of gear, a more active type of additive known as an Extreme Pressure (EP) additive may be needed. EP additives are usually reactive synthetic organic compounds containing phosphorus, sulphur or chlorine, although lead naphthenate is still used in certain gear oils, and molybdenum disulphide is occasionally used.

EP properties are assessed in several different ways, but the most widely used test methods are probably the Timken and Four-Ball machines. One criterion, measured by means of the Four-Ball Machine, is the Mean Hertz load, and this can vary from 30 for a plain mineral oil to 85 for a powerful EP gear oil.

One extreme requirement for EP properties is in certain types of metalworking operation, and the most powerful chlorine and sulphur-containing additives are used in the lubricants for such applications. Powerful EP additives can be corrosive, and should therefore not be used where they are not necessary.

A natural oil may be added to an oil to decrease the friction in boundary lubrication, and thus reduce the power consumption and the heat generated. This may be particularly important in a mechanism which is sensitive to friction, such as a worm gear. The natural oil is then sometimes called a lubricity additive, but the term is ill-defined and generally not recommended.

We have now considered all the properties required to give adequate initial lubrication, but there are two other considerations in ensuring continued satisfactory lubrication.

9.3.5 Stability

An oil will decompose chemically in service because of either heat (thermal decomposition) or a combination of heat and oxygen (oxidation). Thermal stability can be improved in manufacturing if the more unstable components of the oil can be removed, but otherwise the only solution is to keep the temperature down.

In general, however, thermal decomposition takes place at much higher temperatures than oxidation, and it is the oxidative stability which determines the maximum temperature at which the oil can be used. Oxidation can be very effectively reduced by the use of anti-oxidants. If the temperature is high in the presence of oxygen, it may therefore be desirable to use an oil containing an anti-oxidant.

Figure 4 shows the relationship between oil life and temperature for typical mineral oils.

Many of the additives used in oils will also decompose, and the useful life of the oil may depend on the depletion of such additives. Anti-oxidants are used up in preventing oxidation, so that where highly oxidative conditions occur the anti-oxidants may be rapidly exhausted and leave the oil unprotected.

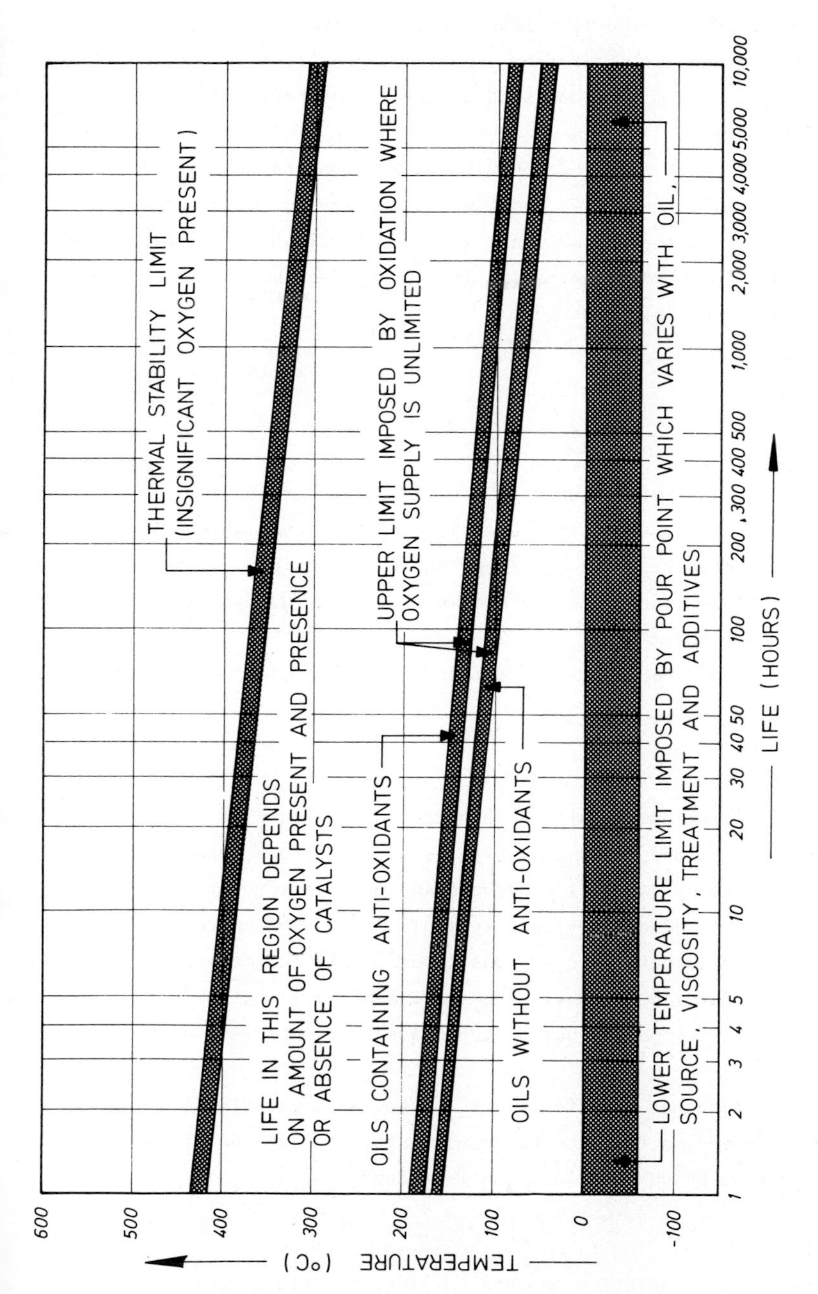

Fig.4 Temperature/life limits for mineral oils.

9.3.6 Contaminants

The quality of a lubricant will also deteriorate in use because of contamination. The contaminants may be solid particles of wear debris or solid decomposition products from a fuel or the lubricant itself, or they may be dissolved acids produced by oxidation, or water from condensation or fuel combustion. They may also enter the system from outside, such as when atmospheric dust or rain enter an oil filler opening.

Some of these contaminants can be removed by filtration and others will settle out in the reservoir, but it is sometimes necessary to use a detergent or dispersant additive to keep the contaminants in a relatively harmless dispersed condition. Ultimately the oil change period in some systems may be determined by the extent of contamination.

9.4 CHOICE OF BASE OIL

9.4.1 Limitations of Mineral Oils

Mineral oils are the most widely used lubricating oils because they are the cheapest, and for many applications they are also the best available. The most important limitations on their use arise from high temperatures, flammability and compatibility problems, and there are a number of other base oils which can then be used as alternatives.

Table 9.2 shows some of the important factors in the selection of alternative base oils.

9.4.2 High Temperatures

More and more industries are requiring lubricants to operate at temperatures too high for mineral oils, and alternative fluids are therefore becoming more and more widely used. Esters were first developed for aircraft jet engines, and are now probably the most common lubricant in applications which are too hot for mineral oils. At higher temperatures various types of silicone can be used, and for even higher temperatures polyphenyl ethers are available, but both silicones and polyphenyl ethers have some serious disadvantages in comparison with mineral oils and esters.

Figure 5 shows the temperature/life limits for several synthetic lubricating oils. These limits depend on the acceptable viscosity, the degree of oxygen access and the extent to which deterioration is acceptable.

9.4.3 Flammability

Some industries, such as aviation and coal mining, have long been concerned with the flammability of lubricants and hydraulic fluids. With increasing plant integration and lubricant system capacities, this concern is now extending

Table 9.2 Important properties in selecting different base oils.

Property / Fluid	Di-ester	Neopentyl Polycl (Complex) Esters	Typical Phosphate Ester	Typical Methyl Silicone	Typical Phenyl Methyl Silicone	Chlori-nated Phenyl Methyl Silicone	Polyglycol (inhibited)	Polyphenyl Ether	Mineral Oil (for comparison)	Remarks
Maximum temperature in absence of oxygen (°C)	250	300	120	220	320	305	260	450	200	For esters this temperature will be higher in the absence of metals
Maximum temperature in presence of oxygen (°C)	210	240	120	180	250	230	200	320	150	This limit is arbitrary. It will be higher if oxygen concentration is low and life is short
Minimum temperature due to increase in viscosity (°C)	-35	-65	-55	-50	-30	-65	-20	0	0 to -50	This limit depends on the power available to overcome the effect of increased viscosity
Density (g/ml)	0.91	1.01	1.12	0.97	1.06	1.04	1.02	1.19	0.88	
Viscosity index	145	140	0	200	175	195	160	-60	0 to 140	A high viscosity index is desirable
Flash point (°C)	230	255	200	310	290	270	180	275	150 to 200	Above this temperature the vapour of the fluid may be ignited by an open flame
Spontaneous ignition temperature	Low	Medium	Very high	High	High	Very high	Medium	High	Low	Above this temperature the fluid may ignite without any flame being present
Boundary lubrication	Good	Good	Very good	Fair but poor for steel on steel	Fair but poor for steel on steel	Good	Very good	Fair	Good	This refers primarily to anti-wear properties when some metal contact is occurring
Toxicity	Slight	Slight	Some toxicity	Non-toxic	Non-toxic	Non-toxic	Believed to be low	Believed to be low	Slight	Specialist advice should always be taken on toxic hazards
Suitable rubbers	Nitrile, silicone	Silicone	Butyl, EPR	Neoprene, viton	Neoprene, viton	Viton, fluoro-silicone	Nitrile	(None for very high tempera-tures)	Nitrile	
Effect on metals	Slightly corrosive to non-ferrous metals	Corrosive to some non-ferrous metals when hot	Enhances corrosion in presence of water	Non-corrosive	Non-corrosive	Corrosive in presence of water to ferrous metals	Non-corrosive	Non-corrosive	Non-corrosive when pure	
Cost (relative to mineral oil)	5	10	10	25	50	60	5	250	1	These are rough approximations, and vary with quality and supply position

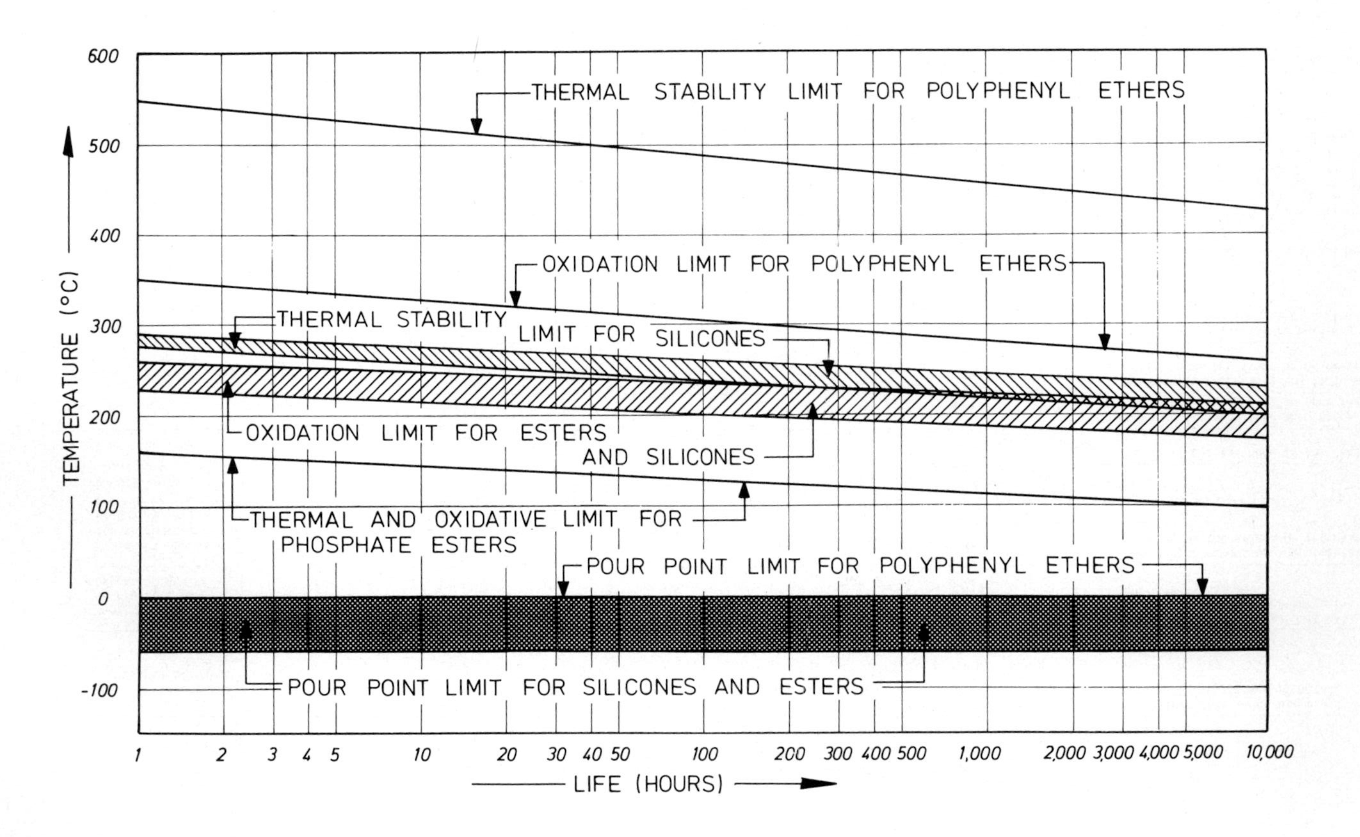

Fig.5 Temperature/life limits for some synthetic oils.

to most industries.

The best fire-resistant liquid is water, but it is a poor lubricant and in its relatively unmodified form is only used as a hydraulic fluid. To improve its lubricating performance for metal-cutting, detergents and EP additives are used. For more general lubricant use, various techniques are employed to increase its viscosity, including thickening with clays and natural organic substances, mixing with polyglycols, and emulsification with mineral oil.

Other fire-resistant liquids with good lubricating properties include phosphate esters, chlorinated hydrocarbons and chlorinated silicones, but these all have corresponding disadvantages.

9.4.4 Compatibility

Problems of incompatibility can arise with any liquid. Mineral oils are incompatible with natural rubber seals and hoses, esters are incompatible with nitriles, and phosphate esters are incompatible with many different rubbers. Mineral oils and most other combustible liquids are incompatible with high-pressure oxygen, and certain special fluorine-containing lubricants are preferred for breathing oxygen systems. Mineral oils are unsatisfactory in contact with red-hot steels because they produce carburisation, and rapeseed oil may be used to avoid this problem.

Most compatibility problems can be overcome by careful selection of suitable base oils and additives, but in some industries such as foodstuffs, pharmaceuticals and chemicals, even the smallest leak of any conventional lubricant may be unacceptable. In such a case it may be possible to use a process fluid as a lubricant. For example, in sugar refining the high viscosity of syrups and molasses enables them to provide effective lubrication of bearings. Where the available process fluids have insufficient viscosity for effective hydrodynamic lubrication, external pressurisation can be used. Such use of process fluids may sometimes be preferred to conventional lubrication, because it may eliminate the need for seals and glands.

9.5 GREASES

A lubricating grease is a liquid lubricant which has been thickened to a semi-solid consistency. The base oil may be mineral oil, ester, silicone, or one of the other synthetic oils, and many of the additives used in lubricating oils are equally effective in greases. In addition, solid lubricants such as graphite or molybdenum disulphide can confer important advantages in greases.

It has been said that more industrial bearings are lubricated by grease than by oil. To understand the reason for this we should consider again the simplest form of lubrication described in Section 9.2, namely the use of a small quantity of plain mineral oil in place in the bearing. There are two disadvantages

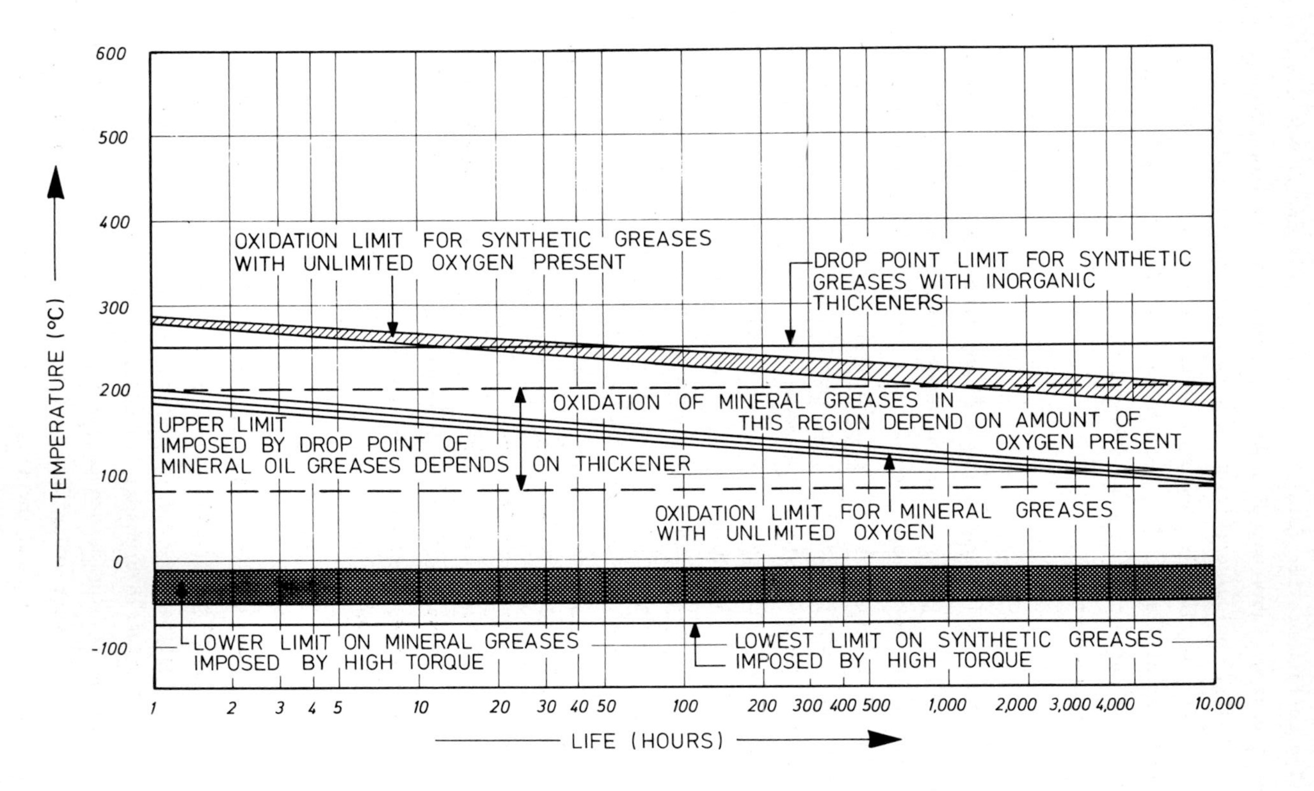

Fig.6 Temperature/life limits for greases.

of that system which are due only to the fluid nature of the oil. The first is that it is difficult to retain the oil in the bearing for any length of time. The second is that it is difficult to stop dirt or other contaminants getting into the bearing.

The use of a grease may overcome both of these problems while retaining the simplicity and economy of the system. The grease in a bearing often distributes itself quite quickly in such a way that a very small proportion is actively lubricating the bearing, while the bulk is outside the critical bearing surfaces, forming a seal against loss of lubricant or ingress of contaminants.

The chief limitation on the use of greases is the lack of coolant flow to remove surplus heat when the bearing is large, fast or heavily loaded, but there are greases which will operate at temperatures as high as 250°C. Figure 6 shows temperature/life limits for some typical greases.

9.6 SOLID LUBRICANTS

Strictly speaking, solid lubricants are solid materials which can be introduced between bearing surfaces to reduce friction or wear, but the title is normally used in a more restricted sense to cover only reduction in friction or prevention of adhesion or seizure. It therefore excludes hard wear-resistant coatings and friction surfaces such as non-skid coatings and brake materials.

Table 9.3 lists a large number of different solid lubricants, but the majority of applications use either molybdenum disulphide, graphite or PTFE (polytetrafluoroethylene). They can all be used in a variety of ways, including free powder, rubbed film, bonded film or sputtered film. Graphite and PTFE may also be used in the form of solid blocks, as may many other polymers, and in this form they are commonly known as dry bearing materials.

The important properties of molybdenum disulphide, graphite and PTFE are listed in Table 9.3.

The advantages of the common solid lubricants are their wide temperature ranges, chemical inertness, low volatility, and the facts that they do not need sealing, do not need feed systems and generally do not cause contamination. Their disadvantages compared with oil or grease are relatively high friction, lack of heat removal, failure to prevent corrosion, and steady wear in service. They are therefore only used where their advantages over oils and greases outweigh their disadvantages, such as at high or very low temperatures, in vacuum, where very long life is required without relubrication, or where contamination is critical.

9.7 GAS LUBRICATION

Gases can be used as lubricants in the same way as oils, and the physical laws governing hydrodynamic oil lubrication apply also to gases. The very low

Table 9.3 Properties of some solid lubricants.

Solid Lubricant	Temperature Limit (^{o}C)	Typical Friction	Specific Wear Rate (mm^3/Nm)	Form in which used
(a) Lamellar solids				
Molybdenum disulphide	350 (in air)	0.1	10^{-6} - 10^{-5}	Powder, bonded film, sputtered coating.
Graphite	500 (in air)	0.2	10^{-6}	Powder
Tungsten disulphide	440 (in air)	0.1		Powder
Calcium fluoride	1000			Fused coating
Graphite fluoride		0.1	10^{-6}	Burnished or sputtered film
Talc		0.1	High	Powder
(b) Thermoplastics				
PTFE (unfilled)	280	0.1	10^{-4}	Powder, solid block, bonded film
Nylon 6.6	100+	0.25	10^{-6} - 10^{-5}	Solid block
Polyimide	260	0.5	10^{-6} - 10^{-5}	Solid block
Acetal	175	0.2	10^{-6}	Solid block
Polyphenylene Sulphide (filled)	230	0.1	10^{-4}	Solid block or coating
Polyurethane	100	0.2		Solid block
PTFE (filled)	300	0.1	10^{-7}	Solid block
Nylon 6.6 (filled)	200	0.25	10^{-7}	Solid block
(c) Others				
Molybdenum trioxide	800			Powder
Phthalocyanine	380			Powder
Lead	200		10^{-6}	Burnished or sputtered film

viscosity of gases means that the film thickness is also very low, so that hydrodynamic gas bearings ("gas-dynamic bearings") are limited to conditions of high speed, low load, small clearances and very closely controlled tolerances. For this reason externally pressurised gas bearings are more commonly used which can carry higher loads, are less critical on clearances and tolerances, and can be used at lower or even zero speeds.

Gas lubrication can be used at higher or lower temperature than oil or grease, but their main advantages are in giving high stiffness in high speed precision bearings, such as in dental drills and precision grinding spindles, and in avoiding sealing and contamination problems where a suitable process gas is available to act as the gas lubricant.

10 LUBRICANT ADDITIVES
Their Application, Performance and Limitations

D.M. SOUL, Lubrizol International Laboratories,
Hazelwood, Derby, U.K.

10.1 INTRODUCTION

The lubrication of industrial plant and equipment covers the whole range of requirements for machinery including the internal combustion engine and the gas turbine. Discussion of such a wide subject in one chapter would be impossible but a very wide field of additive applications exists apart from those for engine and transmission systems. These include hydraulic and gear lubricants, metal cutting and forming fluids, turbine oils, compressor and refrigerator oils, fire-resistant fluids, greases, and a miscellany of specialised products.

Lubricant additives are chemicals incorporated into either a liquid base (mineral oil, synthetic fluid or water), a semi-fluid, or a grease. In general, the additives are soluble or uniformly dispersed throughout the carrier medium. In this respect this discussion does not cover solid lubricant additives such as graphite, molybdenum disulphide and polytetrafluoroethylene.

Additives are used to improve the performance of the base carrier material to provide a desired physical, chemical, or mechanical property. They are manufactured to strict quality control standards to assure consistent performance in similar applications, in many environments throughout the world, for both the base material and the additive system. Many lubricant specifications covering a wide range of industrial processes can only be satisfied by the use of these additional chemicals and certain tests are designed specifically to assess such factors.

10.2 BASIC PROPERTIES OF LUBRICANTS

All types of lubricating fluids display certain basic properties. These may be used to advantage in selecting the most appropriate lubricant for any application. In applications where hydrodynamic lubrication films are formed, an untreated base oil of the correct viscometric properties can provide a perfectly adequate lubricating medium. The properties with appropriate test references for standard evaluation tests are given in Table 10.1.

Mineral oil properties are dependent on the crude oil source and the refinery processes of separation and treatment. Lubricating oil fractions of varying viscosity are produced. By selection of the processes used in the refinery, properties such as pour point, colour and viscosity index may be controlled or modified so as to produce better final products.

10.3 LUBRICANT ADDITIVES

In Table 10.2 is a list of the additives normally employed in lubricants. This table gives the general name of each type together with a summary of its purpose and a suggested mechanism of action.

10.3.1 Detergent and Dispersant Additives

10.3.1.1 Detergent Types

A variety of metal containing compounds called detergents have been described in the literature. Their function appears to be one of effecting a dispersion of particulate matter and neutralising acids rather than one of cleaning up dirt and debris. Some of these materials also function as rust inhibitors and emulsifiers.

The compounds are generally molecules having a large hydrocarbon tail and a "polar group" head. The tail section serves as a solubilizer in the base oil, while the polar group serves as the functional part of the molecule, which attracts particulate contaminants in the lubricant.

The most widely used members of the organometallic class are the sulphonates and phenates. These include both the neutral barium, calcium and magnesium salts, and the alkaline or highly basic products prepared from these organic substrates by incorporating metal in excess of the stoichiometric quantities into those compounds. Phosphonates and salicylates are also used commercially as substrates for metal containing additives.

Commercial interest in barium compounds is declining due to cost of production and environmental considerations.

The meaning of the terms "neutral" and "basic" metal salt used above:

Neutral salt - is a salt of an acid which contains the stoichiometric amount of metal required for neutralization of the acidic group present.

Basic salt - is a salt of an acid which contains more metal than is required for the neutralization of the acidic groups present. Such excess metal may be present

Table 10.1

Standard Lubricating Oil Tests

Test	Scope of Test	Practical use of Data	I.Pet. Method Number and Related Methods
Specific Gravity	Quality of oil. Naphthenic high. Paraffinic medium. Aromatic low. Batch control.	Volume to weight conversions	160/68, 59/72, 189/79
Viscosity	Relation of 'body' oil to temperature. Normally evaluated at two temperatures. Also various methods for low temperatures and varying shear rates.	Viscosity Index. Refining processes. 'Flow' of oil in machinery.	71/79
Flash Point	Volatility, classification of fluids, inflammability.	Storage and operating conditions of oil. Taxation and shipping of Petroleum products.	34/75 (Pensky-Martin) 170/75 (Abel) 36/67 (Cleveland)
Colour	Depth of colour related to refinery processes.	Contamination.	15/72
Pour/Cloud Point	Refining processes - dewaxing. Effect of low temperatures.	Low temperature viscosity and operation.	15/67 (Pour) 219/67 (Cloud)
Acidity	Trace residues from refinery processes.	Normally low for new oil. Deterioration of used oil.	1/74
Carbon Residue	Carbon residue after burning off oil.	Performance of high quality base. Paraffinic - high. Naphthenic - low.	13/78 (Conradson) 14/65 (Ramsbottom)

due to a true basic salt structure, but a more likely explanation is that the excess metal is present in the form of dispersed metal compounds.

10.3.1.1.1 Sulphonates

Normal metal sulphonates derived from "mahogany" acids (the mahogany-coloured petroleum sulphonic acids obtained as a by-product during white oil manufacture) were first employed as detergent additives in commercial crankcase oils during World War II. Almost without exception, the metal present in such sulphonates was calcium or barium. Petroleum sulphonates were not only superior to earlier additives with respect to detergency, but were much less corrosive to sensitive bearing metal alloys and responded well to corrosion inhibitors. They can be represented by the general formula:

R_xArSO_3M

where R_xAr represents complex alkylaromatic radicals derived from petroleum and M is one equivalent of a polyvalent metal.

Supplementing the supply of natural petroleum sulphonates are the synthetic sulphonates derived from long-chain alkyl substituted benzenes (e.g. polydodecyl benzene bottoms) obtained as by-products in the manufacture of household detergents or which are manufactured specifically for this use.

Highly basic sulphonates contain from 3 to 10 or 15 times as much metal as the corresponding normal sulphonates. Called "overbased", "superbasic" or "hyperbasic" sulphonates, these products are manufactured by heating a mixture of certain promoters or solvents with a neutral sulphonate and a large excess of metal oxide or hydroxide and carbonating with carbon dioxide to convert the metal base to colloidally-dispersed metal carbonate. Overbased sulphonates possess the ability to neutralize acidic contaminants formed in lubricating oils and thus reduce corrosive wear of engine components.

10.3.1.1.2 Phosphonates and/or Thiophosphonates

These detergent additives can be represented by the general formula:

$$R - \overset{X}{\overset{\|}{P}}(XM)_2$$

where R is a large aliphatic radical of at least 500 molecular weight. X is oxygen and/or sulphur, and M is one equivalent of a monovalent or polyvalent metal. They are prepared by first heating a polyolefin such as polybutenes of 500 - 2,000 molecular weight with a phosphorus reagent, generally phosphorus pentasulphide, to form a complex organic phosphorus-sulphur compound which is then neutralized with a metal base. By steam treatment or prolonged hydrolysis prior to or during neutralization, a portion or substantially all of the sulphur

TABLE 10.2

Lubricant Additive

Additive	Chemicals	Purpose of Additive	How Additive Works
Antioxidant	Hindered phenols Amines Organic sulphides Zinc dithiophosphate	Minimizes the formation of resins, varnish, acids, sludge and polymers	Reduces organic peroxides terminating the oxidation chain. Reduces formation of acids by decreasing oxygen taken up in the oil. Prevents catalytic reactions.
Corrosion Inhibitor	Zinc dithiophosphates Sulphurized terpenes Phosphosulphurized terpenes Sulphurized olefins	Protects bearing and other metal surfaces from corrosion	Acts as anticatalyst. Coats metal surfaces which protect against acid and peroxide attack.
Rust Inhibitor	Amine phosphates Sodium,calcium and Magnesium sulphonates Alkyl succinic acids Fatty acids	Protects ferrous metal surfaces against rust	Polar molecules are absorbed preferentially on the metal surface and serves as a barrier against water. Neutralizes acids.
Detergent	Normal or basic, calcium, barium,magnesium phosphonates, phenates and sulphonates	Reduces or prevents deposits in engines operated at high temperatures	Controls buildup of varnish and sludge by reacting with oxidation products to form oil soluble material which remains suspended in the oil.
Dispersant	Polymers such as nitrogen containing polymethacrylates, alkyl succinimides, and succinate esters high molecular weight amines and amides	Prevents and retards sludge formation and deposition under low temperature operating conditions	Dispersants have a strong affinity for dirt particles and surround each with oil soluble molecules which keep the sludge from agglomerating and depositing in the engine.
Friction Modifier	Fatty acids Fatty amines Fats	To increase oil film strength to prevent oil film rupture	Highly polar molecules are absorbed on the metal surface and remain in place to cushion and keep metal surfaces apart.
Antiwear	Zinc dialkyl-dithiophosphate Tricresyl phosphate	Reduce rapid wear in steel-on-steel applications	Additive reacts with the metal to form a compound which is deformed by plastic flow to allow a new distribution of load.

Extreme Pressure	Sulphurized fats, olefins Chlorinated hydrocarbons Lead salts of organic acids. Amine phosphates	Prevents seizure and welding between metal surfaces under condition of extreme pressure and temperature	EP agent reacts with metal surfaces to form new compounds having lower shear strength than the base metal and is sheared preferentially to the base metal.
Viscosity Index Improver	Polyisobutylenes Polymethacrylates Polyacrylates Ethylene propylene copolymers Styrene maleic ester copolymers Hydrogenated styrene butadiene copolymers	Reduces the rate of change of viscosity with temperature	Polymer molecule assumes a compact curled form in a poor solvent (cold oil) and an uncurled high surface area in a better solvent (hot oil). The uncurled form thickens the oil.
Pour Depressant	Wax alkylate naphthalene Wax aklylated phenols Polymethacrylates	Lowers the pour point of the oil	Retards the formation of full-size wax crystals by coating or co-crystallization with the wax.
Antifoam	Silicone polymers Polymethacrylates	Prevents the formation of stable foam	Appears to attack the oil film surrounding each bubble reducing interfacial tension. The small bubbles liberated combine to form large ones which float to the surface.
Emulsifier	Sodium salts of sulphonic acids, sodium salts of organic acids, fatty amine salts	To make mineral oil miscible with water	Emulsifier is absorbed at the oil-water interface to reduce interfacial tension resulting in an intimate dispersion of one liquid in the other.
Tackiness	Soaps, polyisobutylene and polyacrylate polymers	To provide the oil with greater cohension	Increases viscosity. Materials themselves are tacky and stringy.
Antiseptic	Phenols, chlorine compounds, formaldehyde bases	Increases emulsion life and prevents odour	Prevents and reduces microorganism growth.
Metal Deactivator	Triaryl phosphites Sulphur compounds Diamines. Dimercapto thiadiazole derivatives.	Stop the catalytic effect of metals on oxidation and corrosion	A protective film is absorbed on metal surfaces which prevents contact between corrosive agents and base metal.

present in the thiophosphonate group, $-P(S)(SM)_2$, can be substituted with oxygen to yield a phosphonate.

The history of these detergent additives closely parallels that of sulphonate additives in that basic and overbased salts have replaced normal salts in all but a few commercial applications. The manufacture of such basic and overbased salts is carried out by the use of methods like those described in the SULPHONATES section.

10.3.1.1.3 Phenates

Phenates and phenate-sulphides have played an important role as detergent additives ever since their introduction during World War II. Among the earliest additives of this type to gain commercial acceptance were:

Calcium and barium phenates of tertiary-octylphenol sulphide and tertiary-amylphenol sulphide having the general formula:

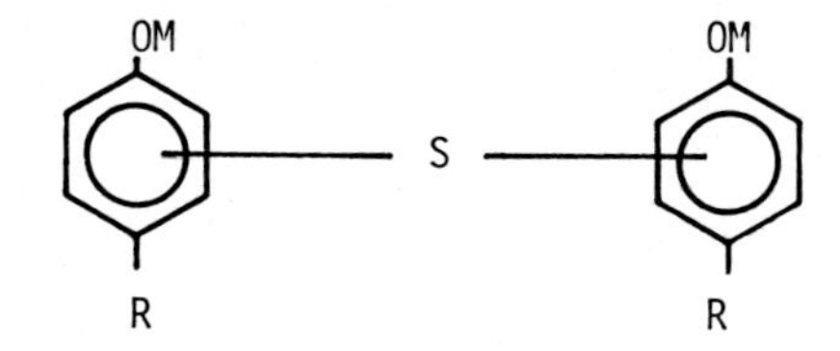

Calcium phenates of tertiary-amylphenol -formaldehyde condensation products.

OH OH

$-CH_2$

R R

Calcium and barium phenates of paraffin wax substituted phenol having the general formula:

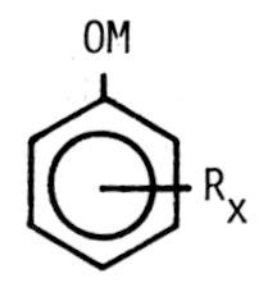

In addition to performing a detergent role, phenates - and especially phenate sulphides - exhibit substantial anti-oxidant properties and are particularly useful for high temperature fluids.

Like the other families of detergent additives discussed, basic and overbased phenates have replaced normal phenates in most applications. Manufacturing techniques for such products are similar to those employed for basic sulphonates or phosphonates.

10.3.1.1.4 Alkyl Substituted Salicylates

These additives can be represented by the general formula:

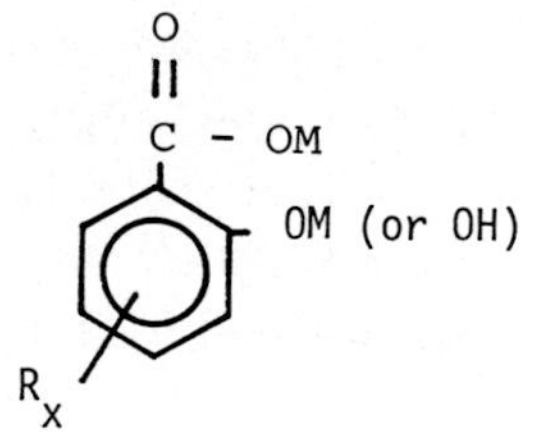

O
||
C - OM
OM (or OH)
R_x

where R is an oil-solubilizing organic radical and M is one equivalent of a polyvalent metal.

One of the first additives of this type to see commercial use was the zinc carboxylate of di-isopropyl salicylic acid. More recently, calcium carboxylates of long-chain alkyl salicylic acids have been used. The manufacture of such additives involves carboxylation of a metal phenate with carbon dioxide. As with the other detergent additives discussed, overbasing techniques have been employed to prepare improved products.

10.3.1.2 Mode of Action of Detergent Additives

Although the mechanism of detergency in non-aqueous media such as mineral oils is not fully understood, researchers have found evidence for the existence of "soap micelles" in non-aqueous solvents. There is reason, then, to believe that detergent additives in mineral oil solution can act in a manner similar to aqueous soap solutions. Basic and overbased detergents also possess the ability to neutralize harmful inorganic and organic acids which accumulate in crankcase lubricants during service. They can also act as high temperature stabilisers as means of reducing thermal decomposition of other additives by neutralising small amounts of acidic products which could cause catalytic decomposition reactions to occur.

10.3.1.3 Dispersant Additives

The term "dispersant" is presently used to designate additives which are capable of dispersing the "cold sludge" formed in engines operated for the most part at relatively low bulk crankcase oil temperatures. Unless maintained in fine suspension in the lubricating oil, this sludge deposits on oil filters, valve train components, and oil control rings where it interferes with good engine performance.

Since known metal-containing detergents did not appear to offer a solution to the cold sludge problem, researchers turned their attention to metal-free organic compounds in the hope that an "ashless detergent" would provide the answer.

Such products can also be used to effect dispersion of insoluble material in oil, i.e. colloidal dispersions, and also to disperse water in oil to produce stable invert emulsions containing up to forty percent water.

The compounds which are useful for this purpose are again characterized by a polar group attached to a relatively high molecular weight hydrocarbon chain. The polar group generally contains one or more of the elements: nitrogen, oxygen and phosphorus. The solubilizing chains are based on polyisobutylene.

Dispersants may be divided into various chemical families.

10.3.1.3.1 Copolymers

Copolymers which contain a carboxylic ester function and one or more additional polar functions such as amine, imide, hydroxyl, ether, epoxide, phosphorus ester, carboxyl, anhydride, or nitrile generally have dispersant properties. Some of these polymers exhibit viscosity modifying properties and thus find application as multi-functional additives. Three types of dispersant VI improver are in used today: polymethacrylates, styrene-maleic ester copolymers and ethylene-propylene copolymers.

10.3.1.3.2 Substituted Succinimides

Hydrocarbon polymers may be introduced into molecules by chemical reaction. Typical products of this type are obtained by treating a polyolefin, with a molecular weight in the range 500-2000, with phosphorus chloride and phosphorus sulphide and then with reagents such as urea, ethylene oxide and boric acid.

The N-substituted long chain alkenyl succinimides contain the characteristic succinimide grouping

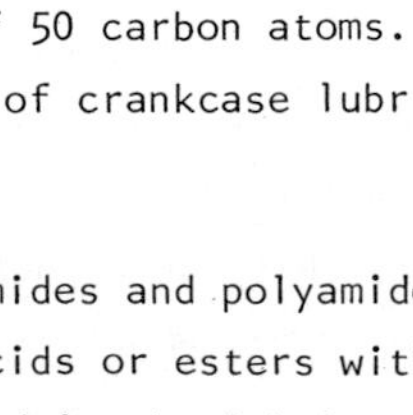

```
                 O
                //
R - CH - C
     |      \
     |       N
     |      /
    CH2 - C
           \\
             O
```

where R contains upwards of 50 carbon atoms. This group of products is used extensively in many types of crankcase lubricant.

10.3.1.3.3 Amides

High molecular weight amides and polyamides are usually prepared by the reaction of higher fatty acids or esters with polyalkylene amines. Products of this type are generally used in the lubrication of two-cycle engines and little in automotive crankcase lubricants.

10.3.1.3.4 Other Chemicals

Extensive research into other classes of detergent chemistry continues. Products may be produced from the following:-

- Poly and benzyl amines
- High molecular weight esters which may be used for emulsification purposes
- Amine salts of high molecular weight acids

10.3.1.4 Mode of Action of Dispersant Additives

Based on microscopy and chromatography studies, several groups of researchers have concluded that ashless dispersants function by adsorption on contaminant particles present in oils, thus keeping them in suspension. Although these researchers have contributed much to an understanding of the mechanism of dispersancy, there is wide agreement that it is necessary to evaluate the products in practical trials since there is little in the way of laboratory testing which can compare products with this function.

10.3.2 Antioxidants and Bearing Corrosion Inhibitors

Antioxidants are probably employed in a wider variety of lubricants than any other kind of additive. In addition to their use in crankcase oils and steam turbine oils, they will be found in gas turbine lubricants, automatic transmission fluids, gear oils, cutting oils, greases and hydraulic fluids.

The function of an oxidation inhibitor is the prevention of deterioration associated with oxygen attack on the lubricant base fluid. Inhibitors function either to destroy peroxides or the free radicals derived from peroxides. The most widely used antioxidants in the lubricant field are the phenolic types, sulphurised polyolefins and the zinc dithiophosphates. The phenols are considered to be the chain-breaking type whereas the latter two are believed to be peroxide destroyers.

The corrosion of bearing metals in internal combustion engines is generally considered to be due largely to acid or active sulphur attack on the bearing metals. The acids involved in the attack originate either from products of incomplete combustion of the fuel which find their way into the lubricant as blow-by gases or from the oxidation of the lubricant. Oxidation inhibitors can reduce or eliminate the latter materials and, hence, reduce bearing corrosion. Generally, in most industrial applications, oxidation is the major cause of acid release.

In most environments in which a lubricating oil is employed, it comes in contact with air often at high temperatures and in the presence of metals or chemical compounds which promote oxidation of the oil. The oil undergoes a complex series of oxidation reactions and the harmful results include, principally, an increase in viscosity of the lubricant, the formation of acidic contaminants such as "petroleum oxyacids" and the development of carbonaceous material.

Among the more effective chemicals employed as commercial antioxidants today are:-

10.3.2.1 Dithiophosphates

Zinc diorganodithiophosphates (also effective in the role of corrosion inhibitors). Other metals may also be used.

Ashless dithiophosphates, replacing metal by amine can be equally effective.

10.3.2.2 Hindered Phenol

Hindered phenols (i.e. phenols in which the hydroxyl group is sterically-blocked or "hindered"):

2, 6-di-tertiary-butyl-4-methylphenol
4, 4'-methylene bis (2,6-di-tertiary-butylphenol)
4, 4'-thiobis (2-methyl-6-tertiary-butylphenol)

10.3.2.3 Nitrogen Bases

Amines such as:

N-phenyl-alpha-naphthylamine
N-phenyl-beta-naphthylamine
Tetramethyldiaminodiphenylmethane
Anthranilic acid
Phenothiazine and alkylated derivatives

10.3.2.4 Sulphurised polyolefins, where the sulphur acts in a similar manner to naturally occurring sulphur chemicals found in crude oil but removed during refining.

Of all these antioxidants, the most widely used are the zinc diorganodithiophosphates. These have a dual function of behaving also as antiwear agents to protect cam and tappet scuffing when used in crankcase oils. These compounds are also used extensively as mild load-carrying additives in gear oils and in hydraulic oils. Performance requirements today generally dictate the use of more than one antioxidant.

Hindered phenols are favoured for hydraulic and turbine oils because they give excellent oxidation lives for long field use. The phenyl naphthylamine antioxidants, especially the beta form, or those containing traces of it, are now regarded as carcinogenic. They are therefore not used to a great extent and are more specific to some types of grease.

10.3.2.5 Mode of Action of Antioxidants

There is general agreement among independent investigators that the oxidation of a lubricating oil involves a chain oxidation reaction in which initially-formed organic peroxides attack unoxidized oil and are subsequently regenerated by oxygen in the air to continue such attack. According to this widely accepted "peroxide theory", an effective antioxidant is a chemical compound which reduces organic peroxides and consequently causes the chain reaction to cease.

10.3.3 Corrosion Inhibitors

In the additive industry, the term "corrosion inhibitor" is applied to a material which protects corrosion-susceptible non-ferrous metal components, principally bearings, against attack by acidic contaminants in the lubricating oil. A different term - rust inhibitor - is used to designate materials which protect ferrous metal surfaces against rust.

Among the earliest types of corrosion inhibitors to see commercial application were organic phosphites. Most of these were not pure chemicals but rather mixtures of mono-, di-, and tri-organo phosphites obtained from the reaction of alcohols or hydroxyesters (e.g. methyl lactate, trimethyl citrate) with phosphous trichloride.

The major classes of corrosion inhibitors in commercial use at the present time are:

10.3.3.1 Dithiophosphates

Metal diorganodithiophosphates, especially zinc diorganodithiophosphates. They possess the structure:

$$(RO)_2P(=S)-SM$$

where R is an aliphatic or aromatic radical and M is a polyvalent metal such as zinc or nickel. Their manufacture involves first heating an alcohol or phenol with phosphorus pentasulphide to form diorganodithiophosphoric acid, then neutralizing such acid with a metal base.

10.3.3.2 Dithiocarbamates

Metal diorganodithiocarbamates, especially zinc diorganodithiocarbamates. They are described by the general formula:

$$R_2N-C(=S)-SM$$

where R and M are as defined in 10.3.3.1. Their manufacture involves the reaction of an organic amine, carbon disulphide, and a metal base.

10.3.3.3 Sulphur Products

Sulphurized terpenes, for example, sulphurized dipentene. These products are manufactured by heating elemental sulphur with a terpene hydrocarbon and then optionally washing the crude product with aqueous alkali or alkali metal sulphide to remove dissolved or "corrosive" sulphur.

10.3.3.4 Phosphorus-Sulphur Products

Phosphosulphurized terpenes, for example, phosphorus pentasulphide-treated turpentine. These additives are manufactured by heating phosphorus pentasulphide with a terpene hydrocarbon.

Of the four major classes of corrosion inhibitors listed above, metal dithiophosphates - particularly zinc dialkyldithiophosphates - have achieved the widest commercial acceptance. Many of the compounds listed above may not be true corrosion inhibitors but function by inhibiting oxidation and so reducing the formation of corrosive carboxylic acids.

10.3.3.5 Triazoles and Chelating Agents

Benzotriazole and its derivatives are used to form a surface layer on copper and silver based alloys by chelation. This process passivates the surface and reduces its ability to act as a catalyst towards fluid degradation by preventing solubilisation of small quantities of the metal.

There are versions of the chemical which are not soluble and other which are soluble in both synthetic base fluids and aqueous systems.

10.3.3.6 Dimercapto Thiadiozole Derivatives

These are prepared by reacting hydrazine and carbon disulphide followed by various reactions to make the material oil soluble. They are effective in reducing corrosion.

10.3.3.7 Mode of Action of Corrosion Inhibitors

It can be said that compounds like 10.3.3.5 and 10.3.3.6 are true corrosion inhibitors. They function by reacting chemically with the non-ferrous surface of the metal component: (e.g. in an engine, copper-lead or lead-bronze bearings) to form thereon a corrosion-resistant, protective film. This film must adhere tightly to the bearing surface lest it be removed by dispersants or detergents and expose the underlying metal surface to attack by acidic components in the lubricating oil.

10.3.4 Rust Inhibitors

The presence of water in lubricants varies between very small levels to values as high as forty percent (40) in invert emulsion hydraulic fluids and ninety-five

(95) percent in emulsion fluids. Inhibiting ferrous surfaces against rusting is therefore a requirement in all types of oil.

A range of rust inhibiting chemicals is necessary to cope with different environments, and their chemical or physical action is important. The strong surface adsorption exhibited by these chemicals to restrict the contact with water means that careful choice is necessary to ensure that other surface active chemicals, notably the antiwear and extreme pressure members, are able to perform their function.

Typical chemicals used for oil soluble systems include the following:-

(i)	Alkenyl succinic acids and derivatives	Turbine, hydraulic and circulating oils
(ii)	Alkyl thioacetic acids and derivatives	Turbine, hydraulic and circulating oils
(iii)	Substituted imidazolines	Gear oils
(iv)	Amine phosphates	Preservative oils
(v)	Sulphonates, neutral or low base	Engine preservative oils. Storage, etc.

Additional film strength is achieved by the addition of fatty materials such as lanolin in the case of thick film storage composition.

10.3.5 Viscosity Improvers

Viscosity improvers are materials which improve the viscosity temperature relationship of a lubricant. They are generally oil soluble polymers with molecular weight ranging from approximately 50,000 to 1,000,000. The polymer molecule interacts with the oil to effect the final oil viscosity. The higher the tempature of the system, the larger the polymer volume, the greater the thickening effect, and hence the less the "thinning" tendendy of the oil due to increased temperature.

In addition to viscosity characteristics, the performance of these polymers is dependent on the shear stability or resistance to shear and on their chemical and thermal stability. With a given polymer, the shear stability decreases with an increase in molecular weight. The loss due to shear is reflected in a decrease in viscosity of the lubricant in the mechanical system.

Viscosity Index or "V.I." is an arbitrary number - calculated from the observed viscosities of a lubricant at two widely separated temperatures - which indicates the resistance of the lubricant to viscosity change with temperature. The higher the V.I. value, the greater the resistance of the lubricant to thicken at low temperatures and thin out at high temperatures.

The significance of V.I. is much less with current oil specifications. Measured low temperature viscosity is now most important. For motor oils such values are reported using the Cold Cranking Simulator and for automatic transmission fluids and gear oils using the Brookfield Viscometer.

The use of hydrocarbon polymers of low molecular weight makes it possible to formulate multigraded axle oils. Reduction in oil drag and consequent fuel savings may be achieved with lower viscosity index values but a high degree of shear stability.

All important viscosity improvers are manufactured by processes of polymerisation. The groups of polymers, soluble in oil, are:-

(i) Polyisobutenes

(ii) Alkyl methacrylate and acrylate copolymers

(iii) Rubber type chemicals such as olefine co-polymers (OCP) and butadiene-styrene copolymers

10.3.5.1 Mode of Action of V.I. Improvers

V.I. improvers exert a greater thickening effect on oil at high temperatures than they do at relatively lower temperatures. The result of such selective thickening is that the oil suffers less viscosity change with changing temperature. It is believed that the selective thickening occurs because the polymer assumes a compact, curled form in a poor solvent such as cold oil, and an uncurled high surface area form in a better solvent such as hot oil.

Polymers whose solubility in oil changes very little with temperature act as thickeners, but are not as effective V.I. improvers as are those polymers whose solubility is poor at low temperatures but good at higher temperatures. V.I. improvers are more effective in increasing the viscosity order of low viscosity oils and become progressively less effective as the viscosity of the base oil increases.

V.I. improvers undergo temporary viscosity reduction under shear because of the alignment of the polymer molecules in the direction of flow. This temporary viscosity reduction has the effect of reducing friction in high shear zones and gives a significant advantage for the V.I. improver-treated oil over a base oil of the same viscosity. Another advantage is that in low shear zones the viscosity of the treated oil remains high, thus minimizing oil consumption. Polymer-treated oils also exhibit lower bearing wear than their comparable mineral oil counterparts. These factors have been important in the acceptance of multigrade engine oils.

Polymers used as viscosity-index improvers must be relatively stable to chain scission under high shear rates. Molecular weight is drastically reduced by shearing, thus causing a deterioration of the properties which improve viscosity index. The shear stability of a polymer type is dependent on molecular weight and molecular weight distribution and increases with decreasing molecular weight. It can be seen that shear stability and viscosity-index improvement require divergent molecular weights. The molecular weights of commercial polymers are based on a compromise between these two properties. In many cases, this has resulted in the use of larger amounts of lower-molecular-weight polymers for a given application.

Chemical and thermal stability are important properties of a good viscosity-index improver. For good stability, it is important that catalyst residues be completely removed during the manufacturing process. Some commercial polymers are formulated with inhibitors to enhance their stability. Several of the commercial viscosity-index improvers are multi-functional in that they may be effective pour-point depressants and possess dispersency performance.

10.3.6 Pour Point Depressants

Ever since lubricating oils were prepared from crude oils, refiners have experienced difficulty with congelation of these products at low temperatures. Part of the difficulty arises from a natural thickening of the hydrocarbons comprising the bulk of the oil; something which can usually be corrected by the use of a solvent such as kerosene to reduce viscosity. The rest of the difficulty - the more serious part - arises from crystallization at low temperatures of the paraffin wax present in almost all lubricating oil fractions. Upon crystallization, this wax tends to form interlocking networks which adsorb oil and form a voluminous gel-like mass which restricts the flow or "pour" of the oil.

Pour point depressants are chemicals which modify the wax crystallization process in such a manner that the oil will pour at low temperatures. Although some monomeric compounds such as tetra (long chain alkyl) silicates, phenyl tristearyloxysilane, and pentaerythritol tetrastearate have been shown to be effective, all commercially important pour point depressants are polymers:

(i) Alkyl methacrylate polymers and copolymers
(ii) Vinyl carboxylate-dialkyl fumarate copolymers
(iii) Alpha-olefin polymers and copolymers
(iv) Friedel-Crafts condensation products of chlorinated wax and aromatic compounds such as naphthalene or phenol (some investigators do not class type (iv) products as polymers; others maintain that they are relatively low molecular weight polymers having a plurality of aromatic rings and paraffin wax radicals).

The molecular weight range of polymers effective as pour point depressants is generally below that of polymers used as V.I. improvers, and is usually in the area of 500 to 100,000.

10.3.6.1 Mode of Action of Pour Point Depressants

Pour point depressants function by adsorbing on or co-crystallizing with the precipitating wax, thus inhibiting lateral crystal growth. This promotes growth of smaller crystals than the platelets formed in the absence of pour point depressants. This change diminishes the ability of the wax crystals to overlap and interlock to form large conglomerates of wax which would impede the flow of the oil.

There is evidence that aliphatic polymers function by a co-crystallization mechanism, and that alkylaromatic types such as chlorinated paraffin wax-aromatic compound condensation products function by adsorption on the nascent wax crystal.

10.3.7 Extreme Pressure Additives

These additives, commonly called "E.P." agents, are chemicals which are added to lubricants to prevent destructive metal-to-metal contact during lubrication. Plain mineral oils provide good lubrication when a film of oil is maintained between the moving surfaces (hydrodynamic lubrication) but fail to provide adequate lubrication when pressure and rubbing speeds are such that the film of oil is squeezed or wiped out. The latter kind of lubrication, called "boundary lubrication", is governed largely by parameters of the contacting surfaces such as surface finish, metal shear strength and the coefficient of friction between the contacting metal surfaces. Unless such parameters can be chosen to meet expected pressures and rubbing needs, destructive metal-to-metal contact will take place.

10.3.7.1 Commercial E.P. Additives and their Application

Virtually all commercial E.P. additives are organic compounds that contain one or more elements or functions such as sulphur, halogen, phosphorus, carboxyl, or carboxylate salt which can react chemically with the metal surface under conditions of boundary lubrication. The ease with which an E.P.additive reacts with the metal surface, i.e. its "activity", determines to a large extent whether it would be used in a lubricant such as a cutting oil, a hypoid gear oil, hydraulic oil, or a steam turbine oil. An assignment of likely fields of application of commercial E.P.additives based on their relative activity is shown in Table 10.3

Table 10.3 Fields of Application of E.P. Additives

High Activity Additives	"Moderate" or Intermediate Activity Additives	"Mild" or Low Activity Additives
Straight cutting oils	Hypoid gear oils	Worm gear oils
Drawing compounds	(e.g. multi-purpose gear lubricants)	Spiral bevel gear oils
Metal-forming lubricants	Industrial gear oils	Manual gear box oils
Some hypoid gear oils	(e.g. "open gear" oils)	Motor oils
	Industrial gear oils for general application	Steam turbine oils
		Jet aircraft turbine oils
		Gas turbine oils
		Automatic transmission fluid
		Hydraulic oils (fire resistant emulsion type)
		Industrial gear oils for closed or circulating systems

10.3.7.2 Automotive E.P. Gear Oils

These oils are used to lubricate the worm, spiral bevel, or hypoid gear drives of automotive vehicles. Since hypoid gears require the greatest measure of E.P. protection of all commercial gear drives, most additive treatments are designed to give satisfactory performance in this environment.

Historically, E.P. additives for use in gear service fall in the following general categories. This information should not be interpreted as indicating that early treatments are obsolete; in fact, all of the listed treatments are in commercial use today, many on industrial applications. The more recently developed additives, however, enjoy most of the market.

Earliest successful treatment	- Lead soap (e.g. lead naphthenate plus an active or "corrosive" organic sulphur compound.
Late 1930's	- Chlorine and "Moderately active" sulphur present in the same or different organic molecules
World War II period	- Sulphur, chlorine and phosphorus in suitable organic carriers
Most recent types (1960-)	- Sulphur and phosphorus in suitable organic carriers

Typical E.P. additives used commercially in formulating gear lubricants have included:-

Chlorinated paraffin wax (40% - 60% chlorine)

Chlornaphtha xanthate (reaction product of chlorinated naphtha and alkali metal xanthate)

Chlorinated paraffin wax sulphides (reaction product of chlorinated paraffin wax and alkali metal sulphides)

Sulphurised fatty oils (e.g. sulphurised lard oil, sulphurised fish oil, sulphurised sperm oil)

Sulphurised hydrocarbons such as polybutenes

Sulphurised synthetic esters (e.g. sulphurised methyl oleate of fatty acids)

Sulphur chloride-treated fatty oils (e.g. S_2Cl_2 - treated fish oil)

Aliphatic and aromatic polysulphides (e.g. benzyl disulphide, chlorobenzyl disulphide, butyl disulphide)

Phosphosulphurized fatty oils (e.g. lard oil heated with sulphur and phosphorus pentasulphide)

Organic phosphites (obtained by treating alcohols with PCl_3)

Alkarylphosphates (octylphenol treated with P_2O_5)

Alkyl phosphates (alcohols teated with P_2O_5)
Lead naphthenate
Zinc and lead di-organo dithiophosphates
Zinc and lead di-alkyl dithiocarbamates

An evaluation of the practical effectiveness of a hypoid gear lubricant must be carried out in full-scale equipment in the laboratory and in the field. Bench test rigs have not been able to perform this task, although they are useful in screening likely candidates for full-scale evaluation. A list of these typical bench tests is given in Table 10.4.

10.3.7.3 Antiwear Additives

Although discussing primarily gear systems, the advent of high performance engines for passenger cars posed new lubrication problems. Engine inspection began to reveal unmistakable evidence of excessive wear and scuffing of valve train components; high rotational speeds and increasing pressures between cams and lifter foot surfaces had apparently combined to shift lubrication requirements into the boundary region. The first and perhaps most effective E.P. additive for controlling or eliminating wear and scuffing in the valve train area was found to be that versatile additive - zinc dialkyldithiophosphate.

Other additives found useful for the control of valve train wear include:

Tricresyl phosphate
Dilauryl phosphate
Didodecyl phosphite
Sulphurized terpenes
Sulphurized sperm oil
Chlorinated compounds
Zinc dialkyl dithiocarbamate

It has been mentioned earlier that the zinc diorganodithiophosphate plays a very important role in antiwear hydraulic oils and in mild EP gear oils. The choice of the correct type of compound depends upon the desired performance characteristics. A zinc dithiophosphate may be prepared using a variety of alcohols. Primary and secondary alkyl groups are used for applications such as gasoline engine and hydraulic systems and aromatics (substituted phenols) for diesel engines where the major role is as a high temperature antioxidant.

It is essential therefore that the performance of these materials is related to the application. Generally, good antiwear performance means low thermal stability, but a good antioxidant. For hydraulic system use the additive must be resistant to hydrolysis at temperatures typical of hydraulic circuits.

An indication of the major performance properties of a range of zinc diorganodithiophosphates is given in Table 10.5.

Table 10.4

Lubricant, Friction, Wear and Gear Test Machines

Test Machine	Type of Contact	Type of Measurement	Type of Loading	Range of Loading (kg)	Speed m/s	Other characteristics
Almen-Weiland	Conforming area, pin/bearing shells	Friction, wear, load-carrying	Mechanical steps	0-2000	0 - 0.2	Also used for corrosion test on specimens after running
Falex IP 241/77T	Multiple line/area	Friction, wear, load-carrying	Mechanical continuous	0-2000	0.1-0.25	Also for corrosion test
Shell 4-Ball IP 239/79 T	Multiple point hardened steel balls	Friction, wear, load-carrying	Dead weight steps	0-1800	1500 rpm	Also with different loads speed and specimens for wear tests. "Rolling" unit for bearing studies
Timken IP 240/76 T	Block and ring. Line	Friction, wear, load-carrying	Dead weight continuous or step	0-50	0-400 (0-800rpm)	Also for grease studies
Niemann-FZG IP 334/77 T	Gear. Two tooth forms. Pinion teeth 16. Wheel teeth 24. Spur/Case hardened	Load-carrying wear rate	Dead weight step, start under load	1600 'Normal' load	7.3	Also runs at 2175 r.p.m. Oil temperature 90°C
I.A.E. IP 166/77	Gear-one tooth form pinion teeth 16. Wheel 16. Spur/ Case hardened	Load-carrying	Dead weight step, start under load	Up to at least 70 Lever	7.9 4000 rpm	Also runs at 2000 and 6000 rpm Oil temperature 2000-60°C, 4000-70°C
Ryder FTMS 791a-6508	Gear-one tooth form pinion/wheel teeth 28	Load-carrying	Hydraulic applied when running	-	10000 rpm	Oil temperature 74°C Start up at no load

TABLE 10.5 Relative Performance of Zinc Dithiophosphates

Substrate	Performance				
Alcohol	Thermal STability	Anti-Oxidancy	Hydrolytic Stability	Anti-Wear	Bearing Protection
Secondary-1	4	2	2	2	2
Secondary-2	5	1	1	1	1
Primary-1	2	4	4	4	4
Primary-2	3	3	3	3	3
Aromatic-1	1	5	5	5	5
Aromatic-2	1	5	5	5	5

These additives are typical of those used for lubricants

Rating 5 = worst condition
1 = best condition

10.3.7.4 EP additives for Turbine Oils

Advances in the design of steam turbine, gas turbine, and jet aircraft turbine engines and their associated gearing have introduced problems of boundary lubrication. To solve these problems, the additive industry has developed E.P. additives of the "low activity" variety which provide protection against excessive wear and scuffing of turbine engine components. Such additives must withstand high temperatures - and in the case of steam turbines, water contamination - without promoting corrosion of turbine engine components. For use in gas turbine and jet aircraft turbine engines formulated with synthetic ester-base fluids different chemicals may be required.

For steam turbine oils additives based on phosphorus and hydrolytically stable chlorine containing chemicals have been used. Certain sulphur-phosphorus additives are also known to have been used. In this case, the sulphur is inactive towards copper at normal temperatures and becomes released only at high operating temperatures.

For gas turbine applications phosphorus compounds, both phosphites and phosphates, are used. In certain cases polyglycol fluids may act in the same way - presumably by strong adsorption of the hydroxy groups to metal surfaces.

10.3.7.5 EP Additives for Cutting Oils

Very high pressures and temperatures are developed locally between the work and the cutting tool in machining operations. The so-called "straight cutting oils" widely employed to cool and lubricate the work and cutting tool are mineral oils which have been blended with suitable E.P. additives. Since the lubrication of a metal-cutting operation is almost completely within the boundary region, E.P. additives of high activity are required for best results. In the case of an organic sulphur compound, its activity should be such that a dilute solution (e.g. 1 or 2 percent) of the compound in a mineral oil will completely blacken a copper strip within one hour at 100°C. Sulphur compounds having this degree of activity are known in the industry as "corrosive sulphur" additives and are used extensively in the compounding of commercial cutting oils. Examples of E.P. additives developed for use in cutting oils include sulphurized mineral oil, sulphurized fatty oils, sulphur chloride-treated fatty oils, sulphurized olefins, sulphur chloride-treated olefins, benzyl polysulphides, chlorinated paraffin wax, and chlorinated mineral oils.

However, metal working and forming operations require many types of chemical additive apart from the active or corrosive sulphur type. Generally, the active sulphur additives are used in metal removal operations. The list below indicates the increasing degree of severity:

(i) Turning and Milling
(ii) Drilling and Reaming
(iii) Tapping and Threading
(iv) Broaching

As very high temperatures are reached, the choice of suitable additive depends also on the material being machined, so that the best cooling, surface finish and tool life are retained.

The chemical activity of sulphur compounds may be chosen to give particular performance in cutting operations. When such materials are incorporated into fluids having a dual-purpose nature, such as machine lubrication, then the choice has to be made with the requirements of the hydraulic or circulating oil circuit metallurgy matched against the metal working operation.

Additives are also used in metal forming operations such as rolling, drawing, stamping and forging. For rolling oils the greatest requirement is high surface finish, which means that the additives must not be chemically active. Generally, fatty alcohols and esters are preferred.

Drawing and forging operations require highly specialised products. Fatty-based systems are used containing a variety of additives including solid lubricants such as graphite and molybdenum disulphide.

10.3.7.6 Mode of Action of E.P. Additives

E.P.additives function by reacting with relatively moving surfaces under boundary lubrication conditions to form an adherent film which has lower shear strength than that of the metal surfaces themselves. This film acts as a sort of solid lubricant, and takes over the task of lubrication when the oil is no longer able to provide a separating and protective lubricating film. With an appropriate E.P. additive, there is little or no formation of such "solid lubricant" under conditions of hydrodynamic lubrication. It forms only at the elevated temperatures which develop locally between metal surfaces under conditions of boundary lubrication. Because temperature has been shown to be the most influential parameter in the function of E.P. lubricants, F.P.Bowden and co-workers have suggested that the term "Extreme Temperature" lubricants and additives might be more appropriate.

10.3.8 Emulsifiers

Generally, an emulsifier is a chemical for dispersing either water in oil or oil in water. The former, the more difficult system, produces fire-resistant hydraulic fluids, rock drill lubricants, and many types of wire-drawing media. The emulsifier may be of the alkenyl succinimide type, fatty esters, or others containing fatty residues. For metal working and hydraulic media of the soluble oil (oil in water) type, a large range of emulsifiers is available. The choice depends upon whether it is desirable to have a chemical with ionic characteristics, a salt, or one with non-ionic properties such as esters, phenol ethers and other oxygenated chemicals. A short list is given below of some of the chemicals:

Sodium sulphonates
Tall oil amides
Ethanol amines
Quaternary Ammonium salts
Polyalkylene phenol ethers and associated oxygenated products
Ethoxylated fatty acids
Salts of fatty acids

Emulsifiers are classified according to a numerical value, the HLB number, obtained by estimating the emulsion stability when prepared in a standard way.

For oil-in-water soluble oil emulsions, the HLB range for emulsifiers is 12-15, for water-in-oil invert emulsions it is 4-6 when checked in a naphthenic base stock.

10.3.9 Friction Modifiers

Certain deficiencies are sometimes observed with lubricants with respect to their frictional characteristics. Where metal surfaces are designed to slide, two opposing requirements may occur:

(i) Smooth sliding with no vibration and minimum coefficient of friction.

(ii) No sliding maximum coefficient of friction for engagement of clutch surfaces or friction locking devices.

Additives for (i) are generally fatty-based, naturally-occurring products such as fatty esters and amides. They find use in sliding motion systems such as machine tool slideways.

In the case of type (ii), the clutch or friction mechanism should engage or disengage smoothly without vibration. Two types of application are difficult to control. A choice of different products may be made:-

(i) Anti-chatter in limited slip axles, or other metal/metal friction locking units. It is necessary to reduce the stick-slip action of lubricated steel/steel contacts.

Amide - metal dithiophosphate combinations
Amine dithiophosphates

(ii) Anti-squawk additives reduce vibration which gives rise to audible noise in clutches of dissimilar material (bronze on steel, asbestos on steel, etc.). Chemicals used for this purpose include:

N-acylsarcosines and derivatives
Sulphurised fats and esters
Organophosphorus acid and fatty acid mixtures
Esters of dimerised fatty acids

Formulations which meet the complex performance desired in automatic transmission fluids, universal tractor engine-transmission oils or machine tool systems are the result of careful matching of friction modifiers in the whole additive system.

10.4 CONCLUSIONS

The chapter will serve to indicate the complex nature of the chemicals used in lubricant formulations. Mainly, the discussion has been about additives for mineral lubricating oils although there has been reference made to synthetic fluids and water containing hydraulic and metal working fluids.

In formulating lubricants of any type, the presence of these additives requires that not only should each individual product carry out its desired function but that interactions between more than one chemical must not cause deterioration of performance. Neither must mixtures react together to form oil insoluble by-products. The final oil formulation must be stable over the range of operating and storage temperatures.

Mixtures of additives may be formulated and used as a package at a given treatment level, requiring only the carrier fluid. Again, there must be no deterioration after manufacture or storage and the combination must be selected with great care so that maximum performance is achieved at an economic cost.

The author wishes to acknowledge the assistance of his colleagues in the Lubrizol Corporation for reviewing this chapter and suggesting additional data;

also to the Lubrizol Corporation for permission to compile the information.

REFERENCES

1 Smallheer and Kennedy-Smith, Lubricant Additives, The Lezius-Hiles Co., Cleveland, Ohio.
2 C.V.Smallheer, Lectures on Lubricant Additives, Imperial College, London, March 1970.

11 CONSUMPTION AND CONSERVATION OF LUBRICANTS

A.R. LANSDOWN
Director, Swansea Tribology Centre, U.K.

11.1 CONSUMPTION

Apart from the relatively small quantities of vegetable and animal oils, almost all modern lubricants are derived from petroleum, either by fractional distillation or chemical conversion. It is now generally recognised that the world's remaining supplies of petroleum have only a limited life.

The various estimates of the remaining life of petroleum reserves depend on many assumptions. The rate of future consumption is itself dependent on several assumptions, such as the availability and competitiveness of alternative fuels, improvements in efficiency of utilisation, and fiscal incentives. The estimates of remaining reserves also depend on several assumptions, such as the rate of discovering new reserves and the potential for using low-grade sources. For energy purposes this last factor reaches its limit when the energy required to exploit a source is equal to the energy ultimately obtained from the source.

There is a general tendency to discuss the future of petroleum purely from the energy standpoint, and to relate it to the phasing in of alternative energy sources. There are however many products from petroleum which will be less readily replaced from non-petroleum sources, and lubricants are in this category.

A further limitation on future lubricant supplies is that not all crude oils can be used to produce lubricants without expensive, and energy-expensive, chemical processing. The United Kingdom's North Sea oil is in fact generally unsuitable for lubricant manufacture.

We are thus faced with the situation that within the working life of some of our younger lubricant technologists petroleum-based lubricants will cease to be plentiful and may become extremely expensive. Conservation of lubricating oils should therefore already be a matter of serious concern, and will inevitably be recognised as such within one, or at most two, decades.

Table 11.1 shows a breakdown of the total United Kingdom consumption of petroleum for the calendar year 1977, and it can be seen that lubricating oils and greases represent just over one million metric tons, or 1.17% of the total.

Accurate figures for the breakdown of lubricant types are more difficult to obtain, but Table 11.2 gives estimates of U.K. lubricant consumption in some of the more important categories. The biggest single category consists of automotive engine oils, comprising perhaps one third of the total lubricant consumption.

Table 11.1 United Kingdom Consumption of Petroleum Products 1977
(Figures from Institute of Petroleum "Petroleum Statistics")

Product category	Consumption (thousand tonnes)
Motor spirit	17,336
Gas, diesel and fuel oils	47,920
Aviation fuels	4,218
Other fuels	9,639
Naphtha/Light distillate	5,179
Lubricating oils and greases	1,029
Other products	2,835
Total	88,156

Table 11.2 Estimated Lubricant Consumption by Types

End Use	Type	Consumption (%)
Aviation		0.4
Marine		8.5
Tractor		3.0
Motor	Oils	35.0
Motor	Greases	0.1
Industrial	Hydraulic oils	16.0
Industrial	Bearing oils	22.5
Industrial	Metalworking	5.5
Industrial	Process oils	5.5
Industrial	I.C.E. oils	1.5
Industrial	Greases	1.0
Industrial	Fuel as lubricant	1.0

Overall lubricants account for only a little over 1% of the total petroleum consumption, but the importance of this 1% is increased by the two factors previously mentioned, namely the greater difficulty of finding substitute sources for lubricants and the limited range of crude oils which can be used to produce lubricants.

The ultimate fate of the various lubricants is also difficult to assess accurately. Table 11.3 shows some estimates made in 1973 of the fate of automotive engine oils. From the conservation point of view there are two interesting aspects of these estimates. The first is that some 185,000 tons were apparently potential polluters of land and water. The second is that some 258,000 tons could theoretically be reclaimed.

An article published in the AA "Drive" magazine in January, 1974 suggested that the 58,000 tons of engine oils changed each year by motorists at home was possibly disposed of as shown in Table 11.4. The larger quantity changed in garages was, however, generally disposed of more legally.

Since 1974 there have been more detailed surveys, but the increasing awareness of the need for conservation, has also led to a tendency to longer oil-change periods and to a greater contribution of re-refining to the disposal problem.

Both of these factors, reduction in consumption and re-refining, will inevitably assume greater importance in coming years.

Table 11.3 Estimates of the Fate of Motor Oils

Burned (exhaust)	25%	95,000 tons
Changed (garages)	50%	190,000 tons
Changed (home)	15%	58,000 tons
Leaked	5%	19,000 tons
Scrapped with vehicle	1%	4,000 tons
Spilled	1%	4,000 tons
Railway (?)	1%	5,000 tons

Table 11.4 Estimates of Oil Disposal by "Do-it-yourself" Car Owners

Buried in garden	20%
Poured down drains	8%
Burned	18%
Taken to garage etc. for proper disposal	17%
Otherwise disposed of	37%

11.2 REDUCING CONSUMPTION

There are of course purely mechanical ways to reduce consumption, such as eliminating leakage and improving serviceability of engines. Improvement may sometimes also be possible by the use of a more viscous oil, but this must obviously be done with caution. Not only must care be used to ensure that a suitable viscosity is used, but as a general rule the use of a higher viscosity oil will mean higher power consumption, and the resulting energy wastage will probably more than offset any lubricant saving.

The best prospect for reducing consumption lies in ensuring that oil changes are not carried out any more frequently than they need to be.

Even in identical systems the rate of oil degradation can vary considerably. The following are some of the factors which lead to rapid degradation.

Dusty or dirty environments
High temperatures
Very low temperatures
Temperature fluctuations, leading to condensation
High altitude
Poor filter maintenance
Low oil levels
Frequent stop-start operation
Short journey lengths in a vehicle
Contamination by chemicals
Contamination by unburned fuel
Contamination by combustion products
Contamination by wear debris

Within the oil formulation the following factors can also lead to rapid degradation.

Poor quality base oil, containing unstable molecules
Inadequate anti-oxidant content
Insufficient dispersant or detergent additives
Insufficient anti-wear or EP additives
Insufficient basic additives with sulphur-containing fuel
Excessively reactive or unstable additives

Because of all these variables, the required oil change period in two identical engines may vary by a factor of ten in different operating conditions.

In an era of relatively cheap lubricants, recommended oil change periods will tend to be short enough to ensure satisfactory quality in the worst set of conditions, because the economic penalty from using an oil fill too long will be far greater than the value of the oil saved.

The solution is oil quality monitoring, which enables the oil remaining in an individual system to be assessed periodically, and only changed when its condition is approaching an unsatisfactory level. As a bonus, oil monitoring will also give valuable information about the condition of the engine or other lubricated system.

Oil monitoring is an important part of machinery health monitoring, which is described in detail in Chapter 18. It is therefore unnecessary to describe the various techniques at length here, but there are two aspects which should be mentioned.

The first of these is the need to tailor the monitoring techniques to the size and importance of the oil system being monitored. A large critical system will justify the effort and cost involved in frequent spectrographic oil analysis or ferrography. Even a small system, such as a car engine, may well justify the use of a magnetic plug to monitor wear debris, examination of a drop of oil from the dipstick by the filter paper technique, or viscosity by the Flostick method.

The second factor is the need to establish for any system a criterion for deciding when to change the oil. This may be a level of acidity in the Total Acid Number, a degree of viscosity change, or the visible appearance of contaminants on a filter-paper, but an objective criterion or criteria will be essential if oil change is to be determined by oil quality. It is a salutary thought that the incidence of inflight engine failures in certain United States Air Force aircraft decreased when not only routine oil change, but routine engine overhaul periods were discontinued in favour of scheduling in accordance with spectroscopic oil analysis.

11.3 RECLAMATION AND RE-REFINING

Even when a lubricant has deteriorated so that it is no longer fit for service, the greater part of it will still be unchanged. Much of the degradation is by contamination, while a further factor is depletion of additives. Only a very small proportion of the base oil will usually have been degraded, and this will consist of the most unstable molecules, usually oxidised to aldehydes, ketones or carboxylic acids.

It is theoretically possible to remove all the contaminants and the degraded additives and base oil molecules, to add fresh additives, and thus to produce a lubricant which differs little if at all from the original. The processes used are described as reclamation or re-refining, depending on the extent of treatment involved.

In the simplest case the use of an in-line filter to remove solid contaminants is a reclamation technique. Another example is that of a transformer oil, which degrades in service to generate a small but unacceptable level of of electrical conductivity and can be cleaned up in situ to recover the required insulation level. Such techniques are commonly known as "laundering" and have been widely used for many years.

A well-established system for more intensive re-refining is the acid-clay system, which basically consists of the five components shown in Fig.1. The strainer at the inlet to the waste oil storage tank removes major solid contaminants, while water and sludge are drained off from the bottom. Contaminating fuel or other volatile materials are removed in the steam stripper. The clarified oil from the intermediate storage is treated with hot sulphuric acid, which reacts with most of the reactive compounds present and removes them as acid sludge. Treatment with heated active "earth" or "clay" removes the remaining polar compounds, and the solids are finally removed in a filter-press. The cleaned oil obtained by this process can be made suitable for use as lubricant base oil, but is often blended into heavy fuel oils.

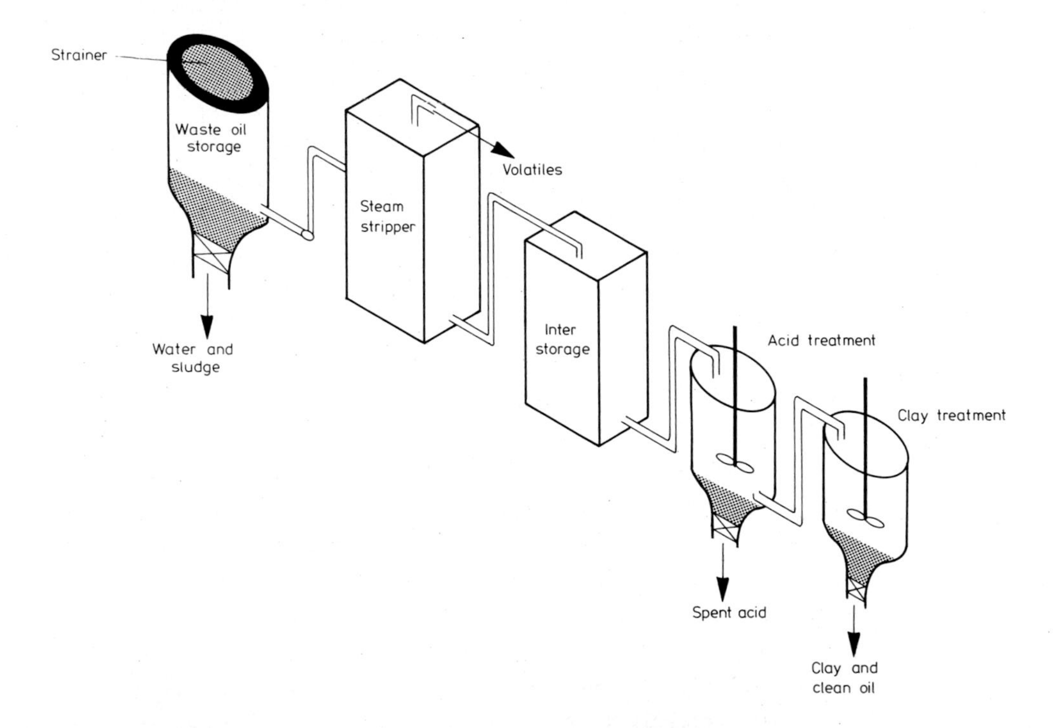

Figure 1 Acid-clay re-refining process

The main disadvantage of this process is the problem of disposing of large amounts of acid and acid clay. An alternative process developed by the Institute Francais du Petrole uses liquid propane precipitation to remove the degraded materials and residual additives. This considerably reduces the generation of waste by-products, but at present the process is probably not economic.

A Matthys process consists essentially of two distillation stages. The first is at atmospheric pressure and 180°C and removes water and light hydrocarbons. The second takes place at 340°-360°C and is designed so that the undesirable materials form a coke which is carried off in the liquid products and removed by centrifuging.

Most re-refining processes require a finishing stage to produce fractions for re-use as lubricants and this may be clay treatment or a conventional hydrofinishing.

The recovery of oil from dilute emulsions or oily plant waste-water is difficult, but is important not only for oil conservation but for elimination of pollution. The standard procedure is to crack emulsions by chemical treatment and allow the product to settle in holding tanks until the oil can be skimmed off. The oil fraction then tends to contain a high proportion of dirt, water and the treatment chemical. It is likely to be uneconomical for re-refining and is often blended into burner fuel.

In recent years, techniques have been developed which use polymeric membranes to filter out particles and droplets from the water, either by direct ultrafiltration or by reverse osmosis. These techniques are reported to give a much cleaner oil fraction as well as a water fraction suitable for discharge to sewers.

One problem associated with re-refining of engine oils is that there may be a slight build-up of polynuclear aromatics, which are carcinogenic in higher concentrations. It may therefore be desirable to treat re-refined oils by a finishing process such as solvent extraction which will remove polynuclear aromatics.

11.4 ECONOMICS

The economics of optimising oil-change periods depends mainly on tailoring the monitoring techniques to the size and importance of the system. Systems which are either very large or of critical importance are already being monitored by sophisticated techniques. Smaller and less critical systems may already justify simple monitoring techniques, and as lubricant availability decreases, the balance will inevitably shift in favour of increased levels of monitoring.

The economics of re-refining also depends on availability of lubricating oil. During the second World War large quantities of lubricating oils were re-refined in most countries, but the industry declined in the following thirty years. The extent of the decline varied in different countries, but nowhere did it completely cease. The reason for the decline may have been partly psychological, in that re-refined oils were considered to be inferior in some way, but the main reason was probably financial. The economics of re-refining did not permit more than a token reduction in retail prices of re-refined oils.

It has been possible for many years to produce re-refined oils of similar quality to new oils, but in a period of relative affluence and plenty most people have been happier to pay the marginally higher price for "new" oil.

The main economic problem has been the cost of collecting and transporting waste oil, but a second factor has been the difficulty of persuading operators to keep wastes of different qualities clean and separate.

In the Federal Republic of Germany a deliberate move was made to reduce lubricant consumption and encourage re-refining by the introduction in 1968 of the "Law on Measures to Ensure the Disposal of Waste Oil". This law enabled organisations disposing of waste oils by approved methods to claim an allowance to cover costs involved in disposal. The approved methods included re-refining. The funds were raised by a special duty on imported lubricant grade oil, and this had the effect of introducing a price margin in favour of re-refined oils.

As a result, it was estimated in 1973 that up to 30% of total lubricating oil consumption was being re-refined in Germany, compared with only 5% in Britain.

With declining availability of petroleum lubricants, the proportion of re-refined lubricants will probably approach more and more the theoretical limit of 60-70%.

12 HEALTH AND SAFETY ASPECTS OF LUBRICANTS

A.R. EYRES, M.A., M.Sc.
Environmental Health & Product Safety Advisor, Mobil Europe Inc.

12.1 INTRODUCTION

The majority of mineral oil based lubricants and greases are relatively harmless to man. Their use normally involves no unusual hazards provided that reasonable care is taken to avoid excessive skin contact or inhalation of mists and vapours. A small number of products may, because of compositional requirements to meet particular technical performance needs, present a higher degree of hazard. Because lubricants are mainly composed of organic chemicals which have some solvency for natural components of the skin, it is unlikely that provision of completely safe products could be possible. Synthetic lubricants generally are similar in hazard potential to mineral oil products. In order to review the hazards of lubricants, it is first of all necessary to look briefly at their composition.

12.2 COMPOSITION OF LUBRICANTS

Mineral oil based lubricants are prepared from base oils manufactured from naturally occurring crude petroleum oils. Crude oils occur in many parts of the world and their composition varies according to source. In addition to complex mixtures of paraffinic, isoparaffinic, naphthenic (cycloparaffinic) and aromatic hydrocarbons, some compounds of sulphur, oxygen and nitrogen will be present plus traces of a number of metals. Distillation of crude oil, followed by various other refining processes such as solvent extraction, hydrogenation or acid treatment yields various fractions in the broad categories shown in Table 12.1.

A lubricating oil fraction typically contains several thousand individual hydrocarbon compounds. A base stock is usually described as paraffinic or naphthenic depending on the predominant type of hydrocarbon compound present. This is a function of the crude source. Crude oil also normally contains polycyclic aromatic hydrocarbons, some of which (4 to 6 ring compounds) are known to be carcinogenic, eg. benz(a) pyrene. Some refining processes, such as solvent or severe acid treatment, remove most of these polycyclic aromatic compounds so that very few are present in the final lubricant base stock. However,

Table 12.1

	Molecular Size	Boiling Range, °C
Gases	C_1 - C_4	≦ 0
Light Naphtha	C_3 - C_8	0 - 100
Gasoline	C_4 - C_{12}	20 - 200
Heavy Naphtha/Kerosine	C_9 - C_{16}	100 - 270
Gas Oils	C_{10}- C_{26}	170 - 400
Lubricating Oils	C_{17} upwards	> 300
Residuum	C_{26} upwards	

it should be noted that the polycyclic compounds are present in the extracted material. Because of their boiling points, the 4 to 6 ring polycyclic compounds are not normally present in distillate fractions boiling below about 370°C.

To obtain lubricant performance characteristics which could not be provided by the base oils themselves, various additives such as antioxidants, detergents/dispersants, emulsifiers, biocides, anti-corrosives and anti-wear/extreme pressure compounds are incorporated. Potential health hazards of such additives therefore also need to be considered.

For special applications, a variety of synthetic lubricants have been developed. These are based on synthetic hydrocarbons and esters, polyglycols, silicones and phosphate esters. Additives of similar types to those used in mineral oil based lubricants are generally incorporated.

12.3 MINERAL BASE OIL FACTORS

12.3.1 Acute Toxicity

All types of mineral oil base stocks have a low order of acute (short-term) toxicity. Acute oral LD_{50}'s (the dose to test animals resulting in 50 per cent mortality) for the rat are well above 10g per kg of body weight. Extrapolated to man this equates to ingestion of more than one litre of oil to cause death. Low viscosity products such as kerosine present a somewhat greater hazard because of the danger of aspiration into the lungs, followed by pneumonitis, if vomiting occurs.

Mineral oils are also essentially non-toxic by absorption through the skin. Dermal LD_{50}'s for the rabbit are well above 10g per kg body weight, a level generally considered as harmless.

12.3.2 Dermatitis

Dermatitis is undoubtedly the major potential health problem with mineral oil products, resulting from repeated or prolonged skin contact and inadequate skin care. Primary irritation and defatting of the skin can occur to varying extents, depending on the type of product and the degree of exposure. The lighter petroleum oils with final boiling points below about 350°C, for example, kerosine, tend to be direct skin irritants. Since the chemical nature of mineral oil ensures that there will be some solvency effect on the natural fats of the skin it is unlikely that potential problems can be totally eliminated by product development. However, adoption of simple measures to prevent repeated and prolonged contact, together with good personal hygiene practices and care of the skin, can virtually eliminate dermatitis problems. Dermatitis from lubricants is an avoidable occupational disease in the vast majority of cases.

12.3.3 Oil Mist

Although it has often been suggested that inhalation of oil mist over extended periods may lead to an increased risk of lung cancer, evidence from animal testing and exposed human populations does not support this. For example, animal exposures to white oil mist at 5 and 100 mg/m^3 showed no indications of increased lung cancer risk. In further work, no injury or indisposition from inhalation of oil mist was observed amongst animals exposed for 18 months to mist from a sulphurised solvent-extracted naphthenic base oil at 50 mg per cubic meter [1,2,3]. Decoufle of the U.S. National Cancer Institute [4,5] has published two epidemiological studies of the cancer mortality of workers exposed to cutting oil mists. The results indicated that exposure to oil mists does not pose a hazard in terms of respiratory cancer and fatal non-malignant respiratory disease, but may be associated with a slight increase in cancer of the gastrointestinal system. A study of workers in metal machining plants in Germany by Draschs [6] showed no adverse respiratory effects among 443 employees in 17 factories. About 63 per cent of these employees were reported to have been exposed to high oil mist levels in the range 40 to 80 mg/m^3. Comparison of smokers and non-smokers in exposed and non-exposed populations suggested in fact that inhalation of oil mist may provide some protective effect against the harmful effects of smoking on the bronchial system.

The current Threshold Limit Value (the atmospheric concentration to which it is believed most workers can be exposed for 8 hours daily without adverse effects on health) published by the American Conference of Gernmental Industrial Hygienists is 5 mg/m^3. This has been set on the basis of preventing nuisance and unpleasantness to workers rather than on health effect aspects and it is essentially an index of good industrial hygiene practices. It is believed that the 5 mg/m^3 limit provides a safety factor of at least ten against even relatively minor changes in the lungs.

As with many chemicals, inhalation of very high concentrations of oil mists (or vapours of more volatile products such as kerosine) may cause irritation of the lungs and may lead to a chemical pneumonia.

12.3.4 Oil Vapours

At normal temperatures, typical lubricating oils do not produce any significant levels of vapour in the working atmosphere. Saturated vapour concentrations for a typical lubricating oil have been calculated by Sanderson [7] to be 0.016 ppm at 20°C and 1.4 ppm at 100°C - these are so low that any adverse health effects will not occur. Lighter products such as diesel oil and kerosine do have the potential for evaporation of light ends to produce significant vapour concentrations. For example, the saturated vapour concentration for kerosine at 20°C is about 2000 ppm. These lighter products may produce irritation of mucous membranes. In metal machining operations, significant vaporisation of lubricating oils may occur at the high tool/workpiece temperatures, but on cooling in the surrounding atmosphere, this vapour will be condensed to droplets of oil mist.

12.3.5 Skin Cancer

It has been known for many years that some types of mineral oil can cause skin cancer with repeated and prolonged exposure over long periods of years. It should be noted however that skin cancer is normally less serious than other forms of cancer and is usually curable with early treatment.

In 1922, Leitch [8] reported that Scottish shale oil distillates caused cancer of the skin when painted on to animals. It is not possible to review all of the extensive studies carried out since that time, but among the most important were those reported in 1966 by Bingham and Horton [9], sponsored by the American Petroleum Institute. It was shown that base stocks prepared by solvent refining, which removes polycyclic aromatic hydrocarbons (PCAH), did not cause tumours in mice whereas typical acid refined base stocks did. In 1968, the UK Medical Research Council published a report "The Carcinogenic Action of Mineral Oils: A Chemical and Biological Study" [10]. The salient conclusions from this report and other studies can be summarised as follows:

- Some fractions of certain crudes from which lubricating oils are refined have been shown to produce tumours on skins of rabbits or mice.
- Solvent extraction methods of refining which remove aromatic compounds markedly reduce the carcinogenic activity of the refined lubricants.

- Carcinogenicity of mineral oils appears to be related to the presence of PCAH's, some of which are known to be carcinogens.
- It has not been possible to define any simple analytical parameters which correlate with carcinogenic activity.

An Ad-hoc Committee of the UK Institute of Petroleum studied all available evidence and advised member companies in 1968 that oils which have been solvent refined or treated in other ways to appreciably reduce the content of polycyclic aromatic compounds, were less likely to promote skin cancer than oils which had not been thus refined.

The difficulty of defining analytical parameters to correlate with carcinogenic activity is illustrated by results reported by Scala [11], shown in Table 12.2.

Table 12.2

OIL	A	B	C
Pyrene, ppm	3.5	18.3	4.4
Benz (a) anthracene, ppm	6.6	7.9	2.7
Benz (a) pyrene, ppm	4.5	1.2	0.2
% C_A (carbon in aromatic rings)	15.4	15.2	12.4
Cancer activity	+	-	+

The reason for this is the unpredictable effects of cocarcinogens, inhibitors and accelerators. For example, it has been shown that sulphur and some organo-sulphur compounds can increase carcinogenicity, as also can certain types of hydrocarbons.

So far, despite extensive research, no entirely satisfactory method for evaluating the potential carcinogenicity of an oil has been devised. The best method to date is the long term painting of mice skins. Such tests take two years to complete and require high standards of experimental techniques; they are therefore very costly. There is also the difficulty of translating results of animal tests to the human exposure situation. It is therefore impossible to test each individual formulation. However, a large number of tests have been done and these form the basis for current recommendations to minimise hazards.

Cases of occupational scrotal cancer have been reported from many countries including France, Sweden, the UK and the USA. The incidence in the UK is equivalent to about 5 cases per million males per year whereas in Sweden it is lower, at about 1 case per million males per year. Wahlberg [12] has reported that only 21 per cent of Swedish cases had had definite exposure to mineral oil.

This contrasts markedly with a figure of 86 per cent reported for UK cases. No obvious reason for this difference has been identified, but factors such as plant and personal hygiene may well be significant. As these are improved, incidence of scrotal cancer can be expected to decrease eventually. The long latent period from first exposure to diagnosis of the cancer (10 to 43 years) means that any changes in incidence cannot be observed quickly.

12.3.6 Eye Irritation

In common with a multitude of commonly used materials, eg. soapy water, many lubricants may cause some irritation if splashes enter the eye. With the majority of products this will not be more than very slight, but some such as neat soluble oils, may, because they contain appreciable amounts of surface active materials such as soaps, be somewhat more irritating.

12.4 ADDITIVE FACTORS

Before use in lubricants, additives are screened for toxicity and skin or eye irritancy. This is essential to assess potential hazards and determine any required handling precautions during blending of the finished lubricant. If it appears likely that an additive may lead to any increased hazard in the blended lubricant, further testing may be done to define this. The extent of possible increased hazard in a finished formulation must be assessed to decide whether or not the additive should be rejected.

In the majority of products, additives are minor ingredients and the potential hazards are essentially those associated with the base mineral oil. Where additive contents are higher, eg. in neat soluble oils and in some engine lubricants, the final product may be more irritating than the base mineral oil. If use of these additives is essential for technical performance reasons, the hazards must be controlled by the implementation of adequate handling and use precautions.

A wide range of additives is used to achieve improvements in lubricant performance. For each additive type, various chemical compounds have been found to be effective. Very few of these have been found to present any significant potential hazards. Among those which have been the cause of some concern are lead compounds, ortho isomers of phosphate esters, chlorinated naphthalenes, sodium nitrite in combination with amines, sodium mercaptobenzothiazole and trichloroethylene.

12.4.1 Lead Compounds

One of the early effective methods of improving antiwear and extreme pressure properties of a lubricant was to incorporate lead soap. Although there are no reported cases of sufficient lead absorption to cause adverse health effects, Van Peteghem and Vos [13] reported slight increases in blood lead levels in steel mill employees with frequent or prolonged skin contact with this type of lubricant. New additive technology has enabled this type of formulation to be largely replaced over the last few years by unleaded lubricants.

12.4.2 Ortho Phosphates

Absorption of ortho tricresyl phosphate has been shown to cause central nervous system damage leading to neuromuscular problems and various stages of paralysis. The para isomer does not have this effect and is essentially inert providing the content of ortho isomer is at a very low level. Suppliers of tricresyl phosphates have for many years restricted the ortho content to less than 1 per cent in order to avoid the possibility of central nervous system effects.

12.4.3 Chlorinated Naphthalenes

Chlorinated naphthalenes were used for a short period many years ago as effective extreme pressure additives in cutting oils. Their use was discontinued when an association with chloracne of the skin was found. The types of chlorinated additive now used, eg. chlorinated paraffins, do not cause this effect.

12.4.4 Sodium Nitrite and Amines

Sodium nitrite in combination with tri and diethanolamines has been used for many years to provide satisfactory anticorrosion properties in aqueous grinding fluids, and, at lower concentrations, in some soluble cutting oils. Recently, small amounts of nitrosamines, a type of chemical of which some are known to be carcinogenic, have been found in both concentrates and diluted versions of such products [14,15]. Nitrosamines are also found in many foods, drinks and cosmetics, are present in the atmosphere of city streets and are also formed within the body itself. Assessment of any increased hazard from grinding fluids is difficult because of the problems of estimating exposure and extent of absorption into the body. Taking the worst possible case, it appears that absorption could approach that from foods, but typically it is likely to be very much less than this. As a precaution however, suppliers have eliminated the combination of nitrite and amine except in some critical applications where it has not yet been possible to meet the technical requirements with alternative formulations.

12.4.5 Sodium Mercaptobenzothiazole

This compound was used at one time as a very effective corrosion inhibitor in aqueous lubricants. It was later identified as a strong skin sensitizer and is no longer used in applications where skin contact is likely.

12.4.6 Trichloroethylene

Trichloroethylene was used widely as a non-flammable diluent in open gear lubricants to enable a highly viscous lubricant film to be easily applied to gear teeth surfaces. Exposure to trichloroethylene vapour above the time weighted average TLV of 100 ppm may be hazardous, causing depression of the central nervous system with visual disturbances and lack of co-ordination, plus the possibility of damage to the liver and kidneys. Addiction to sniffing trichloroethylene vapours has also occurred. Such exposures on a continuing basis are unlikely to occur except perhaps in confined and poorly ventilated spaces. However, to provide a greater margin of safety, trichloroethylene can be substituted with a slightly more expensive safer alternative, 1.1.1 trichloroethane, which has a TLV of 350 ppm.

It has also been indicated that trichloroethylene is a carcinogen in animal tests. However, the doses given by ingestion into the stomach were so massive compared to possible human exposure that the significance of these test results can be seriously questioned.

12.5 BACTERIA AND BIOCIDES

Water based lubricants and mineral oil lubricants contaminated with water, eg. marine engine oils, can support the growth of bacteria, yeasts and fungi. Growth does not normally occur in products which do not contain water. As supplied to users, products are normally free of bacteria, but contamination occurs from a number of possible sources such as water from engine cooling systems, the water used as diluent, residual bacteria in plant circulation systems, refuse such as cigarettes thrown into the coolant, employees spitting into the product or even from bacterial contamination of the air in the plant.

The bacteria, yeasts or fungi which grow in aqueous coolants or lubricants contaminated with water are not normally harmful to humans. Although concern has been expressed that bacterial contamination may lead to increased respiratory or skin infections, industrial medical advisers responsible for large metal machining plants have reported that they can find no evidence for this. The types of bacteria found are almost invariably the non-pathogenic type which are harmless to humans although very occasionally a potentially harmful pathogenic type may be identified.

For technical reasons, it is desirable to control bacterial growth. With aqueous coolants, this is achieved either by inclusion of small amounts of biocides in the original product or by addition of biocides during use. By their nature, biocides are moderately to highly toxic by ingestion and may be skin or eye irritants in the concentrated form. Therefore, their use must be carefully controlled to avoid increased health hazards. In the concentrations normally used, and provided appropriate handling precautions are observed, biocides should present no hazard to health. However, use of excessive concentrations in "topping-up" may cause skin irritation. With marine engine oils, bacterial problems can be controlled by correct oil selection and appropriate operating procedures [16].

12.6 SYNTHETIC LUBRICANTS

A variety of chemical types are used as synthetic lubricants to meet operational requirements which cannot be satisfied adequately with mineral oil products. Types include various esters of organic fatty acids, silicones, synthetic hydrocarbons such as polyolefines, polyglycols and phosphate esters. There are no unusual health hazards associated with these. Defatting of the skin, similar to that with mineral oil, is possible in most cases if repeated or prolonged contact occurs. In the case of phosphate esters, the use of the ortho isomer should be avoided as indicated in 12.4.2.

12.7 USED AND RECLAIMED OR RE-REFINED OILS

There is strong evidence that PCAH content of mineral oil based lubricants increases during use [17]. The extent of the increase appears to depend on the type of application, being up to about ten-fold for cutting oils and diesel engine oils, but perhaps one hundred-fold or more for gasoline engine oils and quenching oils. Non-engine industrial lubricants such as hydraulic, gear and bearing oils would not be expected to show any significant PCAH increase during use because of the limited temperature increases to which they are subjected. Much of the increase in engine oils appears to arise from gasoline combustion products.

The significance of these increases in PCAH content in relation to any increased skin cancer risk is not clear at present for the reasons discussed in 12.3.5. In the case of cutting oils, calculations based on oil mist concentrations at the TLV of 5 mg/m^3 indicate that PCAH levels will be of the same order of magnitude as background atmospheric levels. Provided oil mist levels are maintained below the TLV, it appears there should be no significant increase in risk.

For environmental conservation reasons, there are attractions in reclaiming or re-refining used lubricants for further use. In specific situations there may also be economic justification for reclamation. A variety of processes may be employed from simple centrifuging and earth filtration to acid treatment, redistillation and solvent extraction. Until further information is available on the potential hazards, it is considered it would be prudent to limit use of such oils to applications in which there is little skin contact unless it is certain that the oil has only been used where PCAH increase is unlikely, or has been treated by a process which will remove PCAH, eg. solvent extraction of aromatics.

Used gasoline engine oils can contain up to about 1 per cent of lead [18]. This originates mainly from lead additives in gasoline with perhaps a minor contribution from wear of engine parts. Repeated or prolonged skin contact with these oils may result in some increased absorption of lead into the body.

Used cutting oils usually contain small metal chips or swarf which present an additional hazard to the skin. Many of these metal particles are needle-like in shape and can cause micro-lesions of the skin, leading to a general irritation. It is also believed that entry of swarf into the skin may destroy an electro-negative barrier beneath the surface over about one square inch around the site, thus allowing other materials such as the cutting fluid to penetrate into the skin.

12.8 HEALTH AND SAFETY PRECAUTIONS

12.8.1 Supplier's Responsibilities

Section 6 of the Health and Safety at Work Act etc. 1974 places responsibilities on suppliers "to ensure so far as is reasonably practicable, that the substance is safe and without risks to health when properly used" and to make available information on any relevant tests and "about any conditions necessary to ensure that it will be safe and without risks to health when properly used." Suppliers are also required to eliminate or minimise risks to health and safety, as far as is reasonably practicable. Suppliers cannot be expected to produce completely safe products; indeed, it can be said that there is no such thing - there are only safe ways of using a product. Even pure water can be harmful if one drinks too much of it and many people will suffer skin problems if their hands are immersed in water for several hours daily.

For many years, reputable lubricant suppliers have been assessing the potential hazards of products. Of particular importance is the assessment of toxicity of possible additives and rejection of those which may lead to significantly increased risk. As indicated in Section 12.4, a number of additives which may present hazards under some conditions of use have been replaced in recent years by alternative safer materials. Care must always be exercised however, to ensure that a material with a known risk is not replaced by a new material with unknown risk which may in fact be much greater.

The Health and Safety at Work Act does not require suppliers to disclose details of product compositions to users. Whilst most suppliers will make broad compositional information generally available, or even detailed information on a confidential basis to a user health professional where necessary, formulations are proprietary information to the individual manufacturer in a competitive business world. Considerable research expenditure may have been incurred in developing new additives and in selecting the best combination of additives to provide a performance benefit in a particular application. Suppliers are naturally reluctant to increase the possibility that this information may pass to their competitors. In any case, such information is usually of little value to the user in assessing potential hazards. Of much greater value is the toxicological information and the recommended handling precautions based on this information.

Most suppliers also provide users with booklets or leaflets recommending safe handling precautions and reviewing potential health hazards. Typical recommendations, together with practices of large user companies, have been reviewed by the Institute of Petroleum and incorporated in a Code of Practice for Metalworking Fluids, published in July 1978 [19].

12.8.2 Skin Protection

The basic requirement for avoiding skin problems is to minimise contact. With the majority of lubricants, occasional skin contact for short periods will cause no problems. Compliance with the normal recommendation to avoid prolonged or repeated skin contact will be sufficient to prevent dermatitis and skin irritation problems. The precautions needed to avoid dermatitis and skin irritation will also prevent skin cancer. Modern solvent refined types of mineral oils (or equivalent) also minimise the risk of skin cancer. With the more irritant types of product, such as kerosine and neat soluble oils, occasional very short contact is unlikely to cause problems, but if any appreciable contact is likely, suitable protective measures should be employed.

Contact can be minimised by using suitable protective gloves and clothing, barrier creams, and the proper installation and use of splash guards on cutting machines. Protective clothing which becomes contaminated with oil should be changed frequently and cleaned by any laundering process (dry or wet, or a combination of both) which produces visually clean garments. If clothing becomes grossly contaminated, eg. by spray or spillage, it should be changed immediately.

Special aprons are available, consisting of an impervious back with a detachable absorbent front which can be easily removed for cleaning. The use of this type of protection by toolsetters, who are particularly likely to be exposed to heavy contamination in leaning over oily machines, is strongly recommended. Use of this type of apron removes the temptation to stow oily rags or tools in trouser pockets, a practice which could result in the skin in the groin area being in prolonged contact with oil-soaked clothing. Sleeves of clothing should be short or rolled up to avoid continual friction between oil-soaked cuffs and the skin of forearms and wrists. The golden rule is - "Do not wear oil-soaked clothing". It should be remembered that this applies also to underclothing which may become contaminated from oil-soaked overalls. Underclothes should also be changed frequently. To help minimise clothing contamination, separate locker facilities for work and street clothes are desirable in changing rooms.

Employees who come into contact with oil should wash exposed skin at the end of any work period, using warm water and soap, mild detergent or proprietary skin cleanser. Strong soaps and detergents and abrasive type soaps or cleansers should be avoided. Kerosine, petrol and other degreasing solvents should not be used for cleaning the skin. Hands should also be washed before eating, drinking or smoking and before and after using the lavatory. Easily accessible washing and toilet facilities should be available and should be well maintained. Ideally, employees should shower at the end of the work shift to remove all traces of oil from the skin. If rags are used to wipe oil from the skin or machinery, a plentiful supply should be available so that they can be changed frequently. Preferably they should be of a disposable type to avoid the possibility of accumulation of metal chips and swarf which may cut or scratch the skin.

Barrier creams are often used on the assumption that they protect the skin from direct contact with oil or coolant. Although their effectiveness is somewhat uncertain, they do have a part to play in minimising dermatitis problems by increasing awareness of the need for care of the skin. They can also make eventual washing of the skin more effective, particularly where "dirty" work is involved. Use of creams from reputable suppliers is recommended since long experience has enabled them to avoid use of components which may harm the skin. It is important to use the correct type of cream for the type of oil involved since a water resistant type intended for use with aqueous coolants will not give effective protection against mineral oil based lubricants.

Use of a skin reconditioning cream after work is also important to help replace the natural fats and oils removed from the skin by exposure to lubricants and by washing. This is a very important part of a skin care programme to avoid dermatitis problems. It is particularly important with older employees who tend to have drier skins, and in the winter when low temperatures and humidity also tend to cause dryness and cracking of the skin. Male employees are often reluctant to use a skin cream because of their association with the "soft feminine touch" of consumer advertising. There is no doubt that the common association of masculinity with tough and rough hands which do not need skin creams has contributed to numerous cases of occupational skin disease. Efforts to persuade more men to use reconditioning creams regularly would undoubtedly be well rewarded in reduced absence from work, or need to transfer to other jobs, because of dermatitis. Suitable creams are available from reputable barrier cream suppliers.

Finally, an important part of any skin protection programme is to ensure that all employees who use or are exposed to any type of lubricant or metalworking fluid keep a careful watch on all areas of their skin and obtain medical advice at the first sign of any abnormality. Medical attention should be obtained for any cuts and scratches as well as discoloration, soreness, itching, swelling or warty growths. Awareness of skin disorders and skin care can be promoted by displaying and distributing posters and leaflets published by the Health and Safety Executive and by the oil suppliers. Employees should be aware of and observe any special instructions on product package labels or in the supplier's product literature. Under the Health and Safety at Work Act, it is the responsibility of the employer to inform his employees of any known or potential hazards to health and to instruct them on the appropriate precautions to be followed. Constant reminders and proper supervision are necessary to ensure that the contempt bred of familiarity does not override prudence, or the "problems only happen to other people" syndrome does not become predominant.

12.8.3 Oil Mist and Vapour

Although exposure is unlikely to create a health hazard, concentrations of oil mist and vapour in the plant atmosphere should be minimised to avoid an unpleasant environment. The oil mist concentration should be maintained below the Threshold Limit Value of $5mg/m^3$ and preferably below $2.5mg/m^3$. Oil mist is determined by use of sampling pumps to collect mist on filter papers for analysis by weighing or other methods. The Occupational Hygiene Sub-Committee of the Institute of Petroleum Advisory Committee on Health has published details of suitable techniques [20]. As a general guide, if mist can be seen in the plant atmosphere, it is likely to be above $5mg/m^3$.

If the oil mist level is excessive, the first step is to try to reduce the amount generated. This will require an assessment of the way in which the mist is produced. There are two basic mechanisms by which mist is formed:

(1) In some operations oil may be atomised in small droplet form. This may occur in some high speed cutting operations or from mist lubrication systems through

over application of mist or poor reclassification of the lubricant. In these cases, mist concentrations may be reduced by adjustments to the method of application or by modification to the composition of the lubricant. In metal machining the rate, volume flow and point of application can affect the degree of misting. Proper positioning of splash guards can control the escape of oil mist into the general plant atmosphere. Specially formulated anti-mist cutting oils have successfully reduced plant mist levels in some applications. Mist lubricant formulations require a careful balance between adequate misting and reclassification properties. The rate of application of the mist lubricant and the design of the application system are also important.

(ii) Vaporisation of lubricant may be followed by condensation to form small droplets of oil mist. In this case it may be possible to provide additional cooling by increasing the volume flow rate of oil applied. It is often thought that high velocity jet of oil is the most effective means of cooling whereas in fact a low velocity high volume flow will result in less oil vaporisation. With a high velocity jet there may also be a greater tendency to formation of oil mist.

If formation of oil mist and vapour cannot be effectively controlled by application or formulation changes, local exhaust ventilation should be used. Systems incorporating filters to remove oil so that clean air can be returned to the plant atmosphere are commercially available. For maximum effectiveness, the exhaust hood should be located as close to the point of mist generation as possible.

Oil mist is sometimes generated by the use of air jets to remove swarf from machined parts. If there is no other way of performing this operation, construction of an enclosure with local exhaust ventilation may be needed.

12.8.4 Skin Cancer

If good quality solvent refined oils (or those treated adequately in other ways to reduce the aromatic contents) are used in all applications where any significant skin contact is likely, skin cancer should not be a problem in future, provided that precautions discussed under Skin Protection in sec.12.8.2 are followed. Because of the long latent period from initial exposure to occurrence (often more than 20 years) some cases associated with prolonged or repeated exposure to poorly refined mineral oils many years ago, or with unsatisfactory hygiene practices in the past, can still be expected to arise. However, following the introduction of solvent refined oils in cutting fluid formulations and the implementation of better hygiene practices, there should be fewer cases as time passes.

Because changes occur in composition of used oils, particularly with gasoline engine lubricants and quenching oils, it would be prudent to exercise care with these products and avoid skin contact as much as possible.

12.8.5 Bacteria and Biocides

Bacteria in aqueous coolants and mineral oil lubricants contaminated with water need to be controlled for technical performance reasons. As indicated in Section 12.5, there is no evidence of health hazards associated with the occurrence of these bacteria. However, addition of biocides to coolant systems may present hazards in handling these materials. Biocides are normally irritant and toxic. Therefore, the supplier's handling recommendations should be carefully followed.

12.9 CONCLUSIONS

Provided that users are aware of potential hazards and follow recommended handling practices in combination with good personal and plant hygiene standards, lubricants should present no undue health risks. The major points which need to be repeatedly stressed can be summarised as follows:

- Use good quality solvent refined (or equivalent) mineral oils if there is to be significant skin contact.
- Ensure that adequate information is available to enable products to be used safely.
- Develop proper awareness of hazards through training, cautionary notices, supplier publications etc.
- Avoid repeated or prolonged skin contact.
- Encourage good personal hygiene with proper skin cleaning practices.
- Provide suitable protective clothing.
- Encourage proper use of barrier and reconditioning creams.
- Keep oil mist concentrations well below $5mg/m^3$.
- Establish good oil and machine maintenance practices.
- Obtain early medical advice for any skin problems.
- Ensure good practices are maintained with diligent supervision.

REFERENCES

1 Wagner,W.D., et.al., Am. Ind. Hyg. Assn. J. 1964, 25, 158.
2 Lushbaugh,C.C., et. al., Arch. Ind. Hyg. 1950. 1, 237.
3 Wagner,W.D., et. al., Unpublished results, USPHS, 1014 Broadway, Cincinatti, U.S.A.
4 Decoufle,P., Ann. New York Acad. Sci. 1976, 271, 94.
5 Decoufle,P., J. Nat. Cancer Inst. 1978, 61, 1025.
6 Drasche,H., et. al., Zentbl. Arb. Med. Arbschutz 1974, 10.
7 Sanderson,J., Oil Mist - Recent Interests in Europe. Presented at Esso Symp. on Oil Mist and Nitrosamines, Stockholm, March 1977.

8 Leitch,A., Brit. Med. J., 1922, 2, 1004.
9 Bingham,E., and Horton,A.W., 'Advances in Biology of Skin', Vol.VII - Carcinogenesis, 1966, Pergamon Press, New York.
10 Medical Research Council, 'The Carcinogenic Action of Mineral Oils: A Chemical and Biological Study'. Special Report Series No. 306, 1968, H.M.S.O. London.
11 Scala,R.A., J. Occ. Med. 1975, 17, 784.
12 Wahlberg,J.E., Acta.Derm. (Stockholm), 1974, 54, 471.
13 Th. Van Peteghem and H. De Vos, Brit. J. Ind. Med. 1974, 31, 233.
14 Zingmark,P.A. and Rapp,C. Ambio, 1977, 6, 237.
15 NIOSH Technical Report 'Control of Exposure to Metalworking Fluids', February 1978, (Publication No. 78-165).
16 Technical Bulletin, 'Microbiological Degradation of Lubricating Oils'. Mobil Oil Co. Ltd., 1977.
17 Thony,C., et al., Arch. Mal. Prof de Med Trav et Sec. Soc. (Paris), 1975, 36, 37.
18 Clausen,J., and Rastogi,S.C., Brit. J. Ind. Med. 1977, 34, 208.
19 Institute of Petroleum "Code of Practice for Metalworking Fluids", 1978, Heyden & Son Ltd., London.
20 Institute of Petroleum Occupational Hygiene Sub-Committee, Ann. Occup. Hyg. 1975, 18, 293-297.

13 EFFECTIVE CONTAMINATION CONTROL IN FLUID POWER SYSTEMS

J.B.SPENCER, Manager, User Support Division,
Sperry Vickers.

13.1 INTRODUCTION

The selection of a filter and its proper location in a hydraulic system needs as much care and the same level of expertise as the selection of other components such as pumps, valves and cylinders. Many system designers look no further than the hydraulic equipment manufacturer's catalogue for guidance, but unfortunately it is still common for hydraulic equipment manufactuers to specify one general level of filtration such as 25 micrometre, without regard to the working pressure, environment, or duty cycle. With certain types of equipment a lower standard may be acceptable; an example of this being the earlier designs of pumps, many of which give long trouble-free service protected only by a 0.13 mm strainer. On the other hand, more modern equipment such as the miniaturized controls with smaller clearances than many servo valves, will need much higher standards of protection.

Usually the next step is to decide on the location of the filter and, again, the generalized recommendations of the filter manufacturer are often accepted without regard to the particular system requirements. Finally, the size of the filter is fixed and sometimes the overriding consideration is simply a desire to match the filter port size to the diameter of the adjoining pipework. This may well ensure that recommended velocities are not exceeded but much more important and often overlooked is the filter efficiency and dirt holding capacity.

It must be admitted that the hit-or-miss approach described above often appears to achieve an acceptable result, but with over 70% of hydraulic system failures known to be due to poor fluid condition, there is a clear need for a more systematic approach to contamination control.

The need has been accentuated by the increasingly arduous conditions under which systems operate. For example, a pressure of about 70 bar was common in industrial hydraulic systems for many years; today, 140-210 bar systems are commonplace and much of the contaminant formerly washed away is now forced into the clearances where it does considerable damage. Smaller oil reservoirs mean more rapid circulation and less opportunity for particles to settle out. Higher operating temperatures result in thinner oil, which in some situations may give less protection against wear giving rise to increased contamination.

In the face of these trends, the hydraulic equipment user wants improved reliability and it must be easier and less costly for him to achieve this if he works with clean oil. It is not difficult to keep the oil in good condition provided the machine design is right, and the rewards for doing so are better reliability and longer life from both the equipment and oil.

13.1.1 A Systematic Approach to Filtration

To work towards the most effective protection consistent with economy, we must first define our aim. It is not, as is widely assumed, simply to separate out particles larger than a certain size chosen arbitrarily. Instead, we must achieve stable levels of contamination acceptable and appropriate to various parts of the system.

For a stabilised contamination level, 'dirt in' must equal 'dirt out' (collected by the system filters). 'Dirt in' is made up of in-built contaminant, contamination in the initial charge of oil, and contamination drawn in from the atmosphere through the air breather and cylinder seals - all of which contribute to the generation of particles by the process of wear.

The effect of the in-built contamination must be carefully considered. Inevitably it will be high, even when care is taken in the preparation of pipework and manifold blocks. Flushing will remove some initial contaminant, but there are many systems where this is not done, and at the first start-up high pressures are generated with high contamination levels present. The result is usually rapid pump wear and valve malfunction, the first of which will almost certainly go undetected at this time. All hydraulic systems should be run in an off-load condition until the desired contamination level is attained.

Figure 1 shows a typical relationship between the design contamination level and the actual level prior to start-up. This convenient method of presentation follows naturally from the logarithmic distribution of particle size that occurs in practice. The relative slopes of the initial and acceptable contamination lines are a clear pointer to the type of filtration needed.

The methods used to determine and control contamination levels will be discussed in more detail later, but at this stage we can summarize the practical and performance requirements of the filtration system as follows:-

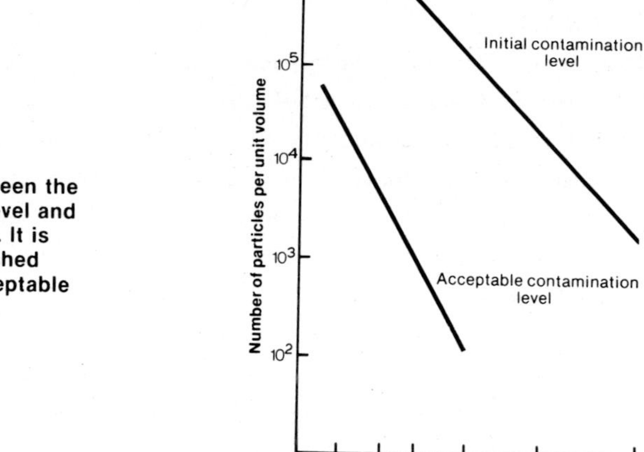

Fig. 1: Typical relationship between the desired design contamination level and the actual level prior to start up. It is essential that the system is flushed and run at no-load until the acceptable contamination level is achieved.

(i) It must be capable of reducing the initial contamination to the desired level within an acceptable period of time, without causing premature wear or damage to the hydraulic components.

(ii) It must be capable of achieving and maintaining the desired level, including a suitable factor of safety to cater for a concentrated ingress which could occur; for example, when a system is 'topped-up'.

(iii) The quality of maintenance available the end user location must be acknowledged.

(iv) Filters must be easily accessible for maintenance purposes.

(v) Indication of filter condition to suit the end user's requirements must be provided.

(vi) In continuous process plant, facilities must be provided to allow changing of elements without interfering with plant operation.

(vii) The filters must provide sufficient dirt holding capacity for an acceptable interval between element changes.

(viii) The inclusion of a filter in the system must not produce undesirable effects on the operation of components, e.g. high back pressures on seal drains.

(ix) Sampling points must be provided to monitor initial and subsequent levels of contamination.

13.2 DIRT INPUT - TYPES AND SOURCES OF CONTAMINATION

13.2.1 New Oil

Although oil is refined and blended under relatively clean conditions, it is usually stored in drums or in a bulk tank at the user's factory. At this point it is no longer clean, because the filling lines contribute metal and rubber particles and the drum always adds flakes of metal or scale. Storage tanks can be a real problem because water condenses in them to cause rusting and contamination from the atmosphere finds its way in unless satisfactory air breather filters are fitted.

If the oil is being stored under reasonable conditions, the principal contaminants on delivery to a machine will be metal, silica, and fibres. With oils from reputable suppliers, sampling has shown average counts of 30,000 to 50,000 particles above 5 micrometre per 100 ml, with a relatively low silt level. Using a portable transfer unit or some other filtration arrangement, it is possible to remove much of the contamination present in new oil before it enters the system and is ground down into finer particles.

It must be said in passing that contamination arising from delivery and storage varies with the industry. For example, aircraft operation generally needs high standards of cleanliness and fairly quick turnover of stores, whereas much longer storage periods are the rule for marine systems and the environment may be more difficult to control.

13.2.2 Built-in Contamination

New machinery always has a certain amount of built-in contamination. Care in assembly and in flushing the system reduces this, but never eliminates it. Typical built-in contaminants are burrs, chips, flash, dirt, dust, fibre, sand, moisture, pipe dope, weld splatter, paints, and flushing solutions.

The amount of contaminant removed during flushing depends not only on the effectiveness of the filter used but also the velocity of the flushing fluid. Unless high velocities are attained, much of the contaminant will not be dislodged until the system is in operation, with component failure the almost certain result. Irrespective of the standard of flushing, an off-load running-in period should be regarded as essential. Some built-in contaminant, such as weld scale, often remains intact until high pressure oil is forced between it and the parent metal, thus loosening it.

13.2.3 Environmental Contamination

Contaminants from the immediate surroundings can be introduced into a fluid power system. On large installations such as those within steelworks, it is relatively easy to ascertain the environmental conditions, though they vary considerably. For example, a coke oven system operates in conditions very different from a cold mill. Sometimes the best solution is to protect the hydraulic equipment by providing a clean room where maintenance can be carried out under controlled conditions. Unfortunately, it is not uncommon to see hydraulic power sources exposed to the worst possible environment, while alongside the electrics are protected by pressurised and temperature controlled cabins.

In most machine shops the relatively large contaminant particles of 10-15 micrometres do not demand a high standard of air filtration, although grinding machines without effective extraction equipment can result in localised problems. On the other hand, foundries and stone quarries demand a very high standard of filtration because of the airborne abrasive particles.

The mobile equipment field presents special problems because the original manufacturer usually sells a standard machine to operate in a wide variety of environmental conditions.

13.2.4 Entry Points for Environmental Contamination

13.2.4.1 Air breathers. Very little information appears to be available on what the filter will actually achieve and purely nominal ratings are usually specified. There have been instances where the element has shrunk leaving a free passage for the air, which highlights the need for more rigid engineering standards on this type of product. The amount of air passing through the filter will depend on draw-off, which means for example that single acting cylinders in bad environments must result in a greater ingress of contaminant.

It is encouraging to see some manufacturers now offering better grades of filter paper and that in certain areas the combined filler/breather has become unacceptable. A separate breather is more efficient and helps to release the air while filling takes place through a suitable gauze strainer.

13.2.4.2 Power unit access plates. In some plants it cannot be assumed that access plates will always be replaced, though happily this state of affairs is not as common as it once was. In power unit design, good sealing is vital, and in bad environments such items as strainers should not be positioned inside the reservoir if access requires the refitting of removable plates. Other removable items will allow ingress during maintenance and good design practice should minimize this.

13.2.4.3 Cylinder seals. Wiper seals cannot be 100% effective in removing very fine contaminant from the cylinder rod.

If they were, they would remove the oil film from the piston rod, producing a result that is usually diagnosed as a leaky seal. In any case, a completely dry rod would quickly wear out the seals. Where cylinders remain extended in a heavily contaminated atmosphere considerable quantities of fine particles can get into the system unless protection such as a bellows is provided.

It has been shown that cylinder piston rod seals naturally ingress about one particle over 10 micrometre for each square centimetre of swept rod area. Wear of seals or wipers can increase the ingression rate considerably. Thus in bad ambient conditions a 50 mm diameter rod in a 100 mm bore cylinder, cycling at a speed of 12 metres per minute, could ingress about 20,000 particles over 10 micrometres every minute, and this quantity could increase by a factor of 100 for every 100 hours of running.

13.2.5 Generated Contamination

Contamination is created internally by the operation of a hydraulic system. The generated contaminants are products of wear, corrosion, cavitation, and fluid breakdown, i.e. decomposition, oxidation, etc. Experience shows that in a system which has been carefully flushed and has filtered oil added to the sealed reservoir (incorporating an effective breather), the contamination will be mainly system generated.

If the initial level is not satisfactory, this induces wear which greatly accelerates the build-up of generated contaminant.

13.3 EFFECTS OF TYPES AND SIZES OF PARTICLES

We know that contaminant particles are of all shapes and sizes and that the finer they are, the more difficult it is to count them and to determine the material of which they are composed. However, we can say that the majority are abrasive and that, interacting with surface protrusions, they plough and cut fragments from a surface. This wear accounts for about 90% of failures due to contamination.

Failures arising from contamination fall into three categories:

13.3.1 Catastrophic Failure occurs when a large particle enters a pump or valve. For instance, if a particle causes a vane to jam in a rotor slot, the result may well be complete seizure of the pump or motor. In a spool type valve a large particle trapped at the right place can stop a spool closing completely. Another example of catastrophic failure occurs when the pilot orifice of a valve is blocked by a large particle. Fine particles can also cause catastrophic failure, for instance if a valve fails to operate due to silting.

13.3.2 Intermittent Failure is caused by contaminant on the seat of a poppet valve which prevents it from reseating properly. If the seat is too hard to allow the particle to be embedded into it, the particle may be washed away when the valve is opened again. Later, another particle may prevent complete closure only to be washed away when the valve opens. Thus a very annoying type of intermittent failure occurs.

13.3.3 Degradation Failure follows wear, corrosion, and cavitation erosion. They cause increased internal leakage in the system components, but this condition is often difficult to detect.

The eventual result, particularly with pumps, is likely to be catastrophic failure. The particles most likely to cause wear are clearance-size particles which just pass through the clearances between moving parts (Fig.2).

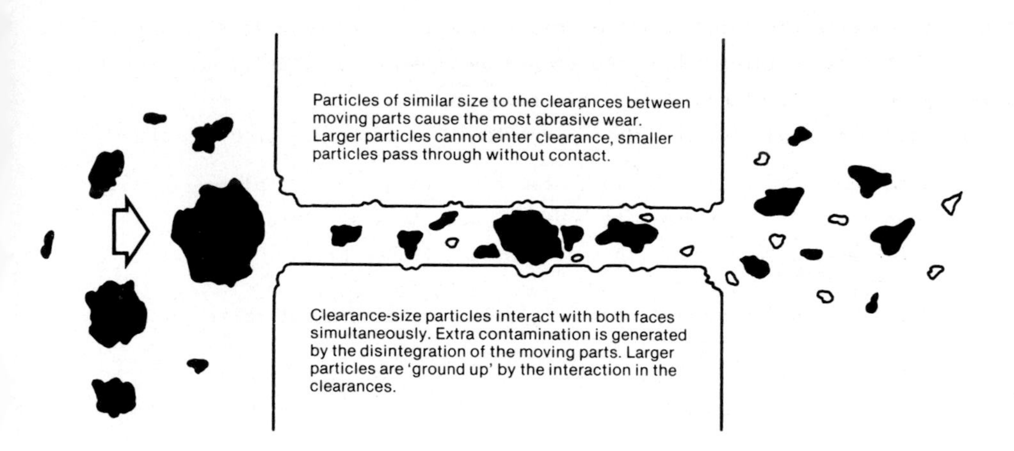

Fig.2 Interaction of moving parts.

Manufacturing clearances within hydraulic components can be divided into two principal zones, i.e. around 5 micrometres for high pressure units, and 15-40 micrometres for low pressure units. The actual clearance may vary considerably depending on the type of unit and operating conditions it sees. Good component design is important, thus minimising the effect of small clearances.

We should therefore look at the factors affecting critical clearances and also at the type of failure occurring in various groups of components.

13.3.4 Pumps

All hydraulic pumps have component parts which move relative to one another, separated by a small oil-filled clearance. Generally, these components are loaded toward each other by forces related to pressure, and the pressure always tends to force fluid through this clearance.

As the finite life of most pumps is determined by a very small quantity of material being removed from a few surfaces, it follows that if the fluid within the clearance is heavily contaminated, rapid degradation and eventual seizure could occur. With low pressure units the design permits relatively large clearances and contamination has less effect. Also at the lower pressure there is less force available to drive particles into critical clearances. Increasing the pressure therefore is of major significance in determining the effect of contamination on a pump.

Another factor affecting clearances is the oil film thickness, which is also related to fluid viscosity. An optimum viscosity is used for design which provides good film thickness to support loads hydrodynamically but which is also low enough to allow adequate filling of the pump without cavitation. It is generally found in practice that filtration requirements become less critical where higher viscosities are used, and for this reason the maximum viscosity which is compatible with the inlet conditions should be chosen. Similarly, good temperature control will be of benefit in this respect.

The areas in pumps particularly subject to these clearance problems are:-

Vane pump	- Vane tip to cam ring, rotor to side plate.
Gear pump	- Tooth to housing, gear to side plate.
Axial piston pump	- Shoe to swashplate, cylinder block.

Figures 3, 4 and 5 illustrate the critical areas diagrammatically.

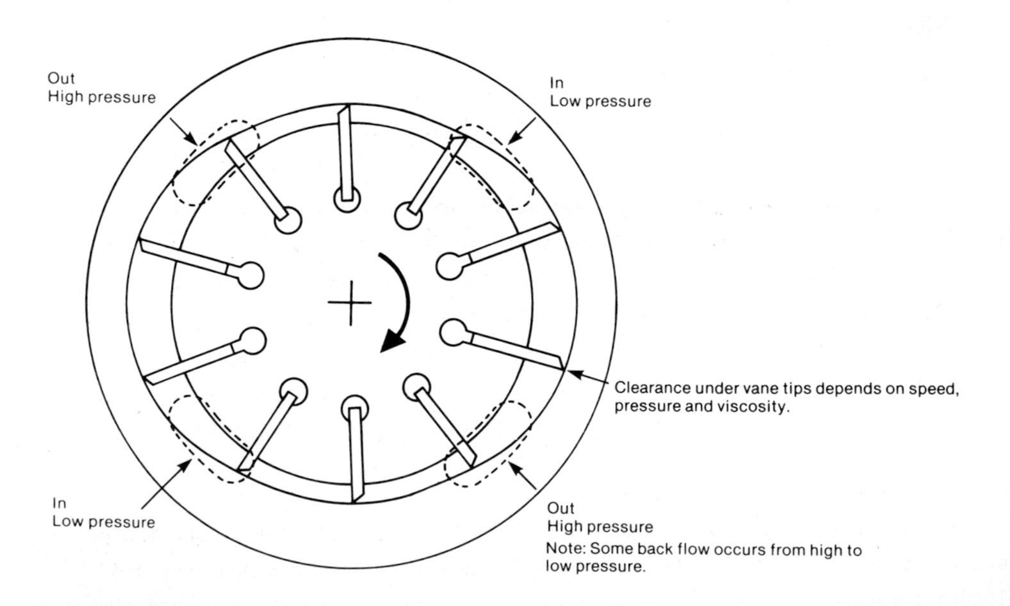

Fig.3 Critical clearances in a vane pump.

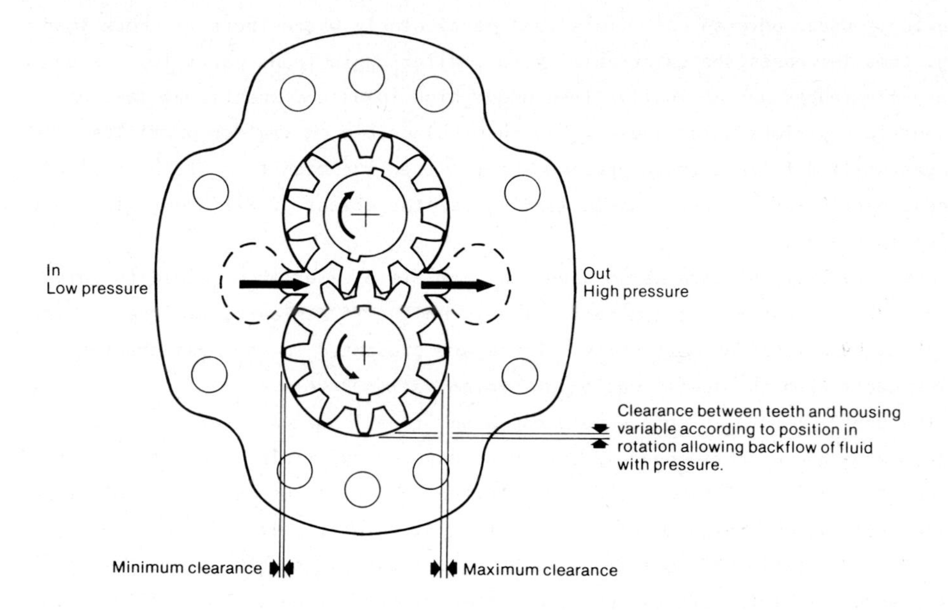

Fig.4 Critical clearances in a gear pump.

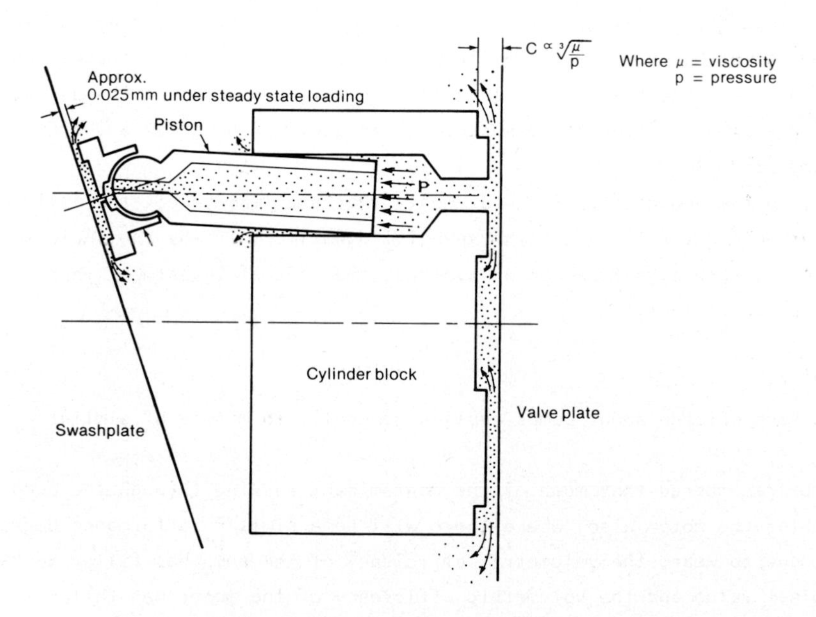

Fig.5 Critical clearances in an axial piston pump. Although piston clearance is nominally fixed, actual clearance varies with eccentirc ity due to load and viscosity.

In many of the foregoing cases the clearances are effectively self-adjusting under operating conditions, i.e. with increasing pressure clearances become smaller. Under adverse conditions, and particularly where there is shock loading, this increases the vulnerability to smaller contaminant particles. Even where clearances are nominally fixed, under high loads components may take up eccentric positions which again makes them vulnerable to smaller particles. It is extremely difficult to be precise about either the magnitude of these clearances, particularly under dynamic loading, or the effect of different size particles in the gaps.

However, from the data we do have and from field experience gained to date, we are able to suggest contamination levels which, if achieved, will result in an acceptable life for most pumps. These are presented in the next chapter, which deals with the specification of contamination levels.

The useful life of a pump should end when it no longer delivers the required output at a given shaft speed, discharge pressure, and fluid temperature. As a guide, 15-20% loss of flow would indicate the end of the useful life. All too often degradation goes undetected until, finally, catastrophic failure occurs with vast quantities of contamination being released into the system. If, following such a failure, the system is not then properly cleaned, the life of the replacement pump will be reduced.

In the interest of the end user, the system designer should specify the minimum acceptable flow rate from the pump to achieve satisfactory machine performance. Means should be provided for monitoring pump output by inserting suitable instrumentation, either temporarily or permanently, so that routine checks can be carried out to reduce the risk of catastrophic failures. With piston units, it is usually a simple matter to measure case leakage, which can be a useful guide to pump condition.

Remember, to the end user it is total costs that are important; the failure of a low-cost pump may well result in expensive downtime. If, by the inclusion of a flow meter, such a failure can be avoided, the initial investment in a flow meter would be fully justified.

13.3.5 Motors

What has been written about pumps applies generally to motors of similar design.

It must be remembered that much of the contaminant passing through the pump may be reaching the motor also, where there will be a similar performance degradation. If, due to wear, the volumetric efficiency of the pump has fallen to 85% of its original value and the volumetric efficiency of the motor has fallen to, say, 90% of original, then the overall volumetric efficiency of the pump and motor will be down to $0.85 \times 0.9 = 76.5\%$ of the original value. For this reason contamination control is particularly important in hydrostatic transmissions to provide the necessary level of fluid cleanliness.

13.3.6 Directional Valves

The radial clearance specified between bore and spool in most directional valves is in the range of 5 to 13 micrometres. As is well know, the production of perfectly round and straight bores is exceptionally difficult, so it is unlikely that any spool will lie exactly central in the clearance band. In a nominal $\frac{1}{8}$ in. valve, a good spool is likely to have less than 2.5 micrometres clearance.

In an electrically operated valve, the forces acting on the solenoid are shown in Fig.6. They are: Flow forces + Spring force + Friction + Inertia.

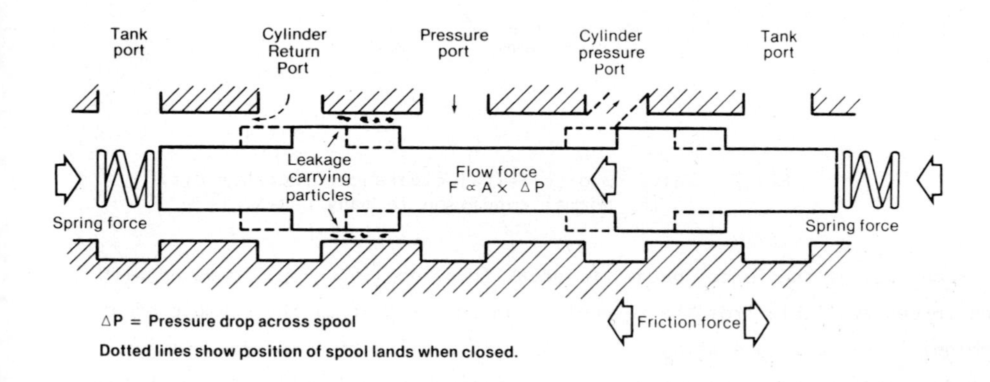

Fig.6 Valve spool critical clearances (with flows and forces).

Flow, spring, and inertia forces are inherent factors but friction forces are, to a great extent, dependent on filtration. If the system is heavily contaminated with particles similar in size to the radial and diametral clearances, higher forces will be needed to move the spool.

An even worse situation results from silting, where contaminant is forced into the clearances under pressure, eventually leading to breakdown of the oil film and spool hang-up (Fig.7).

This situation occurs where valves subjected to continuous pressure are operated infrequently. Such valves should preferably have local filtration of a very high order in the adjacent pressure line but due account should be taken of possible pressure surges generated during component operation. The use of high efficiency filters as a special protection for single units or groups of units can result in the need for a very high dirt capacity if the general level of filtration in the system is much lower.

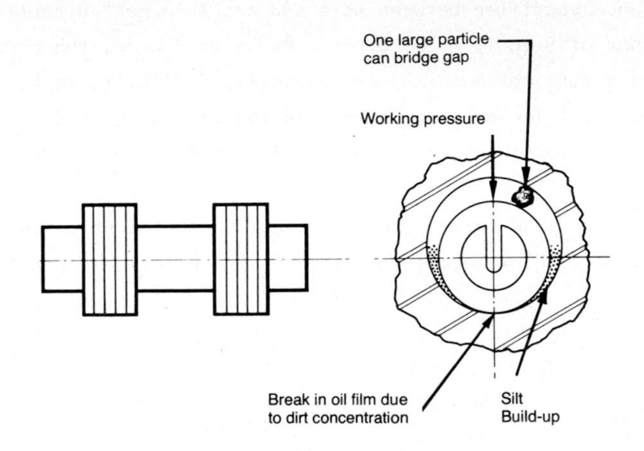

Fig.7 Valve spool critical clearance. Eccentricity is the normal condition in many cases.

Some idea of the forces needed to break this spool hang-up, compared with the forces available from the solenoid, can be gained from the example of a nominal $\frac{1}{8}$ in. valve operating at 210 bar. If a valve of this type remains selected in the spring offset or energised position for a lengthy period of time, silting takes place between spool and bore to produce total immobility. The force needed to overcome this state has been found by experiment to be of the order of 135 Newtons, but both spring and solenoid could exert only 45 Newtons. Thus the effect of the silt is to cause total system failure.

13.3.7 Pressure Controls

Highly abrasive particles in high velocity streams of oil erode the surfaces with which they come into contact. This situation is common to pressure controllers, particularly relief valves which are subjected to maximum system pressure drop and velocities of the order of 30 m/s. Pilot control stages generally see low volumes at high velocities and heavy contamination affects both their stability and repeatability.

13.3.8 Flow Controls

The contamination tolerance of flow control valves will depend very much on the orifice configuration.

Figure 8, for example, shows two orifices which are of entirely different shape, although having equal areas. The groove type (a) will tolerate a high contamination level except when used at low setting, whereas type (b) is much more prone to silting at all settings.

With all types of pressure-compensated flow controls, the performance of the pressure reducing element can be considerably affected by contamination, irrespective of valve setting. Damage to the metering orifice can also occur, which will become particularly apparent at lower settings.

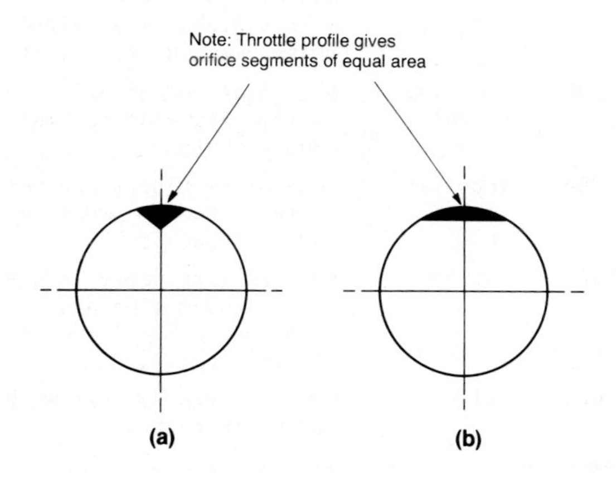

Fig.8 Flow control valve throttle sections. Profile (b) is more prone to silting.

Generally speaking, all spool-type control valves are affected by contamination in the system, especially at high pressures. The effects are likely to be magnified if precise axial positioning of the spool is necessary as, for example, in pressure reducing valves where limited forces are available to operate the spool. On the other hand, poppet valves, though affected by large particles of contamination, tend to be far more tolerant of silt due to the self-clearing action of the seat. However, erosion is still likely to occur.

13.3.9 Summary

It can be seen from the foregoing that an individual large particle arriving at the wrong place at the wrong time can cause catastrophic failure.

Surfaces within components should be separated by an oil film, the thickness of which may be continually changing. When this gap is bridged by contaminants, wear will occur thereby generating further particles which may well be ground into many more smaller particles. Fine particles individually or in small

Fig.9 Suggested acceptable contamination levels for various hydraulic systems

Target Contamination Class to CETOP RP70H 5µm	15 µm	Suggested Maximum Particle Level 5µm	15µm	Sensitivity	Type of System	Suggested Filtration Rating βx > 75
13	9	4 000	250	Super critical	Silt sensitive control system with very high reliability. Laboratory or aerospace.	1-2
15	11	16 000	1 000	Critical	High performance servo and high pressure long life systems, i.e. aircraft, machine tools, etc.	3-5
16	13	32 000	4 000	Very important	High quality reliable systems. General machine requirements	10-12
18	14	130 000	8 000	Important	General machinery and mobile systems. Medium pressure, medium capacity.	12-15
19	15	250 000	16 000	Average	Low pressure heavy industrial systems, or applications where long life is not critical.	15-25
21	17	1 000 000	64 000	Main	Low pressure systems with large clearances.	25-40

Note: This graph assumes viscosity to be within recommended range.

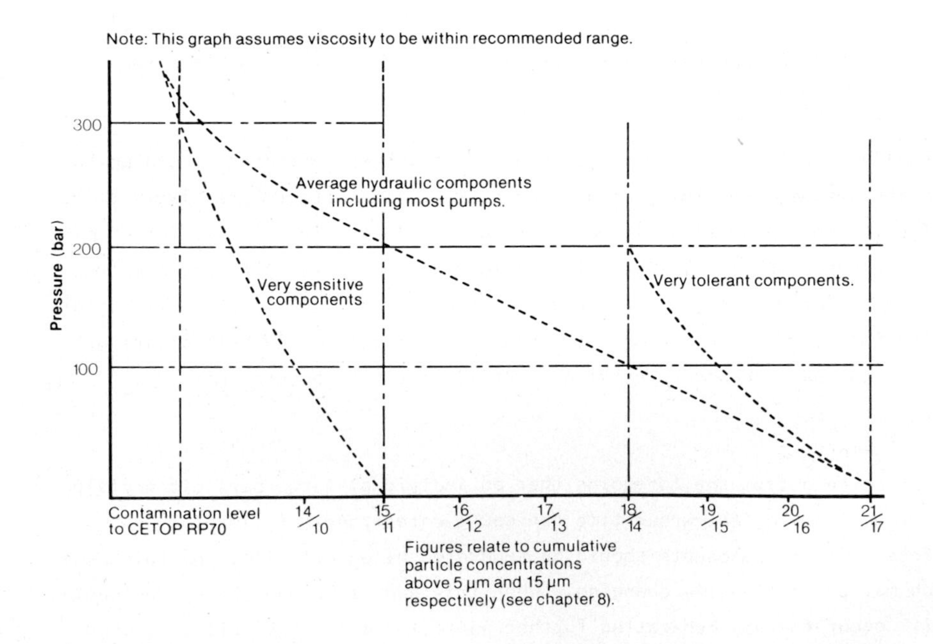

Fig.10 Suggested cleanliness level for good life

quantities may not cause damage, but if present in larger concentrations may lead to failure through silting.

The objective must be to obtain the most economic combination of contamination control and contamination tolerance for a given reliable system life under known performance and environmental conditions.

13.4 SPECIFYING CONTAMINATION LEVELS

As stated previously, 25 micrometres is a typical general level of filtration, specified without regard to working pressure, local environment or duty cycle. We know from experience that under similar environmental and operating conditions the effectiveness of a 25 micrometre filter will vary depending on its location in the system. Also, under steady flow conditions as the pores tend to clog the filtration performance may actually improve. Under varying or intermittent flow, however, the result can be very different because contaminant particles are dislodged from the pores allowing more fine particles to pass through.

It is obvious that the conditions of use have the greatest effect on the contamination level obtained with a specific filter, which means that it is generally unwise to offer without qualification or accept without question a blanket recommendation of, say the use of a 25 micrometre filter. For the user it means putting his investment at risk and for the manufacturer it makes the administration of warranties extremely difficult.

From a wide spectrum of field data, Fig.9 summarises the levels of contamination which are considered acceptable for most hydraulic systems.

The decision of whether a system is critical will depend largely on the type of components used and the system design pressure. Figure 10 has therefore been produced to give the recommended contamination levels with knowledge of the system pressure and having analysed the dirt sensitivity of the system components.

13.5 SELECTING THE FILTER

Before a choice of filter may be made the various ratings used by manufacturers must be examined.

13.5.1 Nominal Rating

Specifications MIL-F5504A and MIL-F5504B were established for determining nominal ratings. Version A defines a 10 micrometre filter as being able to remove 98% by weight of all particles of the elected contaminant (AC fine test dust) larger than 10 micrometres at a certain high concentration. Version B defines a 10 micrometre filter as being able to remove 95% by weight of 10-20 micrometre glass beads at a high concentration. Although little use has been made of these particular specifications, many manufacturers use similar tests to provide nominal ratings for their filters.

Such tests have two major limitations. Firstly, they do not limit the maximum size of particle allowed to pass through the element and from tests it has been found that filters meeting these requirements can pass particles up to 200 micrometres. Secondly, the high concentration of contaminant added is not typical of conditions experienced in a normal system. In practice, the particles approach the filter in small concentrations and those particles that are smaller than the mean pore size pass readily through the filter as long as the filter medium remains reasonably clean.

For the reasons given above there is a good case for discontinuing the use of nominal ratings.

13.5.2 Absolute Rating

The NFPA Fluid Power Glossary of Terms defines the absolute rating as being the diameter of the largest hard spherical particle that will pass through a filter under specified test conditions. This is an indication of the largest opening in the filter element.

13.5.3 The Bubble Test

This is a test used by manufacturers to determine the area of greatest porosity. It is achieved by applying air pressure to the inside of the filter element, which is submerged in a liquid such as alcohol, which wets the filter media. The operator rotates the filter element at each pressure level and records the pressure at which the first stream of bubbles emitted from the filter element. The test can be continued to measure the pressures of the second, third, fourth, etc. largest hole. By continuing to slowly increase the pressure, a point is reached called the 'open bubble point' at which air bubbles appear over the entire surface of the filter element; this is a simplified method of measuring the mean pore size.

It is claimed that precise results are not obtainable from the bubble test, which is unfortunate because such a simple test would be invaluable. As it is, its main use is in the quality control of elements to ensure there is no damage to the media or a bad seal.

13.5.4 Mean Filtrating Rating

This is a measurement of the average size of the pores of the filter media. This is a very significant rating, since it is a measure of the particle size above which the filter starts being effective. It can be measured using the 'open bubble point' method just described.

13.5.5 Multipass Filter Test

This test, designed to provide a means of describing the performance characteristics of a filter, involves the continuous injection of a controlled contaminant into a test system. As the contaminant can be removed only by the test filter, it will continue to circulate in the system unless it is captured. The separation capability of the test filter is monitored by analysing upstream and downstream fluid samples. The dirt holding capacity is measured by the amount in grammes of test contaminant which can be added to the system before a specified terminal pressure drop across the filter is reached.

The mathematical relationship which describes the test is developed from the following expression:

$$\begin{array}{l} \text{Number of particles} \\ \text{downstream of size} \\ > x\mu m \end{array} = \begin{array}{l} \text{Number of particles} \\ \text{originally of size} \\ > x\mu m \end{array} + \begin{array}{l} \text{Number of} \\ \text{particles} \\ \text{injected of} \\ \text{size} > x\mu m \end{array} - \begin{array}{l} \text{Number of} \\ \text{particles removed} \\ \text{of size} > x\mu m \end{array}$$

13.5.6 Beta Ratio

The separation characteristics are given by the Beta ratio, which is defined as follows:

$$\beta x = \frac{\text{Number of upstream particles larger than } x\mu m}{\text{Number of downstream particles larger than } x\mu m}$$

A Beta ratio of 1 indicates that no particle contamination is removed. A figure of less than 1 is clearly impossible unless the filter is unloading contaminants.

For a filter exhibiting a Beta ratio greater than 1, the downstream concentration of particles above a given size will stabilize to give an almost constant contamination level.

13.5.7 Practical Classification of Filter Performance

Whatever format is used by the filter manufacturer to give performance information on his product, the degree of filtration provided will basically fall into one of three categories, depending on the degree of silt control. Typical data corresponding to these categories are given in Fig.12 though it should be the manufacturer's responsibility to state into which of the three classifications his products fall. At present there is no universally recognised standard classification but work being carried out by various bodies should eventually lead to an internationally agreed definition.

Two adverse factors affect the actual performance of filters in service, namely pulsating flow, and the sometimes uncertain performance of internal seals and bypass valves.

Fig.12 Definition of practical classification categories

Category	Nominal Rating µm	Absolute Rating	Beta Ratio
Silt Control	½ to 1	3 to 5	β3-5 >75
Partial Silt Control	3 to 5	10 to 15	β10-15 > 75
No Silt Control (Chip removal)	10 to 15	25 to 40	β25-40 > 75

13.5.7.1 Pulsating Flow forces through the media those fine particles which would otherwise lodge among the fibres and between larger particles already intercepted. The effect may be compared with a sieve holding a mixture of stones, some larger and some smaller than the openings in the mesh; when the sieve is stationary, many of the small stones are retained but they fall through when the sieve is shaken. Pulsating flow therefore increases the proportion of silt particles in the system downstream of the filter, and this is reflected in the performance curve of the filter (Fig.13).

13.5.7.2 Bypassing by internal or external valving is acceptable for many systems since, in a number of passes, all the fluid eventually goes through the filter.

For filters other than those designed for permanent bypassing it would normally be expected that the bypass be operative only when the element is reaching the end of its useful life. Therefore the effects of premature opening of the bypass valve or a faulty internal seal need to be considered. Since bypassing, whatever the cause, does not discriminate between fine and coarse particles, the effect is to weigh the contamination profile heavily at the coarse end. Figure 14 shows the effect of increasing percentages of bypass flows, and it will be noted that even at 0.1% bypass the maximum particle size has almost doubled. At 1% it has more than trebled, and at 10% it has increased by a factor of five.

The practical significance of bypassing and the need to make provision for it depend on the location of the filter in the system. For example, on inlet filters a bypass valve is mandatory in order to protect the pump from cavitation as the filter becomes blocked.

Pressure filters are intended to protect the system if there is catastrophic failure of the pump. If such a failure occurs when the filter is bypassing, then the protection is non-existent. There is obviously a case for fitting non-bypass pressure filters where the need for this type is established. An element must be fitted that will withstand the total system pressure, which increases the cost of the filter. Clogging of the element causes system performance to fall off and a reliable indicator is essential to give early warning of this.

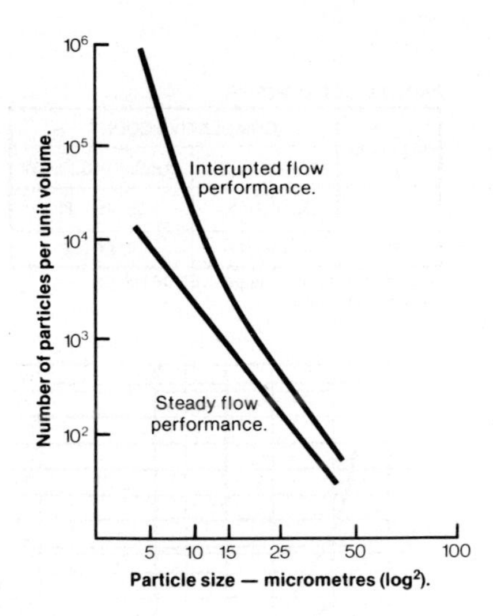

Fig.13 Filter performance deteriorates under intermittent or pulsating flow. The effect is more marked with finer particles which are forced through the media.

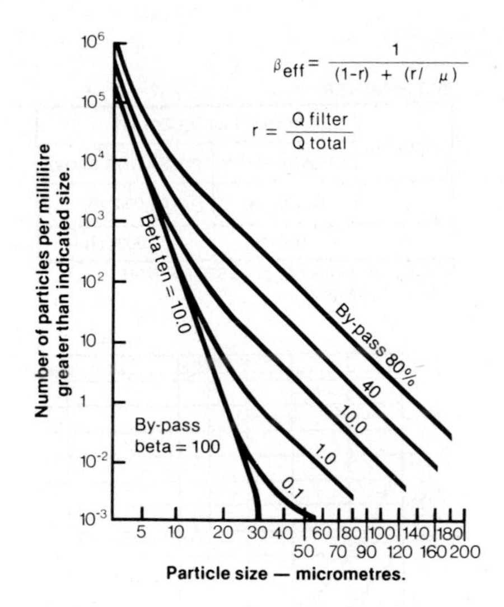

Fig.14 Influence of by-pass leakage on the filter performance.

Where a bypass is judged to be permissible, the system designer must weigh carefully the effects of premature opening due to cold starts and surge flows; the elimination of surges, even those at low pressure, will reduce the tendency for the bypass to open under normal operation.

At present there is little information available on the extent of the adverse effect that pulsating flow has on filter performance. Hopefully, Beta ratios will eventually be quoted which do relate to conditions which line filters are subjected to in actual practice. Naturally, where steady flow exists relating more closely to laboratory test conditions, more accurate predictions of filter performance can be made. However, using the limited knowledge at our disposal, likely performance levels for each of our categories have been produced in Fig.15.

On each graph a cumulative count of both 5 and 15 micrometre size particles has been shown. The reason for using these figures is because of the standard set by the CETOP RP70H.

For ease in relating to this document,the range numbers are also quoted. In addition to fixed figures at the two values the lines are extended to show expected trends for typical filters.

SILT REMOVER

SIZE μm	CUMULATIVE COUNT	
	STEADY FLOW	PULSATING FLOW
5	16,000 (14)	64,000 (16)
15	1,000 (10)	2,000 (11)

N.B. Range number as per CETOP RP70H shown in brackets.

PARTIAL SILT REMOVER

SIZE μm	CUMULATIVE COUNT	
	STEADY FLOW	PULSATING FLOW
5	32,000 (15)	25,000 (18)
15	2,000 (11)	4,000 (12)

N. B. Range number as per CETOP RP70H shown in brackets.

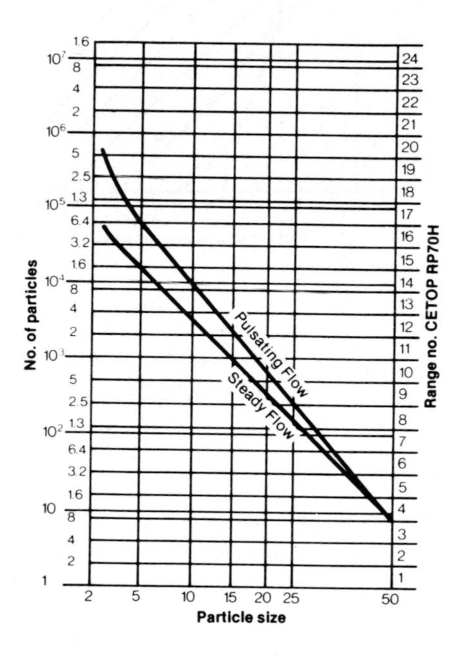

Fig.15A Assumed mean pore rating 3 micrometre. Curves show deterioration of filter performance with respect to particles below this size under pulsating flow.

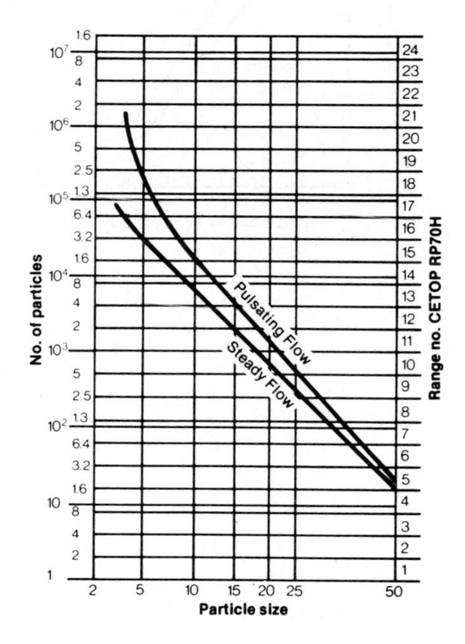

Fig.15B Assumed mean pore rating 7 micrometre. Curves show deterioration of filter performance with respect to particles below this size under pulsating flow.

CHIP REMOVER

SIZE μm	CUMULATIVE COUNT	
	STEADY FLOW	PULSATING FLOW
5	250,000 (18)	16,000,000 (24)
15	16,000 (14)	64,000 (16)

N.B. Range number as per CETOP RP70H shown in brackets.

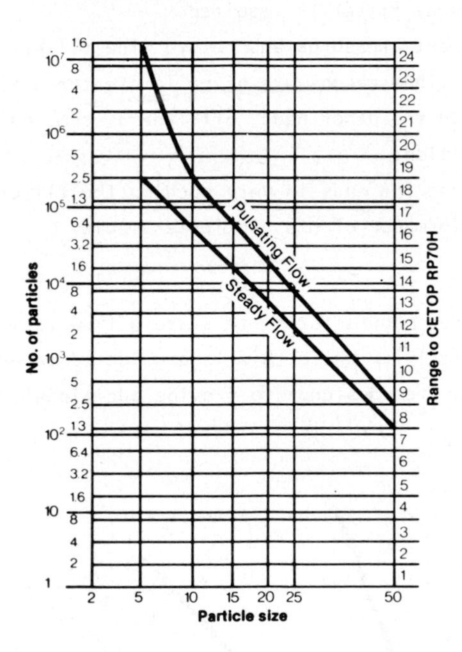

Fig.15C Assumed mean pore rating 15 micrometre. Curves show deterioration of filter performance with respect to particles below this size under pulsating flow. There is little or no control of 5 micrometre particles.

Assuming these graphs relate to performance of the filter supplied by the particular manufacturer chosen it is necessary to show that the actual requirement when plotted stays below the line drawn for steady or pulsating flow.

As an example, if we have a system with a working pressure of, say, 150 bar using components with average dirt tolerance, our cleanliness level graph (Fig.10) tells us that the desired contamination level should be somewhere between 15/11 and 18/14.

Referring now to Fig.15a and b, we see that under steady flow conditions this could be achieved by our partial silt control filter, but if pulsating flow is present a silt removal filter is required.

It can be assumed that most pressures and return line filters are subject to pulsating flow conditions. The reasons are given in greater detail in the section on filter location. On the other hand, off-line systems have the benefit of a steady flow rate condition.

It must be emphasised that the onus is very much on the filter manufacturer to identify the likely performance of his particular media.

13.5.8 Filter Sizing

It is current practice for manufacturers to state a flow rating at a specific clean pressure drop (see Fig.16). While this is a guide to capacity, it may well be necessary for the system designer to provide additional dirt capacity so

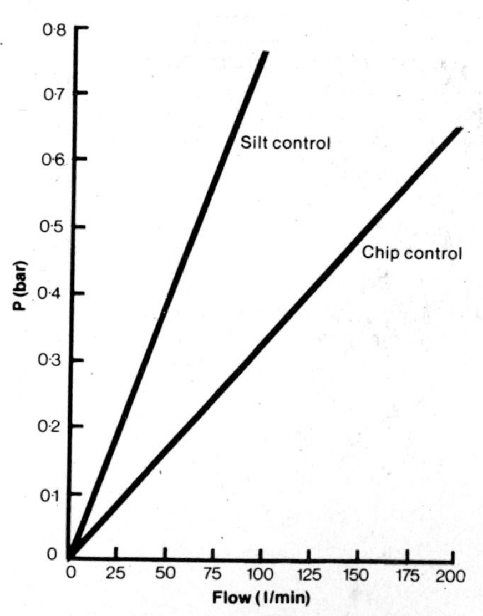

Fig.16 Typical pressure drops for clean silt control and chip control elements of similar size.

as to ensure that the end user obtains an acceptable element life. All too often filters sized purely on flow rate have a short element life. In choosing larger filter capacities greater initial expense may be incurred, but this is almost certain to be recovered in reduced running costs, i.e. fewer element changes, reduced labour costs and reduced downtime.

Correct filter sizing necessitates relating the dirt entering the filter to the effective element area and the maximum allowable pressure drop. The relationship of area to pressure drop is not simple, however, and filter inlet dirt levels are rarely known.

There is a laboratory test (the 'comparative life' or 'dirt capacity' test) which is designed to compare the dirt holding capacities of hydraulic filters. An artificial contaminant is added at a constant rate to a continuously recirculating oil system and the resultant increase in differential pressure is plotted against the weight of contaminant added, as shown in Fig.17. The resulting curve has a characteristic form which is constant for a given filter media.

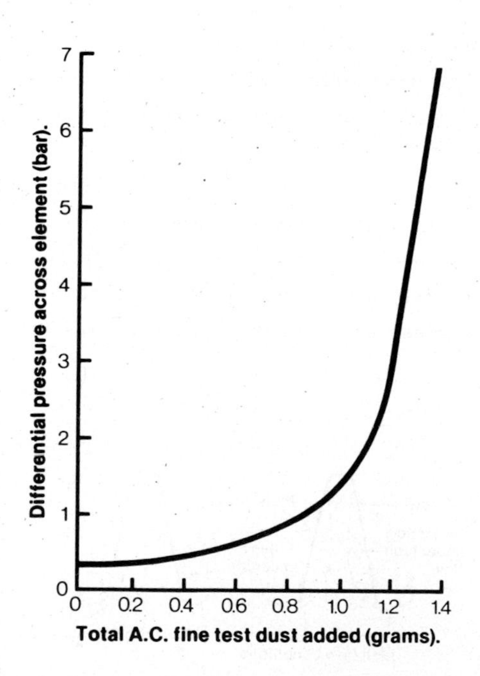

Fig.17 Typical dirt capacity curve for hydraulic filter element.

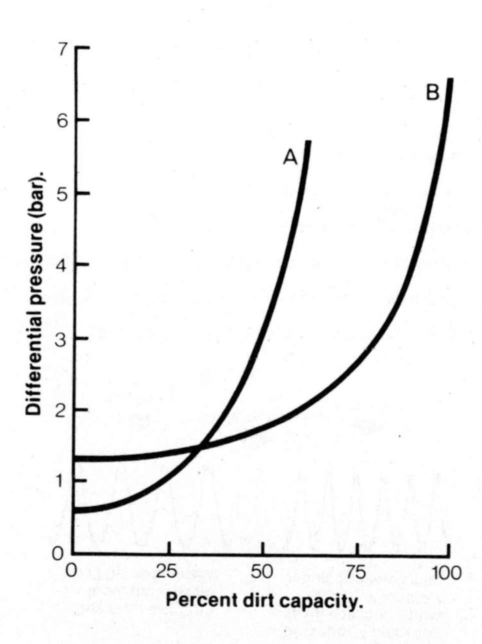

Fig.18 Comparison of dirt capacities and differential pressures for two different filters.

In the first stages of dirt addition it can be seen that the pressure drop increases slowly, whereas later the pressure drop increases very rapidly. This illustrates why very little is gained in terms of element life by allowing the filter to operate at a high differential pressure after the 'knee' of the curve has been passed. The curve also illustrates the irrelevancy of a system designer concerning himself solely with clean pressure drops; the more significant factor is the differential pressure across the filter after a specified amount of dirt has been added.

Except with non-bypass filters, the maximum pressure drop across the element is usually determined by the bypass valve setting. The system designer must also check that the system performance is maintained with the pressure drop at its maximum value. If filters are oversized, the bypass may pass a substantial flow without indicating.

It is commonly thought that to obtain longer service life from a given filter element without sacrificing filter efficiency, it is only necessary to pack more media area into the filter envelope. This is not so since an optimum area exists for a given envelope and it is detrimental to exceed this area.

The comparative life test previously described is used to compare the dirt holding capacity of different filter elements, and in Fig.18 we compare the dirt capacity of two filters of identical envelope size. The filter A has a lower clean pressure drop than filter B because A has more area. However, the optimum area has been exceeded and therefore filter A would have a shorter life than B for a given pressure differential.

How close packing reduces the effective area is illustrated in Fig.19 where the pleats close up under pressure and the small angles between them clog rapidly. Fatigue failures can also occur when pressure is applied. As data on the results of dirt capacity tests are not readily available, we must revert to the manufacturer's flow rating at a specific clean pressure drop and use this as a basis for assessing dirt capacity.

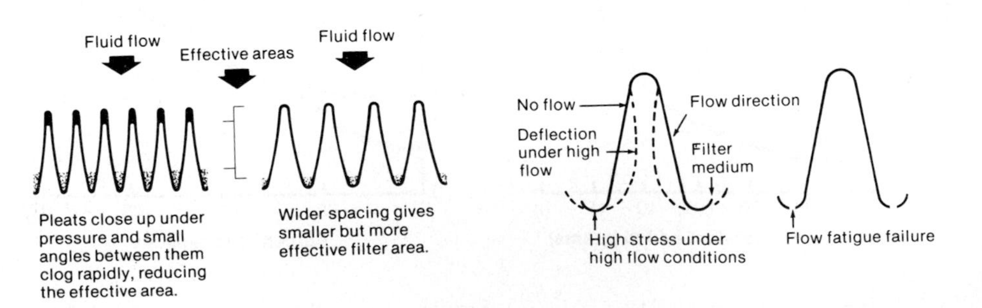

Fig.19A Increasing filter area within a given package may actually reduce the dirt holding capacity.

Fig.19B Changes in flow and pressure drop cause the sides of element corrugations to flex and the root to stretch, giving rise to fatigue stresses.

Recapping, our objective is to achieve a 'dirt in equals dirt out' condition. The desired dirt capacity of the filtration system will therefore depend to a large extent on our ability to control the dirt input. The dirt input is a product of inbuilt contamination and ingressed contamination which, in turn, produces system generated contamination. Let us first examine each source in detail and the factors controlling its input (Fig.20).

	Contamination source	Controller
Dirt input =	Inbuilt: in components pipes, manifolds,etc.	Good flushing procedures, system not operated on load until acceptable contamination level obtained.
	plus	
	Present in initial charge of fluid	Integrity of supplier. Fluid stored under correct conditions (exclusion of dirt, condensation,etc.) Fluid filtered during filling.
	plus	
	Ingressed through air breather	An effective air breather with rating compatible with degree of fluid filtration.
	plus	
	Ingressed during fluid replenishment	Suitable filling points which ensure some filtration of fluid before entering reservoir. This task undertaken by responsible personnel. Design should minimise the effects.
	plus	
	Ingressed during maintenance	
	plus	
	Ingressed through cylinder rod seals	Effective wiper seals or, if airborne contamination, rods protected by suitable gaiters.
	plus	
	Further generated contamination produced as a result of the above and the severity of the duty cycle.	Correct fluid selection and properties (viscosity and additives) maintained. Good system design minimising effects of contamination present on system components.

Fig.20 The practical steps which control contamination in hydraulic systems.

Based on Fig.20 we will now grade the cleanliness level of a system between 1 and 7. An example of a grade 1 clean system would be a clean workshop with effective control over all contamination ingress. A grade 7 dirty system would be, say, a foundry with little or no control over contamination ingress and a system operating several exposed cylinders.

Figure 21 will assist in making a numerical assessment between these values based on the environment and the degree of control over contamination.

Environmental Conditions			
Good	Average	Bad	Degree of Control
3	6	7	Little or no control over contamination ingression (many exposed cylinders)
2	4	5	Some control over contamination ingression (few cylinders).
1	2	3	Good control over contamination ingression (gaitered cylinders).

Fig.21 Assessment and classification of system cleanliness level into 7 grades.

We must now relate this to the effective element area and the maximum allowable pressure drop. The relationship between area and pressure drop is not simple, but by using very broad approximations of these values and assuming that the manufacturers flow rating at a specific clean pressure drop is a good guide to dirct capacity, the following selection guides can be used:

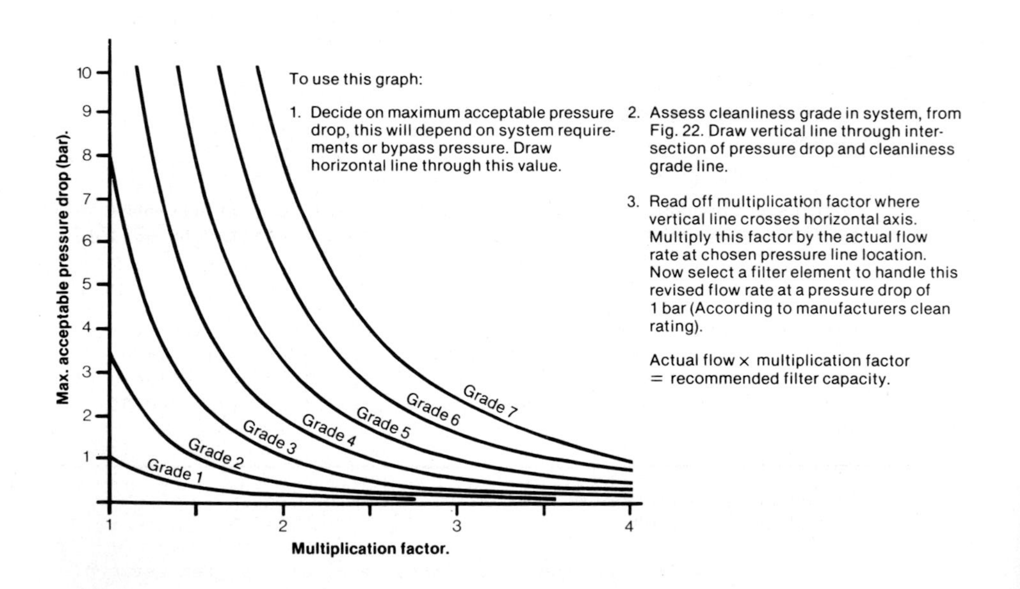

Fig.22 Pressure line filter selection guide.

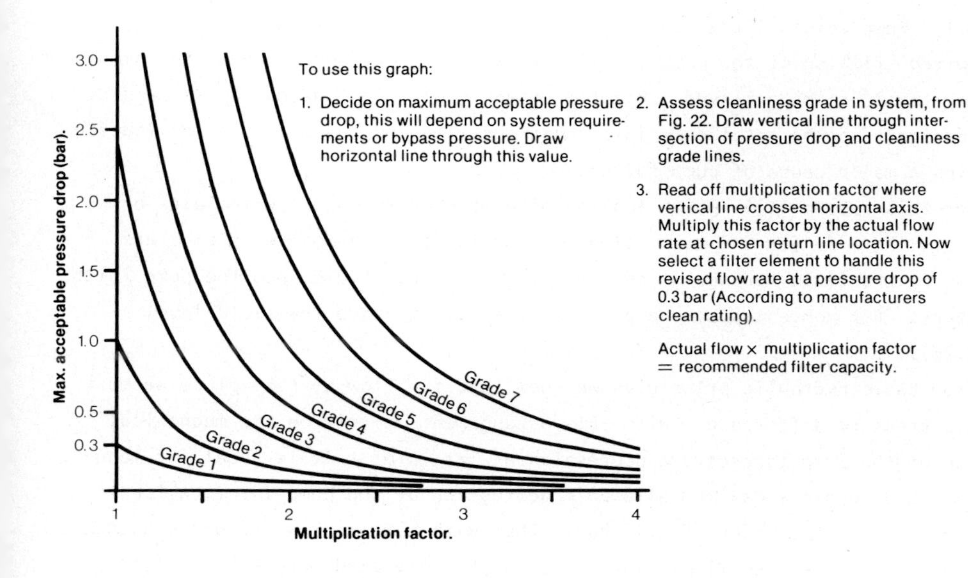

Fig.23 Return line filter selection guide.

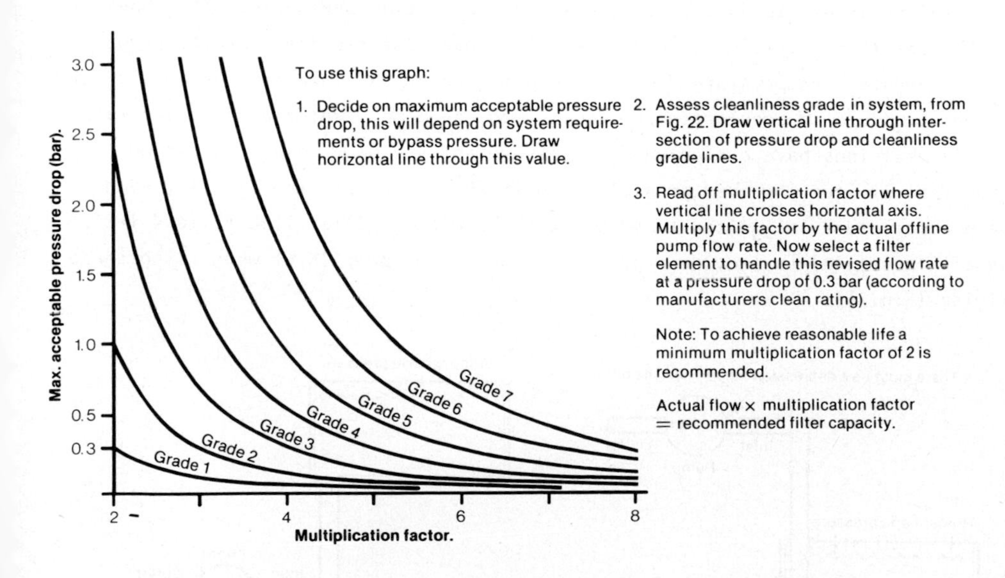

Fig.24 Offline filter selection guide.

13.6 LOCATING THE FILTER

13.6.1 Pump Inlet Filtration

Correct filling of the pump is vital if satisfactory operation of the hydraulic system is to be achieved. All too often insufficient attention is paid by the designer to the total pump inlet configuration and, as a result, cavitation remains a major cause of pump failure.

For this reason it is worth looking once more at the basic principles of pump filling. By far the most common method is to use atmospheric pressure acting on the fluid surface of the reservoir to force fluid into the pump inlet chambers. For convenience, the pump is often sited above the fluid level (Fig.25).

From basic hydraulic principles we know that for flow to take place we must have a pressure difference. With this arrangement we rely on the mechanical action of the pump to create a depression at its inlet. It is usual for manufacturers to quote a maximum allowable depression at the pump inlet, which is likely to be of the order of 0.17 bar. Thus with normal pressure drops accounted for, only a very small pressure drop can be tolerated across the filter. For this reason the size and cost of inlet filters is often greater than, say, filters in the return line. Furthermore, such low pressure drops make silt removal virtually impossible.

Fire-resistant fluids are very sensitive to suction pressures. They have higher specific gravities than mineral oils, particularly the synthetic types. This increases the pressure drop to the pump and at the same time demands a higher pressure to accelerate the fluid into the pump. Water glycols and water-in-oil emulsions have a high vapour pressure and the pump inlet depression should be limited wherever possible to half the value for mineral oils, even when temperature is limited to 50°C. With or without inlet line filters it is usually essential to provide a positive head at the pump inlet when using these fluids (Fig.26).

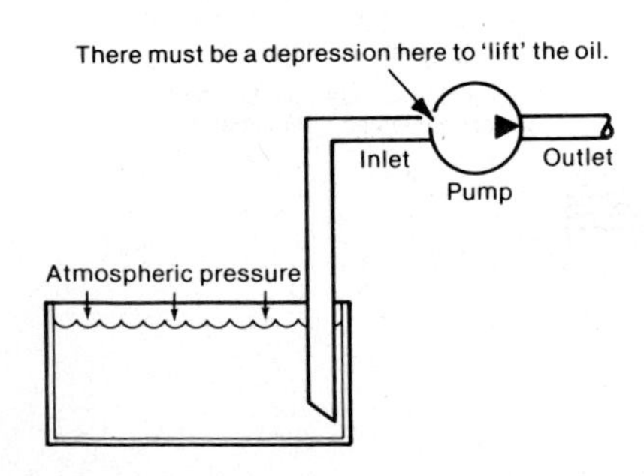

Fig.25 Negative head tank.

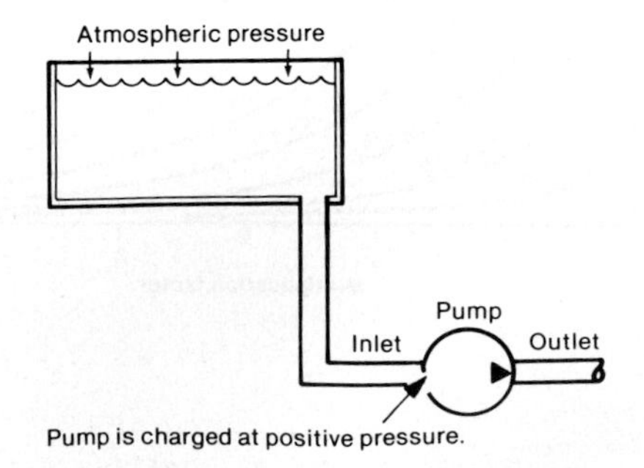

Fig.26 Positive head tank.

Irrespective of the type of fluid, a positive head will improve inlet conditions because it increases the force available to create the required flow.

For both negative and positive head inlet arrangements, when sizing inlet filters we must ensure that any filter (housing and element) and associated pipework should: 1) pass the full pump volume within the permitted inlet depression for that pump, and 2) permit a bypass flow that is still within that limit when the filter element is blocked. (This requirement often necessitates the operation of the bypass feature at pressures as low as 0.085 bar, a level at which operation is seldom consistent.)

All calculations should take into account the effect of higher viscosity fluid, e.g. at cold starts, otherwise cavitation will occur.

The usual micrometre rating for inlet filters is 75 or 150. However, units are available incorporating elements down to 10 micrometre. This means large housings and probable oversizing of the inlet configuration.

The 75 and 150 micrometre mesh elements will remove most of the particles above their rating but are relatively ineffective in removing anything smaller. The addition of magnets will remove some of the fine metallic particles but the location of the magnets within the filter must be such that under no circumstances can accumulated contaminant break away, thereby passing a conglomerate into the pump.

An advantage often claimed for inlet filters is ease of servicing. However, incorrect re-assembly of access covers can result in air ingress, which often goes undetected and can be harmful to the system as dirt.

Inlet filters are generally used in systems where maintenance procedures do not prevent quantities of large particles entering the reservoir, for instance during topping-up. Providing bypassing is not occurring, they protect the pump from this type of ingression. However, good reservoir design that includes a mesh screen or baffle can provide an equally acceptable contamination level at the pump inlet. Under these circumstances, strainers and filters and their associated fittings can be omitted from the inlet line, thus improving pump filling conditions.

It is an encouraging sign that more designers are providing separate filling arrangements, usually through some form of coarse filter to reduce the need for last-chance pump inlet protection.

If the filling requirements of the pump are critical and supercharging is necessary, as is the case quite often with large variable-displacement piston pumps, it is common to site a filter between the supercharger and main pump. For such applications, the same guidelines given for the sizing of pressure or return line filters can be used. However, depressions can occur when a variable pump moves from zero to full displacement, and the likely effect on filter performance must be carefully considered.

13.6.2 Pressure Line Filtration

Let us start by discussing the location of the pressure line filter in relation to pump and relief valve.

Figure 27 shows the pressure filter located downstream of the relief valve. For the non-bypass type the arrangement shown in Fig.28 is mandatory. The actual flow seen by the filter during the operating cycle depends on the system demand and during off-load periods there is leakage flow only if block-centre directional valves are employed. Naturally, if off-loading is achieved through open-centre directional valves the filter will see full pump output for this period.

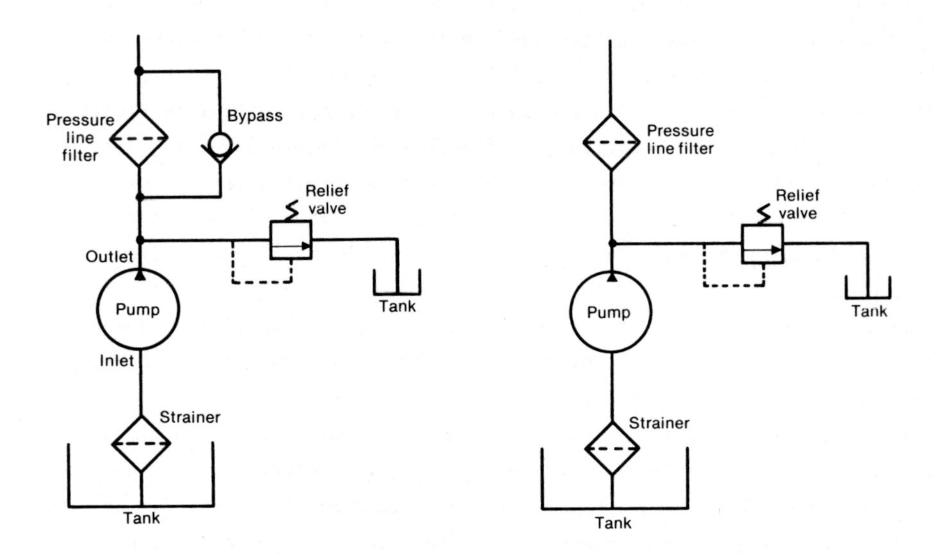

Fig.27 Pressure line filtration with bypass filter.

Fig.28 Pressure line filtration with non-bypass filter.

To increase the flow across the filter it has become common practice to locate the pressure filter between pump and relief valve (Fig.29). The advocates of this arrangement point to the fact that the relief valve is protected from pump generated dirt. A valid point, however, is that this generated dirt is caused by pump wear which, in turn, is directly proportional to the contamination level of the fluid entering the pump inlet. The pump is often a costly component and we should therefore direct more attention to reducing the contamination entering the pump to a level which will minimize wear. For this arrangement, a bypass is mandatory and there must be an assurance from the filter manufacturer that any filter malfunction will not result in excessive pressure at the pump outlet.

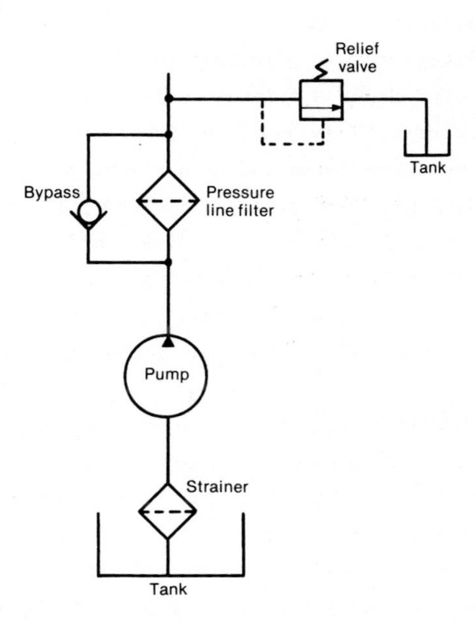

Fig.29 Locating pressure filter before relief valve gives constant flow through filter.

Where variable-displacement pumps are employed, careful analysis of the flow seen by the filter must be made. Take the classic case of a pressure compensated pump which is compensating (i.e. low displacement at maximum system operating pressure), where a low flow rate exists although the pump is still operating at pressure. With a pressure line filter dirt removal is limited. Even when flow demands are made these are limited to cylinder displacement.

As illustrated by Fig.15, the filter efficiency depends upon the type of flow it sees, although currently no standard test exists for evaluating filters when they are subjected to abrupt flow and pressure changes and mechanical vibration. Practice has shown there is a reduction in efficiency but its extent depends not only on the quality of the actual filter media but also on how well this media is supported. We must hope that standard test parameters can be agreed and a Beta ratio quoted which relates more closely to conditions seen by a filter in real hydraulic systems. In the future it will also be essential that the system designer has information on the performance of integral bypass valves under system operating conditions.

In addition to mechanical vibration and abrupt flow and pressure changes when valves are operated, a pressure line filter is also subject to pump pulsations.

These effects were demonstrated recently on a hydraulic component test rig where the so-called 10 micrometre nominal filter proved almost totally ineffective in removing sub-5 micrometre contaminant.

Because pressure filters have to withstand the full system pressure with adequate margin of safety, there is a tendency to make these small, thus limiting their dirt holding capacity. It is usual for manufacturers to quote a rated flow at around 1 bar for bypass types and at a slightly higher pressure for non-bypass types which incorporate high pressure differential capability elements.

In summary, it can be said that pressure filters are subjected to conditions far removed from that found in a laboratory filter test rig, and because of this the designer's task in assessing the resultant contamination is extremely difficult. The levels given in Fig.15 are based on field experience using good quality elements.

Pressure filters may well be used in certain applications to protect the system should catastrophic failures of the pump occur, or to provide special protection to a single unit or group of units. An example of the latter would be a servo valve, where failure might be extremely expensive, though it should be noted that a pressure filter does not protect a servo valve from dirt ingressed through cylinder rod seals. In such cases a non-bypass filter should be considered with the assurance that total element collapse cannot occur. If bypass types are employed, some means of indication should be provided to give warning of a partially blocked element.

From a maintenance point of view, changing elements involves stopping the system unless external bypass valving is provided. This operation often allows free air into the system which must be cleared before satisfactory machine performance can be obtained.

13.6.3 Return Line Filtration

The usual return line filtration arrangement shown in Fig.30 has all return lines passing through the filter. Drain lines from pumps, motors, and certain valves should not be subjected to pressure surges emanating from the system return lines and should return separately to tank.

Where there are high surges (e.g. due to uncontrolled decompression or the rapid acceleration of the fluid column in the relief valve tank line when this valve operates) it may be undesirable to pass these through the filter. To prevent collapse of the element due to high-viscosity oil, e.g. at a cold start, or when the element is loaded with dirt, an internal bypass should be provided.

When bypassing occurs under minimum flow and surge conditions the circulated fluid should not be contaminated by dirt already retained in the filter. The type of surge experienced in most return lines will reduce the filter efficiency and in the absence of precise data the filter should be selected using the pulsating flow condition from Fig.15.

Full flow return filtration should be of sufficient capacity to handle the maximum return flow (including any in excess of pump flow where, for example, unbalanced cylinders are used) without the bypass opening.

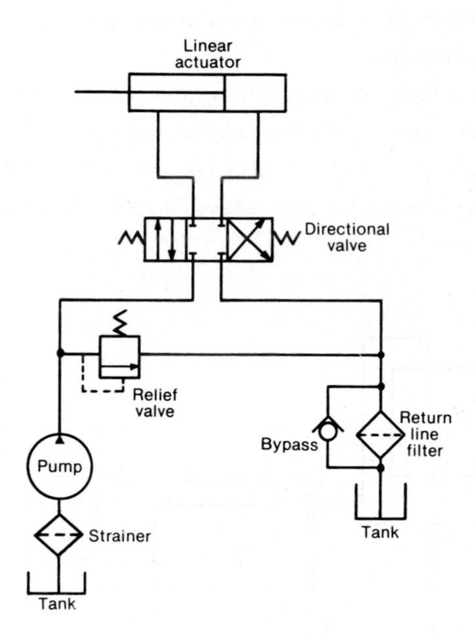

Fig.30 Basic arrangement of return line filtration.

Careful assessment of the flow across the filter is necessary. The comments relating to the use of variable displacement pumps with a pressure filter again apply, since with pressure compensated pumps the filter sees only the volume displaced by the cylinder.

For certain systems partial return line filtration is acceptable with the bypass (internal or external) always passing a percentage of the flow. With this arrangement much depends on the continuous rate of flow which we can get across the filter. Satisfactory performance can sometimes be achieved with as little as 10% passing across the actual element.

To summarize, a return line filter does not protect the system from environmental dirt entering the reservoir via breathers or during topping-up. However, if we start off with a clean reservoir and take precautions to prevent the ingress of environmental contamination, then experience has shown that effective filtration can be achieved economically with return line filters.

13.6.4 Off-Line Fitration

It has been stated that the effectiveness of filters sited in both pressure and return lines is reduced by shocks, surges, pulsations, vibrations, etc. to an extent which depends on the type of media and how well it is supported. Steady flow relatively free of pressure fluctuations provides optimum filter performance. The simplest way of achieving this is to remove the filter from the main system and place it in an independently powered circulating unit where its performance is more predictable.

Though connected to the system reservoir, a typical arrangement is shown in Fig.31. The likely contamination level from our three filter ratings under steady flow conditions is given in Fig.15, and our ability to provide the correct flow rate will determine whether or not these levels can be achieved and maintained.

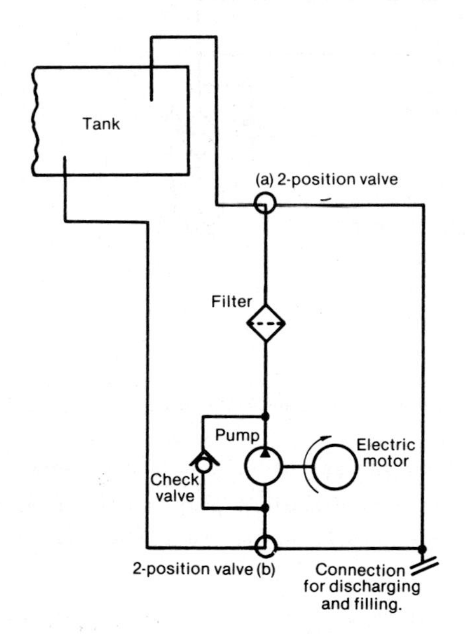

Fig.31 Layout of off-line filtration system. Valves (a) and (b) can be positioned so that the unit will provide a filtered fill or discharge.

Assuming reasonable standards of engineering in the design and build of the system, the most relevant factors in selecting the flow rate through the off-line filtration are likely to be environment and tank size. From field experience to date the guidelines given in Fig.32 can be applied and the flow rate derived can be used to determine the size of the filter by the procedure previously outlined.

Environment	Flow rate (1/min) as percentage of tank capacity (litres)
Good	5%
Average	10%
Bad	20%

Fig.32 Suggested flow rates for off-line filtration.

With off-line filtration the designer's task is simplified because he is not governed by the flow and pressure characteristics of the main system. It enables him to select the best filter and the best flow through it and then select the size to achieve the desired frequency of maintenance.

Should the desired contamination level not be achieved, then corrections are easily made to the flow rate or type of filter without in any way affecting the design of the main system. Furthermore, the off-line installation can be run prior to starting the main system in order to clean the oil in the tank and reduce the contamination level the pump will see at start-up. By the addition of simple valve gear it can be employed to filter the initial charge of fluid and any used subsequently in topping-up. Ideally, it should be left running continuously to provide a complete tank of clean fluid ready for every start-up.

Unlike line filters, the off-line installation will continue to clean up the fluid when variable delivery pumps are running at minimum displacement.

Being independent of the main system off-line filters can be placed where they are most convenient for servicing. When element changes are necessary the main system is not affected; the operation can be carried out at any time without stopping or introducing air into the main system, thus making the very minimum of servicing skills acceptable.

Whether or not off-line filtration can be the sole means of filtration depends on many factors relating to the character, quantity and origin of the contamination. As it is partial filtration we must decide if it is necessary to protect individual or groups of components from stray particles likely to cause a catastrophic failure.

The foregoing has attempted to show that there is a place for inlet, pressure, return and off-line filtration, which is borne out by filter manufacturers who once told us that pressure filters answered all our problems and who now offer a range of return line filters, and manufacturers of return filters who now offer ranges of pressure filters.

Surely what is most important is to realise that like all branches of engineering, system filtration engineering involves taking a risk. A single particle

Fig. 33: A pictorial representation of the basic filtration equation.

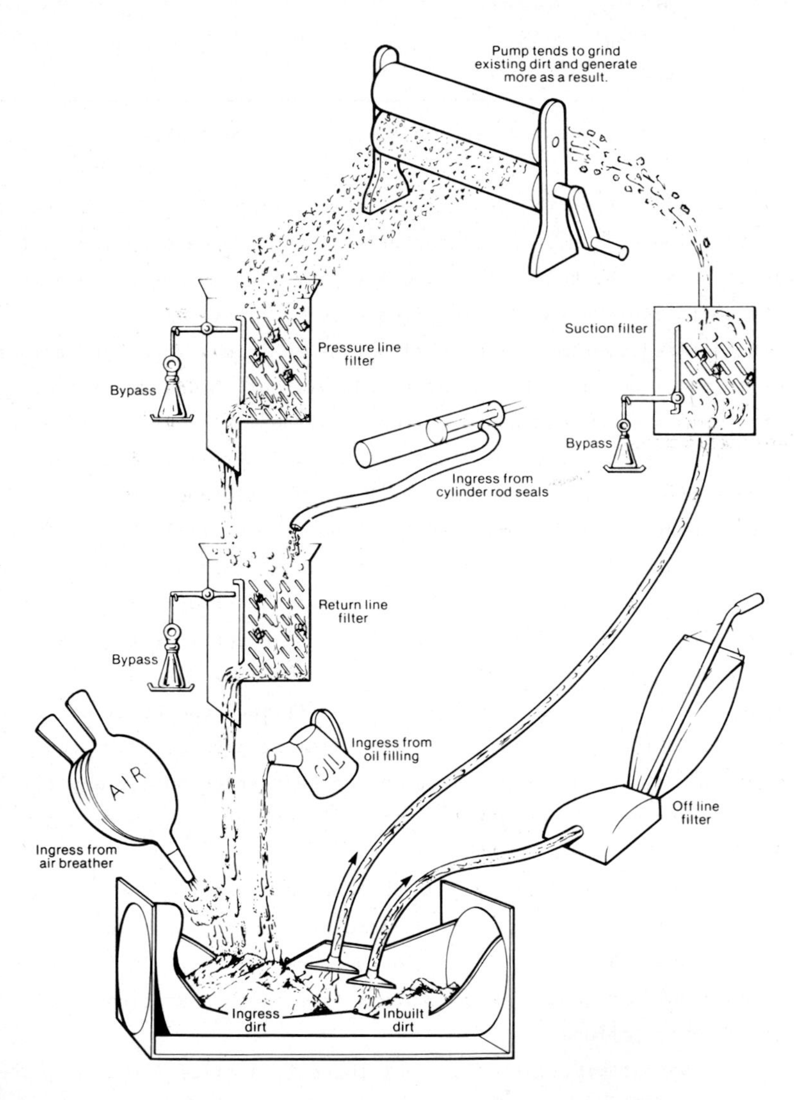

$$\text{Total system dirt} = \text{Inbuilt in component} + \text{Inbuilt in oil} + \text{Ingress from breather} + \text{Ingress from oil} + \frac{\text{Ingress from cylinder}}{\text{Return line filter factor}} + \frac{\text{Generated dirt}}{\text{Pressure \& return line filter factor}} - \text{Dirt removed by line filters} - \text{Dirt removed by off line filters}$$

say, in the 5 to 10 micrometre size range, could cause system malfunction if it arrived at the wrong place at a critical time. The machine tool designer may well have the benefit of a prototype on which to evaluate the filtration system performance and make changes prior to the first production batch. The designer of heavy equipment, say, for steelworks is afforded no such opportunity; he must be right first time and may well have to use all the filter selection procedures available to him in order to minimize the risk of expensive stoppages later. That risk will be with us until (a) we have more realistic data on filter performance, and (b) maintenance practices are greatly improved.

The message of the preceding chapters can best be summarised diagrammatically (Fig.33). This highlights the fact that control over the inbuilt contamination by applying good installation practices and using all means available to limit the amount of ingressed contamination, enables us to control the system generated contamination to an acceptable level.

The level of contamination entering the pump is a critical factor and should be so controlled that it prevents the sort of action depicted by a mangle. The difficulty in controlling this level by inlet filters has been clearly stated. In the main, they should be considered only for preventing large particles entering the pump and causing catastrophic failure.

Downstream of the pump the ability of a pressure filter to trap particles will be influenced to a large extent by any abrupt changes of flow and pressure, which have the effect of driving particles through the filtering media. We know from practice that bypass valves can malfunction under certain circumstances, and it is essential that the design of the filter prevents any migration of contamination if this should happen.

Ingress through cylinder rod seals, although in the smaller particle range, can nevertheless add up to significant quantities. The performance of any return line filter will, again, depend upon the magnitude of changes it sees in terms of flow and pressure.

Off-line filtration enables optimum filter performance to be accomplished, thus enabling us to control contamination levels more effectively. Our ability to remove contamination depends not only on the filter micrometre rating but also on the flow rate across it, and we must provide sufficient dirt holding capacity to ensure an element life acceptable to the end user.

13.7 SUMMARY

Any examination of the subject of contamination involves four groups of people:

(i) The fluid manufacturer or supplier.

(ii) The hydraulic equipment and filter manufacturers.

(iii) The manufacturer of the machinery which uses the hydraulic equipment.

(iv) The end user of the machinery.

Each of these has a commercial responsibility to supply equipment which will perform its duty satisfactorily at a reasonable price, and each must have some knowledge of the cleanliness of the working fluid.

The fluid supplier will supply fluid as clean as required and will charge accordingly.

The hydraulic equipment manufacturer must advise the user to whom he is selling his equipment on the type of fluid and its degree of cleanliness best suited to the hydraulic equipment he is recommending for a specific application.

The machinery manufacturer, who is responsible to the end user for supplying equipment, must be aware of all the conditions to be met, including supply of fluids, servicing facilities, and the type of reliability which has to be maintained. It is his prerogative to offer his customer the best commercial proposition and this could mean, for example, either an inexpensive throwaway unit which must be replaced relatively frequently, or a more expensive item for which the only service needed is the occasional exchange of a relatively inexpensive component, such as a filter cartridge.

The end user has to make the final judgement. He needs to be able to compare the real value of each installation offered to him; each user will place different emphasis on the value of each type of equipment offered. To the end user the value of a pump is its fitness for the purpose, how long it will perform its duty satisfactorily, and what the cost of servicing will be. He is not interested in how many 10 micrometre particles are contained in 100 ml of hydraulic fluid.

The user's interest is in the least expensive filter that will provide the required degree of cleanliness. In making this assessment, the original or capital cost of the equipment has to be balanced against the cost of service. He may have to compare, for example, the advantages offered by a very expensive pump that will operate on 'dirty' oil with those of a low-cost pump plus filter.

If it is to maintain its present high integrity, the hydraulic equipment industry needs to establish more meaningful specifications for different types of filter. The specifications must allow the end user to buy performance without necessarily having to know how this performance is achieved. The people who need to write these specifications are those who know the actual requirements. The responsibility must be divided between manufacturers of filter elements, who should know what is possible, and the manufacturers of hydraulic equipment, who should know what is needed. These two groups should be able to communicate in meaningful terms.

At present, there are no adequate techniques covering all aspects of contamination measurement in hydraulic fluids. This should not deter us from making a start in the right direction. The cleanliness level chart, for example, will not be 100% right first time; there will be a continual need to review it in the light of new experience.

The escalating cost of equipment downtime and maintenance must encourage the end user to consider running costs more carefully, for both new and existing plant. To be able to do this he must appreciate more fully the part that contamination control plays in determining those costs for the hydraulic installation. It is hoped that this volume has given a useful insight to the real nature of the problem and has pointed the way to a systematic approach and more cost-effective solutions.

This chapter has been edited by kind permission of Sperry Vickers from a more comprehensive paper "Effective Contamination Control in Fluid Power Systems", written by J.B. Spencer and C. Balmer, published by Sperry Vickers.

14 SEALS FOR FLUID POWER EQUIPMENT PART ONE

B.D.HALLIGAN, C Eng,MIMechE AMPRI, Technical Manager
(Product Applications) James Walker & Co.Ltd.

14.1 INTRODUCTION

The fact that a seal manufacturing industry exists at all originates from the commercial or physical impracticability of achieving perfect and permanent mutuality of adjacent surfaces - be they in dynamic or static relationship.

Sealing devices, which are usually of an essentially deformable nature, permit the fluid power designer to work within economically sensible conditions of fit, surface finish, tolerance and fluid cleanliness across the range of temperatures and pressure conditions with which he is likely to contend.

The designer of fluid power equipment has, in fact, influenced trends in the sealing industry by going beyond the fundamental requirement of introducing an environmentally tolerant component for preventing leakage. The quest to minimise production costs has accelerated the movement towards seal designs embodying integral bearings and anti-extrusion devices capable of supporting seals against substantial extrusion gaps at significant pressures. Overall in the context of those seals specified for linear dynamic motion, a much axially shorter seal has evolved which allows hydraulic cylinders, for example, to be produced to more compact proportions with the consequent saving in raw material costs and in the bulk of fluid power sub-assemblies.

None of this is bad. However it does mean that a proliferation of seal designs and materials exists and those responsible for selection are required to make their choice with discretion to ensure that the most cost-effective solution is found. Picking a seal which facilitates the achievement of the cheapest production costs for the component into which it fits is seldom the most reasonable basis for guaranteeing minimum leakage and maximum serviceability.

It is an unfortunate fact that most fluid leakages are attributed to a faulty seal, whereas investigation will frequently show that the number of occasions of malfunction due to a faulty product are few. The real source of difficulty usually stems from incorrect seal selection, working conditions which differ from those considered to apply, seal containments outside specified tolerances in terms of dimension or finish, faulty fitting, contaminants in the fluid to be sealed or, importantly, personal interpretation of 'leakage' in quantitative terms.

To understand the limitations of fluid seal performance it is necessary to know a little of the materials most typically used, the origins of fundamental seal designs, how they function, and their refinement to the present state of the art. Fortunately, the seal is no longer an after-thought but more often receives the attention it properly deserves as a vital interactive element in any complete fluid power system.

14.2 MATERIALS

Within the scope of this chapter, attention will be centred on solid elastomeric seals, typified by the 'O' ring for static connections, elastomer containing seals such as cylinder packings based on proofed textile, and those designs which employ plastic components for various functions.

The compounding of elastomers is regarded by many as being a black art and it is true to say that widely differing properties can be developed by varying the quantity and type of rubber chemicals added to the stock rubber in either internal mixers or on mixing mills. The raw material supplied to the seal manufacturer must usually be processed to tailor the physical characteristics required for the sealing function. Parameters such as tensile strength, elongation at break, compression set, brittle point, and behaviour in control fluids are typical of those reviewed when assessing performance requirements and laying down compound specifications.

In addition to application suitability, processability in mixing, extrusion and moulding sequences must be assured as must satisfactory post-moulding operations such as de-flashing.

To maintain consistency of seal elastomers requires extensive laboratory support and continuous standard testing of the compound itself on a batch-to-batch basis.

Tables 14.1 - 14.3 are taken from the British Rubber Manufacturers' Association 'Guide to Elastomeric Seals' to give broad reference to most commonly used elastomers, their properties, fluid compatibility and temperature resistance.

TABLE 14.1

Polymers-types and General Properties

NATURAL POLYISOPRENE	(Natural Rubber-NR)
SYNTHETIC POLYISOPRENE	(Polyisoprene-IR)
Very good	General physical properties, in particular tear resistance, resistance to glycol-ether brake fluids and vegetable oils.
Poor resistance to	Heat, weather, ozone and mineral oil.
CHLOROPRENE (NEOPRENE-CR)	
Very good resistance to	Ozone, weathering and abrasion
STYRENE-BUTADIENE (SBR)	
Very good resistance to	Abrasion, glycol-ether brake fluids and vegetable oils.
Poor resistance to	Mineral oil and ozone.
ACRYLONITRILE-BUTADIENE (Nitrile-NBR)	Low, medium, and high nitriles are available based on increasing acrylonitrile content which significantly affects low temperature and fluid swell properties.
Very good resistance to	Mineral oil, compression set and abrasion.
ISOBUTYLENE-ISOPRENE (Butyl-IIR)	
CHLORO-ISOBUTYLENE-ISOPRENE (Chloro-Butyl)	
Very good resistance to	Tear, weather and gas permeation.
Poor	Tensile properties and mineral oil resistance.
POLYBUTADIENE (Butadiene-BR)	
Very good resistance to	Tear, abrasion, low temperature, glycol-ether brake fluids and vegetable oils. Offers high resilience.
Poor resistance to	Water, ozone and mineral oil.
POLYSULPHIDE (Thiokol-T)	
Very good resistance to	Ozone, mineral oils, petroleum fuels and weather.
Poor	All other properties.
ETHYLENE PROPYLENE (EPR-EPM)	
ETHYLENE PROPYLENE TERPOLYMER (EPT-EPDM)	
Very good resistance to	Weather, ozone, heat, water, steam, glycol-ether brake fluids and vegetable oils.
Poor resistance to	Mineral oil.
CHLOROSULPHONATED POLYETHYLENE (Hypalon-CSM)	
Very good resistance to	Water, ozone, abrasion, acid and weather.

TABLE 14.1 (contd.)

METHYL-VINYL SILOXANE (Silicone-VMQ)	
PHENYL-METHYL-VINYL SILOXANE (Silicone-PVMQ)	
Very good resistance to	High and low temperature.
Poor resistance to	Abrasion, tear and tension.
TRIFLUOROPROPYL SILOXANE (Fluorosilicone-FMQ)	
Very good	Mineral oil and fuel resistance. All other properties as VMQ.
POLYURETHANE DI-ISOCYANATE (Urethane-AU)	
Very good resistance to	Mineral oil, abrasion, tear, ozone and weather. Offers high modulus and tensile properties.
Poor	Moist heat resistance.
FLUORINATED HYDROCARBON (Fluorocarbon-FPM)	
Very good resistance to	High temperature (in air and most oils), weather and petroleum fuel.
Poor	Tear strength.
POLYACRYLATE (Acrylic-ACM)	
Very good	Heat, weather, mineral oil and ozone resistance.
Poor	Water resistance.

Of major interest to the fluid power industry are:-

Acrylonitrile-butadiene (Nitrile - NBR)

Probably upwards of 80% of seals supplied to the fluid power industry are based on nitrile compositions. The balance of properties available from NBR in terms of good compression set qualities, abrasion resistance, mineral oil compatibility, ease of processing and low initial cost favour this choice.

Nitrile compositions are not, however, compatible with fire-resistant fluids of the phosphate ester type.

Isobutylene-isoprene (Butyl - IIR)

Satisfactory in service with phosphate ester fluids butrather lifeless from a physical point of view. Not suitable for mineral oil service, neither should butyl seals be smeared with mineral oil or grease on fitting.

Ethylene propylene (EPR - EPM)

or Ethylene propylene terpolymer (EPT - EPDM)

Same service function and limitations as butyl but generally better physical properties.

Fluorinated Hydrocarbon (Fluorocarbon - FPM)

The only usual selection where compatibility with mineral oil and phosphate ester is required.

Polyurethane (Urethane - Au)

Widely favoured as a dynamic seal material in the US and in Germany, polyurethane exhibits excellent abrasion resistance and is very durable in situations which might adversely affect nitrile compositions or even proofed fabrics, e.g. passage of seal lips over ports in a cylinder wall. Poor low temperature flexibility, hydrolysis in hot water and a lower operational temperature ceiling are limiting factors.

Other seal component materials worthy of mention are:-

Textiles - woven cloth such as cotton, asbestos, terylene and nylon are used as the substratum in elastomer proofed fabric packings for medium and high pressure duties.

Non-woven materials such as polyester and polyamide are also available.

The proofing elastomer can be varied to suit but will frequently be based on NBR and/or chloroprene (Neoprene-CR).

PTFE - employed as a back-up ring material for '0' rings. Poor creep properties restrict its use as a seal material in its own right unless energised by an elastomeric component or spring.

Nylon and acetal - main function is heel support for dynamic seals or as component bearing ring material.

Hytrel - the registered trade name of a Du Pont range of thermoplastic polyester materials which form a bridge between elastomers and thermoplastics. Has similar properties to polyurethane but exhibits much better low temperature flexibility. Used currently as an anti-extrusion element in certain spool-type piston head seals.

When referring to rubber technology as a 'black art' there is a two-fold interpretation. On the one hand it is certainly a manufacturing area influenced by many variables. On the other, the end product used in a sealing context is invariably black in colour due to the use of carbon black as a filler necessary for toughening the finished article. Seal identification is therefore a major problem unless housekeeping is of a high order with seals positively separated and bagged in small numbers having full specification of material, size, and part number clearly shown.

Colour coding is one solution but the real answer lies in the use of self-coloured elastomers. At the moment a good deal of development work is being

TABLE 14.2 R-Recommended A-Acceptable N-Not recommended

	NATURAL RUBBER	NEOPRENE	NITRILE	BUTYL	EPDM	FLUORO CARBON	SILICONE	ACRYLIC	URETHANE	FLUORO SILICONE	SBR	THIOKOL	BUTADIENE
Air or Oxygen	N	A	A	R	A	R	R	R	R	R	N	R	A
Dilute Acid	R	R	A	R	R	R	N	N	N	A	R	R	A
Dilute Alkali	R	R	A	R	R	A	N	N	N	A	R	R	A
Water	R	A	R	R	R	A	A	N	N	R	R	R	R
Lower Alcohols	R	R	R	R	R	R	N	N	N	R	R	A	A
Commercial Petrols	N	N	A	N	N	R	N	A	A	A	N	R	N
Fuel and Diesel Oils	N	N	R	N	N	R	N	R	A	A	N	R	N
LUBRICANT OILS:													
(a) Mineral Based	N	R	R	N	N	R	A	R	R	R	N	A	N
(b) Synthetic Based	N	N	A	N	N	N	N	N	N	A	N	A	N
HYDRAULIC OILS:													
(a) Mineral Based	N	N	R	N	N	R	N	R	R	A	N	A	N
(b) Ester Based (non-flam)	N	N	N	A	A	R	A	N	N	N	N	A	N
(c) Water Glycol Based	N	N	R	R	R	A	A	N	N	N	N	A	N
(d) Chlorinated	N	N	N	N	N	A	N	N	N	N	N	N	N
(e) Silicone Based	N	N	A	N	N	A	N	N	N	N	N	N	N
(f) Glycol-ether brake fluids and vegetable oils	R	A	N	A	R	N	N	N	N	N	R	N	R

TABLE 14.3

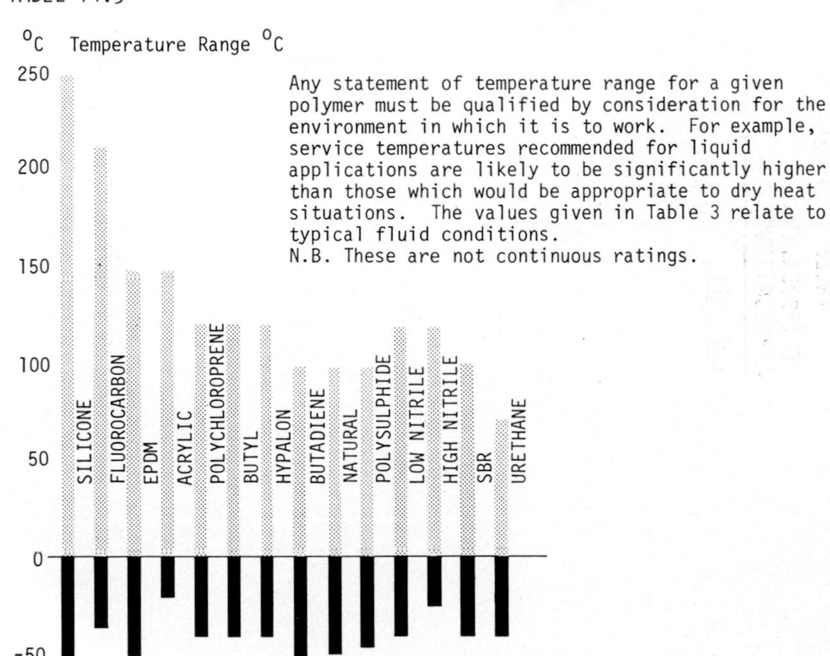

Any statement of temperature range for a given polymer must be qualified by consideration for the environment in which it is to work. For example, service temperatures recommended for liquid applications are likely to be significantly higher than those which would be appropriate to dry heat situations. The values given in Table 3 relate to typical fluid conditions.
N.B. These are not continuous ratings.

TABLE 14.4 TYPES OF SEALS FOR RECIPROCATING, ROTARY AND STATIC APPLICATIONS

RECIPROCATING APPLICATIONS	APPLICATIONS	AVAILABLE STANDARD MATERIALS
LIP TYPE		
U Ring (Also known as U Packing) 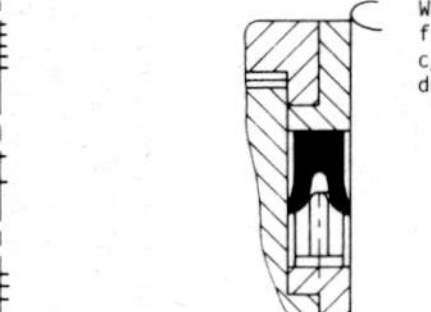	Wide variety of rod and piston head sealing duties from low pressure pneumatic cylinders to hydraulic cylinders, valves, main rams of up-stroking presses, down-stroking presses, drawback rams, jacks and hoists.	
		Solid elastomer—for pressures up to 30 bar—good film wiping and low pressure sealing characteristics—ideal for air cylinder. Prone to abrasive wear, extrusion and fluid cutting at higher hydraulic pressures. Can be fitted in recesses requiring no separate access.
		Solid elastomer with plastic heel—for pressures up to 100 bar—as solid rubber but higher pressure capability due to enhanced extrusion resistance.
		Fabric—for pressures up to 350 bar although robust constructions of less sensitive profile will extend this range. Frequently fitted with an adaptor ring (as shown in the assembly sketches) or an internal lip spreader. Spring loaded adaptors are available to augment low pressure sealing ability. Usually require separate housing access such as removable cover plate but special fabric U-rings for inaccessible grooves are available. Other materials include polyurethane, PTFE and leather.

None of these designs are normally recommended for shock loading, rams subject to lateral thrust, abrasive conditions or high speed reciprocation. U rings seal in one direction only. A double-acting arrangement will require two seals to be rigidly separated and each facing the pressure source.

V Ring (When in set form, more commonly known as Chevron Packing)	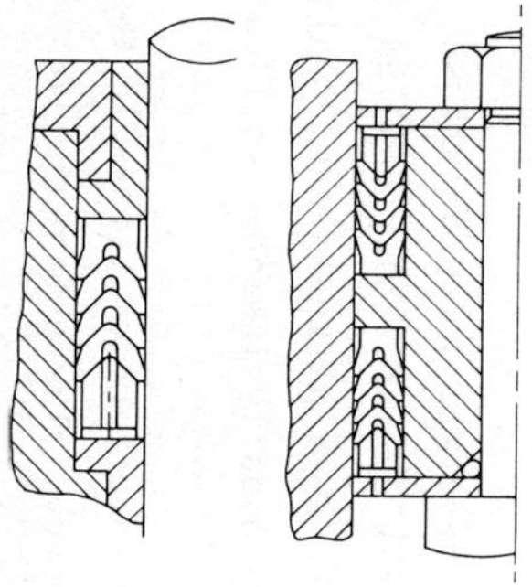	Extremely wide range of hydraulic services—most types of hydraulic press, reciprocating steam, oil and water pumps, sludge pumps, hydraulic cylinders, oil pipeline expansion glands, hydraulic valves.	Normally proofed fabric. Others include elastomer, PTFE or leather
Heavy Duty Lip Packings	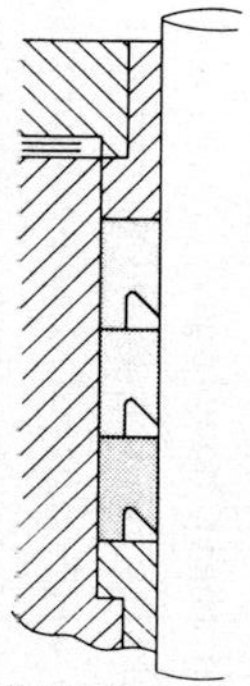 (a) Single Lip (b) Two-Lips Single Lip 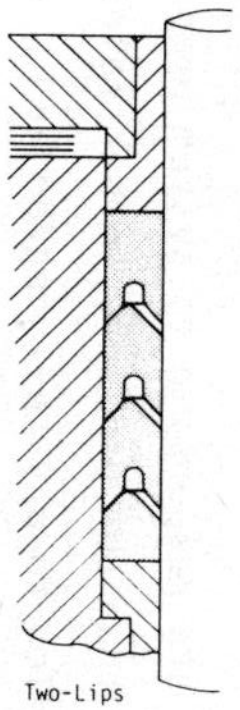 Two-Lips	(a) Up-stroking slow moving press rams, reciprocating pump rams, lifts, cranes, hydraulic accumulators. (b) Large diameter rams, up-stroking or down-stroking (where frictional resistance is not of paramount importance) even where subjected to shock loads, e.g. in forging presses; horizontal rams of any diameter including those of extrusion presses, rams subject to lateral thrust; high pressure reciprocating pumps; those heavy duty applications where ram packing housing and busnes are not in first class condition.	Proofed fabric.

carried out to achieve this end without any significant loss in physical properties. Coloured fluorocarbon rubbers are already available.

14.3 SEAL DESIGNS

Within the boundaries set by fluid power systems the majority of seals employed can be categorised as lip or squeeze types.

Figure 1 illustrates the simplest forms utilised for static or dynamic reciprocating situations - the 'U' ring and the 'O' seal - and indicates how both rely on interference stress for at least a component of their functional operation.

Fig.1 Lip and Squeeze Seals

Both are responsive to system pressure and will generate a radial stress greater than the pressure to be sealed. Neither rely on externally applied compression to any real degree as in the case of the soft packed gland. This is particularly relevant to squeeze seals which are usually intended to float axially in their housings.

The quality of the sealing contact area and the interference stress in the same zone will control low pressure sealing efficiency. These factors plus overall seal geometry and composition will set the point at which time hydraulic response takes over from the manufactured interference condition. The hydraulic component may become significant at pressures of 40 bar for an elastomeric 'O' ring or as high as 200 bar for a rigid proofed fabric packing.

Sealing slack oil and low pressures is generally a function of seal integrity and is the more difficult condition. Preventing leakage at high pressures is a feature of seal containment, i.e. preventing extrusion or rapid wear if in a dynamic duty.

Tables 14.4 and 14.5 list typical lip and squeeze types in common and traditional usage. (Courtesy of BRMA).

Some particular points to note on each category:

(i) Lip seals - are single-acting to applied pressure and must be separated by a fixed component in double-acting duties so that the hydraulic load from the element under pressure is not passed on to the trailing element.

For maximum service life multi-ring packing sets are preferred where the succession of sealing edges ensures that breakdown is not sudden. The use of split rings is also entirely feasible with most designs of this sort without sacrificing sealing performance to any critical degree. Savings in downtime will be obvious.

(ii) Squeeze seals - have the advantage of being double-acting and are usually housed more economically than their corresponding lip brethren. The lineage from the humble 'O' ring to the unit seals in contemporary use can be seen from Table 14.5.

Most seals in this category use a single sealing zone and damage in service will be followed by more immediate breakdown than in the case of multi-lip packings.

The squeeze seals are rarely used in split form although they can be engineered so to do.

Tables 14.6 and 14.7 depict the extension of lip and squeeze seals to rotary and static functions.

More advanced seal designs are detailed in the next section.

14.4 TRIBOLOGICAL CONSIDERATIONS

It is heartening to find the study of seal behaviour an essential part of tribology seminars, receiving as much attention, indeed, as bearings, lubricants and surface topography. In recent years much company research and independent study by organisations such as BHRA has been expended on defining sealing mechanisms - particularly in relationship to reciprocating motion.

14.4.1 Film Conditions

All dynamic seals rely on a coherent fluid film under their contact area if they are to function consistently and predictably. Such films may stem from boundary lubrication in some modes and be truly hydrodynamic in others. The film will vary in thickness according to seal profile, interference stress, pressure, speed, surface finish, type of fluid employed and its temperature. On reciprocating duties it might typically vary from 0.25 - 3.0 microns. On rotary shafts values of 0.6 - 1.0 micron would usually apply.

Leakage from rotary shaft lip seals is seldom evident unless seal wear down or under-lip cracking has taken place - all other features being equal. However,

TABLE 14.5

RECIPROCATING APPLICATIONS	APPLICATIONS	AVAILABLE STANDARD MATERIALS
SQUEEZE TYPE		
O-Seal 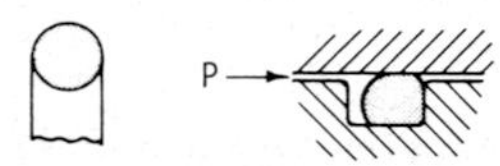	Light and medium duty, pneumatic and hydraulic reciprocating services, e.g. small cylinders, valve spools and stems, rod film wiping.	Principally elastomer but polyurethane and PTFE are also manufactured.
Lobed Seal 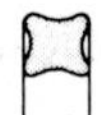 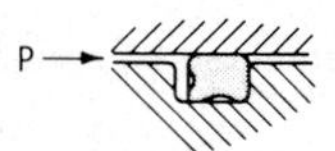	As O-seal but better resistance to spiral twist.	Elastomer.
Energized Sleeve Seal 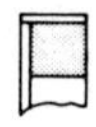 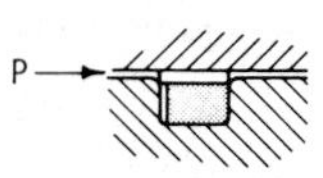	Pneumatic and hydraulic cylinders particularly where sensitivity is essential, e.g. weighing machines, testing equipment, etc. Primarily for piston heads but also available for gland duties.	Elastomer and PTFE (plain or reinforced) or elastomer and proofed fabric.
Supported Single-Acting Seal 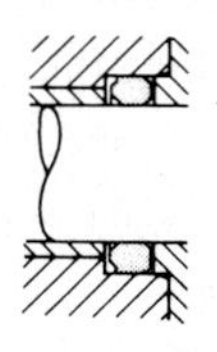	Basically developed for hydraulic cylinder rod and piston head duty particularly in mobile and industrial hydraulic environments.	Elastomer/Plastic, Elastomer/Fabric, Elastomer/Fabric/Plastic can also be made in polyurethane.
Supported Double-Acting Seal 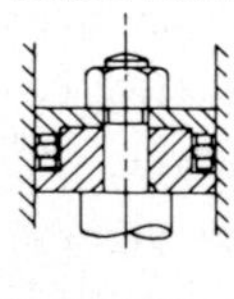 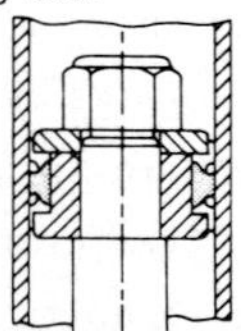	Developed for piston head duty in hydraulic cylinders associated with mobile/ industrial hydraulic industries.	Elastomer/Fabric, Elastomer/Fabric/Plastic, Elastomer/Plastic.

TABLE 14.6

ROTARY APPLICATIONS	APPLICATIONS	AVAILABLE STANDARD MATERIALS
LIP TYPE		
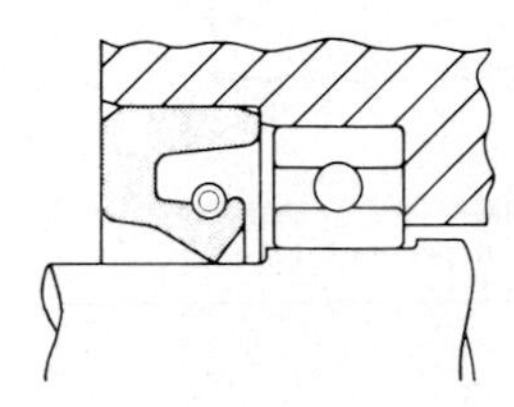	Retention of lubricant and exclusion of foreign matter from bearings and critical surfaces.	Elastomer/metal, Elastomer/fabric, Spring loading to lip normally provided by toroidal spring or finger spring.
SQUEEZE TYPE	The application of squeeze-type seals to rotary duties requires specialised attention. Consultation with the seal manufacturer is recommended.	

TABLE 14.7

STATIC SEALS	APPLICATIONS	AVAILABLE STANDARD MATERIALS
GASKETS AND JOINTINGS		
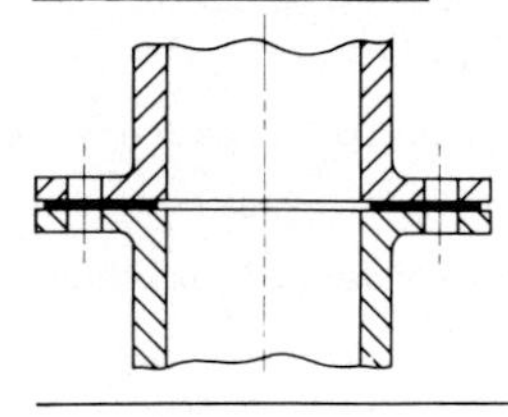	The most widely used form of static connection where external mechanical load is available.	Elastomer, reinforced elastomer, proofed fabric, elastomer bonded cork, compressed asbestos fibre and a wide variety of non-elastomer based compositions.
LIP TYPE U Ring (Also known as U-Packing) 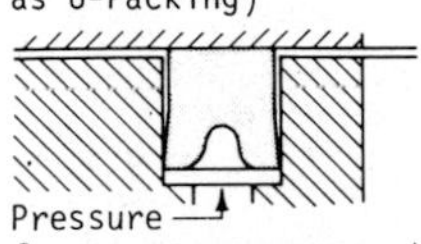 Pressure from common or separate source	A wide range of static sealing duties. Cylinder end caps, autoclave doors, pressure vessels, couplings, etc.	Elastomer, proofed fabric, leather, Polyurethane and PTFE.
SQUEEZE TYPE O-Seal 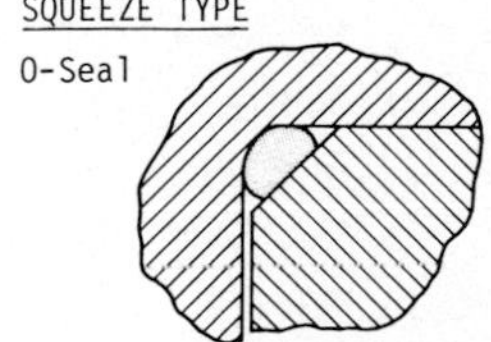	All static sealing duties which will permit the use of the selected O-seal material and which are of a design which will contain this material within the confines of the seal recess at operating pressure.	Elastomer and PTFE.

in reciprocating applications passage of the shaft through the gland will carry the oil film to the atmosphere side of the seal.

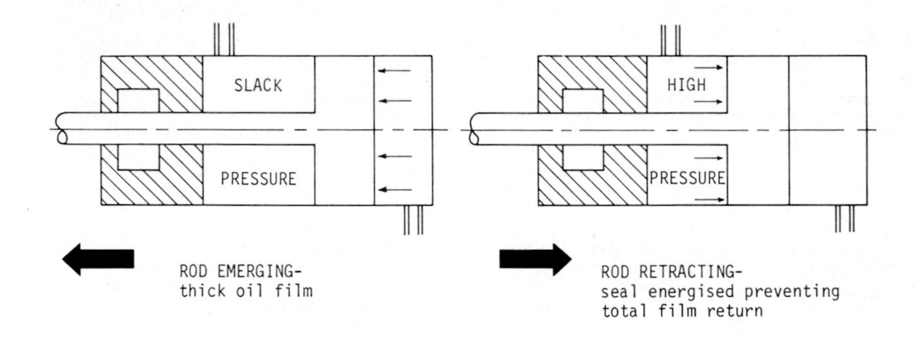

Fig.2 Condition of Maximum Collected Oil Film.

Figure 2 is a diagrammatic representation of the worst film transport situation involving an emerging cylinder rod with slack pressure to the gland which retracts with the sealing element under load. In this mode the heel of the seal will be energised and can prevent the return of the total quantity of film carried by the rod. The fitting of an effective wiper can aggravate the condition.

To combat the emergent film, careful attention is required to pressure side seal geometry. Designs such as that shown in Fig.3 with knife-cut sealing edges and a specific relationship between contact edge and groove heights have proved very successful despite their short axial length.

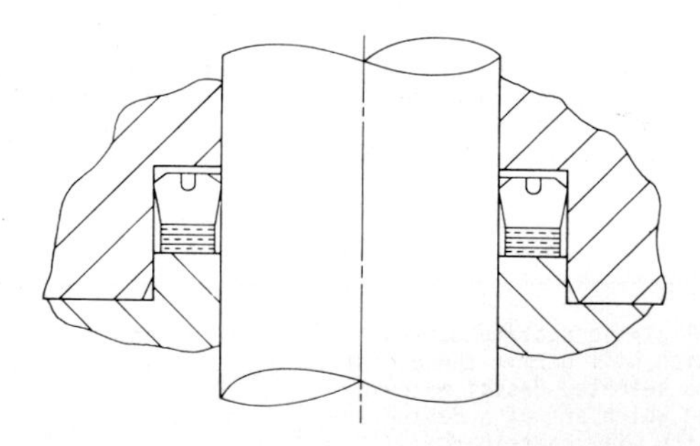

Fig.3 Minimum Film - Shallow Design

Equally efficient and having the advantage of more than one sealing edge is the concept shown in Fig.4 which is a marriage of lip and squeeze sealing principles.

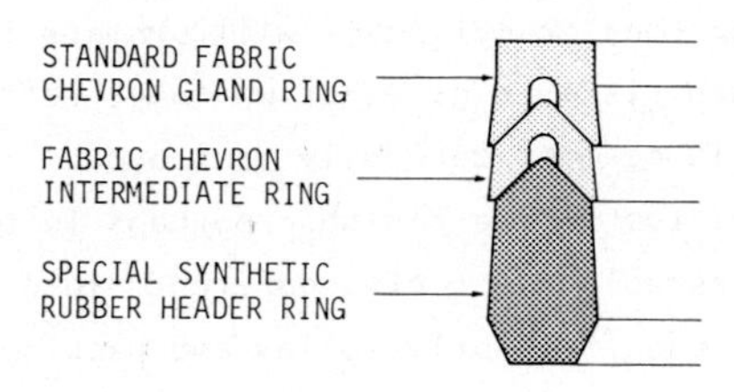

Fig.4 Minimum Film - Multi-Lip Design

Figure 5 illustrates typical improvement in performance against standard V-ring packing under offset load conditions and shows its adaptability even in split form.

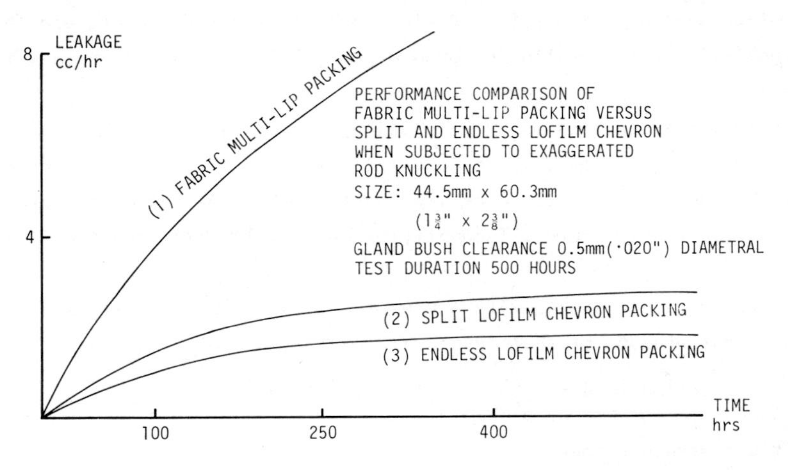

Fig.5 Comparative Seal Performance

14.4.2 Surface Finish

It is not only the numerical value of surface roughness which is relevant to seal performance but also the manner in which such a finish is achieved.

In the context of fluid power equipment, honed or roller burnished barrel finishes are recommended with an average value of between 0.4 to 0.8 μm Ra. Larger cylinders should be finished to 0.8 μm Ra or better, if possible, but in any event should be no worse than 1.6 μm Ra.

Rods should in all circumstances be of a surface roughness 0.8 μm Ra or better. Many of the smaller sizes will be available from proprietary rod suppliers to a standard finish of 0.2 μm Ra. Type of finish will depend on material, function, and the designer's experience of similar equipment.

All static housing areas may be finished in the range 0.8 to 1.6 μm Ra for proofed fabric packings and, preferably, 0.4 to 0.8 μm Ra for smaller housings appropriate to 'O' rings, etc.

Any longitudinal marking on rods or cylinders will promote leakage and manufacturing processes posing such risks, e.g. as-drawn tube, retraction marks on roller-burnishing heads, should be most carefully monitored.

Equally, the achievement of too fine a finish - perhaps in the area 0.01 - 0.05 μm Ra - can prevent the establishment of a coherent fluid film under the seal. An erratic performance can frequently follow and packing life can be curtailed. Several practical cases have been demonstrated where the deliberate introduction of a coarser finish has restored an adequate quality of sealing. In any event, on most hydraulic installations the advantages to be gained by improving finishes below 0.2/0.3 μm Ra are disproportionately costly without offering tangible performance improvement.

14.4.3 Seal Friction

In estimating seal drag loads for critical applications, designers are faced with the practical problem of finding even a general idea of actual values from seal manufacturers.

There are real difficulties in extrapolating results based on laboratory equipment and applying values to much larger plant. Equally, type of lubrication, choice of packing, and degree of lip interference, whether as-moulded or as a result of compression, are all quantities which will influence the final result.

Work carried out in this area with a view to giving a notional allowance for friction takes due account of seal length and expresses seal drag load as :

$$\text{Friction load} = \frac{DL}{25.8} \times [142 + (0.8 \times R)] \text{ kgf}$$

where R = fluid pressure (bar)

D = seal contact diameter (cm)

L = seal contact length (cm)

It is often desirable to add a contingency allownace of 15%. As a result, it may be seen that a multi-lip packing of 1000 mm contact diameter and 50 mm deep would require 0.2% approx. of the thrust developed on this full area diameter at a pressure of 200 bar.

Figure 6 makes the point that any elastomeric and therefore flexible seal construction will suffer an increase in contact band width with increasing system pressure. Even in the case of relatively tough, high modulus materials such as proofed fabric or polyurethane, pressures of 200 bar or over may be sufficient to promote total axial contact of the seal with the dynamic wear surface.

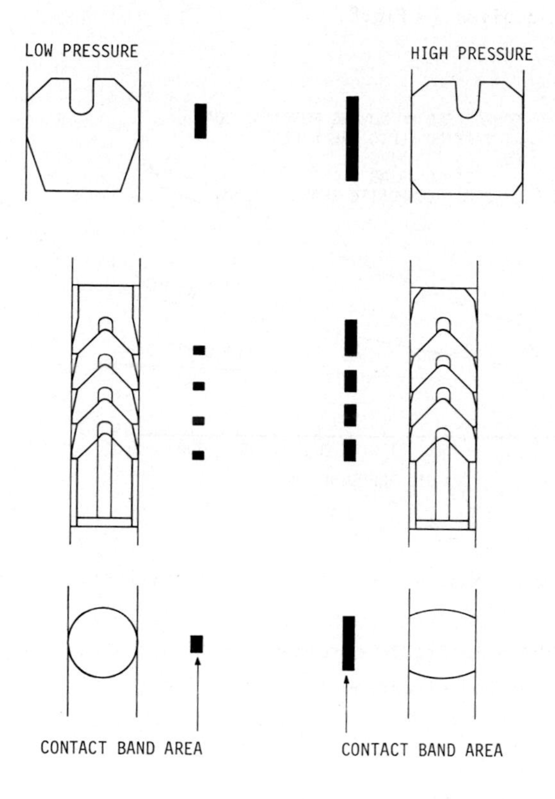

Fig.6 Seal Contact Band Width

Interposing a loaded PTFE sleeve is an effective means of limiting seal drag dramatically despite substantial hydraulic pressure increases. Figure 7 shows one form of the composite principle utilising a rectangular section elastomeric energiser.

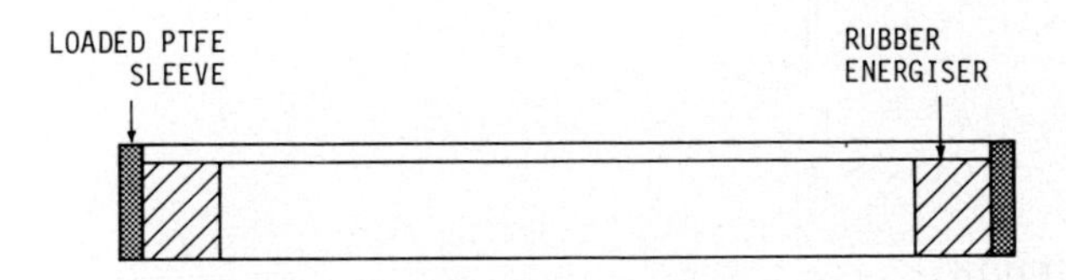

Fig.7 Energised PTFE Sleeve Seal

The order of friction control available by comparison to comparably tested 'O' rings and 'U' rings is given in Fig.8.

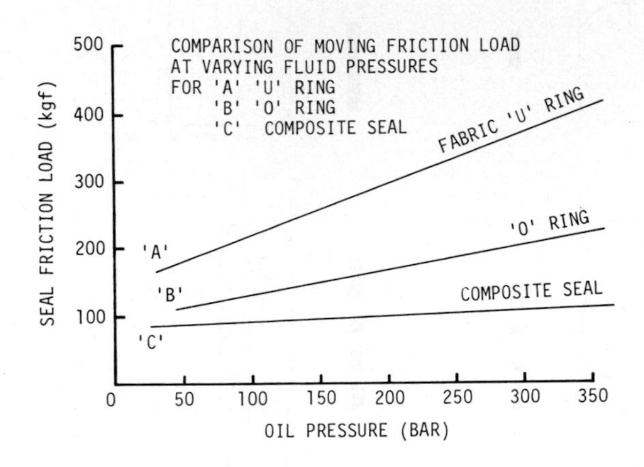

Fig.8 Comparative Friction Characteristics

The patented design shown in Fig.9 combines friction control with maximum film wiping ability for the rod situation.

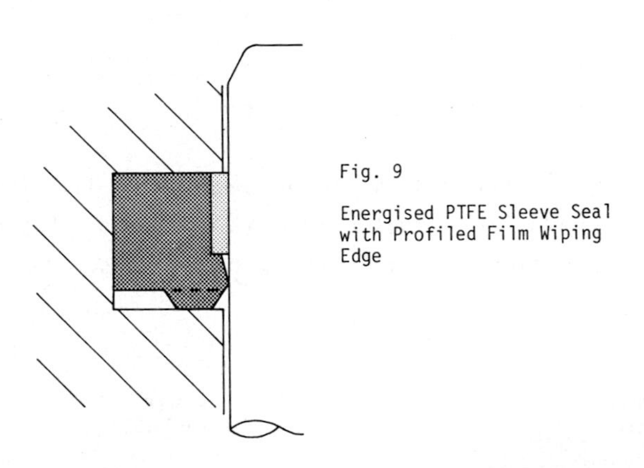

Fig. 9

Energised PTFE Sleeve Seal with Profiled Film Wiping Edge

14.4.4 Type of Fluid

As far as dynamic seals are concerned, the vast majority of mineral-based or phosphate ester fluids do not present lubricity problems.

One of the most demanding types of duty for any dynamic seal involves plain water at high pressure whether by design or by total loss of soluble oil content in a nominally lubricated system. High speeds are particularly difficult to sustain unless a minimum soluble oil level of 2% is guaranteed.

Tests conducted on three-throw mining pumps operating at 250 bar, 0.6 m/s, have shown that for even a 2% soluble oil content, average seal life will be extended by a factor of four by comparison with untreated water. Equally, a separate lubricant feed will produce similarly dramatic improvements.

The temperature of hydraulic fluid should not exceed 60°C if at all possible, as significantly faster swell and softening of proofed fabrics and straight polymers will occur above this value. To illustrate the point, the following data is based on immersion testing for seven days in Shell Tellus 27 mineral oil.

Nitrile proofed fabric -	room temperature	+ 0.7% volume increase
	60°C	1.2%
	90°C	2.6%
	120°C	3.3%
High nitrile elastomer -	60°C	0.8%
	90°C	2.3%
	120°C	3.8%

Softening of moulded fabric material usually reduces intrinsic strength and will normally diminish service life.

14.4.5 Filtration

For the bulk of elastomer proofed fabric and solid elastomer seals fluid filtration of 25 microns should be perfectly adequate from the sealing performance aspect. Special seals such as those embodying PTFE wear faces will benefit from 10 micron filtration or better. In either event, filtration equipment offering a finer cut-off will probably be specified to suit control valve functions on a given press or hydraulic component.

The majority of conventional hydraulic seals which are tested in-house work in conjunction with no more than a coarse wire strainer in any hydraulic system. However, modern filtration aids must be considered an advantage, particularly if the operation of the plant involves produces aggressive residues.

14.4.6 Air Entrainment

As cycle speeds become faster due to increasing work demands, system pressures must fluctuate more quickly. In many situations, full working pressure must be exhausted in milliseconds, e.g. die-casting machines, plastics injection moulding presses, etc. If air is entrained in the hydraulic fluid such rapid decompressions can be exceedingly dangerous if no automatic venting is available.

If one considers that in 10:1 of hydraulic oil at 200 bar and 10°C it is possible to dissolve nearly 200 1 of air, some indication of the magnitude of risk will be apparent.

The main problem relates to piston head situations where fluid collects between two opposed seals. Lip packings are largely self-venting but heavy interference polyurethane cup rings are often suspect in this direction as they do not react sufficiently quickly.

Under no circumstances should two squeeze type seals, e.g. '0' rings, be employed on a piston head as the air entrainment contingency can be aggravated by a proven phenomenon known as inter-seal pressure whereby three or four times system pressure can be built up in the annular clearance between the seals. Extrusion of such seals <u>into</u> the applied pressure can be noted in typical cases.

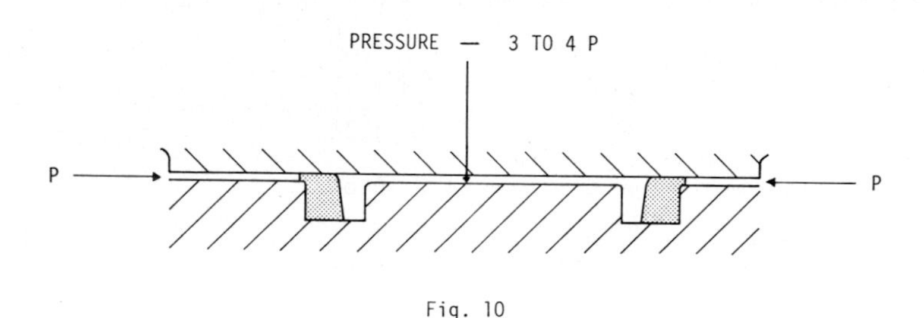

Fig. 10

In the gland situation, the risks are not of the same degree unless there is restricted access of working fluid to the packing via a single port in the neck bush or some similar feature. Adequate fluid access to all seals should be assured unless deliberate attempts are being made to reduce the actual pressure at this point or to dampen pressure variations.

14.5 SELECTION

As so many seal designs will apparently meet given conditions and yet be significantly different in material, size, and price, the fluid power equipment designer can be forgiven for being confused.

Table 14.8 illustrates the variations that could apply to a given actuator of fixed rod, barrel and stroke dimensions and reflects a survey made several years ago when eighteen quotations were sought for a 3" dia. cylinder with double cushioning and a stroke length of 18". Prices received at that time varied between £175 and £24.50, yet all purported to do the same job.

Much will depend on the user's own experience and preference based on his knowledge of the application. Schedule maintenance periods, accessibility, consequence of leakage, initial cost and availability will all play their part in steering the decision towards ultimate security or some other level of cost-effectiveness.

TABLE 14.8

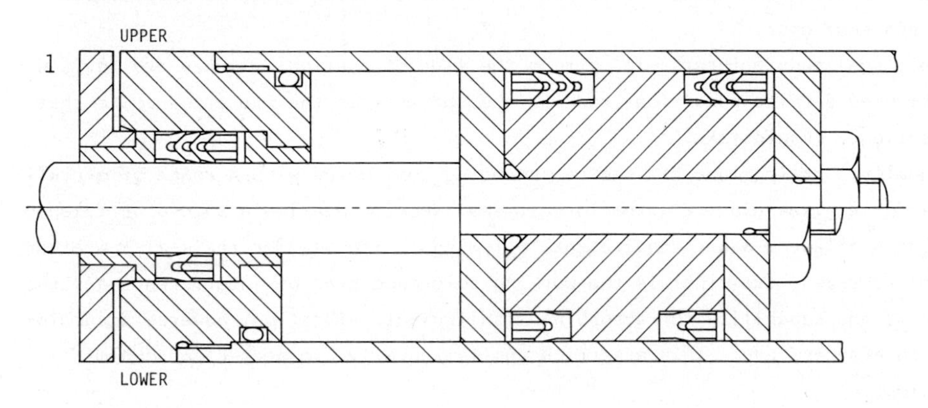

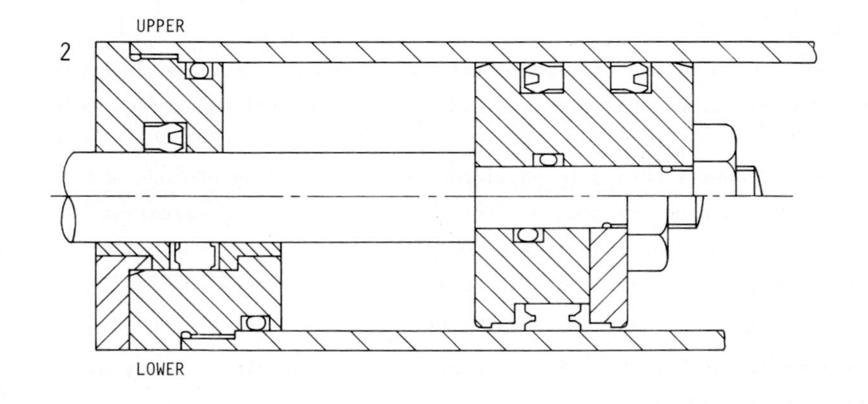

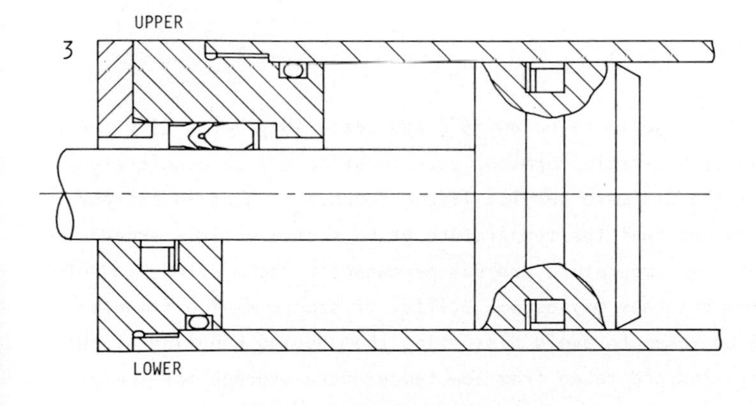

It is significant that British Steel Corporation are now setting their own standards for critical cylinder duties and will use multi-lip packing throughout. Such a solution would be entirely unacceptable to the manufacturer of earth-moving vehicles who would seek a more economic solution in terms of cylinder size and seal cost.

Most seal manufacturers will err on the side of caution if they are advised of intended service conditions and if in doubt this is the preferred route that the designer should take.

Equally, having made his choice and being confronted with a range of overall sizes for a given diameter, the largest seal section available should be taken - not the smallest - other requirements allowing. The smaller the seal - whether lip or squeeze - the finer is the working tolerance band of interference and the lower is the capability for absorbing misalignment, vibration, adverse accumulative tolerances, etc. This assertion applies equally to static and dynamic positions.

14.6 STORAGE

BS 3754:1963 'Storage of Vulcanised Rubber' was prepared under the authority of the Rubber Industry Standards Committee and includes the following recommendations:

> "Most vulcanised rubbers change in physical properties during storage and ultimately may become unserviceable, for example, because of excessive hardening, softening, cracking, crazing or other surface degradation. These changes may be the result of one particular factor or a combination of factors, namely, the action of oxygen, ozone, light, heat and humidity."

The deleterious effects of these factors may, however, be minimised by careful choice of storage conditions.

14.6.1 Recommendations

14.6.1.1 Temperature

The storage temperature should be below 25°C and preferably below 15°C. At temperatures exceeding 25°C certain forms of deterioration may be accelerated sufficiently to affect the ultimate service life. Sources of heat in storage rooms should be so arranged that the temperature of no stored article exceeds 25°C. The effects of low temperature are not permanently deleterious to vulcanised rubber articles but they may become stiffer if stored at low temperatures and care should be taken to avoid distorting them during handling at that temperature. When articles are taken from low temperature storage for immediate use their temperature should be raised to approximately 30°C throughout before they are put into service.

14.6.1.2 Humidity

Moist conditions should be avoided; storage conditions should be such that condensation does not occur.

14.6.1.3 Light

Vulcanised rubber should be protected from light, in particular direct sunlight and strong artificial light with a high ultra-violet content. Unless the articles are packed in opaque containers, it is advisable to cover any windows of storage rooms with a red or orange coating or screen.

14.6.1.4 Oxygen and Ozone

Where possible, vulcanised rubber should be protected from circulating air by wrapping, storage in air-tight containers, or other suitable means; this particularly applies to articles with large surface area to volume ratios, e.g. proofed fabric, cellular rubber.

As ozone is particularly deleterious, storage rooms should not contain any equipment that is capable of generating ozone, such as mercury vapour lamps, high voltage electrical equipment, electric motors, or other equipment which may give rise to electric sparks or silent electrical discharges.

14.6.1.5 Deformation

Vulcanised rubber should, wherever possible, be stored in a relaxed condition free from tension, compression or other deformation.

High quality requirements for storage and periodic inspection such as those specified by the Aero-Space Industry are obtainable from BS 2F.68:1963 "Recommendations for the storage and inspection in store of vulcanised rubber items".

14.7 ASSEMBLY

Although individual applications will involve particular fitting problems there are a number of basic points of good practice which, if observed, will contribute to optimum seal performance:

(i) Check that seal is of correct type, part number or size and material.

(ii) Ensure that seal is in undamaged condition and clean.

(iii) Where permissible, smear the sealing edge of dynamic seals with clean grease. Consult any fitting instruction label provided by the manufacturer to ascertain whether further grease application to inter-seal cavities etc. is recommended.

(iv) Do not treat flat gasket surfaces with any form of jointing paste or lubricant unless instructed so to do, otherwise the ability of the gasket to grip the adjacent sealing faces may be impaired.

(v) Clean all seal housing or gasket seating areas. Check that other surfaces adjacent to the passage of the seal on fitting are also free of dirt, swarf or other contaminants.

(vi) Check seal housing dimensions and surface finish to design recommendations.

(vii) If a seal is likely to contact threads, sharp corners, ports, circlips or similar contingencies during the assembly operation then suitable fitting aids must be provided. The slightest nick or tear on a critical edge of an elastomeric component will reduce sealing integrity. If frequent use is envisaged, non-metallic fitting sleeves can be of advantage since damage to a similar metallic device can duplicate the hazard to the seal.

(viii) Do not leave seal in part assembly for any length of time if sealing edges are subject to misaligned loads; for example rod seal fitted to cylinder with rod in position but no gland bush fitted.

(ix) If appropriate, apply any post-assembly operation recommended by seal manufacturer, for example compression of the prescribed amount in an adjustable gland housing; running at half-speed to assist seal bedding-in on a rotary duty; following up flange bolts after a period at temperature,.... and so on.

15 SEALS FOR FLUID POWER EQUIPMENT
PART TWO ROTARY SHAFT LIP SEALS

B.D.HALLIGAN, CEng, MIMechE, AMPRI.,
Technical Manager (Product Applications)
James Walker & Co.Ltd.

15.1 INTRODUCTION

For the purpose of sealing lubricant within a bearing or excluding foreign matter from bearing surfaces, the spring-loaded rotary shaft lip seal, as shown typically in Fig.1, is widely accepted. The nature of most designs precludes the use of this type of seal from operating at significant pressures unless the sealing lip is adequately supported by a shaped plate. For pressure conditions much in excess of 2.0 - 3.0 bar, combined with rotary movement it would be preferable to consider either a compression packing or a radial face mechanical seal as first choice.

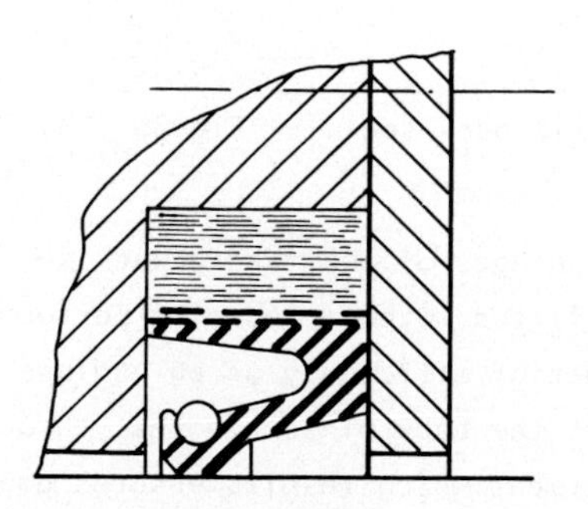

Fig.1 Standard fabric back seal with retaining plate.

15.2 DESIGN

A wide variety of lip seal designs is available in terms of overall construction, material, and lip profile. Some are intended for fitting in housings which have no separate cover-plate and may be supplied with a rigid metal case to which the seal is bonded, the unit being a force-fit in the housing, (Fig.2).

Fig.2 Typical metal-encased seal.

Others have the advantage of flexibility by virtue of having a proofed fabric back (Fig.3a) whilst requiring the provision of a retaining plate to nip axially the back of the seal to prevent leakage and obviate rotation. More recent developments combine the merits of flexibility and self-retention in a metal-supported all-rubber configuration (Fig.3b).

Fig.3a. Standard fabric back seal.

Fig.3b. Moulded-in flexible steel band self-retaining seal.

One of the several advantages of the fabric back seal is the facility with which split seals can be fitted without reducing performance which, in many cases, is of the same order of efficiency as an endless ring. A rubber inlay is frequently moulded into the back of such a seal through which the split is effected. The rubber abutment which results ensures good sealing across the split portion.

The profile of the lip contact area is subjected to each manufacturer's design philosophy. Some have a knife-edge contact band and rely on heavy as-moulded lip interference. Others depend on the spring tension to urge the lip into intimate shaft contact. In practice, a careful balance of interference, spring characteristics and contact band width must be allied to knowledge of the type of material being used, bearing type, and condition and other environmental considerations such as temperature, fluid and, if any, pressure.

A typical range of seal profiles is shown in Fig.4 which also illustrates expedient used when insufficient space is available to accommodate a pair of seals, i.e. the so-called dust lip design.

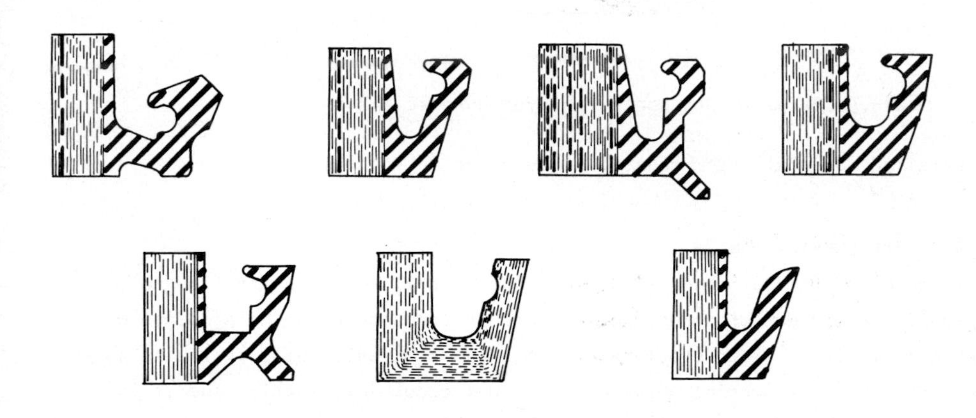

Fig.4 Typical seal profiles.

15.3 MATERIAL

Nitrile-base synthetic rubber compounds are widely employed as standard materials by reason of their compatibility with most lubricating oils and greases and their resistance to wear. There are, however, certain rolling oils and coolants which may cause excessive swell or shrinkage of such polymers and a fluorocarbon rubber, e.g. Viton, may be a necessary selection. This is particularly true of some palm oil solutions and other soluble types of oil.

Consideration for operational temperature is important and, in this respect, it is the condition at the lip of the seal which is paramount - not necessarily the environmental temperature. However, it may be said that nitriles are frequently used with success for intermittent service temperatures of 150°C and are continuously rated for 120°C in lubricated conditions. Above these values either acrylic rubbers, fluorocarbon or silicon compounds would require investigation.

Many silicone and nitrile rubber shaft seals are fitted in automotive applications for crankshaft and gearbox sealing, in which areas much use is being made of grooved sealing surfaces which are designed to produce a hydrodynamic effect in the oil film being sealed, effectively causing the oil to be pumped away from the seal lip (see Fig.5).

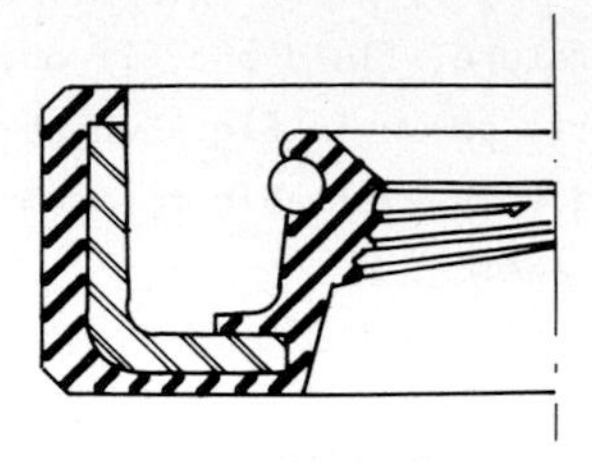

Fig.5 Metal insert seal with hydrodynamic aid.

Detailed material recommendations are given in Appendix 1.

15.4 SEAL LUBRICATION

In common with most other forms of dynamic seal, rotary shaft lip seals depend for their efficient and consistent performance upon the presence of a stable fluid film. The establishment of this film, which will typically be in the order of a few microns thickness, may not occur immediately, and it is not uncommon for a high percentage of wear to occur on rotary shaft seal lips during this period before steady-state conditions are achieved. For this reason, it is always good practice to apply a lubricant to the seal lip and in the seal cavity before fitting.

Where single seals are housed at each end of the housing, there is normally sufficient bearing lubricant in contact to provide adequate lubrication. Again, where two seals are housed together, it is often found that with bearing lubricant on one seal and either rolling fluid or roll coolant on the other, further lubrication is unnecessary. In all cases the liberal application of grease to the seals on assembly will ensure lubrication from the beginning and, in some instances, this will be found sufficient to last from one fitting to the next.

However, there are many sealing arrangements, especially on rolling mill bearings, in which two or more seals are fitted together in the same housing and there is the danger that at least one will run dry unless lubricant is supplied from an external source. This can best be accomplished by drilling a hole through the chock connecting with an annular groove in the back of the seal housing, as shown in Fig.6, and using a special type of seal which has a series of radial ports in the base for passage of lubricant to the seal lip. When it is difficult or impossible to machine an annular groove in the back of the seal housing, this groove may be incorporated in the seals themselves, as shown in Fig.7.

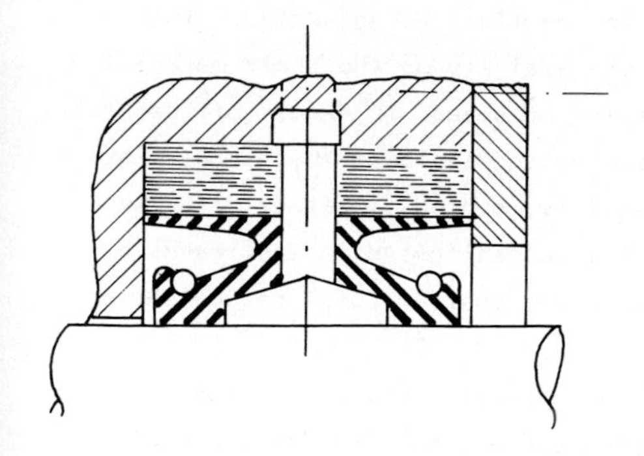

Fig.6 Seal lubrication via ports in seal base.

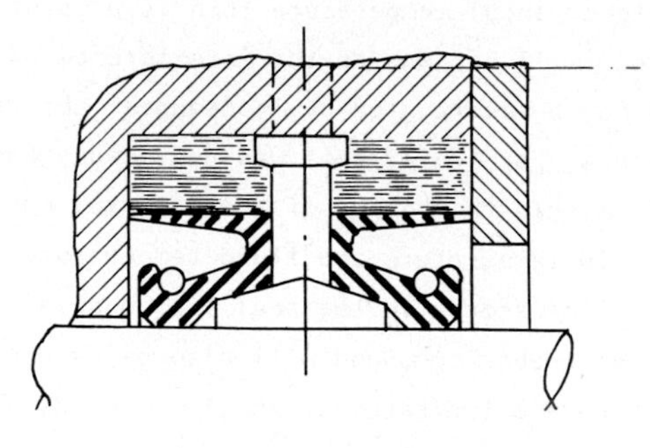

Fig.7 Seal lubrication via ports and annular groove in seal base.

In general, a good quality mineral oil or grease is suitable for seal lubrication, but molybdenised lubricants may be used to advantage where application is infrequent. Care should be taken to ensure that the grease or lubricant with which the seal is to come into contact is compatible. For example, where seals are being used with dilute mineral acids, butyl compositions may well be used. This material is likely to swell in contact with mineral oil or grease and an alternative lubricant will be essential.

15.5 SHAFT SURFACES

The sealing area of the shaft should be a fine ground finish of 0.4 to 0.8 μm (16 - 32 μin.) CLA or Ra for most applications, but for the higher speed range it is recommended that the surface finish be improved to 0.2 to 0.4 μm (8 - 16 μin) CLA or Ra. In all cases it is important that the shaft sealing area be free from machining marks, dents, burrs and scratches.

Where fluorocarbon or acrylic rubber seals are being employed, it is also advisable to use the finer level of finish indicated above in order to eliminate pick-up of the seal material.

If lubrication is adequate and free from abrasive contact, unhardened mild steel shafts will generally give satisfactory results under normal operating conditions. However, a harder shaft material is to be preferred for applications where lubrication is poor, abrasives are present, or speed and pressure conditions are particularly arduous.

A fine machined finish is suitable for the housing bore.

15.6 FRICTION

The rubbing friction of the seal lip on the shaft material inevitably causes a higher local temperature than is present in the fluid being sealed. This effect will be due to the interference of the seal lip on the shaft material and may be aggravated by inadequate lubrication or speeds in excess of that for which a particular seal was designed. Assuming the system temperature to be well within the bounds of the material capability, a high differential between lip tip temperature and fluid temperature will be manifested in many cases by hair-line scores in the seal lip co-axial with the shaft. Local carbonisation of the rubber compound will also be evident.

Figure 8 indicates a set of curves derived by experiment, projecting frictional horse power against seal diameter for a conventional rotary lip seal design operating at 500 r.p.m. in mineral lubricating oil at different system pressures.

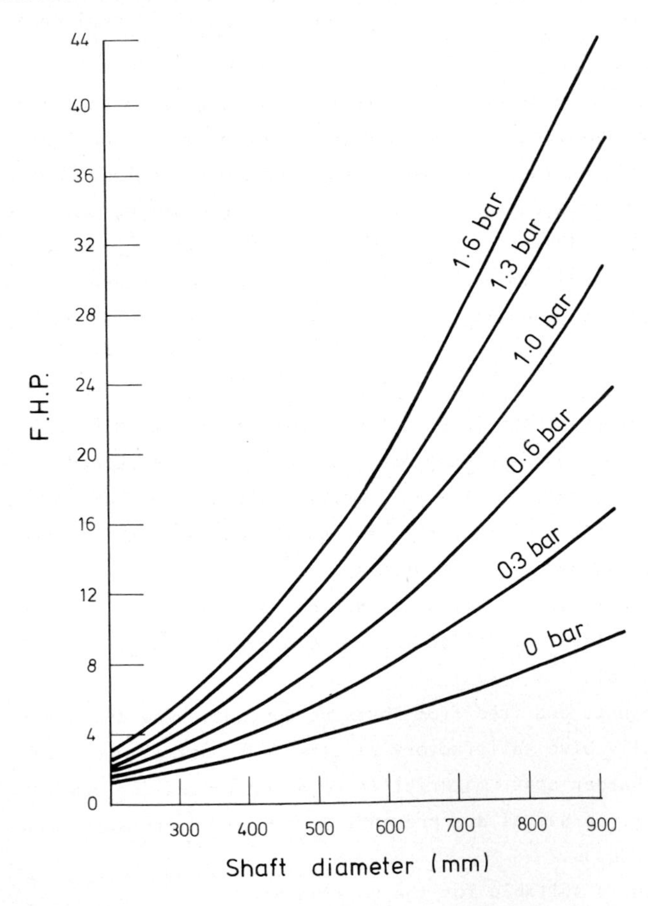

Fig.8 Frictional horse power absorbed by rotary shaft lips seals, having conventional lip interference operating at 500 rev/min.

15.7 SPEEDS

Many of the fabric-backed type of seals together with those of the metal-supported type are working satisfactorily on 400 mm roll-neck diameters at speeds of up to 25 m/s (5000 ft/min) over sustained working periods. There are, however, many factors such as surface finish, eccentricity, and lubrication which can limit the maximum speed for which any seal is suitable, and manufacturers should be consulted in cases of doubt.

15.8 ECCENTRICITY

Where plain metal or synthetic composition bearings are being employed, it is not uncommon to find that the shaft is not truly concentric with the seal housing, owing to bearing clearance and subsequent wear. In this event, it is essential for efficient sealing that the seal lip be capable of following all shaft movement and, indeed, on many large comparatively slow-moving shafts eccentricity values of 2.5 mm have been satisfactorily accommodated. Naturally, the seal performance in terms of eccentricity capability will be speed and shaft diameter dependent.

Where split seals are fitted, then particular consideration is necessary to the problem of shaft eccentricity, since there may be a tendency for a split seal to open at the join.

15.9 PRESSURE

Although few rotary shaft lip seals are specifically designed as standard components to accept significant pressures, the use of metal-supporting plates will extend the usefulness of this type of seal. A typical profile is shown in Fig.9.

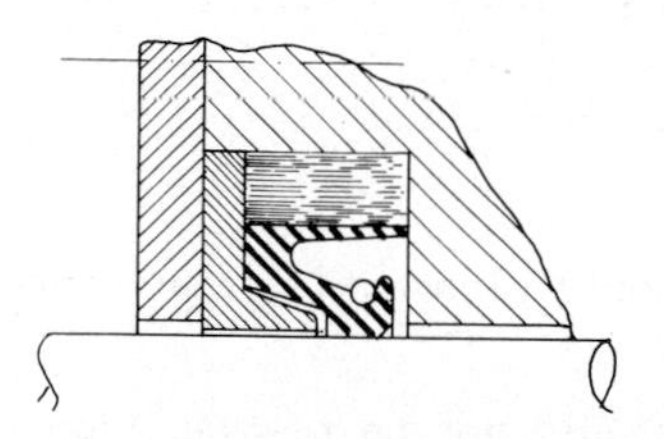

Fig.9 Seal with shaped support plate for pressures up to 3.0 bar.

As a result of experimental work on the sealing of oil-filled marine stern glands and manoeuvring thrusters, a seal lip profile has been developed which satisfactorily sustains pressures up to 4.0 bar without the use of a shaped support plate. The base must be fully supported as indicated in Fig.10.

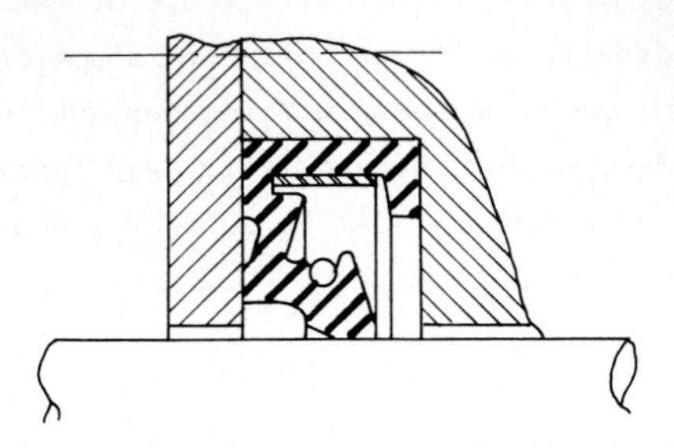

Fig.10 New seal development for pressures up to 4.0 bar without shaped support plate.

Where constant applied pressure is not anticipated, it is frequently suggested that grease-lubricated bearings are equipped with relief holes and that oil-lubricated bearings have drains of adequate size. Drains taken from the ends of the bearing near the seals will help to dissipate any localised pressure build-up. Where possible, steps should be taken in bearing design to prevent escaping high pressure oil impinging directly upon the seals.

In some cases where the loss of a small amount of grease is unimportant, a simple relief system may be formed by facing seals away from the bearing and allowing the seal lip to be lifted under the influence of the lubricant pressure.

15.10 CARE AND HANDLING

Fluid seals should be handled at all times with extreme care since the life of bearings or other costly machine parts may depend upon their efficiency.

Attention to the following vital points will assist in ensuring trouble-free operation during service.

15.10.1 Storage

(i) The store should have a cool, clean and dry atmosphere, free from dust and grit.

(ii) Whenever possible, seals should not be removed from the wrapping in which they were supplied, as this provides protection and identification.

(iii) Avoid untidy stacking as the weight may distort the seals at the bottom of the stack.

(iv) Seals should never be threaded on wire or string as this will damage the lips.

15.10.2 Handling

(i) It must be remembered that seal lips are extremely vulnerable to damage and the smallest nick provides a potential leak path.

(ii) Seal reinforcing inserts, although adequate for their duty, may deform under adverse handling or stacking.

(iii) Seals having metal outside surfaces may damage other seals, especially if the metal edges contact the rubber parts of neighbouring seals.

15.10.3 Fitting

A high proportion of failures and leakage of oil seals is due to incorrect fitting resulting in damage to both seal and sealing surface. Strict attention to the following matters is essential if best performance is to be obtained.

(i) Before fitting, the seal should be examined to ensure that it is clean and undamaged.

(ii) The sealing lip should be smeared with suitable clean lubricant. Seals used as dust excluders should be packed with a compatible grease.

(iii) The sealing lip, normally, should face the fluid to be sealed.

(iv) When fitting, it is important to ensure that the sealing lip is not damaged even by the slightest nick, that the spring is correctly located when in position, and that the seal is properly pressed home into the housing recess.

(v) Examine the shaft which should be free from all roughness and sharp edges and avoid passing the sealing lip over keyways, screw threads, or shoulders. Shaft edges or shoulders should be well rounded or chamfered, and where this is not practicable a fitting sleeve slightly larger than the shaft with a lead-in taper should be used.

(vi) According to the type of assembly, it may be necessary either to first press the seal into the housing and subsequently onto the shaft or, alternatively, to pass the seal over the shaft and then press it into the housing. It is preferable first to mount the seal on the shaft where circumstances permit, since this allows observation of the lip during assembly.

(vii) The assembly should not be allowed to rest for any length of time at an incomplete state of fitting, where the weight of the shaft or housing may be borne by the seal, resulting in damage or distortion to the latter.

(viii) When pressing the seal into the housing, a uniform pressure should be exerted, preferably by means of an arbor press in combination with a suitable tool. The diameter of the tool should be slightly smaller than the diameter of the housing by 0.1 to 0.4 mm. The outside surface of the seal can be smeared with a suitable lubricant in order to facilitate fitting. Care must be taken to ensure that the seal does not enter the housing recess in a tilted position, since this will cause damage to the outer surface.

15.11 SERVICE PROBLEMS AND THEIR SOLUTIONS

A number of fault-finding procedures follow which, if taken in sequence, should analyse the reason for a given difficulty with rotary shaft lip seals.

15.11.1 Unacceptable Leakage

This is almost always associated with oil lubricated bearings, since grease is not a difficult lubricant to seal. The term "unacceptable" can have wide interpretation since an occasional drop of oil might be disastrous if it resulted in contamination of the product being handled by the machine concerned in such spheres as the textile, paper, or food industries, whereas it would probably remain unnoticed in a heavy industrial environment.

Since the vast increase in oil prices there is, however, a greater sensitivity to oil losses in any form and leakage rates that hitherto have been ignored are now becoming regarded as unacceptable.

When dealing with such complaints it is essential to discover the history of equipment concerned, and this broadly falls into three categories:

(I) New equipment recently commissioned, where sealing has been regarded as unsatisfactory from the start.

(II) Equipment that has been in operation for a period of time and only recently has developed leakage problems.

(III) Equipment that was satisfactory during its first term of operation but leakage has occurred after fitting replacement seals during routine maintenance or overhaul.

Since trouble tracing is basically a process of eliminating of substantiating faults, the sequence of checks required would vary with each of the above categories.

In order to simplify the procedures and avoid irrelevant investigations, the recommended sequences for each of the above categories is defined by letter symbols to be used in conjunction with the attached fault-finding chart.

Category (I)

Full checks in order as A, B, C, D and E until fault is discovered.

Category (II)

A, B, C(1), D(2), and D(3). If faults as C(1) or D(2) are exhibited, ascertain period of service with seals. This should be calculated in terms of hours of running and related to speed, temperature, and other environmental conditions.

A moderate speed with good clean lubrication conditions and ambient temperatures would normally anticipate a seal life of around 10,000 hours.

High speeds, poor lubrication, elevated temperatures, or partially abrasive media could reduce this to as little as 2,000 hours.

The problem may therefore be simply that of being due for seal replacement.

If faults as in (D3) are in evidence, obtain details of all media in contact with seal (including any cleaning fluids) as a change of lip material may be necessary to obtain compatability.

If seals display no faults, check for mechanical defects as E(1) and E(2).

Category (III)

A, B, C(1), C(3), D(1), D(3), D(4), E(1), E(3).

Fault-finding Chart

(A) Is leakage actually occurring from the seal or does it stem from such sources as bearing cover flanges and is merely "collected" by the seal housing, giving a false impression? Check by wiping clean all appropriate areas and run machine to ascertain leakage source.

(B) Is leakage from around seal back or from the lip along the shaft? Check by wiping both clean and observing while machine is running.

(C) If O/D leakage - check the following:

(1) Is seal a good fit in housing or is it slack? (On split seals a slack seal will display a gap between seal ends).

(2) If housing bore is correct size then seal O/D dimension is suspect if slack in housing.

(3) If seal is good fit in housing, check for damage on housing bore.

(4) Check housing depth to ensure seal is being axially compressed - if applicable.

(D) If leakage along shaft, check the following:

(1) Shaft size, surface finish, shaft damage at contact area.

(2) If (1) O.K., check condition of sealing lip for hardening and/or cracking. If either in evidence, then speed or temperature conditions are probably incompatible with seal material.

(3) If lip is soft or swollen this is usually an indication of chemical incompatibility with the media in contact with the seal.

(4) If (2) and (3) O.K., check section width of seal with spring fitted.

This should be at least nominal section +1% immediately on removal and increasing to nominal +3% after one hour in free state. Spring may be shortened by up to 5% of its original length if section appears inadequate. Section measurement should be average of 4 equi-distant readings.

(E) If checks (C) and (D) do not reveal any faults the problem may be due to mechanical conditions and the following should be checked:

(1) Shaft to housing concentricity - check by means of calipers between shaft and housing bore at 4 points around periphery. Variations of more than 0.3 mm require further investigation.

(2) If smallest caliper measurement occurs between bottom of shaft and housing, this may indicate bearing wear with resultant dynamic eccentricity. (Applicable to horizontal shafts only).

(3) If bearing is O.K. then housing offset may be responsible. Seal housings are normally centralised with the shaft by means of a machined register with the bearing housing. Where this feature is not incorporated then it may be possible to centralise the housing by slackening the bolts and repositioning.

(4) If (1), (2) and (3) are blameless then the following requires investigation:
If bearings are oil pressure lubricated is there adequate drainage to prevent pressure build-up against the seal?

If equipment operates on a constant oil level principle, are there gearwheels or ball-journal bearings in close proximity to the seal causing oil turbulence or flooding?

Where the latter situation exists, the housing lands should be only marginally larger than shaft diameter in order to form a baffle or, alternatively, a baffle plate fitted between bearing and seal housing.

A temporary remedy can be made by using a 2.5 mm CAF gasket at the bottom of the housing, the I/D of which should be shaft diameter plus 0.5 mm maximum. A further gasket of equal thickness should be fitted underneath the retaining plate to restore the correct amount of axial compression in the case of retained seals (see Fig.11 and Fig.12).

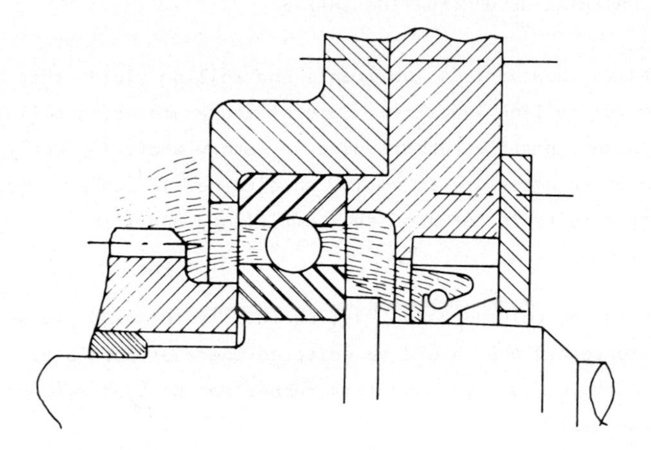

Fig.11 Showing how too large bore diameter of seal housing land permits high velocity oil impingement on seal.

Baffle plate between bearing and seal or close-fitting seal would reduce risk of oil leakage.

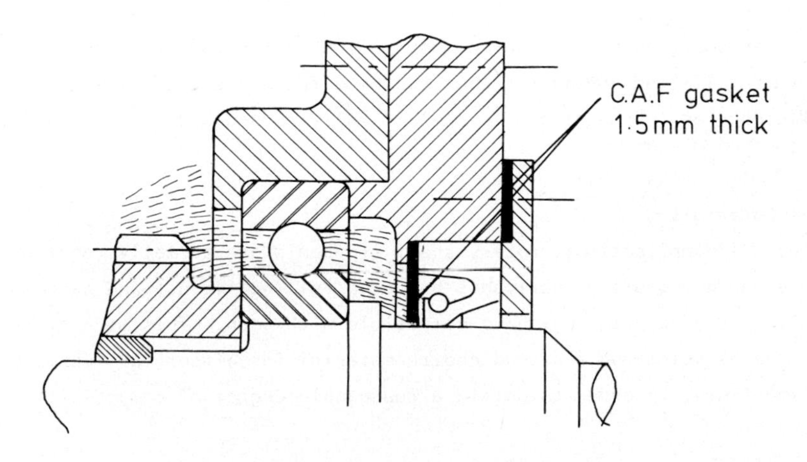

Fig.12 Showing how fitting of C.A.F. gaskets can provide a temporary remedy for situation in Fig.11.

(With acknowledgement to the late Mr.Ray Bladwin - Senior Advisory Engineer, rotary shaft lip seals - who compiled much of the data presented here).

APPENDIX 1

ROTARY SHAFT LIP SEAL MATERIAL RECOMMENDATION CHARTS

The accompanying charts show various lubricants and rolling fluids that are commonly used in the metal rolling industry. Some of these are not chemically compatible with the standard nitrile rubbers used in rotary shaft lip seal production, and it is therefore necessary to employ an alternative rubber compound to achieve satisfactory results in terms of seal efficiency and life.

Seal Lips

The rubber compound in the column headed "1st choice" is the grade least affected by the media concerned and should be selected wherever possible. In most instances a second choice is given and this rubber can be used without serious detriment to the seal.

Where fluorocarbon rubber (denoted by the letter 'A') appears as the only choice, please note that none of the conventional commercial grade rubbers can be used as an alternative. In instances where fluorocarbon rubber appears as the second choice it is usually for economic reasons, indicating that it is only marginally superior to the commercial grade compound given as first choice, and the small advantage would often not justify the additional cost.

Seal Backs

Although with some media the fabric materials used for construction of the seal back suffer limited volumetric change, the effects are not detrimental to sealing efficiency as this part of the seal is fully housed and normally axially restrained.

Selection of Materials

In rolling-mill applications, rotary shaft lip seals are normally required to seal not only against bearing lubricants but also rolling fluids, and care should be taken to select a material which is compatible with both. This may involve some compromise by selecting a second choice material for either the lubricant or the rolling fluid, in order to obtain a reasonable degree of compatibility with both.

In dual sealing arrangements employing back-to-back seals, should only one of the media necessitate fluorocarbon rubber, the opposing seal could be made with a less costly compatible grade of rubber providing that care is taken in identification and fitting.

For convenience, letter symbols are used in the charts to indicate various rubbers, and the key to these is given:

KEY TO MATERIAL CODES

Code	Base rubber compound
A	Fluorocarbon
B	High nitrile
C	Medium high nitrile
D	Nitrile
E	Nitrile with graphite
F	Nitrile with paraffin wax

RECOMMENDED SEAL LIP MATERIALS FOR USE WITH BEARING OILS AND GREASES

Brand Name or Number	Seal lip materials	
	1st Choice	2nd Choice
B.P. HCT 80	C	E
B.P. Spemo 350 HB " 450 HB	C	D
B.P. Energol GR 125-XP	C	D
Castrol 98	C	E
CLP 114	C	E
Calysol Grease	C	A
Duckhams Zero Oil	D	C
EP 69 Oil	C	D
EP 80 Gear Lubricant	D	C
Esso Estic 65	C	A
Esso Nuray 146	C	A
Esso Pen-O-Led EP3	D	C
Fuchs MR 40	C	F
Mobil Exu 66/25	C	D
" Vacuoline 'AA'	C	F
" " 25 x 25R	C	E
" " Heavy	C	E
" " 'EE"	C	E
" Mobilube H.D.90	C	E
" Mobilex EP2 Grease	C	F
OM 100 Oil	C	A
Regal Gear Oil	C	B
Shell Oil 1624 and 1611	B	C
" Carnea	C	B
" Faunus 'B'	B	A
" Macoma	D	C
" Nassa 78	C	A

Seal lip materials for bearing oils and greases (contd.)

Brand Name or Number	Seal lip materials	
	1st Choice	2nd Choice
Shell Telona 945	C	A
" Teressa	C	A
" Tivela 75	C	D
" Vitrea 75	C	B
" Retinox Grease	A	C
" Donax 17	C	B
" Aeroshell Grease No.7	C	A
Ucon Oil	C	D
Alo-Jidac	A	-
Caster Oil (with water)	A	B
" " (with methylated spirit)	C	E
Cimcool	B	A
Cimcool E5	B	C
Castrol Coolage SL	A	B
Croda D4 A	B	A
" Lubrotex AF/MW 7804	A	-
Dasco 900	B	A
Esso Somentor N35	C	A
" " W20	A	B
" " 33 and N60	A	C
" Univis J58	C	B
Emulsion KF B1	A	-
Germ Kinetrol FR3A	A	-
Gulf Mineral Seal	C	D
" Cut Oil	C	D
" 93	C	D
Houghton Permasol	B	A
" Sulfona ED	A	B
Huile de Laminage 102 P & 982 R	B	C
Hydrit	B	D
KF B1 Rolling Oil	A	D
Lubricor 'T'	B	E

Seal lip materials for roll coolants and rolling oils (contd.)

Brand Name or Number	Seal lip materials	
	1st Choice	2nd Choice
Mobil Generex 56, 57 and 404D	A	B
" " 22 and 24	C	B
" " 322	C	D
" Prosol 66	A	B
" " 44	B	A
" " 33	A	-
" Solvac 11	A	-
" " 800	B	A
" Vactra "HH" and "BB"	C	B
Mirobo 415A	B	A
Ocut G	C	A
Palm Oil	B	A
Quaker Tinnol 12	E	D
" " 109	B	D
" Quakerol 41 (with Shell Carnea 31)	B	A
" " 43, 82, 87 M and 88-182 M	B	A
" Qwerl 506	B	D
Rollup 200	B	A
Shell Dromus B	B	A
Sternol PL106 and PL107	C	B
" 1270/1307	B	A
" 1076	D	C
Tayol 316 Emulsion	B	A
Texaco Texatherm 320	A	C
Trellub 12 A	B	A
Wyrol H40	C	A

16 SEALS FOR FLUID POWER EQUIPMENT PART THREE COMPRESSION PACKINGS

B.D. HALLIGAN, C.Eng, MIMechE, AMPRI
Technical Manager (Product Applications)
James Walker and Co. Ltd.

16.1 THE PACKED GLAND

Compared to the finite qualities of ferrous metals for example, the essentially deformable nature of sealing materials has introduced a measure of variability that causes many commentators to look on fluid sealing technology as an art rather than a science. If this is true, and manufacturers of mechanical face seals would be but one area of valid objection, then the field of compression packings is, arguably, the blackest area of that art.

Regarded as an anachronism in a period of high technological achievement, compression packings show no signs of losing significant ground in terms of production quantities as new and improved types proliferate both in Europe and elsewhere. To understand this situation requires some appreciation of the fundamental mode of operation of the adjustable gland or stuffing box shown in Fig.1.

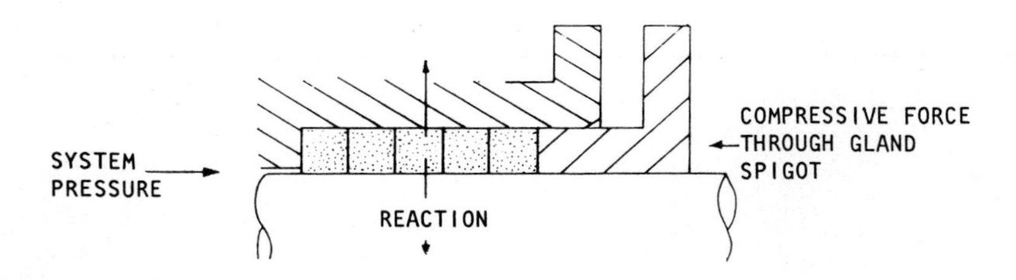

Fig.1 Compression Packing

This may be filled with split packing rings chosen from a variety of materials and constructions, described elsewhere, which are persuaded to react against a shaft, whether rotary or reciprocating, to the extent that the radial force developed exceeds the pressure to be sealed.

Such a principle could not be more elementary and its intrinsic value could be further questioned as packings in this category used for rotating or reciprocating equipment rely on a controlled leakage for long-term lubrication purposes, if they are to survive for an adequate period.

The continued justification for the compression packing might appear obscure against such a background but there can be no doubt that certain areas of application exist where no reasonable substitute is available.

16.1.1 Pumps

Many reasoned and well-researched papers have been published to support mechanical seals against soft packing and vice versa. There is no doubt that the former have supplanted packed glands as original equipment on the majority of rotodynamic pumps for a variety of process and service fluids but they are operating parameters and cost considerations which will frequently dictate the choice of soft packing.

Table 16.1 compares the relative attributes of the two contenders in basic terms.

In general it may be said that, unless zero leakage is an absolute priority, compression packings will retain an important position wherever regular maintenance is available and the following considerations apply:

- simplicity in gland design and ancillary equipment
- ease of fitting
- flexibility of supply and spares for plant utilizing many different types and sizes of pump handling a wide variety of fluids
- frequent ability to cater for adverse conditions without elaborate precautions

16.1.2 Valves

If any doubt exists regarding selection on pumps then a much more obvious choice of soft packing applies to the valve scene. The relative lack of movement, ease of fitting and, in this case, lack of leakage requirement for lubrication purposes plus the most decisive advantage of low cost, are factors which ideally relate to compression packings.

There are areas where moulded elastomeric seals present a reasonable alternative but even the most exotic compounds would seldom be used above 250°C - unless reinforced by asbestos fabric.

TABLE 16.1

Comparison	Soft Packing	Mechanical Seal
Initial Cost	of the order of 10:1 in favour of soft packing depending on size and application	
Reliability	APPROXIMATELY EQUAL ample warning of impending failure with possibilities for correction	little or no warning of end of useful life with possibility of sudden complete failure
Installation	essentially simple - requiring no special skills if correct procedure adopted	skilled fitting required - precisely defined environment and assembly
Maintenance	regular and requiring experience	zero
Spares	facility for stocking length form material or complete pre-formed sets at relatively low cost	spare seal components must be available - cost can be substantial
Shaft Wear	can be considerable; shaft sleeves reduce replacement costs	nil
Operating Costs	friction losses slightly higher with soft packing leakage losses zero with mechanical seals but positive with soft packing as lubrication of sealing rings is essential	

16.2 OPERATING PRINCIPLES

By comparison to the seal types described in the other papers - particularly elastomeric lip and squeeze seals - compression packings respond to applied pressure in inverse proportion to the hardness of their construction and rely on an external force to produce the radial pressure required for effective sealing. The method of generating that force can vary but usually (and preferably) involves a bolted gland spigot as shown in Fig.1 where controlled axial movement is easily achieved by adjustment of the retaining nuts or studs. Spring loading is sometimes used in inaccessible situations but such a provision lacks the fine control demanded by some packing types and has a limited range of load capability.

Whilst the sealing force can be adjusted to cater for service wear, care must be taken to avoid over compression which will lead to excessive friction, shaft wear and premature packing failure.

To increase density and dissipate heat, soft packings invariably contain lubricants, loss of which, through excessive compression or over-heating in service, will result in packing volume loss with subsequent reduction in the effective sealing reaction and correspondingly increasing leakage rates. By limiting compression to a point where slight controlled leakage is obtained, adequate lubrication of the dynamic surfaces is ensured and over-compression of the packing avoided. However, where lubrication is a problem - or a degree of gland cooling is required - a lantern ring can be incorporated into the gland area for the distribution of additional lubricant/coolant (Fig.2a). The position of a lantern ring will depend on the nature of the application but, since the packing rings nearest the gland spigot do most of the work, the additional fluid should usually be introduced near to that area.

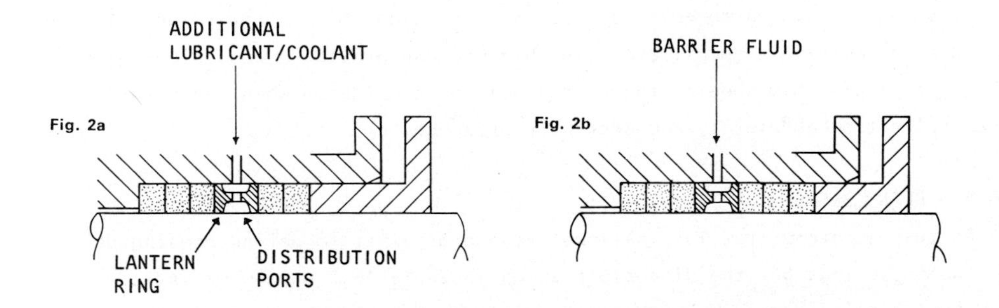

If it is essential that the fluid being pumped does not escape to atmosphere (e.g. a toxic medium) the lantern ring may serve to introduce a barrier fluid at a pressure of 0.5 to 1 bar above that to be sealed (Fig.2b). Similarly, where there is a risk of severe abrasive wear to the packing, a flushing fluid may be introduced through the lantern ring (Fig.2c).

For application with negative pump pressures (i.e. suction) a supply of the medium being sealed can be made through the lantern ring to prevent air-drawing (Fig.2d).

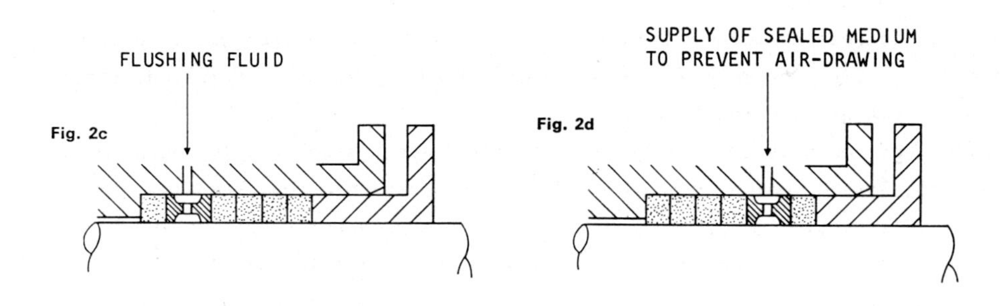

Fig. 2c

Fig. 2d

If extreme temperatures are to be encountered, it is unlikely that cooling through the lantern ring will be sufficient and recourse must be made to internal cooling of the gland housing and shaft to reduce the temperature at the gland to a value within the packing's capabilities. Conversely, when dealing with media which crystallize or congeal when cool (e.g. sugars, tars, etc.), the packing will face rapid destruction unless gland heaters or a steam jacketed arrangement are employed to restore the fluid state before starting up.

It should always be remembered that the inclusion of a lantern ring into the gland area invariably complicates assembly and can provide a possible source of shaft scoring; they should, therefore, only be considered when the nature of the application absolutely demands their presence.

16.3 GLAND DESIGN

At this juncture, few international standards exist to define housing design for soft packings but the dimensions shown in Table 16.2 should be satisfactory for most applications. Housing depths will vary with individual circumstances, such as the inclusion of a lantern ring, but five rings of square section packings are usually recommended for the average, uncomplicated duty.

TABLE 16.2 Suggested housing widths in relation to shaft diameters. All dimensions in mm.

All packings except expanded graphite		Expanded graphite	
Shaft Diameter	Housing Width	Shaft Diameter	Housing Width
Up to 12	3	up to 18	3
above 12 to 18	5	above 18 to 75	5
18 to 25	6.5	75 to 150	7.5
25 to 50	8	150 and above	10
50 to 90	10		
90 to 150	12.5		
150	15		

Other design considerations worthy of note, but often overlooked, can be summarized as:-

(i) The provision of an adequate tapered 'lead in' at the mouth of the gland to facilitate entry of the packing and to obviate the risk of damage in the assembly operation. A minimum of 15° x 6.5mm usually represents good practice.

(ii) The provision of a reasonable surface finish on adjacent metal parts - particularly the dynamic surface. The better the finish the less wear will occur; 0.4μm (16 μ in) CLA or Ra on the shaft and 1.6 μm (64 μ in) CLA or Ra on the stuffing box bore should be ideal for most applications. The use of shaft sleeves can give considerable maintenance advantage when considering the question of surface finish.

(iii) The danger of extreme running clearances at the gland - particularly on the spigot side. In those exceptional cases where excessive clearance is unavoidable, the packing should be protected by an independent ring of suitably robust material or construction which reduces the clearance to a minimum.

(iv) An allowance for entry of the gland spigot well into the gland area; certainly to an extent that exceeds substantially the depth of the tapered lead in. The length of spigot selected must also cater for packing compression, resulting from gland adjustment. Typical entry lengths should be at least two times packing section. For packings of softer construction, maximum length should be provided. With modern packing materials, bevelled glands are seldom an advantage and can actually promote movement of the sealing ring on the spigot side into the live clearance.

(v) The need to avoid excessive shaft misalignment or whip.

(vi) The provision of adequate shaft support. The packing must not be used as a bearing.

16.4 PACKING CONSTRUCTION AND MATERIALS

16.4.1 Fibre Material

Mineral	asbestos		
Vegetable	cotton	Synthetic	aramid
	flax		glass
	jute		graphite filament
	ramic		graphite foil
	sisal		nylon
			polytetrafluoroethylene (PTFE)
			rayon

16.4.2 Lubricants

Dry	graphite	Metals	lead foil and wire
	mica		aluminium foil
	talc		copper foil and wire
	molybdenum disulphite		brass wire
Wet	tallow		monel wire
	castor oil		inconel wire
	straight mineral lubricating oil		stainless steel wire
	petrolatum	Elastomers	natural and synthetic
	solid fractions		
	paraffin wax		
	soaps		
	silicone grease		
	PTFE dispersions		

16.4.3 Construction

The principal forms of constructions for fibrous compression packings are:

(i) Braided Individual yarns are braided tube over tube and squared off. The density of this type of construction is high and ideal for many valve applications.

(ii) Plaited Multiple yarns are interwoven in plaited bundles in such a fashion that the direction of fibre follows the periphery of the packing ring. The natural characteristic of this construction is more suited to centrifugal pump applications than valve service although the inherent flexibility of the form is popular with some users.

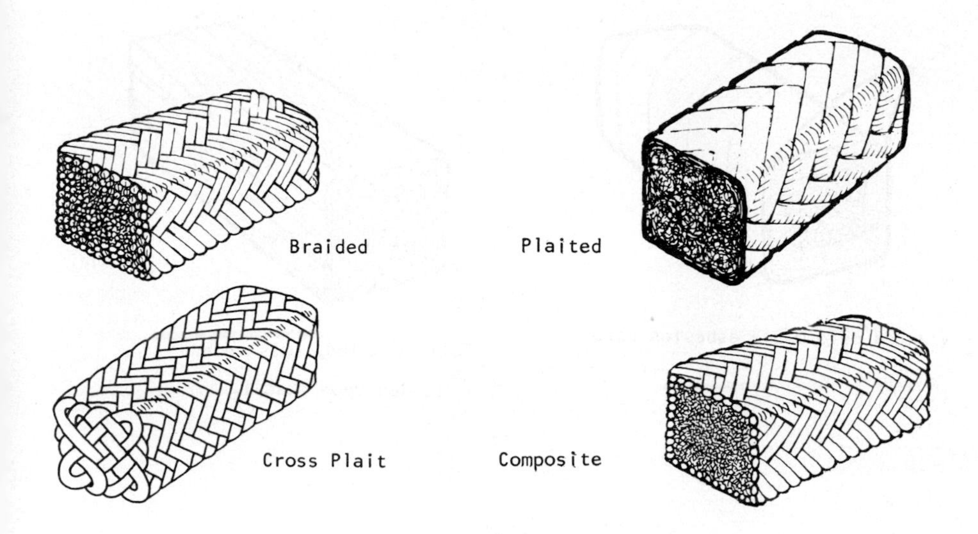

Fig.3 Basic packing constructions

(iii) Cross Plait All the yarns are interlocking and pass diagonally through the packing to provide a firm construction of consistent density and shape. Used extensively for synthetic yarn packings for valves and pumps.

(iv) Composite asbestos plastic This packing category is of fairly recent origin and includes those types based upon a braided and reinforced asbestos jacket enclosing a 'plastic' core. Although inaccurate in the scientific sense, the term 'plastic' conveniently describes those many mixtures of asbestos fibre and lubricant, both mineral and solid, from which readily deformable packing material may be made. This packing is widely accepted for difficult valve sealing duties.

All the fibre-based constructions described here are frequently reinforced with metal. This applies particularly to asbestos based products where the use of metal wire in the yarn can extend the service capability of the packing to 800°C and beyond. But for this feature, even the best quality non-metallic asbestos yarn packing would be restricted to temperatures of about 315°C maximum.

All of the lubricants shown are used in conjunction with fibre packings of different sorts and are applied by dipping, coating, soaking, vacuum impregnation, dusting, etc. The prime object is maximum lubricant retention. Frequently, several treatments and repeat processes are employed to achieve this end.

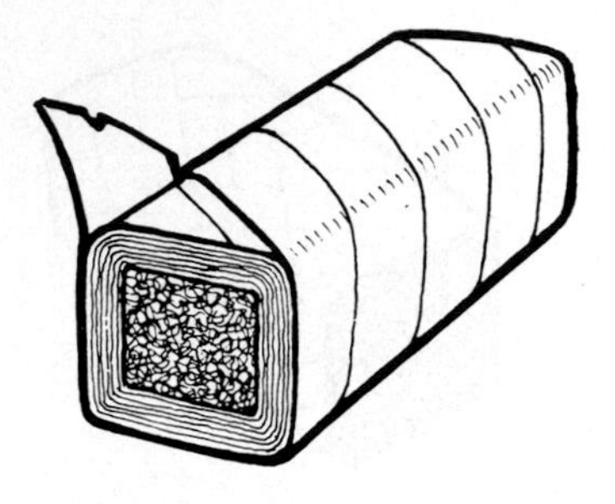

Foil wrapped deformable asbestos core.

Foil crinkled, twisted and folded upon itself.

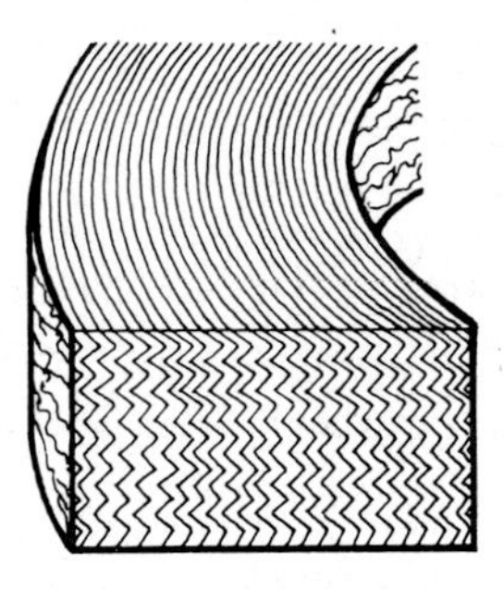

Corrugated foil, concertina wound.

Fig.4 Typical metal foil based packing construction

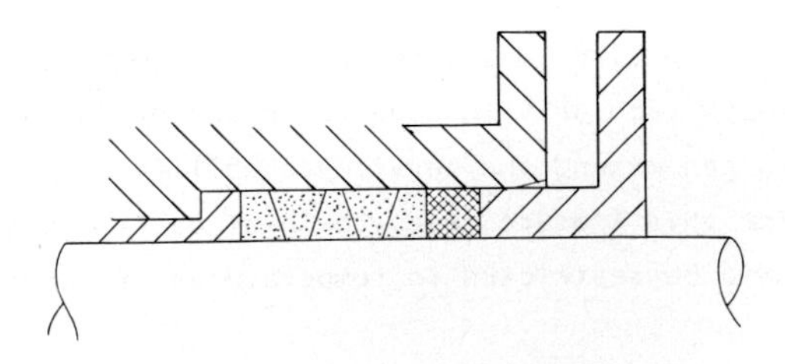

Fig.5 Double-bevelled, elastomer proofed fabric packing for abrasive duties

In the field of compression packings, elastomers are not widely used although some braided packings do employ yarns that are treated with a rubber proofing to render them more suited to difficult wet applications such as condensate duty.

Rings of square or rectangular section compression packing, manufactured from folded, rolled or laminated elastomer proofed cloth, are still popular for relatively slow moving, lower pressure reciprocating pumps handling water of LP steam.

One particular design, with a moulded, double-bevelled section, made from semi-metallic rubberised yarn, is particularly effective on rotary applications dealing with viscous media which solidify when the pump is idle and cause damage to conventional plaited packings on restarting from cold. This moulded packing is also suitable for duties involving solids and abrasives (See Fig.5).

16.4.4 Additional Materials

In high performance rotary and reciprocating packings, three materials which may be regarded as important additions to the seal manufacturers' armoury have become available in the last two decades and play a highly significant part in extending the frontiers of performance of the traditional soft packing:

(i) PTFE Polytetrafluoroethylene yarns provide soft packings for services where corrosive media are being handled or freedom from contamination is an essential requirement.

A semi-rigid fluorocarbon plastic, PTFE is unique in possessing almost complete chemical resistance within its temperature range which, in this field of application, spans the cryogenic area to 250°C. Another major advantage refers to its very low coefficient of friction.

Lubricated plaited PTFE yarn packings are suitable for rotary surface speeds up to 8 m/s and are also finding increasing acceptance on high speed, high pressure, multi-ram reciprocating pumps.

In solid form this material is not acceptable as a compression packing due to poor creep properties and lack of resilience. However, solid junk rings or spacers in PTFE are often used to enhance packing performance on arduous pump duties.

(ii) Aramid fibre Crossplait yarn packings made from aramid fibre, usually of a distinctive yellow colour, are becoming increasingly popular for a variety of pump and valve services hitherto satisfied by PTFE lubricated asbestos packings.

The yarn has high tensile strength, excellent resilience, thermal stability up to 250°C and is resistant to a wide range of chemicals. Aramid fibre packings are suitable for speeds up to 15 m/s and give impressive results with abrasive media.

(iii) Graphite

(a) Yarn packings in this material are a development for rotary pump applications and provide possibilities for extending the range of the packed gland beyond boundaries hitherto established.

A high coefficient of thermal conductivity, low friction and resistance to chemical attack are the useful characteristics of this material. Temperatures up to 400°C may be considered.

If a good performance is to be obtained, then close attention must be paid to mechanical conditions such as shaft run out and finish. Care in fitting and running-in is also mandatory.

(b) Expanded graphite foil is the recent and dramatic application of graphite, particularly in the context of valve applications. Expanded graphite materials combine the well-established thermal and friction characteristics, long associated with the correctly developed use of carbon based products, with a unique flexibility and resilience. The attributes of this exfoliated form of graphite bear recording.

- excellent resistance to compression set resulting in little loss of radial gland force or flange seating stress over long periods (see Table 16.3)
- no loss of volatiles even at high temperature thus minimising frequency of gland adjustment
- high temperature capability particularly in non-oxidising environments
- high thermal conductivity
- low friction properties - self lubricating
- exceptionally low chloride content
- no adhesion or corrosion problems
- fire-safe

TABLE 16.3

Compressibility/recovery ASTM F36-66 Procedure H (Major load 7/mm^2) (1000 lbf/in^2)	Expanded graphite 1mm thick 1.0 g/cc	Expanded graphite 2mm thick 1.0 g/cc
compressibility	39%	33%
recovery	20%	25%
Stress relaxation BS 1832:1972		
Temperature 300°C for 16 hrs Initial stress of 40 N/mm^2		
Residual stress	40 N/mm^2	40 N/mm^2

N.B. 10 N=1 kgf 1 N/mm^2=10 kgf/cm^2 approx.

Fig.6 shows an interesting comparison of performance on a test gland between half rings of expanded graphite and a lubricated asbestos yarn packing. Not only did the former require fewer gland adjustments during the period of testing but the average leakage rate was much less - to the point of running virtually dry for protracted periods.

Rather than use the tape form of expanded graphite which is primarily a useful maintenance expedient, moulded rings to a selected and controlled density should be the first choice for pump and valve glands.

Although more costly than conventional packing materials, economies of radial width and number of rings used are feasible quite apart from the performance advantage likely to be derived from the use of expanded graphite.

One cannot leave materials without special reference to the vital role played by asbestos - a much denigrated mineral fibre without which economic and practical solutions to many sealing problems would not be feasible. Although understandable, in an age of correct awareness of health and safety matters, the over-reaction against asbestos has revealed many inconsistencies. Motor manufacturers may prohibit its use as a plant maintenance material but continue to use asbestos in a brake-lining and clutch-facing role where residual dust is evident. Some users may seek to limit its application in a safe form as a valve packing or gasket but perpetuate its specification for fire-proof positions and roofing where the mineral is cut in a dry form.

Hazards exist but adherence to basic advice on handling asbestos will result in a sensible balance between prohibition and practicality. In this context, users of asbestos based pump and valve packings, gaskets or allied components might heed, to advantage, the statement issued by the Asbestos Information Committee to the British Valve Manufacturers' Association, a copy of which is appended to these notes.

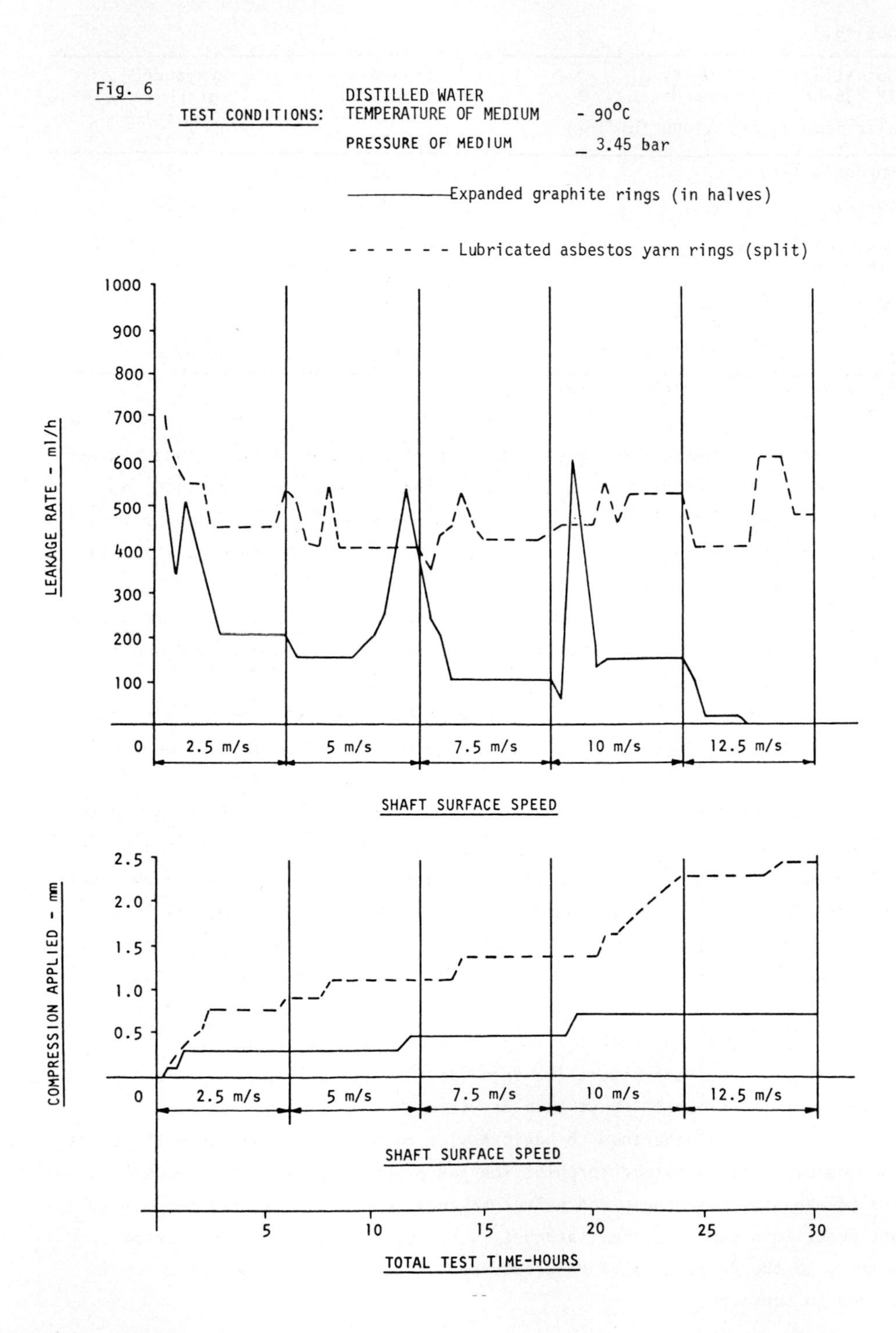
Fig. 6
TEST CONDITIONS:
DISTILLED WATER
TEMPERATURE OF MEDIUM - 90°C
PRESSURE OF MEDIUM _ 3.45 bar
Expanded graphite rings (in halves)
Lubricated asbestos yarn rings (split)
LEAKAGE RATE - ml/h
1000
900
800
700
600
500
400
300
200
100
0
2.5 m/s
5 m/s
7.5 m/s
10 m/s
12.5 m/s
SHAFT SURFACE SPEED
COMPRESSION APPLIED - mm
2.5
2.0
1.5
1.0
0.5
0
2.5 m/s
5 m/s
7.5 m/s
10 m/s
12.5 m/s
SHAFT SURFACE SPEED
5
10
15
20
25
30
TOTAL TEST TIME-HOURS

16.4.5 Selection

The most vexing question as, for many duties, so many reasonable alternatives exist. Much will depend on personal experience, frequency of maintenance, original cost level, contamination considerations, size, etc.

In the case of a manufacturer producing large quantities of valves or pumps to standard dimensions, there is much to be said for purchasing sets or rings rather than length form packing. With the techniques available, packing can be supplied ready for immediate fitting with substantial reduction in that overall cost represented by receiving length form that must be cut to size by skilled personnel. This economy is not confined to the large manufacturers but it is they who will enjoy the greater advantage.

On the other hand, in many instances, the problem of stocking rings or sets tailor-made for an assortment of valves varying in origin, type and dimensions can prove intolerable. For these cases, there is a clear need for the versatility of packing in length form. Comparable with this solution is the expedient provided by those packings of plastic nature that are available in loose form but this advantage must be weighed against the labour cost in the careful fitting required.

The quantity of packing to use and its size for a given application relies largely on the experience of the user/manufacturer in the type of duty being performed, or on liaison with a packing supplier at the design stage. The latter course of action is always favoured if any doubt exists, since an exact knowledge of the capabilities and limitations of the material employed can be found only with those specialists responsible for compounding and production.

Five rings of square section packing are often accepted as a sufficient number for the average uncomplicated duty but there are many pump applications where the presence of a lantern ring or similar consideration may dictate a greater quantity.

The appropriate packing section to use in relation to diameter is open to a degree of individual preference but broad recommendations are shown in Table 16.2.

To give an idea of the capabilities of the various materials and constructions of soft packings which are readily available, reference may be made to:

Table 16.4 - suitability in different media/speed and temperature limits

Table 16.5 - comparative speed performance

Table 16.6 - comparative temperature performance

Table 16.7 - comparative cost indication.

N.B. The statement of speed and temperature limits for a given material should not be construed as meaning that a packing will be suitable for duties where such maxima are jointly encountered.

TABLE 16.4

SYMBOLS

● Recommended
¤ Consult
★ Corrosion inhibitor included

RECIPROCATING, ROTARY PUMPS & VALVES	Max. Rec. Temp °C	Rotary Speed m/s	STEAM	CONDENSATE	WATER	AIR	MINERAL OILS	LIGHT HYDROCARBONS	solvents: HYDROCARBON	solvents: OXYGENATED	solvents: HALOGENATED	REFRIGERANTS	ACIDS	ALKALIS	NON-CORROSIVE SOLUTIONS
Lubricated aluminium foil.	540	7.5		●	●		●	●	●	●	●	●			●
Lubricated braided asbestos.	350	¤	●	●	●	●	●								●
Lubricated plaited asbestos.	315	20	●	●	●	●	●								●
Plaited, lubricated asbestos impregnated with PTFE dispersion.	290	10		●	●	●	●						¤	●	●
Plaited, lubricated asbestos impregnated with PTFE dispersion but with no additional lubricant.	290	8		●	●	●	●	●	●	●	●	●	¤	●	●
PTFE impregnated asbestos and glass fibre yarns with suitable lubricant.	290	7·5		●	●	●	●						●	●	●
PTFE yarn impregnated with PTFE dispersion and inert lubricant.	250	8		●	●	●	●						●	●	●
Soft lead based foil wrapped round lubricated asbestos core.	260	12		●	●		●	●	●	●	●	●			●
Hydrocarbon resistant lubricated plaited asbestos.	200	7					●	●	●						
Lubricated plaited cotton.	90	7			●	●	●								●
Cross plait aramid fibre yarns	250	15	●	●	●	●	●	●	¤	¤	¤	¤	¤	●	●
Pure graphite foil with no volatile additives	¤	¤	●	●	●	●	●	●	●	●	●	●	●	●	●
RECIPROCATING PUMPS & VALVES															
Monel wire reinforced asbestos cover with plastic core. ★	480		●	●	●	●	●	●	●			¤		●	●
Synthetic rubber bonded braided asbestos with brass wire reinforcement.	310		●	●	●	●	●	●	¤						●
Lubricated plaited flax.	90				●	●	●								●
VALVES ONLY															
Constructed from a jacket of asbestos reinforced with inconel wire braided over a resilient asbestos core. ★	650		●	●	●	●	●	●	●	●	●	●		¤	●
Lubricated braided asbestos with monel wire reinforcement. ★	600		●	●	●	●	●	●	●	●	●	●		●	●
Self-lubricating fibrous asbestos with flake graphite or mica.	540		●	●	●	●	●								●
Lubricated braided asbestos and brass wire reinforced. ★	510		●	●	●	●	●	●	●	●		●			●
PTFE yarn impregnated with PTFE dispersion and inert lubricant but with no additional lubricant.	250			●	●	●	●	●	●	●	●	●	●	●	●
Unsintered PTFE cord gland seal for rapid valve packing.	250		●	●	●	●	●	●	●	●	●	●	●	●	●

TABLE 16.5

MAXIMUM ROTARY SPEEDS FOR PUMP PACKINGS

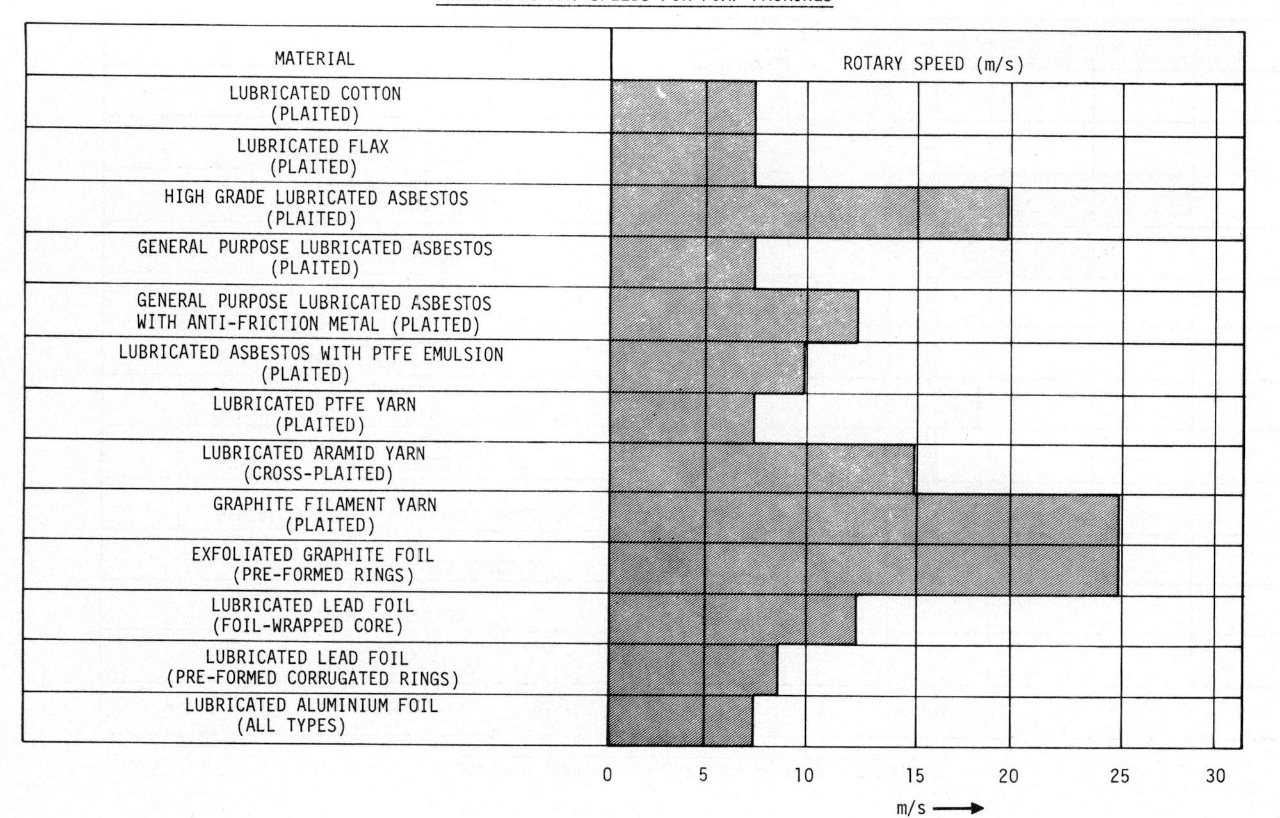

TABLE 16.6

MAXIMUM SERVICE TEMPERATURES OF PUMP PACKINGS

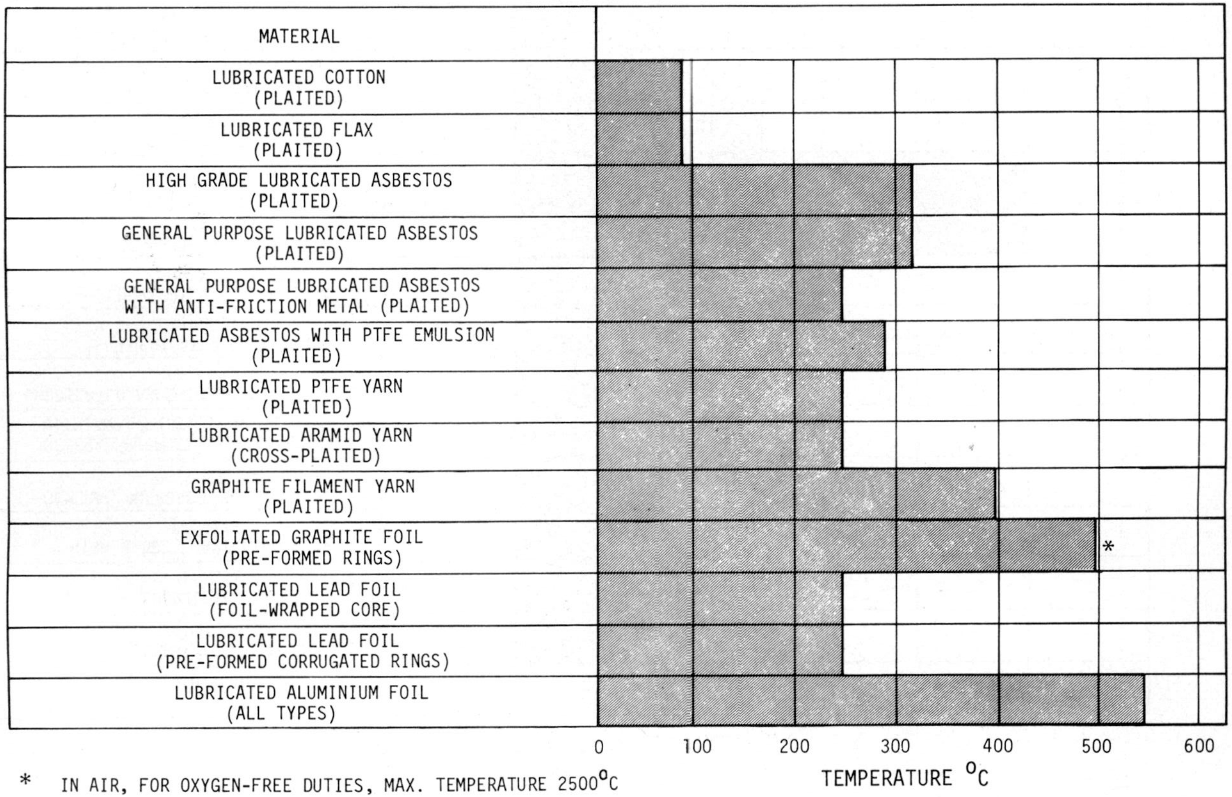

MATERIAL	
LUBRICATED COTTON (PLAITED)	
LUBRICATED FLAX (PLAITED)	
HIGH GRADE LUBRICATED ASBESTOS (PLAITED)	
GENERAL PURPOSE LUBRICATED ASBESTOS (PLAITED)	
GENERAL PURPOSE LUBRICATED ASBESTOS WITH ANTI-FRICTION METAL (PLAITED)	
LUBRICATED ASBESTOS WITH PTFE EMULSION (PLAITED)	
LUBRICATED PTFE YARN (PLAITED)	
LUBRICATED ARAMID YARN (CROSS-PLAITED)	
GRAPHITE FILAMENT YARN (PLAITED)	
EXFOLIATED GRAPHITE FOIL (PRE-FORMED RINGS)	*
LUBRICATED LEAD FOIL (FOIL-WRAPPED CORE)	
LUBRICATED LEAD FOIL (PRE-FORMED CORRUGATED RINGS)	
LUBRICATED ALUMINIUM FOIL (ALL TYPES)	

0 100 200 300 400 500 600

TEMPERATURE °C

* IN AIR, FOR OXYGEN-FREE DUTIES, MAX. TEMPERATURE 2500°C

TABLE 16.7

TYPICAL RELATIVE COSTS OF PUMP PACKINGS

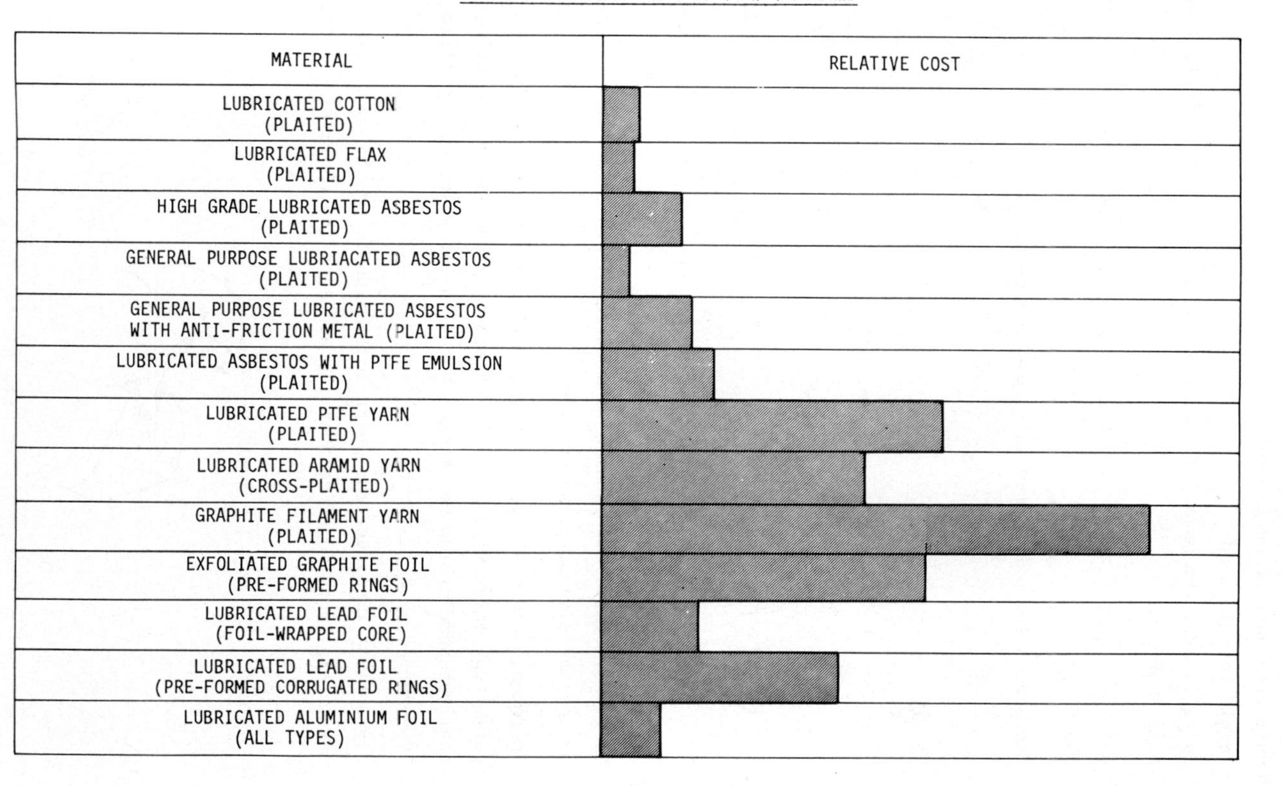

16.4.6 Fitting

It is often assumed that unskilled labour can be used to repack pump glands but this is true only so long as unskilled is not equated with unaware. Familiarity with the following ideal procedure will be more than repaid in terms of trouble free packing performance.

Where length form is used:

(i) Spirally wrap the material around a rod of diameter equivalent to the pump shaft.

(ii) Cut the required number of rings cleanly to obtain good butt-joins. See Fig.7.

(iii) Proceed as for pre-formed split packing rings.

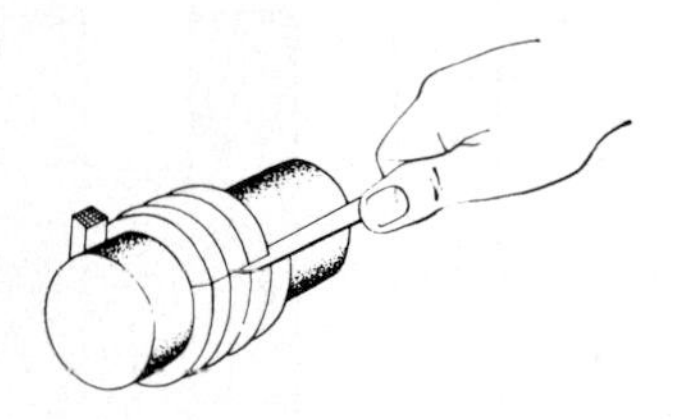

Fig.7

Where pre-formed split packing rings are used:

(i) Carefully remove old packing (including where appropriate the packing on the far side of a lantern ring).

(ii) Thoroughly clean all surfaces that will contact the packing and, where permitted, smear with oil. Gland and neck bushes, shaft surface and bearings should also be checked for signs of wear and rectified as necessary.

(iii) Place first ring over the shaft by opening to an 'S' configuration to ensure that bending effects are spread over the whole ring. See Fig.8.

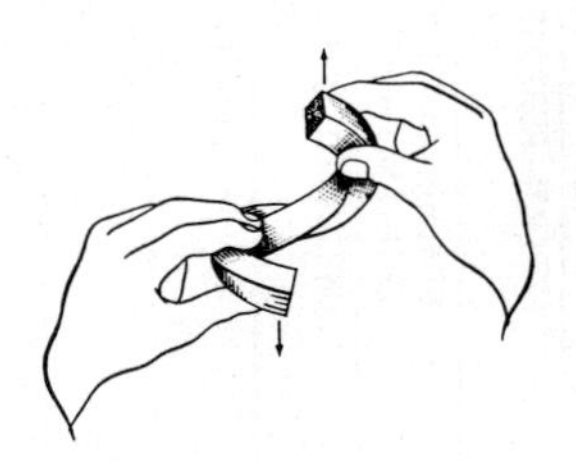

Fig.8

(iv) Insert first ring into stuffing box and lightly bed in with a split (wooden) distance piece and gland spigot. With plaited packing the 'V' formation on the outside diameter of the ring should be pointing in the direction of shaft rotation. See Fig.9.

Fig.9(a) Plaited packing

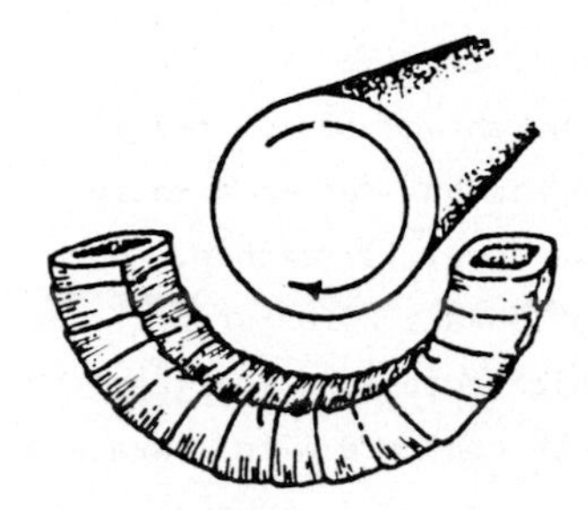

Eig.9(b) Foil wrapped packing

(v) Repeat (iii) and (iv) with remainder of rings ensuring that each ring is firmly seated and that the butt joins are staggered by at least 90°.
N.B. The rings must be fitted individually and under no circumstances should complete sets be fitted as a unit.

(vi) When the requisite number of rings have been fitted, tighten gland nuts until the shaft or spindle torque increases. Then slack off gland and pull up to finger tightness only. (If pump is to be stored before use leave gland slack so that packing resilience is not impaired).

(vii) Running in pumps Prime casting and run pump up to operating speed for 10 to 15 minutes. If pump is not fitted with gland cooling, a cold water spray over the gland housing will avoid excessive heat build-up during this stage. If no leakage occurs, stop pump, vent casing pressure and slacken gland further. Repeat until leakage starts.

The controlled leakage, essential for lubrication purposes, can then be obtained by running the pump and evenly tightening the gland nuts in increments of 2 flats until approximately one drop every few seconds is obtained. Approximately 15 minutes should be left between successive adjustments. DO NOT OVERTIGHTEN.

(viii) Where loose form material is used for valves proceed as (i) and (ii) and tamp packing into a dense homogeneous mass progressively filling the housing to the required degree.

(ix) Because of the danger of corrosion through electrolytic action, packings containing graphite should be avoided on valves or pumps with stainless or chrome steel stems. This risk is most acute when the packing remains in the gland during storage and is particularly aggravated by the presence of moisture.

16.4.7 Fault Finding

A major advantage of compression packings is that breakdown is rarely sudden or catastrophic but rather a matter of a gradual build up of leakage until an unacceptable level is reached. Normally, considerable life can be achieved by controlling leakage with further tightening of the gland nuts (N.B. the seepage of fluid which acts as a lubricant for the packing on rotary applications should not be confused with leakage and the rate of one drop every few seconds should be maintained). However, if other than routine maintenance or just plain 'fair wear and tear' are suspected as the cause of leakage and the need for re-packing, then the following hints could well prove useful.

(i) Confirm that the packing is rated as suitable for the application.

(ii) If one or more rings are missing from the set, check for excessive neck bush clearance allowing extrusion of rings into the system. If the top ring has extruded between the gland follower, anti-extrusion rings could avoid replacement of metal parts.

(iii) If the packing's radial thickness appears diminished in one or more places, check for an undersize shaft or badly worn bearings which could cause shaft whip or spindle wobble.

(iv) If radial section of packing directly beneath the shaft is reduced or premature leakage occurs along the top of the shaft, check for mis-alignment of shaft centre to stuffing box bore.

(v) If packing is worn on the outer diameter, check for loose rings or rings rotating with the shaft due to insufficient gland load.

(vi) If packing rings have bulges on their radial faces, the adjacent ring was probably cut too short, causing packing under pressure to be forced into the gap at the joint.

(vii) If packing nearest gland spigot shows excessive deformation whilst other rings are in fair condition, the set was probably incorrectly installed and subjected to excessive gland tightening.
N.B. Over-tightening is usually the greatest single cause of premature packing failure.

(viii) If the cause of your particular problem is still not apparent, give equal attention to I.D. and O.D. leakage and check for a rough stuffing box bore before seeking specialist advice.

16.4.8 Standardization

In the interests of stock control it is clearly an advantage to rationalise the variety of packings used in any plant to that minimum number which will effectively cater for all the conditions likely to be encountered. If cost is no object, then there are single, sophisticated materials and constructions that will go some way towards satisfying most demands, but it is doubtful if cost effectiveness could be justified. Far better to compromise on a small number of reputable products developed for the areas in question, e.g. pumps, valves, etc.

There are few standards applying to compression packings on a national or international basis, although many individual companies and organizations have domestic standards which have, in many cases, been the subject of collaboration between user and packing manufacturer.

BS 4371 : 1968 specifies minimum standards for lubricated plaited cotton, lubricated plaited flax, lubricated plaited or braided asbestos, dry white non-metallic plaited or braided asbestos, plaited or braided asbestos, metallic wire reinforced, indurated asbestos, and lubricated fibrous asbestos and gives guidance on limiting operating parameters for these constructions.

Where packings are required for service with potable water in the Water Authority distribution system (which covers reservoir to tap), only those materials which have gained a National Water Council Approval may be used. Such products have been tested to establish that they produce no colour, taste or turbidity, are non-toxic and will not support microbial growth.

Statutory Instruments 1978 No. 1927 "The Materials and Articles in Contact with Food Regulations 1978" required that compression packing materials, for example, "... do not transfer their constituents to foods with which they are, or likely to be, in contact, in quantities which could-

(i) endanger human health or

(ii) bring about a deterioration in the organoleptic (sensory quality) of such food or an unacceptable change in its nature, substance, or quality."

Such regulations inevitably restrict the range of available materials and lubricants. Consultation with the supplier is recommended to establish preferred grades.

16.4.9 The Future

The research currently being undertaken as a direct result of the anti-asbestos lobby may produce glass/PTFE/graphite or ceramic constructions which match existing materials. However, many problems remain to be solved in producing a general purpose product that can compete with asbestos in terms of lubricant retention, absorbency or durability at comparable cost.

There can be no doubt that graphite foil products will, because of this situation, see greater acceptance despite high initial cost. Performance reports are extremely encouraging and, irrespective of current market considerations, the material deserves to succeed on its own merits.

APPENDIX 1

COPY OF STATEMENT ISSUED BY THE ASBESTOS INFORMATION CENTRE, 40 PICCADILLY, LONDON W1V 9PA, TO THE BRITISH VALVE MANUFACTURERS ASSOCIATION, 3 PANNEL COURT, CHERTSEY STREET, GUILDFORD, SURREY, GU1 4EU ON 30th APRIL 1980.

SAFETY OF ASBESTOS GLAND PACKINGS AND GASKETS

Crysotile (white) asbestos fibre is a basic constituent of valve packings and gaskets because it combines in one material softness, resilience, absorption properties, strength as a reinforcement and, where required, heat resistance.

Asbestos is only a risk to health if its dust is inhaled. Valve packings and gaskets based on chrysotile (white) asbestos are safe to handle by valve users and by maintenance engineers. A good standard of industrial hygiene should be observed when handling or using products which contain asbestos.

The Advisory Committee on Asbestos in their interim statement and final report published in October 1979 recommend that asbestos dust should be kept to the lowest practicable level and call for a 1 fibre/ml control limit. The report further states that the presence of chrysotile is unlikely to have produced any material increase in the risk of lung cancer in the general population or any appreciable number of cases of mesothelioma, and the same is certainly true of asbestosis. Gland packings and gaskets will not create dust levels in excess of the control limit, given normal usage and maintenance.

'Lubricated' packings are vacuum impregnated with mineral oils and greases, graphite and other lubricants, and do not emit dust.

'Dry' packings are normally treated with a dust-suppressant which significantly reduces dust emission so that in normal use, including maintenance, they would not be expected to present a hazard. They are not commonly used in valve glands.

'Hard' packings and gaskets (compressed asbestos fibre) are made from a combination of asbestos, rubber and other fillers. The asbestos fibre is locked into the rubber matrix and will not normally create dust in use or maintenance. If however, these materials are subjected to grinding or other abrasive processes, precautions should be taken to avoid inhaling any dust which may be emitted.

'Soft' or moulded packings are normally made from rubberised asbestos cloth, and the asbestos is sealed within the rubber coating, so that they also are safe in use and maintenance.

Crocidolite (blue) asbestos

No crocidolite (blue) asbestos has been used in packings and gaskets manufactured in the UK for several years. Some imported packings and gaskets may however, incorporate blue asbestos. When lubricated these packings are safe to handle but respiratory protection will be needed when they are in a dry state.

Removal

Care should be taken when removing old packings which may have lost their lubricants. They should be damped and then removed with the correct tools, i.e. packing extractors.

Labelling

Any packings or gaskets which require special handling precautions carry a warning label.

17 CENTRALISED LUBRICATION SYSTEMS DESIGN

J.G.MERRETT, Managing Director,
Engineering and General Equipment Ltd.

17.1 INTRODUCTION

Whilst we are kept comparatively well informed of the latest developments in oil and grease technology and the vital role it plays in industrial and commercial applications, very little has been written about the equally vital "Centralised Lubrication Systems" and some of the methods available to Engineers by which grease and oil may be transmitted to the point of lubrication.

In our society where energy, machinery and labour are now (1981) expensive commodities, it is clear that in the past insufficient attention has been paid to the direct and indirect losses of energy, occasioned by wear and friction and to the savings of materials. However, in 1977 a government financed American Report suggested that $16.25 billion p.a. (at 1976 values) could be saved by a "Strategy for Energy Conservation through Tribology" [1]. Converted into U.K. (1980) values, this indicates an equivalent saving of energy through tribology in excess of £1½ billion p.a.

In short, correctly selected lubricants and their methods of application by Centralised Systems can effect significant savings, e.g. machine tools, conveyors, cranes, rolling mills, blast furnaces, ball mills, sugar machinery, paper mills, heavy mobile plant, etc. to varying degrees, all require the application of lubricants. A correctly designed and installed Centralised Lubrication System is the engineer's insurance against some of the severe tribological problems, i.e. friction and wear, which would otherwise occur if the plant and machinery were inadequately lubricated.

Unfortunately, all too often in the supply of plant and machinery the selection of the lubricant and Centralised Lubrication equipment are considered at a late stage in the manufacture of the plant which, coupled with the conflicting interests of machinery builders, can result in unsuitable lubrication equipment being selected. Likewise, the lubrication equipment supply companies have a responsibility, not only to know their own product, but also to appreciate the tribological requirements of the plant and machinery which requires to be lubricated. Failure in this area invariably results in the plant user being placed in the unenviable position of having to apply, at an early stage, for additional capital to rectify new plant or, more often than not, to live with the problems and rectify as and when through a maintenance budget - both of which the plant user would be the first to agree is entirely unsatisfactory.

This paper endeavours to deal with some of these factors affecting the choice of lubricant and lubricating equipment, the basic elements of the machinery to be lubricated, and the conditions under which it operates.

17.2 POINTS OF LUBRICATION

Bearings, the essential components of plant and machinery, may be generally grouped into journal, thrust, conveyor chain pin and link, anti-friction, slide-ways and crane rails. Each wear surface must be treated separately with regard to lubricant and lubrication technique.

The lubrication requirement of a plain journal bearing is the provision of an adequate and constant flow of lubricant of specified viscosity to give a fluid film of high-load bearing capacity. The journal bearing has inherently a convergence between the shaft and the bearing. When relative motion takes place a film of lubricant is induced between the surfaces, effectively separating them. Bearings employed to absorb thrust and prevent misalignment of shafts vary greatly in type and lubrication requirement, whilst anti-friction bearings require less lubricant than plain bearings. Most available formulae dealing with the application of grease to these bearings treat speed as an important factor. For small anti-friction bearings such as those employed in lightly loaded fractional horse-power motors, too much grease can be damaging. In such cases, recommended lubrication intervals of up to several years have been established. Chain pins and links present major critical wear points on floor and overhead (including Power and Free) conveyors. In the automobile industry, chain lengths of several hundred metres, having thousands of points requiring lubrication, are commonplace (Fig.1).

Slideway and crane rail lubrication requires the right lubricant and the right applicator (see Fig.10, Section 17.6.1). Too little lubricant results in rapid wear; excessive lubricant can be a hazard to life or limb. Every case is different, yet in every case it is critical that the lubricant is applied in line with the requirements of that component, both with regard to mechanical wear and to energy conservancy.

17.3 SELECTING THE LUBRICANT - OIL OR GREASE?

In modern machinery lubrication, lubricants and the means of the application must be considered together.

The best lubricant will serve no useful purpose if it is not applied at the

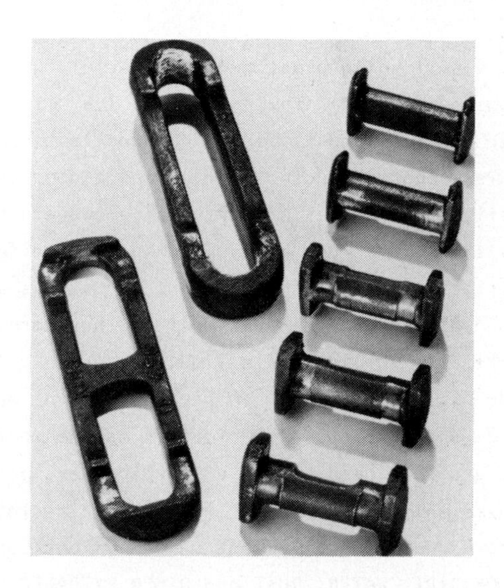

Fig.1 Excessive pin and link wear in a conveyor chain.
A 3 mm pin wear per pin on a 732 m conveyor increases the chain length by 23 metres.

right time, at the right place, and in the correct quantity. Conversely, the best lubrication equipment, applying lubricant quite correctly, will be of little use if the lubricant it feeds is unsuitable for the duty it has to perform.

Lubricants generally, either oils or greases, are supplied by oil companies supported by specialist oil and grease manufacturing companies, who will provide advice on any particular lubricant and application as well as on the choice of lubricants in general. The supplier of the lubrication equipment will also often be able to advise, especially on lubrication design problems, or act as an intermediary.

The subject of lubricants is treated here only in its very broadest terms under three headings:-

"OIL" or GREASE"; when to use one and when the other.
LUBRICATING GREASE; the types and how to select them.
LUBRICATING OILS; types and how to select them.

17.3.1 Oil or Grease? When to Use One and When the Other

Movement between two dry surfaces causes heat and wear. The purpose of introducing a lubricant between the two surfaces is to reduce friction, heat, and wear.

Oil has the following advantages: it flows, it penetrates, it removes heat, and it scavenges. At the same time, it has limited sealing qualities and poor 'staying' power, therefore requires more frequent replenishing than grease.

Lubricating grease, a semi-solid plastic-like material, has excellent sealing characteristics, possesses good 'staying' power, i.e. it adheres to surfaces more readily and longer than oil and it can be an excellent corrosion preventative. Unlike oil, it is a poor conductor of heat and a poor scavenger; it does not flow or spread easily - the latter property can, however, be an advantage where contamination by lubricants must be avoided such as in food and bottling and the textile industries and where the lubricant is used as a sealing medium.

Where an application entails HEAT REMOVAL, oil is therefore generally the choice. It can be applied in the form of a liquid or as an oil mist, i.e. micro-fog comprising an oil-air mixture. In severe cases of heat removal, oil can be recirculated and, during the circulation, it can be cooled and cleaned. Large turbine bearings and fast gear trains are typical examples where oil circulating systems are used and large amounts of oil are circulated.

For applications where heat removal is not a problem, but LUBRICATION or the PREVENTION OF INGRESS OF DIRT, WATER, DUST and OTHER EXTRANEOUS MATTER is - lubricating grease can be used. Its application ranges from heavy mill bearings and slides to textile machinery, etc. It is also the most suitable where application of the lubricant is required at LONG INTERVALS.

Nearly all Ball and Roller Bearings, except those in respect of which heat removal is essential, are generally grease lubricated. Where DIRTY/DUSTY and WET CONDITIONS exist, grease is generally preferable. The convenience of grease lubrication makes grease the 'preferred' lubricant for the vast majority of rolling bearings.

Generally, oil lubrication is employed in the relatively few cases where it is not possible to take advantage of the merits of grease as a rolling bearing lubricant.

Where motion is INTERMITTENT or OSCILLATING, grease is often the more suitable lubricant.

Summarising, where heat conveyance away from rubbing surfaces or penetration of the lubricant is of importance, or the scavenging function is necessary, oil is preferable, whilst lubricating grease can and, in most cases, should be considered as the preferred lubricant for slow moving machinery, long life lubrication, and where dirt and dust has to be kept out of the bearings, sliding surfaces, etc.

17.3.2 Lubricating Grease - The Types and How to Select Them

Greases designed for lubrication are essentially a mixture of mineral oil and thickener, according to the application requirement. In the most widely used modern grease the thickener is a metallic soap, usually of lithium or calcium, with the quantity of oil adjusted to give a solid, semi-solid or semi-fluid consistency.

The soap fibres form a structure that retains the oil, the dimensions and arrangement of the fibres varying according to the metal and the fatty acid from which the soap was made. The quantity, dimensions and distribution of the fibres are the main parameters controlling the stability and flow properties of this lubricating material.

One of the most important physical features of grease is stiffness (for sistency) which is indicated by a test that measures the depth that the cone sinks into a sample of grease. The depth measured is a tenth of a millimetre, and referred to as the degree of penetration. Figure 2 shows one classification system for greases.

Grade Number	Worked penetration at 25°C (0.1 mm)	Description
000	445 - 475	Very fluid
00	400 - 430	Fluid
0	355 - 385	Semi-fluid
1	310 - 340	Very soft
2	265 - 295	Soft
3	220 - 250	Semi-firm
4	175 - 205	Firm
5	130 - 160	Very firm
6	85 - 115	Hard

Fig.2 NLGI classification of the consistency of greases.

Although lithium-based greases today satisfy a vast number of lubrication requirements for cranes, conveyors, forging presses, continuous casting plants, rolling mills, etc., there are a number of situations requiring lubricants with special properties, for example, the ability to withstand high temperatures such as those encountered in power station turbines, bakery oven conveyors, etc. These latter greases include the clay-thickened and other solid-thickened compounds which withstand considerably higher temperatures, especially when employed with high temperature synthetic lubricants, e.g. polyglycol, synthetic esters, and silicones.

The upper temperature limit at which any grease may be used is dependent partly on the type of thickener, partly on the fluid and its required service life. Higher operating temperatures have the effect of shortening the lubricant's service life and reducing permissible operating speeds. It is therefore convenient to express the working limits of a grease in terms of bearing speeds and temperatures, as shown in Fig.3. This diagram shows that lithium, sodium and calcium-based greases have upper temperature limits of 130, 110 and 70°C respectively, and that clay-based and calcium-complex greases can be used at up to 150°C.

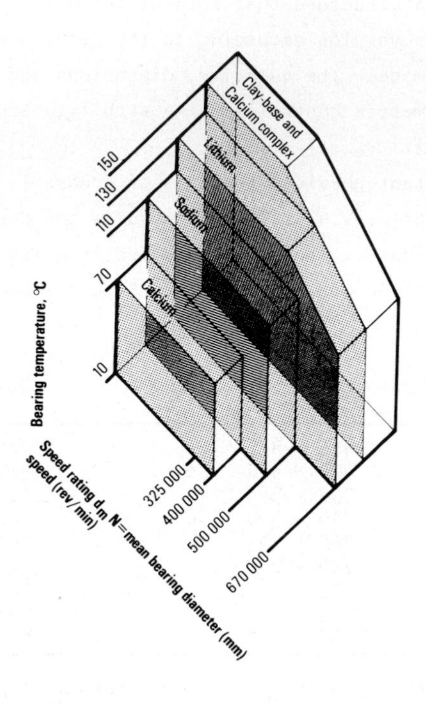

Fig.3 Working limits of mineral oil greases.

17.3.3 Lubricating Oils - Types and How to Select Them

Vegetable and animal oils are excellent lubricants, but have short life as they oxidise and tend to go rancid; as a result, their main use is for 'one shot' applications such as for forging or as additives for mineral oils.

Generally, the origin of the oil determines its use (Fig.4)

Oil Origin	Application
Mineral Oil	e.g. petroleum base for general lubrication of mechanical parts such as engines, gears and general engineering plant.
Vegetable Oil	e.g. castor, palm, and rape seed oils for special applications where high lubricity is desirable such as kilns, bakery ovens.
Animal Oil	e.g. sperm or other fish oils from sheep wool for applications such as kilns, bakery ovens.
Synthetic Oil	e.g. glycol derivatives and diester for extreme high temperature.

Fig.4 Origin of Oils

Figure 5 illustrates how viscosity of oils change with temperature, becoming thinner when they are heated, but they do not change viscosity at the same rate. The rate of viscosity change with temperature is referred to as the 'viscosity index'.

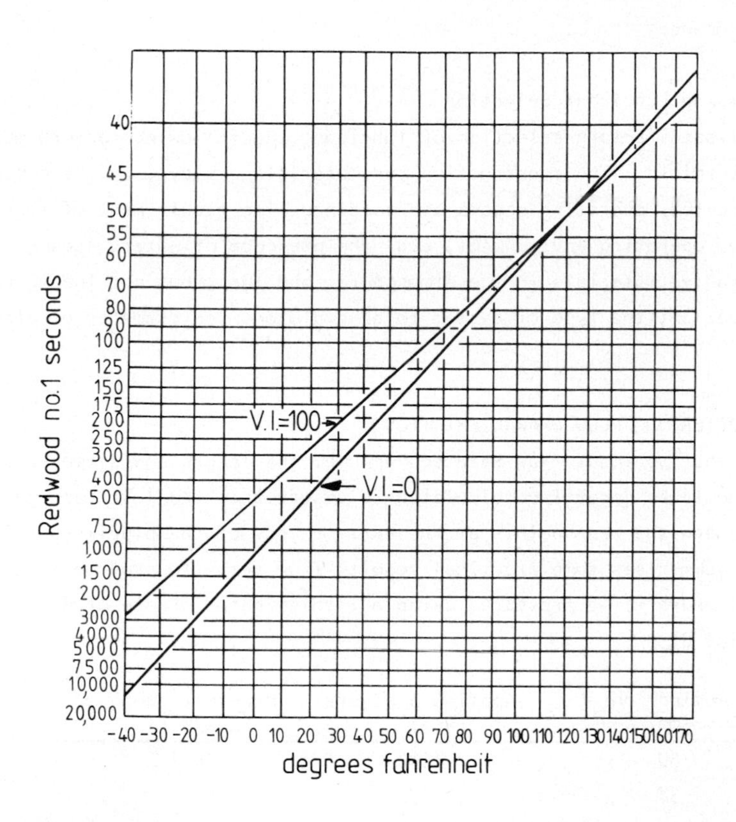

Fig.5 Viscosity vs Temperature for two oils having viscosity indexes of 0 and 100 respectively.

17.3.3.1 The properties of the oil must be carefully considered when designing a lubrication system, e.g. conveyor chains passing through a pre-treatment oven may reach a temperature of 180/200°C. Oils with special inclusions such as graphite or molybdenum disulphide in glycol as used in bakery ovens must have provision for agitation or recirculation within the lubricator storage tank to prevent settling out of the graphite or molybdenum.

Bakery ovens generally use molybdenum disulphide in glycol whilst for extreme pressure loading conditions on Power and Free trolleys, as in the car industry, chains and trolley wheels require special lubricants, having a high adhesion characteristic as well as imparting rust-proofing and water-proofing.

17.3.3.2 A further specialist lubricant is rape seed oil; this is a low acid fatty vegetable oil. Because of its relatively high flash-point it is suitable for high temperature work when refined and given a graphite inclusion. Typical uses are for continuous casting moulds, forging, and as a cutting oil for harder metals. Here again, its special properties must be considered when providing the lubricating means.

17.3.4 Summarising Lubricant Selection

For the final satisfactory selection of lubricant the following factors must therefore be determined. The construction and materials employed in the components to be lubricated, operating speeds and loading, life requirement of lubricant and machine, operating environment, e.g. the presence of water, steam, chemicals or abrasive materials in the atmosphere; and last, but not least, the method of application, the type of system to be employed, the diameter of pipe runs, etc.

17.4 PIPE DIAMETER vs FLOW CHARACTERISTICS

To determine the lubricant flow rate and volume, the length and diameter of the pipelines should be carefully calculated to ensure that the lubricant can satisfactorily reach the wear points at the extreme ends of the pipelines. The following table summarises some practical results from tests to prove optimum measurements and end-of-line pressure, using a lithium-based grease of No.2 consistency at 150°C.

Nominal bore (mm)	Applied pressure (kPa)	Pressure drop (kPa/m)
50	1100	36.1
38	1875	61.5
25	4410	144.7
19	6410	210.3

17.5 LUBRICATION REQUIREMENTS FOR PLAIN BEARINGS

For normal working conditions it has been found that the amount of grease required for plain bearings is equivalent to a layer of 0.1 mm on the developed bearing area (0.1 x d x L) per hour of bearing operation. Figure 6 illustrates a chart for calculating the grease requirements for plain bearings. As an example to calculate the grease required for a bearing of 75 mm diameter by 250 mm long, intersect diameter and length as at * follow line of arrow and where it intersects the top scale, this indicates the amount of grease required, being in this example 6 gm or 0.21 oz per hour.

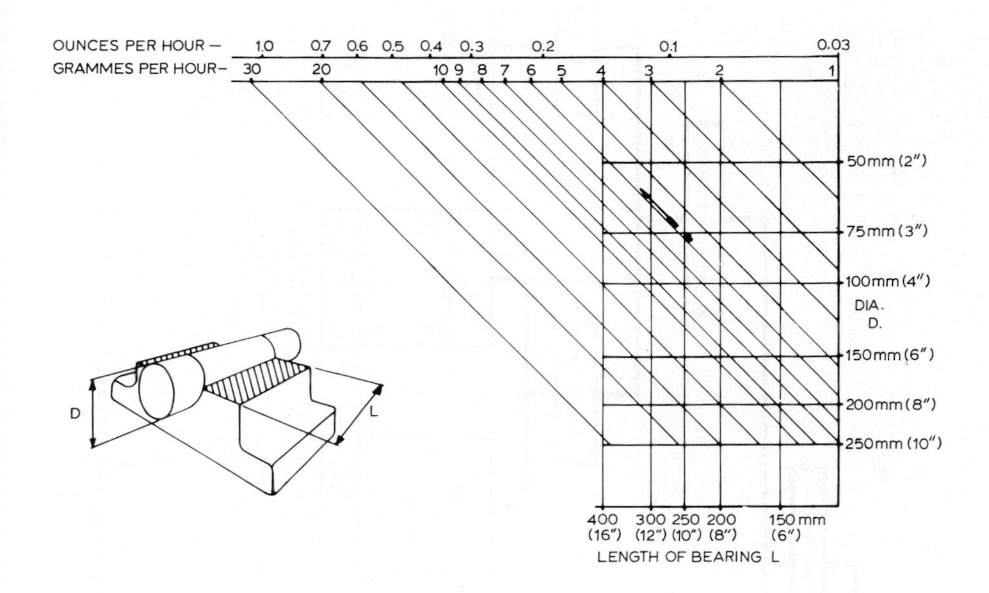

Fig.6 Grease requirements for a plain bearing.

17.6 SELECTING THE RIGHT TYPE OF LUBRICATION SYSTEM

Lubrication systems for plant and equipment, whatever the application, require individual design. They must be functional and correctly engineered to satisfy the bearing requirements and the designers' application specification. With regard to the human element, they must be fool-proof. Systems included in this chapter deal with: Grease, Oil and Micro-Fog.

17.6.1 Grease Lubrication Systems

All Centralised Grease Lubrication Systems are of the non-recirculating type and operate on the total loss principle. They are basically divided into Direct Feeding systems and Indirect Feeding systems, otherwise referred to as Line systems, as illustrated in Fig.7.

17.6.1.1 Direct Feeding Systems are those where the volume output of the Direct Feeding lubricating pump is positive; the pumping plungers and means of metering the output to individual wear points being incorporated in the lubricating pump. Therefore Direct Feeding systems operate on a Positive Volume Principle, i.e. they introduce a metered volume of lubricant into the pipes, and since this volume is not affected by pressure in the pipes, the pumps can work continuously against high back pressures.

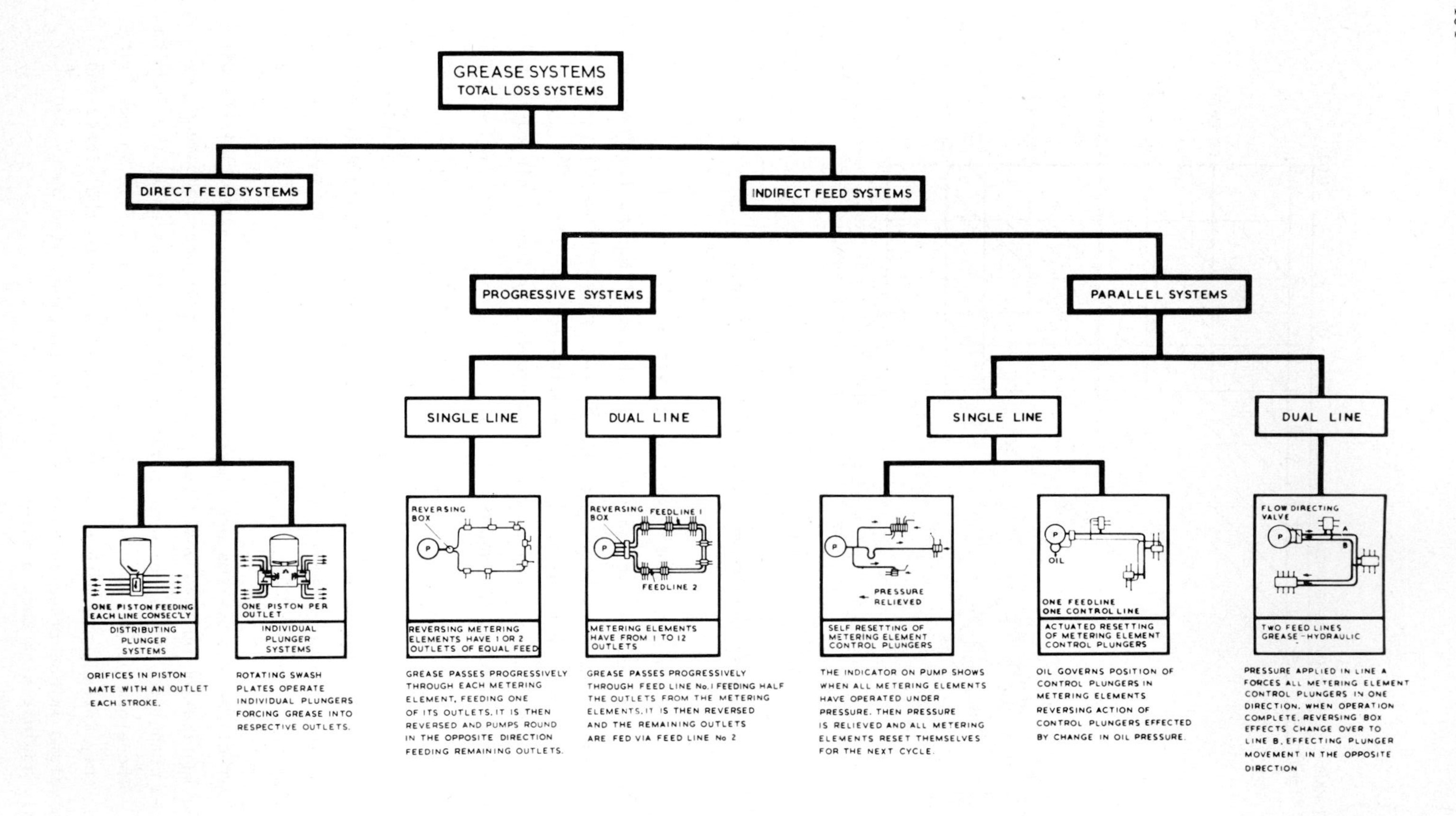

FIG. 7 FUNDAMENTAL DIVISION OF CENTRALISED GREASE LUBRICATION SYSTEMS.

Direct Feeding systems can be divided into those where each outlet has its own metering plunger (Fig.8) and those where a single moving plunger distributes progressively the metered amounts of grease into the various outlets (Fig.9).

Normally, direct feeding lubricators obtain their pumping action either by means of moving plungers, combined with a system of porting, or by the use of plungers in connection with spring-loaded ball valves. The latter type should be avoided in the case of dirty or dusty conditions as any ingress of extraneous matter may lodge in the seating of the ball valve and render the lubricator inoperative.

It follow that where a lubricator is driven by a moving part of a machine, an amount of lubricant required relative to the movement of the shaft in the bearing being lubricated, a direct feeding lubricator should be used (Fig.10) with the lubricator driven via the direct driving elements, e.g. Eccentric Drives, Throw Plates, or Offset Pin Drives.

17.6.1.2 Indirect or Line Systems have the pumping and metering elements geographically separated; they are connected by means of one or several pipelines. The action is hydraulic, the pump serving as a pressure creating unit for pumping grease into a pipeline which is thereby set under pressure. As the pressure increases, so the various metering elements eject their charges either progressively, or according to the back pressures against which they operate.

During normal operations each element, having given up its metered amount, blocks itself and will not pass any further lubricant to the points of application. When all elements have given up their metered charge, a rapid rise in pressure occurs in the main line. Utilizing either this increase in pressure at the pump or at the end of the line, a signal is given indicating that the lubricating phase is completed.

Thereupon the main line has to be depressurized, which is usually effected by opening it to the grease supply unit, e.g. the reservoir. Depending on the type or make of system, the plungers of the metering valves are then reset to permit their further operation, or they are already set for another application phase which moves them back into their original position, thereby completing a system cycle. The method and mechanics of resetting depend on the particular type of system; also whether one complete cycle of the system involves one or two application (lubrication) phases.

The fundamental division of 'line systems' is that between progressive and parallel types of systems. In the progressive system the lubricant must pass through the metering elements or valves progressively, i.e. only after having actuated the first element to feed lubricant to the point of application will the lubricant be passed to the second element, and so forth. This is in contrast to the operation of the parallel system, on which the metering units are

DIRECT FEED LUBRICATOR FOR AUTOMATIC CENTRALISED LUBRICATION

HOW IT WORKS:

The lubricant is fed from the container (5) by the action of the rotating scraper (3), the fixed paddle (4) and the rotating wedge plate (2) through strainer (6) into the main body reservoir (11).

The twelve feed plungers (8) operate in rotation in the pumping body (1). During the suction stroke a feed plunger draws a quantity of lubricant through the holes in the bevel gear (12) into the inlet groove in the distribution cylinder (7) and then into the feed plunger inlet port (14).

When during rotation of the main distribution cylinder its outlet groove (16) connects feed plunger inlet port (14) with the outlet port (15), the feed plunger concerned makes its delivery stroke, thereby forcing the lubricant into the channel leading to its respective outlet pipe (10). This procedure is repeated for each plunger. The stroke of each feed plunger, which is reciprocated by the cam plate (9), can be adjusted individually from zero to maximum, this by means of adjustment screw (13). Such adjustment permits a regulated measure of lubricant to be fed to each point of application.

Fig. 8. Direct Feed Lubricator incorporating its own metering plungers.

PRINCIPLE OF OPERATION

Pulling the hand lever (10) forward raises the plunger (2) and forces lubricant inside the cap nut (6) into the interior of the plunger (2). During the upward movement of the plunger (2) the annular groove, connected to the interior of the plunger by a drilled hole, aligns with the outlet ports at various levels and allows lubricant to be fed to each of the outlets in turn.

Pushing the hand lever (10) backwards returns the plunger to its starting position and draws lubricant into the cap nut (6) so that it may be discharged from the outlets on the next forward movement of the hand lever (10).

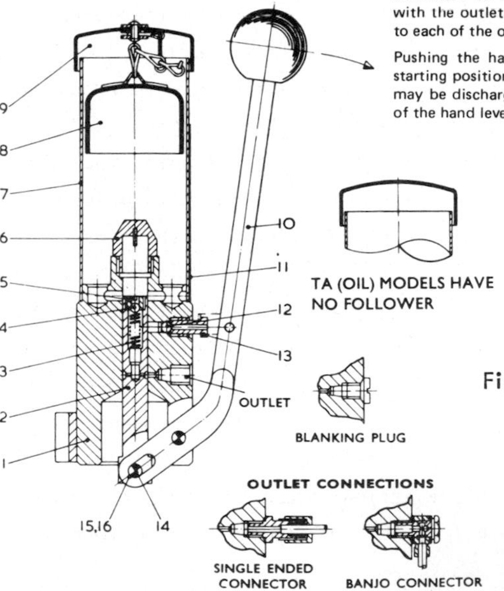

Fig. 9. Direct Feed Lubricator

Distributing lubricant progressively.

1. Pumping body
2. Plunger
3. Spring
4. Ball
5. Valve seat
6. Cap nut
7. Container
8. Follower
9. Top cover
10. Hand lever
11. Instruction label
12. Ball
13. Air-vent screw
14. Plunger pin
15. Circlip
16. Washer

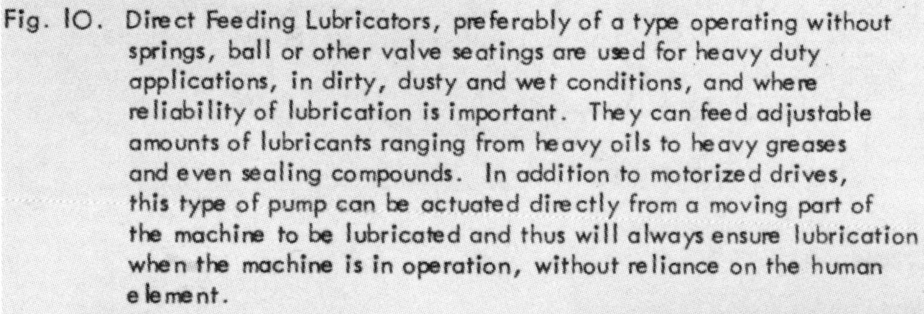

Fig. 10. Direct Feeding Lubricators, preferably of a type operating without springs, ball or other valve seatings are used for heavy duty applications, in dirty, dusty and wet conditions, and where reliability of lubrication is important. They can feed adjustable amounts of lubricants ranging from heavy oils to heavy greases and even sealing compounds. In addition to motorized drives, this type of pump can be actuated directly from a moving part of the machine to be lubricated and thus will always ensure lubrication when the machine is in operation, without reliance on the human element.

Application:
Lubrication of up to 100 points with grease or oil, particularly on presses, machine tools, packaging machinery.

Design:
A main pipe from a simple filling or lubrication pump leads to the distributor; the quantities of lubricant delivered are distributed by the distributor to the outlets in a particular pre-arranged sequence. If required, the lubricant from the distributor can be fed to other distributors for further redistribution.

1st stage:
Lubrication of not more than 20 points by means of a grease nipple screwed into the distributor and a grease gun, which is operated until the flow indicator shows that the lubrication operation has been completed.

2nd stage:
A hand pump and distributors in series, which distribute the lubricant to the lubrication points in a specified manner.

3rd stage:
A pump set with electrical or pneumatic control, including monitoring of the distribution if desired.

Introducing central lubrication in stages:

1st stage

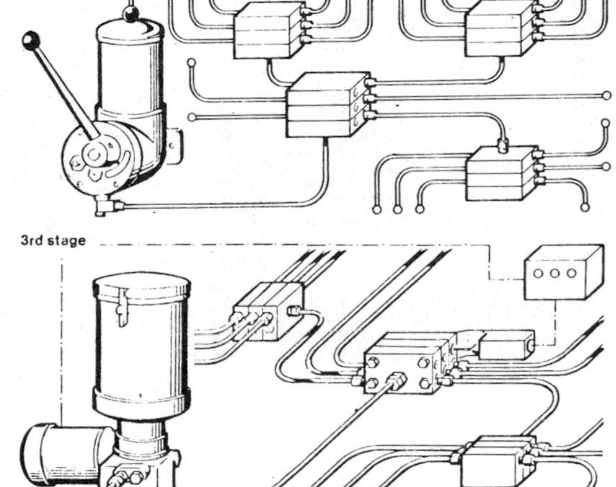

Fig. 11. Examples of Progressive Systems.

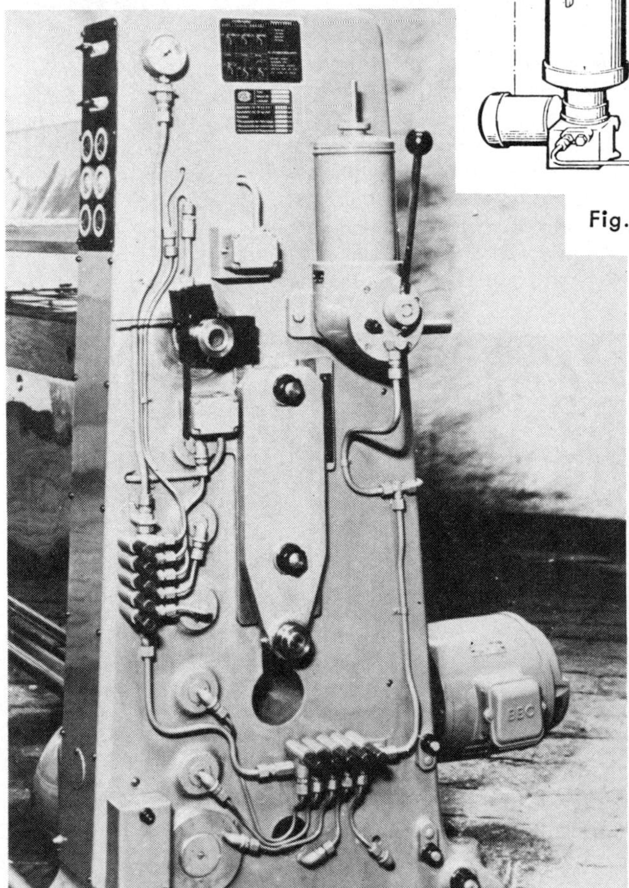

Fig. 12. Progressive Lubrication on a plate shear.

actuated as pressure increases in the main line.

Each group is further divided into systems utilizing one line only and systems utilizing two lines, the former being called single-line systems, the latter dual-line systems. The most popular types of systems used are :

Progressive Systems, operate on pressure/volume cycles; in their case the increasing pressure created by the pump actuates a metering valve which, having given up its set amount of lubricant, allows the grease to pass into the main line leading to the next metering valve. When sufficient pressure has been built up, the valve is actuated and lubricant allowed to flow to the next metering valve, and so on progressively, until it returns to the lubricator or where, when sufficient pressure has been built up, a reversing valve is actuated which reverses the flow of the grease. The selection of progressive systems is dependent upon the number of points to be lubricated. Figure 11 illustrates some options available, and Fig.12 shows a typical installation of a progressive lubrication system on a Plate Shear.

Dual Line Systems (Parallel), operate on the same basic principle, viz: the motorized lubricating pump (Fig.13) forces grease into one of two main feed lines in which are placed a number of dual-line metering elements, each outlet

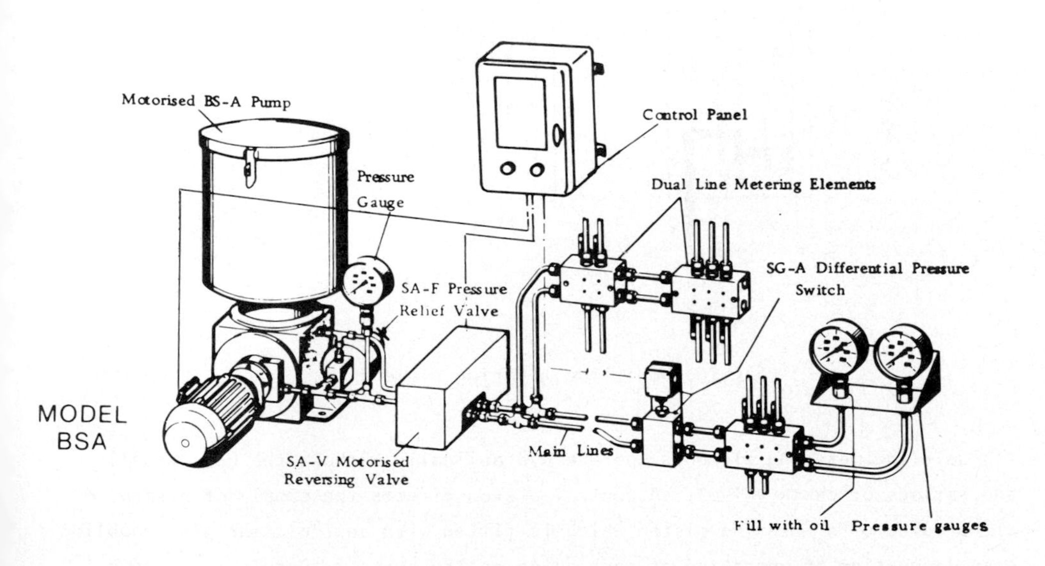

Fig.13 Dual line system.

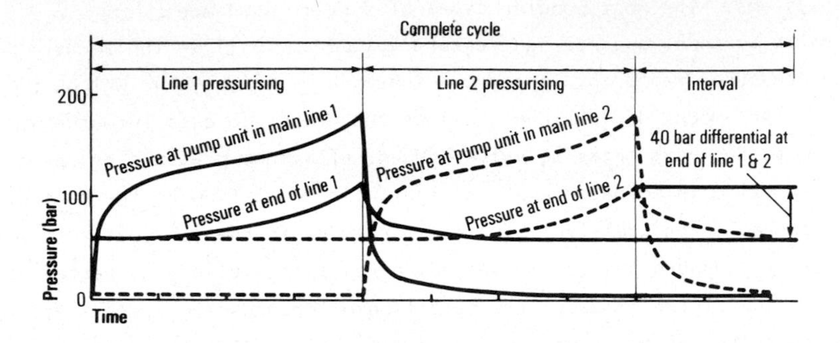

Fig.14 Dual line system pressure.

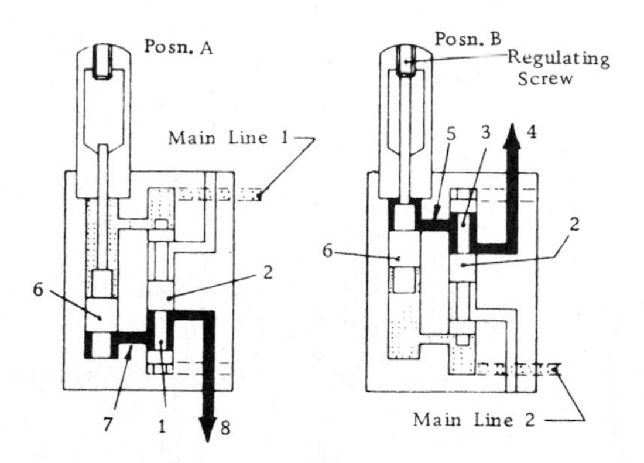

Fig.15 Dual line metering element.

Dual-line metering elements operate hydraulically without the use of balls and springs or check valves. A control piston directs the supply of grease to either side of a metering piston which is fitted with an indicator pin, enabling easy inspection of operation at each point on the system. Regulating screws fitted to each indicator housing permit adjustment down to 20% of maximum output. The position of the control piston (2) and the feed piston (6) are shown in 'A' after the first part of the dual-line cycle. Pressurised lubricant from main line 1 has moved over control piston (2) and then at the upper side of the feed piston (6), displacing it and discharging a measured quantity of lubricant via cross porting (7) and across spool (1) to outlet (8). Position 'B' is the second part of the dual-line cycle and pressurised lubricant from main line 2 has moved control piston (2) and displaced the metering piston (6) thus discharging a measured shot of lubricant via cross port (5) and across spool (3) to outlet (4).

For extremely dirty and abrasive environments such as those found in a Blast Furnace, Pig Caster or Coal Preparation and Washery Plant, it is advisable to have the dual line metering elements housed in toughened glass-fronted protection boxes, similar to that shown in Fig.16.

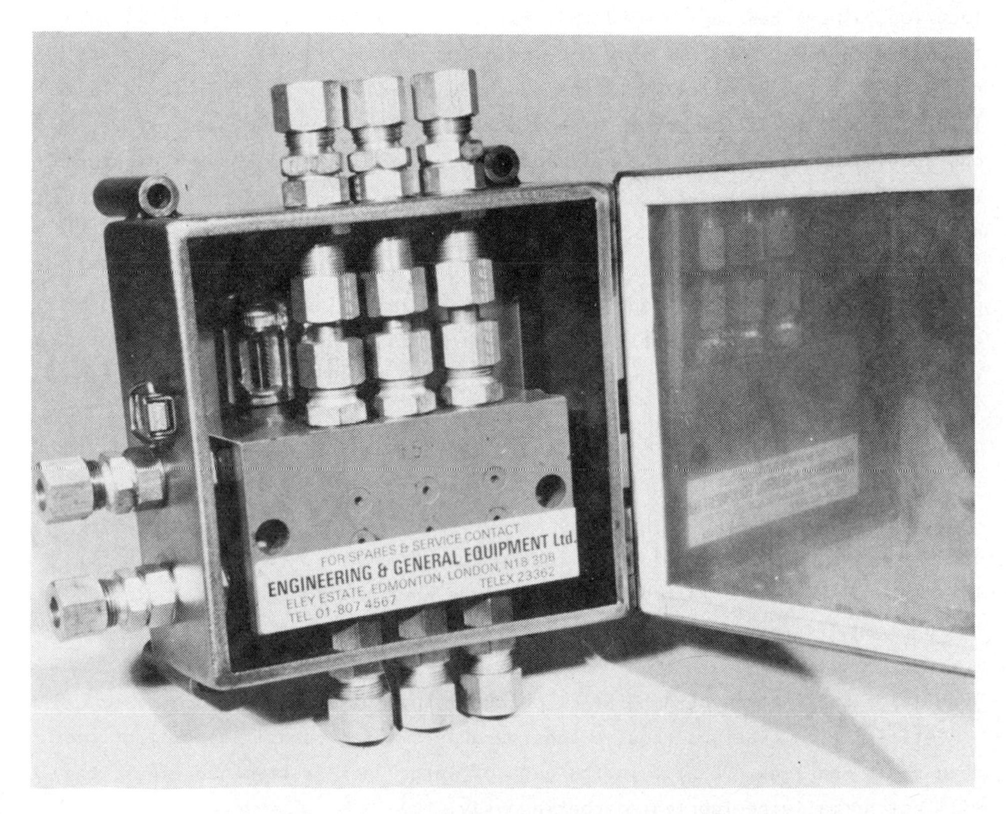

Fig.16 Protection box.

17.6.1.3 Comparison of Parallel Systems - Single-line and Dual-line.

Both systems depressurize the line, but in the case of the Single Line System the resetting of the plungers in the metering elements is usually effected by spring pressure, whereas in the Dual Line system, when main feed line No.2 is pressurised, a second series of dual line metering elements discharge lubricant to the points of application. In doing so, they reset the dual line metering elements in main line No.1, with which they form an integral unit.

17.6.1.4 Selecting Grease Lubrication Systems.

Wherever possible, lubrication systems should be avoided which use equipment incorporating springs and valves, particularly where the lubrication systems are required to operate in arduous and abrasive environments.

In general practice, the majority of Centralised Lubrication Systems used are either of the 'Direct Feed' or 'Parallel-Dual Line' type; both are capable of being operated 'manually' or 'automatically'. The choice of Grease Lubrication Systems is usually dictated by the number of points of application and their location, always bearing in mind that for utmost reliability and minimal maintenance Direct Feed Systems should be selected wherever it is practical. By way of explanation, a Parallel-Dual Line system can cycle and yet fail to deliver lubricant to some of the points of application which is only evident if the operator visually examines the position of every dual line element indicator pin. Quite often this is impractical because of elements being located in hazardous areas only being accessible when the plant and machinery are shut down.

This cannot occur with Direct Feed pumps, for they will only operate effectively providing all the points of application are receiving lubricant; barring, of course, broken feed-pipes (which can apply to both types of system). In practice, a Direct Feed pump can usually cater for up to 40 points of application.

17.6.1.5 Lubrication of Plain or Sleeve Bearings.

These bearings, particularly those over 4" diameter, require as near continuous lubrication as is possible. It is normally recommended that bearings of this type be fed by Direct Feeding lubricators driven from the moving shaft. This ensures lubricant is fed to the bearings when the shaft is in operation and no lubricant is fed to the bearing when the shaft is inoperative.

17.6.1.6 Lubrication of Anti-Friction (Ball and Roller) Bearings.

Anti-friction bearings require considerably smaller quantities of lubricant than plain bearings. Except in the case of large anti-friction bearings, they will not normally be lubricated continuously.

Where anti-friction bearings are close together, a hand-operated or time clock controlled direct feeding pump may be used. Where they are spaced over some distance, a line system, either hand-operated or time clock controlled, is preferred.

For large anti-friction bearings and those installations where the greatest reliability of Direct Feeding systems is desirable but the number of pipes should be kept low, direct feeding pumps with Positive Dividers may be used, splitting volumetrically metered amounts of lubricant independent of varying back pressures.

Very fast operating anti-friction bearings such as those running at 1400 rpm should not be fed continuously. However, over-greasing will do no harm to large slow-running anti-friction bearings where quite often the grease is used as a sealant, preventing dirt and other foreign matter - the greatest destroyer of anti-friction bearings - to enter the bearing. Therefore it is essential that the grease is kept clean at all times, with the lubricating pump container or reservoir bottom filled via a grease keg or bulk grease storage system.

17.6.2 Oil Lubrication Systems

Oil Lubrication Systems serve two purposes: to lubricate and/or cool. On many applications, particularly in the absence of high ambient temperature or where the heat generated in the bearings or the gears is not great, the removal of heat by the oil need not feature as a separate consideration in the selection of the oil circulation system. This can be arranged on the basis of lubrication considerations alone. However, in the case of many other applications, the cooling properties of the oil are of great importance.

Oil systems may therefore be grouped under three main headings:-

- Group 1 Systems designed for lubrication on a total loss basis.
- Group 2 Systems designed for lubrication and with a small amount of heat removal.
- Group 3 Systems designed for lubrication where an appreciable degree of cooling is also required because of operating conditions.

Group 1 and 2 systems vary according to the type of machine and its lubrication requirements. Systems of the total loss type may be operated either manually, mechanically, or motorised, whereas systems of the type which collect the used oil and recirculate it must be automatic. Various combinations of these systems can be employed, and the following are some typical examples:

17.6.2.1 Group 1 - Total loss systems designed for lubrication purposes only.

In this type of system the lubricant, after lubricating the bearings or gears, is not used again. The group consists of manual, mechanical or motor operated pumps. The former generally applies on small items of plant, e.g. machine tools, mechanical handling equipment, jigs and fixtures, presses, etc. These systems may be further sub-divided into Direct Positive Systems or Positive Split Systems.

Direct Positive Systems usually comprise one or more differential plunger type oil lubricators, e.g. Sections 6.1.1 (Fig.9) shows the operation of a manually operated 8-outlet grease pump which is also adaptable for oil, with Fig.17 illustrating a typical application on a press lubricating the slideways and crosshead. For applications where automatic lubrication is required, mechanical lubricators having usually 28 to 32 pumping units may be fitted (Fig.18). These lubricators may be driven either mechanically through a ratchet from the machine being lubricated or by geared electric motor. Each pump unit can be regulated from zero to maximum, to feed minute precise quantities of oil to the points of lubrication application.

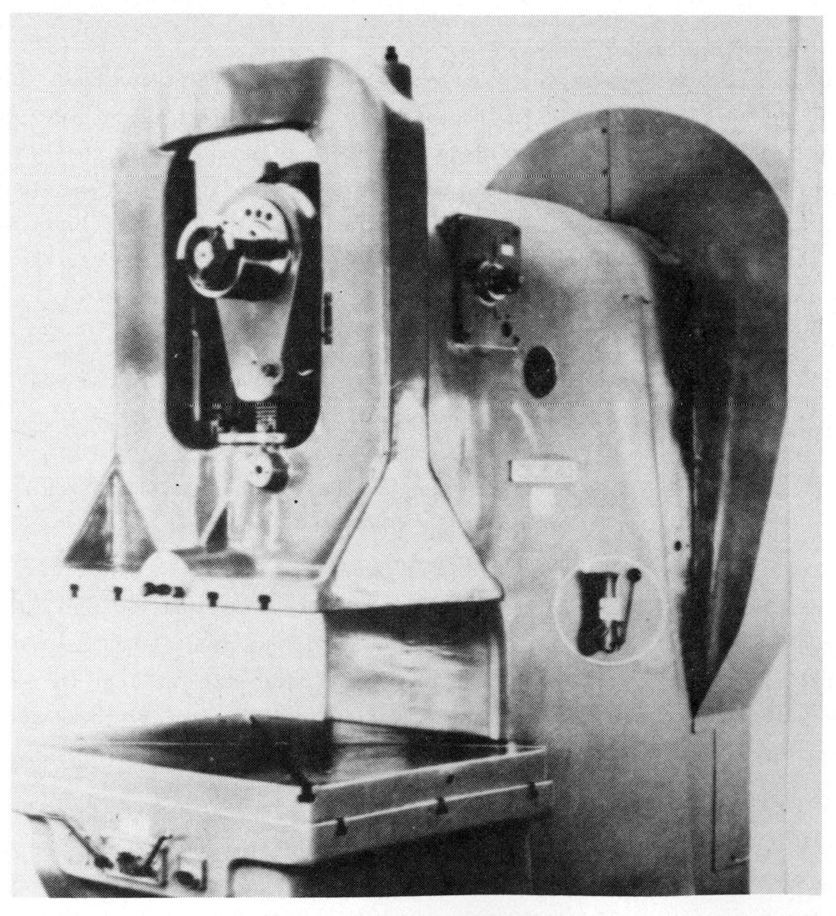

Fig.17 Direct positive system.

Fig.18 Direct positive system.

Positive Split System. This type of system is used where quantities of oil per application point are greater than can be supplied by the previously described direct feeding plunger system. It comprises of one or more small high pressure pumps fitted with integral relief valves, and supplies lubricant to the bearings through volume dividers. The dividers may be used either to increase the number of points or to modify the quantities fed to the lubrication points.

Fig.19 Positive split system.

17.6.2.2 Group 2 - Systems designed for lubrication with small amount of heat removal. This type of system supplies the lubricant to the bearings and returns it under gravity through the return pipes and/or drainways to the supply tank for recirculation. This group can also be sub-divided into Direct Positive Systems or Positive Split Systems. The former consists of the mechanically operated differential plunger type oil lubricator as described in Group 1.

Typical applications include paper machines, large kilns, or generally where a large number of bearings are to be fed positively with an adjustable feed. Systems of this type require a minimum amount of maintenance and attention. The lubrication reservoirs of containers can be kept filled either by a header tank supplying several lubricators, or each lubricator can be fitted with special built-in suction pumps which supply the container with lubricant from the main supply tank. Each lubricator pump unit (outlet) can be connected direct to the lubrication points or to a positive volume divider, depending on the number of feeds and the lubrication requirements of the points.

The lubricant is returned under gravity to the main supply tank through drainways or return pipes (which can be arranged with bearing sump level control devices) for recirculating to the lubricator containers or header tank by means of suction pumps, as described, or to the tank by a float-controlled gear pump (Fig.20).

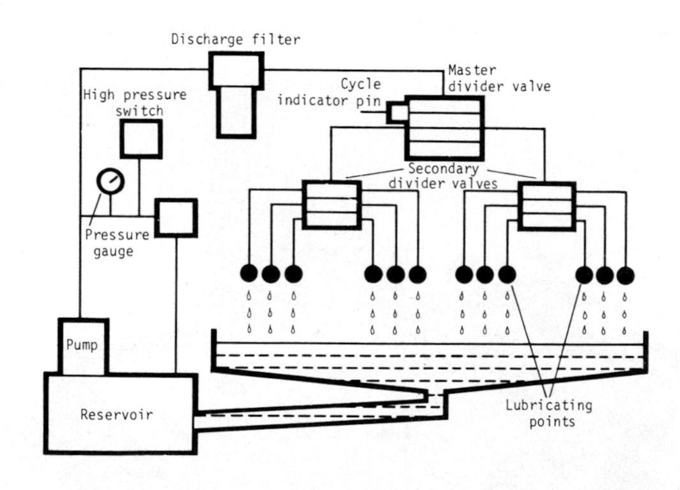

Fig.20 System with small amount of heat removal.

As in the case of the previous group, the lubricator can be driven either direct from the machine being lubricated or fitted with its own motor. Each pump can be fed either from the machine sump or from a separate drain and supply tank. The system is usually installed on machine tools, sugar machinery, gear boxes, printing machines, and special-purpose machinery.

17.6.2.3 Group 3 - Systems designed for lubrication cooling.

Where there is considerable ambient heat or where the power transmitted by the part being lubricated creates a high degree of heat, the cooling function of the lubricant assumes great importance. It is necessary in such cases to apply sufficient lubricant to extract the heat and to maintain the bearings or gears at an optimum temperature. Arrangements must also be made for returning the lubricant to a supply tank for cooling, filtration, and recirculation, between the lubrication equipment manufacturer, the plant designers, and the operators.

Such a system usually comprises a large oil reservoir or storage tank; motor driven pump (normally a gear type pump adjacent to or on the storage tank); coolers; filters; pressure gauges; alarm and flow control equipment; together with the necessary valves and interconnecting pipework. The system may be simple and self-contained with a capacity of 38 cc/sec to 750 cc/sec per minute, or a complex system capable of delivering several litres/sec. These systems can be provided with simple or elaborate flow control, warning devices, and other instrumentation according to the needs of the installation (Fig.21).

Fig.21 Typical example of lubricating and cooling system packaged unit for the lubrication of rubber machinery.

17.6.3 Micro-Fog Lubrication Systems

Aerosol lubrication is the generic term for oil mist or Micro-fog systems which have been used successfully for over twenty years. Compared with Centralisted Grease or Oil lubrication, a Micro-fog system, to perform the same task, requires less lubricant and energy and the initial cost is relatively low. It is also a highly flexible system, readily installed onto existing plant as well as at the new machine stage.

17.6.3.1 Working Principles (Fig.22) outlines the essential and auxiliary elements of the Micro-fog system and provides a guide to some of the more common areas of application. During operation, the system produces continuously a dense concentration of micro particles of oil which are conveyed in a 'dry' fog in a low pressure distribution system. On reaching the point of application the 'dry' fog is passed through reclassifiers, which are really metering and condensing orifices, so as to accurately feed an exact quantity of lubricant to suit the operating conditions.

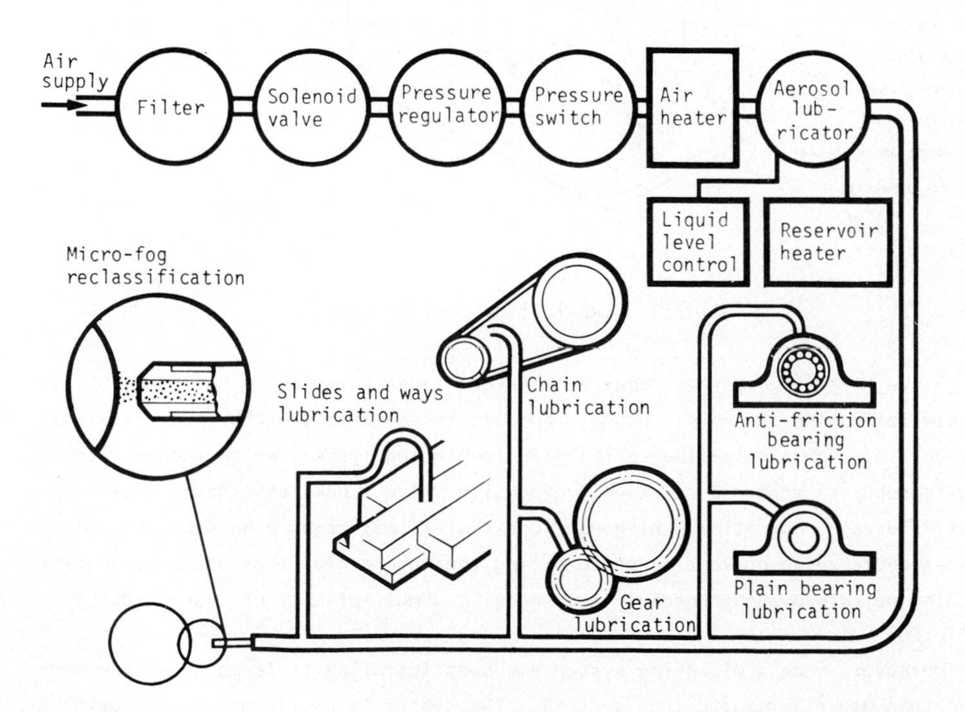

Fig.22 Micro-fog system.

To ensure that the fog reaches the reclassifier dry, the distribution piping is sized to allow oil particles to travel along the piping at a velocity less than 7.3 metres/sec, which is slow enough to prevent condensation. The turbulence in the reclassifiers causes the oil particles to 'wet out' into the line leading to or direct onto the bearing surfaces where they then form a protective film of oil.

In order to create 'dry' fog, oil is first drawn into a compressed air stream as it passes through a Venturi located on top of the lubrication control unit (Fig.23). Oil particles of approximately 0.002 mm in diameter are collected in the air stream and can be transported long distances in the dry condition.

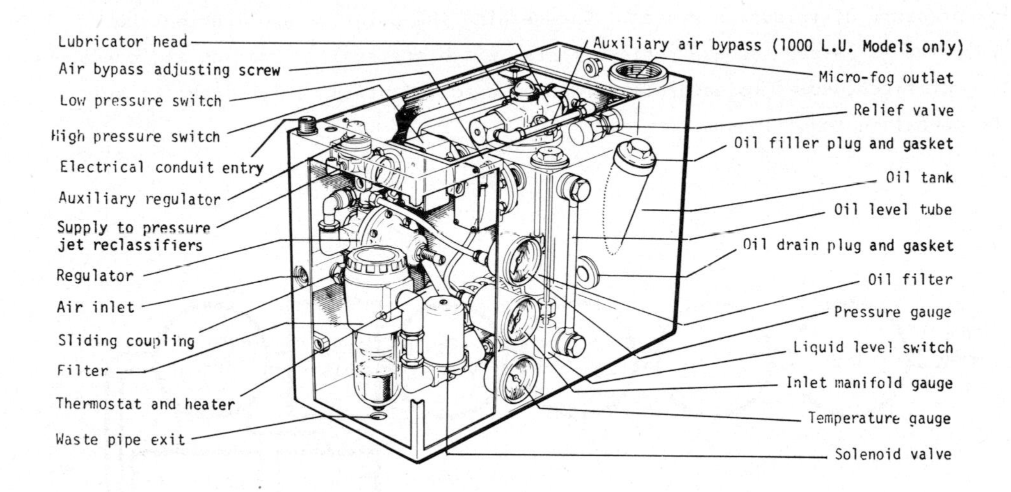

Fig.23 Lubrication control unit.

Since Micro-fog systems require no return lines they can be designed to easy installation and assembly; hence, low cost installation, without the problems associated with single-line series oil circulating systems which incorporate quite sophisticated and relatively expensive valve blocks that have a built-in self-reversing operation. Although such systems may require no separate reverse and recycle valve or venting phase during the lube cycle, they do have, however, a distinct disadvantage because they require a multiplicity of system tubing (Fig.20).

Moreover, once a Micro-fog system has been installed it is much less demanding than other automatic lube systems. The system is easily assembled, using a building block approach, which is designed for easy servicing, repairs and general maintenance.

17.6.3.2 Oil Quality.

Correct oil selection is important as some oils incorporate polymers which suppress aerosol properties, while heavier grade oils may require heating up to between 40°/45°C to attain the viscosity for maximum output. For all normal purposes the control units will perform well when working with oils up to 700 centistokes at 20°C.

Summarised lubricating oil requirements for a satisfactory Micro-fog system are:-

(i) Good aerosol properties.

(ii) Low rate of condensation through pipes.

(iii) Low level of straying by the particles.

(iv) A high degree of rust inhibition.

(v) Absence of clogging tendencies at the venturi nozzle or any polymer precipitation.

17.6.3.3 Compressed Air

Accepting that most industrial compressed air is supplied at 7 bar (100 lb/in^2) in a Micro-fog lubrication system it has to be reduced to about 2 bar (30 lb/in^2). During its passage through the venturi orifice on the control unit, a pressure drop of 0.7 bar (10 lb/in^2) takes place.

17.6.3.4 System Design Considerations

To calculate the lubrication requirements of bearings, an empirical factor referred to as a 'lubrication unit' (L.U.) has been evolved, enabling all moving surfaces requiring lubricant to be converted to their equivalent L.U. rating. The amount of lubricant main and branch line pipe bores and reclassifier nozzles may then be sized to provide the correct amount of lubricant at each lubricating point. In this manner, anti-friction bearings, journal bearings, slides, gears, chains, and other wearing surfaces requiring lubrication can all be converted to equivalent L.U. ratings and served by appropriately sized Micro-fog lubrication systems.

17.6.3.5 Some Typical Applications

Figure 24 illustrates a three Strand Aluminium Foil Mill operating at 1000 to 1500 metres/minute with two 1000 L.U. generating heads (third acts as a standby) serving the mill stack and exit ancillaries, with a separate 300 L.U. generating unit serving the entry ancillaries. The total amount of oil used is less than 2.5 litres per working hour. Figure 25 illustrates the lubrication of vibrator motor gears.

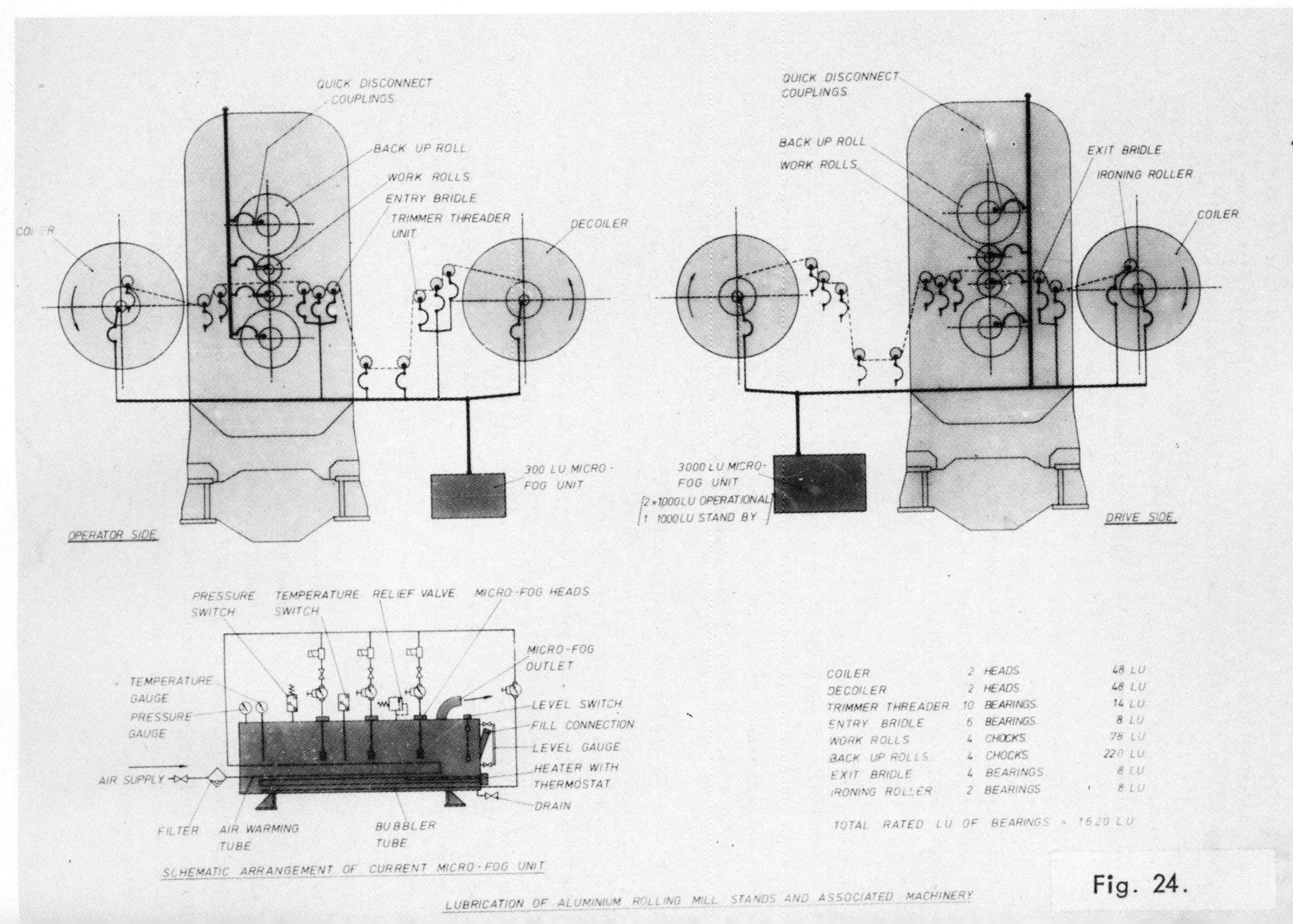

QUICK DISCONNECT COUPLINGS
BACK UP ROLL
WORK ROLLS
ENTRY BRIDLE
TRIMMER THREADER UNIT
COILER
DECOILER
300 LU MICRO-FOG UNIT
OPERATOR SIDE
QUICK DISCONNECT COUPLINGS
BACK UP ROLL
WORK ROLLS
EXIT BRIDLE
IRONING ROLLER
COILER
3000 LU MICRO-FOG UNIT
[2×1000LU OPERATIONAL
1 1000LU STAND BY]
DRIVE SIDE
PRESSURE SWITCH
TEMPERATURE SWITCH
RELIEF VALVE
MICRO-FOG HEADS
MICRO-FOG OUTLET
TEMPERATURE GAUGE
PRESSURE GAUGE
LEVEL SWITCH
FILL CONNECTION
LEVEL GAUGE
HEATER WITH THERMOSTAT
AIR SUPPLY
DRAIN
FILTER
AIR WARMING TUBE
BUBBLER TUBE
SCHEMATIC ARRANGEMENT OF CURRENT MICRO-FOG UNIT
COILER 2 HEADS 48 LU
DECOILER 2 HEADS 48 LU
TRIMMER THREADER 10 BEARINGS 14 LU
ENTRY BRIDLE 6 BEARINGS 8 LU
WORK ROLLS 4 CHOCKS 78 LU
BACK UP ROLLS 4 CHOCKS 220 LU
EXIT BRIDLE 4 BEARINGS 8 LU
IRONING ROLLER 2 BEARINGS 8 LU
TOTAL RATED LU OF BEARINGS = 1620 LU
LUBRICATION OF ALUMINIUM ROLLING MILL STANDS AND ASSOCIATED MACHINERY

Fig. 24.

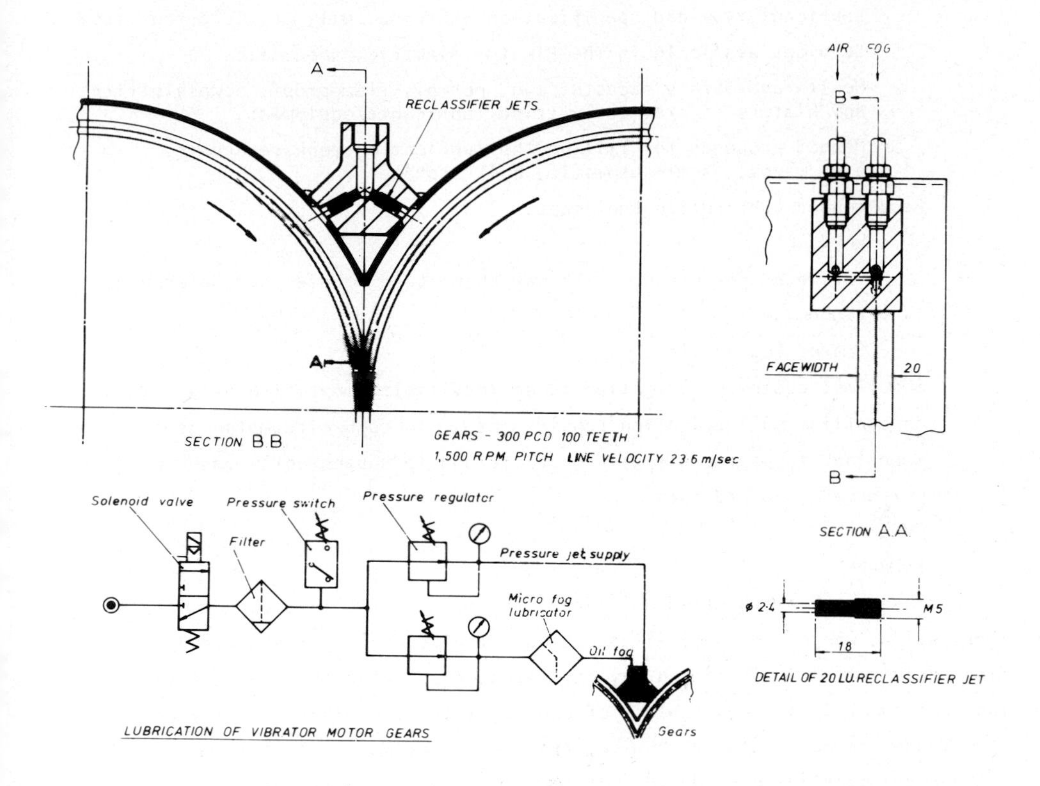

Fig.25 Lubrication of vibrator motor gears.

17.6.4 Check List

Based on the foregoing, prior to actual selection of the appropriate lubricant and associated equipment, it is advisable to draw up a check list of known facts and requirements. The following list, in simplified form, is for guide purposes only and can naturally be elaborated upon to suit the specific needs of the designer or plant engineer.

Specification of Plant to be lubricated:

Type of plant and machinery to be lubricated.

Industrial Application, including operating environmental conditions, e.g. is it dirty, abrasive, wet, hot, etc.?

Surfaces, sizes and speeds to be lubricated, e.g. Bearings (plain), Bearings (roller), Grease (type), Slideways, Chains, etc.

Number and Location of Lubrication Points (Fixed) and (Moving) and proposed site (if known) for lubricator enabling assessment of pipe and flexes:

Frequency plant and machinery operators and whether lubrication needs to be continuous, semi-continuous, or intermittent.

Lubricant type and specification - Grease, Oil, or Micro-fog, etc.

Services available in the Plant - electric, pneumatic.

Health and Safety aspects, e.g. normal, flameproof, accessibility, any history of previously tried lubricants/equipment.

Method proposed for filling the lubrication tank/reservoir, e.g. manual, semi-automatic, bulk storage.

Recommended Lubrication Equipment:

Why?

Estimated performance with any known Case Studies and References.

Economics.

Spares and Service.

Sometimes equipment selection is an inevitable compromise as a result of conflicting lubrication requirements; in such circumstances it is important to appreciate all the facts and to subsequently gauge plant performance accordingly.

17.7 SUMMARY

This chapter has attempted to clarify the more generally accepted methods of lubrication, where necessary illustrating actual examples of plant and lubrication equipment. It is not in any way intended to infer that this is the only lubrication equipment available; of course, there are others readily determined from Trade Journals, etc. Likewise, there are many more select items of lubrication equipment tailor-made to meet the needs of specialist plant.

For example, Overhead and Floor Conveyors which can travel up to 50 metres/minute, unless effectively lubricated (and cleaned where the environment demands) can wear and seize, resulting in costly stoppages.

Initially, conventional static lubricators were used where a mixture of air and oil - and now more recently just oil - was shot over a gap, which resulted not only in the failure to adequately lubricate (Fig.1) but also caused drippage with consequential product contamination and health hazards.

These problems were completely overcome by introducing a range of special-purpose lubricators; (Fig.26) shows one such example.

The conclusion is to determine all the facts regarding the plant to be lubricated and to then evaluate the lubricants and lubrication equipment available.

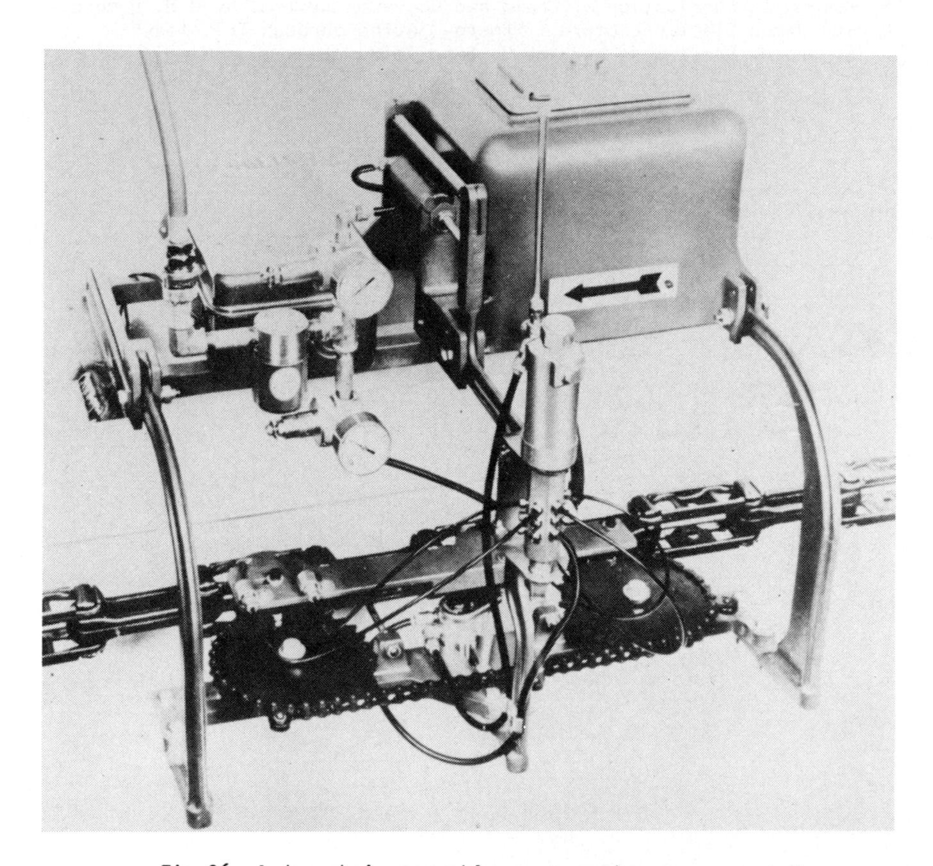

Fig.26 A dog chain assembly engages the conveyor chain which moves the oil-dispensing nozzle plates in and out over the pin links.
Oil is forced through each nozzle.

REFERENCES

1 Mechanical Lubrication of E.O.T. Cranes by Dr.H. Peter Jost and Peter W. Murray.
2 A Fully Automatic Bulk Handling Lubrication System for a Sinter Plant by G. Williams.
3 An Engineering Approach to the selection of Centralised Grease Lubrication Systems by Dr. H. Peter Jost.
4 Modern British and European Steelworks Lubrication Developments by Dr. H. Peter Jost.
5 New Mist Lubrication Concepts for Tapered Roller Bearings used on High Speed Rolling Mill Back-up by Rolls by W.E.McCoy, C.H. West and P.E. Wilks.

6 Aerosol Lubrication Systems - their contribution to savings in operating and maintenance costs by R.E. Knight and J.G. Merrett.
7 Micro-Fog Lubrication for bearing efficiency by J.G. Merrett.
8 Automatic Lubricators and Cleaners Increase Conveyor Life by J.G. Merrett.
9 Automatic Lubrication of Chain and Conveyor Systems by R.M. Dombroski.
10 The James Clayton Lecture - "Energy Saving through Tribology" by Dr. H. Peter Jost and Dr. J. Schofield.

18 ON CONDITION MAINTENANCE

R.A. COLLACOTT, Ph.D., B.Sc(Eng), F.I.Mar.E., F.I.Mech.E.

Director - UK Mechanical Health Monitoring Group

Head - Fault Diagnosis Centre, Leicester Polytechnic.

18.1 INTRODUCTION

Maintenance carried out when required after a significant deterioration in a component as indicated by a sensor or monitored parameter is called on-condition maintenance.

If a person, when visiting the doctor for a health-check was invited to have their body opened up to see whether 'everything was correct' a most unsatisfactory state of affairs would exist. Similarly, to open up a machine in order to check that it is alright is just as unsatisfactory - much more harm than good can be done to a machine in this way. Accordingly it is logical to use diagnostic techniques to assess the 'health' or condition of plant and machinery in just the same way as a medical doctor uses symptoms and aids to assess the condition of human beings - and very often similar or identical equipment is used both by the medical doctor and diagnostic engineer.

18.2 BACKGROUND

On-condition monitoring is already actively and effectively used in many industries - in aircraft, nuclear reactors, steel mills, petroleum refineries, ships, power generation etc. It is a technology which has rapidly evolved during the past 5 to 10 years by using methods and techniques which have been developed since 1945.

Historically, the chain of evolution is shown in Fig.1.

1750 ... steam engines - simple contents gauges, simple governor
1915 ... steam turbines - pressure, temperature, contents gauges etc.
1940 ... steam turbines - automatic controls
1950 ... nuclear reactors - robotry
1960 ... space vehicles - minutiarisation, remote telemetry
1970 ... ... computer minutiarisation, microprocessors

Figure 1 Historical chain of evolution of condition monitoring

18.3 MANAGEMENT OF CONDITION MONITORING

There are three effective stages in the management of an on-condition maintenance system which go hand-in-hand with a change in organisation whereby maintenance involves the running of (i) a diagnostic capability, (ii) a repair team.

The three stages in diagnostic management are:

1) Failure Modes and Effects Analysis - whereby the whole/complete plant is analysed systematically to determine which parts are critical and need to be monitored, also to appraise their typical failure cause
2) Monitoring Technique Selection and Sensor Appraisal - whereby the most effective method is chosen
3) Limit Decision - whereby the amount of deterioration which can be permitted is decided upon

A full account of all three stages is given in reference [1].

18.4 FAILURE MODES AND EFFECTS ANALYSIS

A typical example of the way in which this can be done was explained by Venton and Harvey [2], Bridges [3] and mentioned by Davies [4], in essence it is the use of information to prepare a numerical assessment of the order of essentiality of various components (sub-systems) within a plant or machinery. At its most elementary this may be prepared by a simple 'Delphi' analysis; at its more complex it may involve an integrated data-appraisal method using failure rate data which can be obtained for different components - such failure rate data can be obtained from a number of established data banks [5].

A typical 'starting' appraisal for a marine power plant might be the interdependency chart for all sub-systems as shown in Fig.2. This already shows that for the particular mission for which this appraisal was made, high reliability of electrical generation was important - a state of affairs which was confirmed by a basic PHASE CRITICALITY ANALYSIS and remained unchanged when allowance was made for the inherent safety through a HAZARDS AND RISKS ANALYSIS.

The causes of potential failure need to be established. Again, this must be derived from historical records - it is important to make, maintain and analyse records in order to provide adequate information upon which to choose the correct monitoring methods. In a typical analysis by Wilkinson and Kilbourn [6] the failure rate data of Table 18.1 was used to determine the best arrangement of an electrical generating standby plant to choose the most reliable system.

System	Inter-dependent System (1)	(2)	(3)	(4)	(5)	(6)	(7)	(8)	(9)	(10)	(11)	(12)	(13)
1) Main engine			X	X				X		X	X	X	
2) Transmission	X												
3) Oil fuel					X								
4) Compressed air					X					X			
5) Electrical generation			X	X						X	X		
6) Steering gear					X								
7) Deck machinery					X								X
8) Sea water					X								
9) Bilge and ballast					X								
10) Ventilation					X								
11) Exhaust													
12) Controls					X								
13) Steam			X							X	X		

Fig.2 System Inter-dependency Analysis

As a consequence of past operational experience - either by an individual user or from a manufacturer or some data centre such as that provided by Lloyds Register of Shipping or the UK Atomic Energy Authority (Systems Reliability Service) - it is possible to prepare FAULT TREES such as that shown in Fig.3 [7]. Detailed examination of such fault trees will identify the possible primary and secondary causes of failures and the likely symptoms they will produce.

Failure symptoms are the physical consequences of a malfunction situation. If for example one considers bearing failures there are two quite different physical effects between roller element bearings and plain bearings. With roller element bearings experience shows that failures are associated with local surface defects, consequently operation of defective bearings sets up ultra-high frequency shock waves which can be most effectively diagnosed by shock pulse methods possibly aided by statistical analysis by a method called 'kurtosis'. On the other hand, plain bearings wear in a uniform whole-surface method with the result that clearances are changed and some interference (rubbing) may ensue; for such conditions, methods of clearance measurement by inductive pick-ups or vibration measurement or even debris analysis may be appropriate. For the two classes of bearings different sensor systems will be

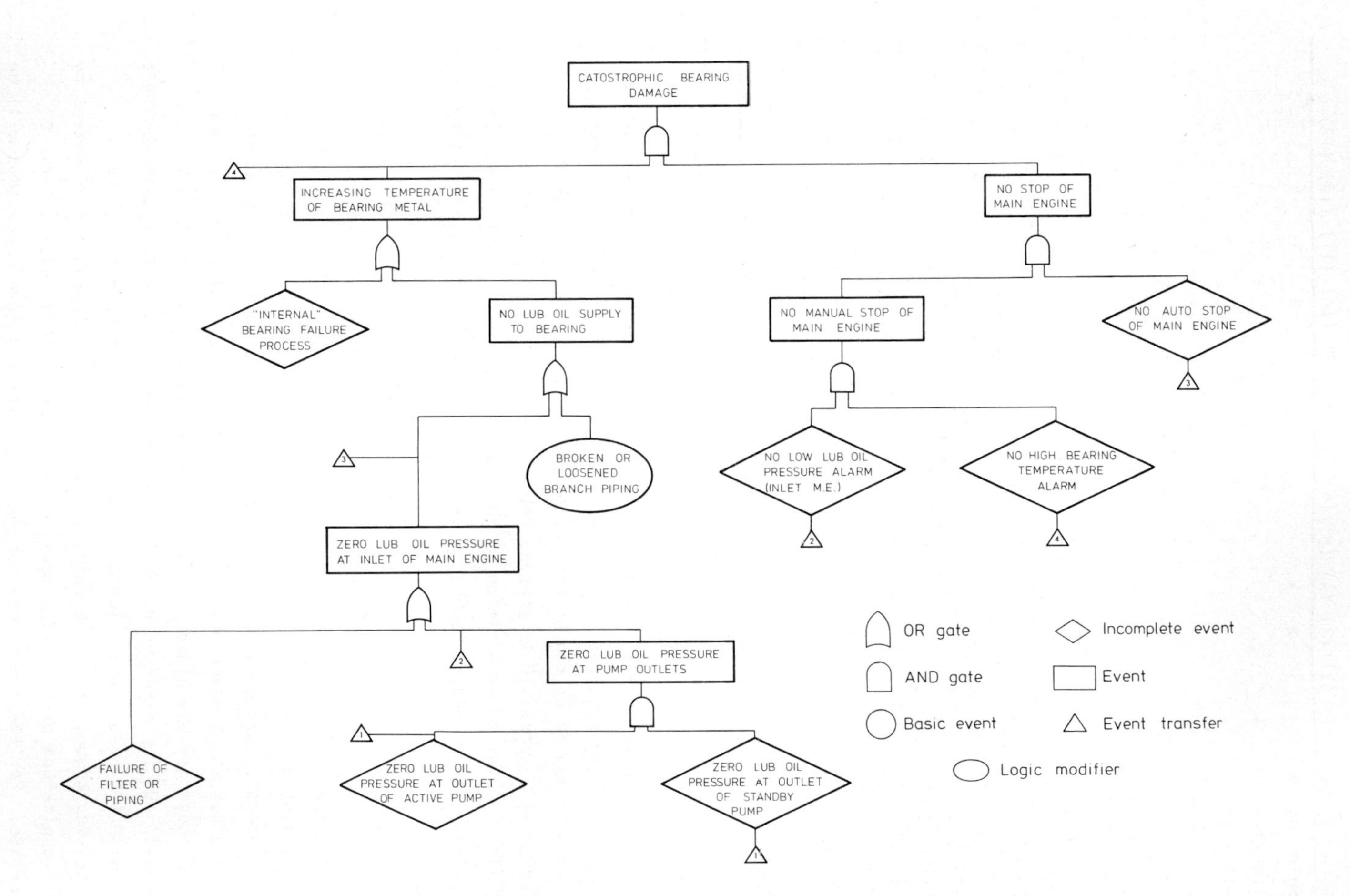

FIGURE 3 Typical FAULT TREE for a bearing failure analysis

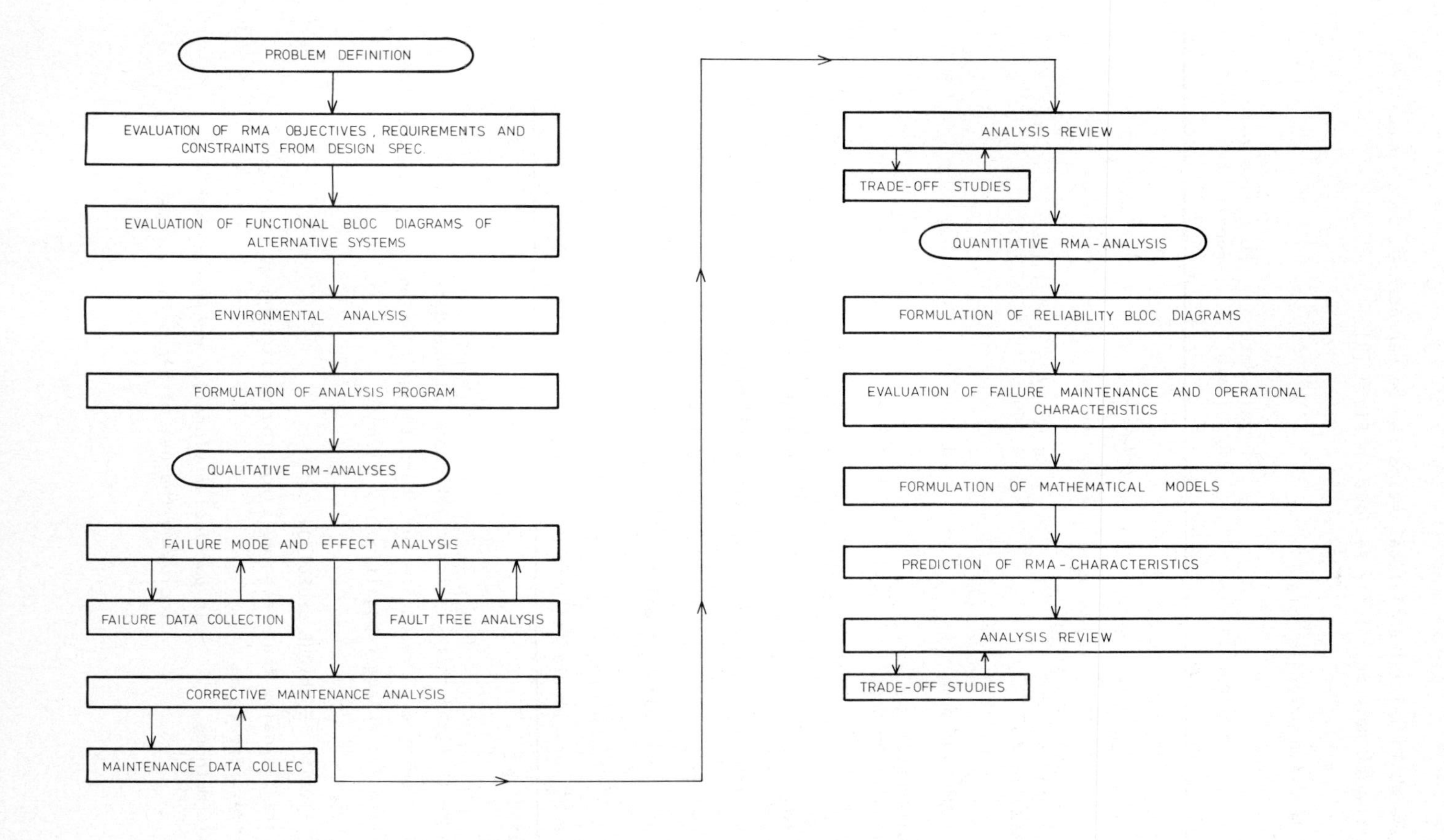

FIGURE 4 Reliability analysis sequence diagram

available.

Thus the whole initial analytical stage can be represented by a number of procedures which have been fully set out in Fig.4 [7].

TABLE 18.1 Failure/repair data

Components	Repair Rate, Repairs/h	Mean Time to Repair, h (2 men)
Turbo-alternator and controls:		
alternator end	0.08197	12.2
turbine end	0.122	8.2
Diesel alternator and controls:		
alternator end	0.08197	12.2
diesel end	0.05051	19.8
Turbo-alternator system sea water circulating pump	0.1	10.0
Diesel alternator sea water circulating pump	0.1493	6.7
Condenser	0.1667	6.0
Air ejector	0.1136	8.8
Extraction pump	0.09434	10.6
Boiler feed pump	0.1429	7.0
Waste heat economiser-type boiler	0.03636	27.5
Oil fired boiler	0.05*	20.0*
Composite boiler circulating pump	0.09259	10.8

* Estimated

To ensure that the characteristics of each failure/defect malfunction it is necessary to be able to recognise the characteristics - which is only obtained through extensive defect recognition experience.

18.5 MONITORING TECHNIQUE SELECTION

There are basically four techniques for the monitoring of plant and machinery deterioration:

1. dynamic methods - in particular vibration monitoring but also including the use of air-borne sounds

2. inspection/integrity surveillance methods - which originated with non-destructive testing techniques but have now extended to include leak-testing; odour identification; corrosion monitoring and stress wave emission

3. contaminant inspection - as a means of identifying wear debris and relating it both quantitatively and qualitatively to its source

4. trends analysis - effectively data logging either as straight sensor outputs as for example temperatures, pressures and speeds or in a coordinated form using such parameters as specific fuel consumption rate or even 'deltas' of variations from the norm

18.5.1 Vibration Monitoring

This is a well established technique ranging in sophistication from the use of judiciously placed dial gauges in conjunction with stroboscopes to broad-band analysis, narrow band analysis, auto-correlation, signal averaging and other highly instrumented techniques [8]. Most monitoring applications are satisfactorily dealt with by means of:

(i) proximity probes and pick-ups with possibly a cathode ray tube (crt) display

(ii) seismic velocity transducers or piezo-electric accelerometers outputting to either
  (a) broad-band (overall) vibration meters
  (b) vibration spectrum analysers (part-octave or narrow-band width)

(iii) waveform analysers

18.5.1.1 Proximity Probes

The orbit moved through a shaft which is loose in its bearings yet subjected to the influence of various forces can be observed by using two proximity probes placed at a 90° relative angle and their outputs led to an X-Y plotter or a CRT. The resulting display, Fig.5, can be used to measure the actual amount of orbital eccentricity (and thus to determine whether 'bearing wipe' is likely to occur) also the shape of the orbit is related to the type of defect force so that the source of trouble can be established.

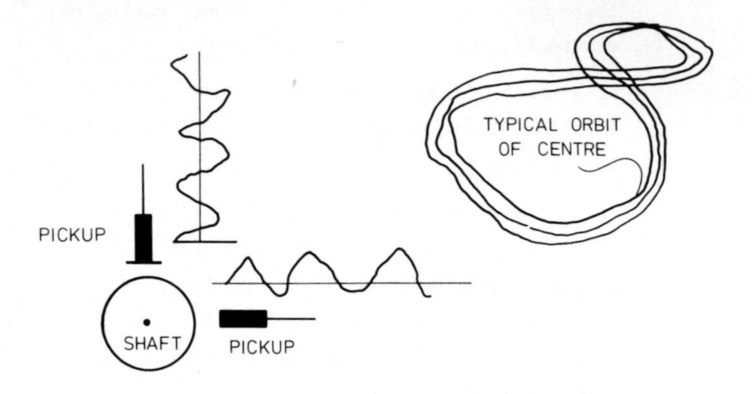

Fig.5 Shaft orbital analysis

18.5.1.2 Seismic Velocity Transducers/Accelerometers/Vibration Meters/Spectrum Analysers

This is the developing area of interest in 'straight' vibration analysis. Velocity transducers are most applicable at the lower frequency ranges; accelerometers are effective at the higher frequency ranges; at intermediate frequencies (around 500 Hz = 500 x 60 = 3000 cpm) either transducer is applicable.

With overall (wide) bandwidth vibration analysis a transducer picks up a signal derived from all the constituent frequencies and this is measured by the meter - which accordingly tells whether the vibration is increasing in strength, ie. the system is deteriorating.

To tell what is deteriorating in a machine it is necessary to measure the vibration signal produced by each constituent component. This is done by recording the frequency 'spectrum' for the machine. Thus each component will generate vibrations at a particular frequency (its 'discrete' frequency) and when plotted as in Fig.6 produce a 'spike' on the graph at that frequency. If the 'spike' increases with succeeding spectra it will mean that a defect is developing in that particular component. To know the values for the discrete frequencies for different components it is necessary to make frequency calculations of various kinds - in any event machine designers make these calculations when machines are designed and such information can be obtained when the machines are being purchased. Typical frequencies are given in the following Table.

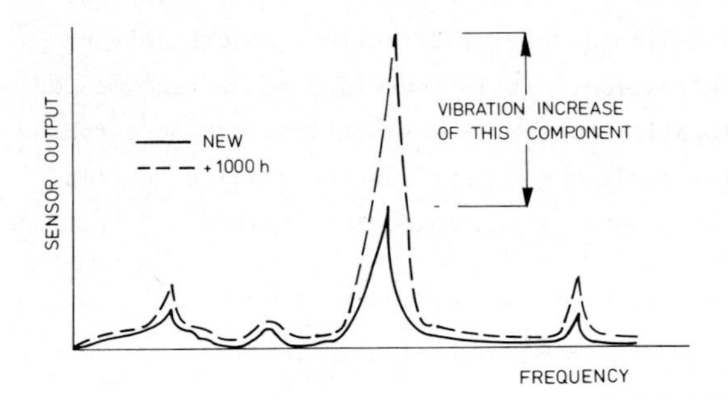

Fig.6 Typical frequency spectrum

TABLE 18.2 Discrete Frequency Calculations

Vibration Type	Frequency Equation
Simple harmonic	$f = \frac{1}{2\pi} \frac{\Delta}{M}$
Pendulum (simple)	$f = \frac{1}{2\pi} \frac{g}{L}$
Pendulum (compound)	$f = \frac{1}{2\pi} \frac{gh}{(k^2 + h^2)}$
Bar, uniformly loaded, fixed both ends	$f = \frac{3.57}{\ell^2} \frac{EI}{\omega}$
Shaft, torsional oscillations, single flywheel	$f = \frac{1}{2\pi} \frac{\tau}{I}$
Ball bearing - malfunction of outer race	$f = \frac{nN}{2} (1 - \frac{d}{D} \cos \beta)$
Ball bearing - malfunction of inner race	$f = \frac{nN}{2} (1 + \frac{d}{D} \cos \beta)$
Ball bearing - defective ball	$f = N(\frac{D}{d} - \frac{d}{D} \cos^2 \beta)$
Gear teeth - tooth defect	$f = N_1 t$

where,

Δ = stiffness of system
M = mass of system
g = acceleration of gravity
L = length of pendulum
h = distance,c.g. to pivot, compound pendulum
k = radius of gyration about c.g., compound pendulum
ℓ = length of bar
E = modulus of Elasticity (Young's Modulus)
I = second moment of area of section about neutral axis
ω = load per unit length applied to bar
τ = torsional stiffness of shaft $\frac{GJ}{\ell}$
G = shear modulus
J = polar second moment of area per shaft
I = polar moment of inertia of flywheel
n = number of balls in bearing
N = shaft speed (rev/min)
d = ball diameter
D = ball pitch circle diameter
N_1 = gearwheel speed (rev/min)
t = number of teeth on gearwheel.

When using vibrations sensors particular care must be taken in selecting their mounting position and in methods by means of which they are attached. Some vibration defects only show up in the radial direction, others only produce major effects in the axial direction. It is therefore to be recommended that 3 sensors be used at each location so that the vibrations can be established in two perpendicular radial positions as well as the axial direction.

A study of the effects of various defects indicates the following vibrational characteristics:

TABLE 18.3 Vibration Characteristics of Defective Components

Defect	Occurrence	Direction
unbalance ...	occurs at shaft speed	in a radial direction
misalignment ..	" at shaft speed (sometimes 3 or 4 x)	in a radial or more dominantly in an axial direction
plain bearing loose in housing ..	occurs at 1/2 or 1/3 shaft speed	in a radial direction
worn gears ...	occurs at tooth mesh frequency = rpm x no. teeth ..	in a radial or axial direction
faulty belt drive ...	occurs at belt frequency	in a radial direction
mechanical looseness ...	2 x shaft speed	
electrical induced ...	at synchronous frequency	should disappear off power

18.5.1.3 Waveform Analysis

The shape of the vibration waves and the general interaction of superimposed vibrations can be studied by passing the vibration signal through a time-domain recorder such as an ultra-violet (UV) recorder. Such methods are in general limited to very low frequency (structural) vibrations and some transient studies.

18.5.2 Inspection/Integrity Surveillance

Methods to determine the presence of flaws which have been adopted from non-destructive (ndt) testing methods include the following:

(i) Dye penetration - which reveals cracks as small as 0.025 μm to the naked eye

(ii) Flux testing - magnetic materials magnetised to reveal the presence of cracks when the surface is spread with magnetic particles/powder

(iii) Electrical resistance - by which two probes are moved over the surface with an electrical p.d. between them; cracks alter the resistance and therefore the passing current

(iv) Eddy current testing - currents induced in a material (not necessarily magnetic) and cracks located by a search coil

(v) Ultrasonic testing - whereby cracks are located by the reflection of ultrasonic waves propagated into the material; this method is developing into one of the primary ndt techniques

(vi) Radiographic - by which X-rays penetrate the surface and locate hidden cracks.

If one adds to this catalogue of techniques the newer ones which have been developed exclusively for inspection/surveillance purposes and add to this leak detection and corrosion monitoring, a vast array of possible methods is presented. An indication of the range of these is given in the following table:

TABLE 18.4 Classifications of Integrity Surveillance Techniques

Acoustic/ultrasonic	- holography; stress wave emission; ultrasonic;
Electrical	- capacitance; corona discharge; corrosion probe; eddy current; microwave; resistance
Magnetic	- hysteresis; particles; prints
Radiography	- beta-ray backscatter; X-ray; gamma-ray; neutron
Thermal	- infra-red; surface impedance; thermographic compounds
Visual	- borescopes; C.C.T.V.; dye-penetrant; holography photo-electron emission;

TABLE 18.5 Some further descriptions

Beta-ray Backscatter	-	to determine layer thickness; assess the composition of aggregates
Capacitance	-	crack detection, bond defects in non-metallic materials
Corrosion-probe	-	material loss by corrosion measurement
Dye Penetrant	-	crack penetration using visible or fluorescent dyes
Eddy Current	-	surface defects detected by electromagnetic induction
Flux Sensors	-	uses residual of induced magnetic flux perturbations
Magnetic Hysteresis	-	measures magnetic changes due to the presence of faults
Magnetic Particles	-	solid magnetic particle migration in the presence of an applied magnetic field locates surface defects
Magnetic Paints	-	a strippable paint film under magnetic field effects which illustrates surface defects in ferromagnetic materials
Microwave	-	infra-red inspection of non-conducting materials
Optical Aids	-	borescopes, fibre optics, CCTV
Photoelectron Emission	-	the detection of spontaneous electron emissions from plastically deformed surfaces
Radiography	-	penetration to show up defective and burnt-out parts
Resistivity	-	crack or bond testing by electrical resistance measurement
Stress Wave Emission	-	a method for locating to position and severity of defects in metal surfaces and structures
Ultrasonics	-	which uses transmission characteristics of ultra-high frequency sonic waves to locate defects.

18.5.3 Contaminant Analysis

Moving contact between the metallic components of any mechanical system is accompanied by wear, which results in the generation of wear debric particles. In a lubricated system these particles are in suspension in the circulating oil. Under normal conditions the rate of wear is low and particles formed are very small. The size and rate of generation of these particles increase as the rate of wear increases. By identifying and measuring these metallic particles, the surface from which the particles were worn can be identified and the rate of wear can be determined to be normal or abnormal.

Techniques frequently used in oil condition monitoring are

(i) Magnetic plug inspection
(ii) Spectrometric Oil Analysis Procedure (SOAP)
(iii) Ferrography
(iv) Particle Counting (for hydraulic fluids)
(v) Patch Testing

A typical magnetic plug, Fig.7, incorporates a non-return valve so that it can be inserted into a pipe-line and/or withdrawn without loss of fluid. Debris which has been trapped by such a magnetic plug can be measured with a magnetometer to determine the amount that has been collected; this may then be recorded to determine the collection trend. At the same time, examination under a simple microscope and comparison with Debris Recognition Drawings/Photographs makes it possible to tell the component from which the major amount of wear material has arisen, as shown in Fig.9.

When spectroscopy was first introduced as a chemical analytical instrument 100 years ago, it brought about a revolution in chemistry. Its advantages were not merely that spectroscopy was a sensitive detector but that the instrument could detect and measure the quantity of an element present in the sample independently of how the element was incorporated in a compound. The flame or spark of the emission or absorption apparatus broke down the compounds and each element displayed its individual set of spectrum lines, Fig.10. Nothing was occluded, this was the fundamental difference from all previously existing methods of chemical analysis.

When the advantages of oil analysis were first realised at the time of the introduction of diesel locomotives to the railroads, the emission spectrograph was adapted to oil analysis, the theory being that a rapid increase of a metallic element in the lubricating oil would imply that a part made of that element was wearinq rapidly. Air dirt, assembly or repair debris, system wear metals, corrosion products and coolant water inhibitors are some of the materails which may be detected.

Element	wavelength (Angstrom units)
Aluminium	3092
Copper	3247
Chromium	3579
Iron	3720
Lead	2833
Sodium	5890
Tin	2354

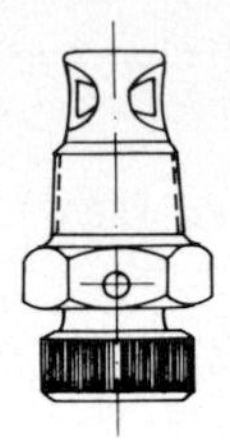

Fig.7 Magnetic plug (chip detector)

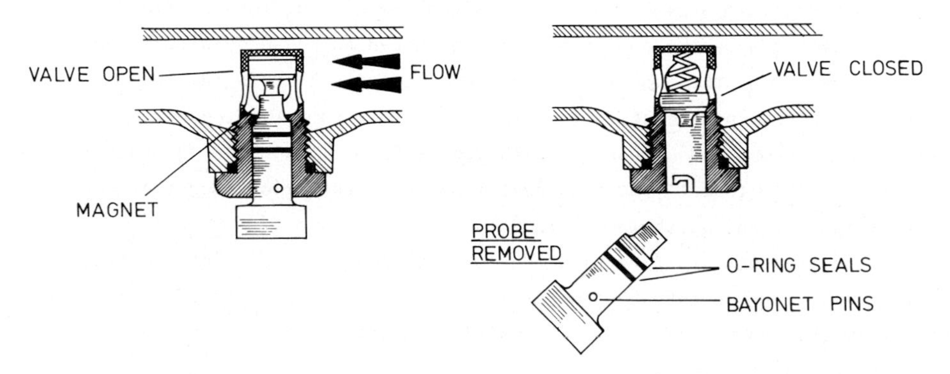

Fig.8 Operating principle of the TEDECO/Muirhead Vactric Magnetic Plug/ Chip Detector

(a) Ball Debris - rounded 'rose-petal' radially split shape

(b) Track Debris - rounded, surface break up, criss cross scratches

(c) Roller Debris - generally curled and rectangular, parallel lines across width

(d) Gear Tooth Debris - irregular shape, grey surface as splash of solder

Fig.9 Magnetic plug debris source recognition

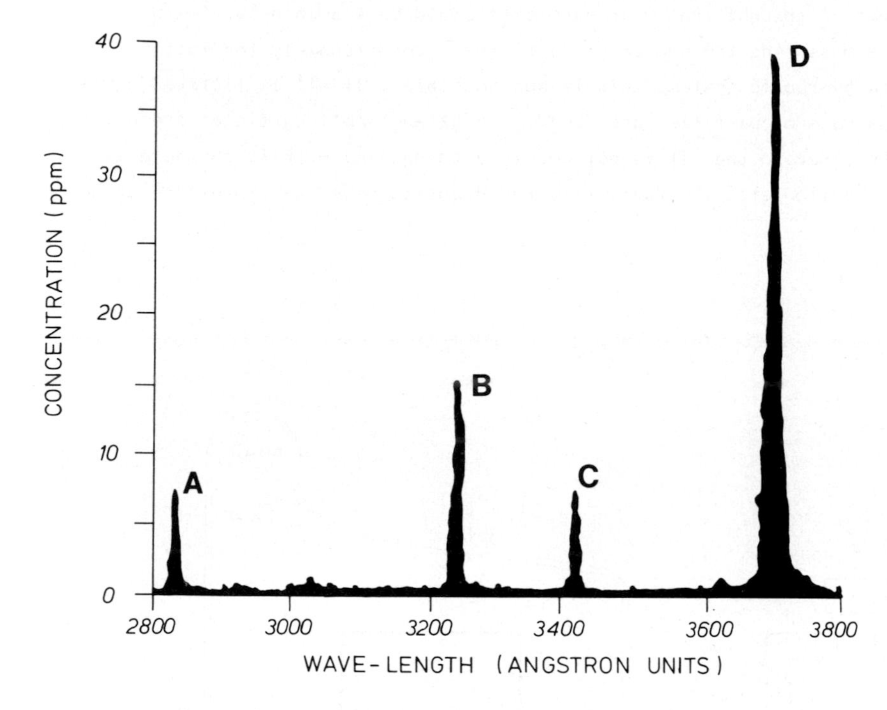

Fig.10 Typical spectroscopic spectrum

TABLE 18.6 Element detected and frequent source

Element	Sources
Aluminium, Silicon	Dust and Airborne dirt
Borax, potassium, sodium	Coolant inhibitor residues
Calcium, Sodium	Salt water residues
Chromium, cooper, iron, lead tin, zinc	Wear, corrosion or residual assembly debris
Barium, calcium, magnesium, phosphorus, zinc	Engine oil additives

Some manufacturers incorporate special elements into different parts of an engine to act as tracers so that their presence in a sample provides an unambiguous indication of the source of trouble.

SOAP has developed to the point where the majority of military services, railroads and airlines operate spectrographic oil analysis as a standard procedure for detecting problem areas, schedule overhauls and routine oil monitoring.

It might be thought that wear particles could be examined by viewing a filter and measuring the number of particles. Unfortunately for most cases other than hydraulic systems this is not possible. If oil is filtered a miscellaneous mass of particles are found. Large and small particles are piled one on the other so that it is not possible to determine their characteristics, size distribution etc. In fact, filtered deposits give the impression of being "just dirt".

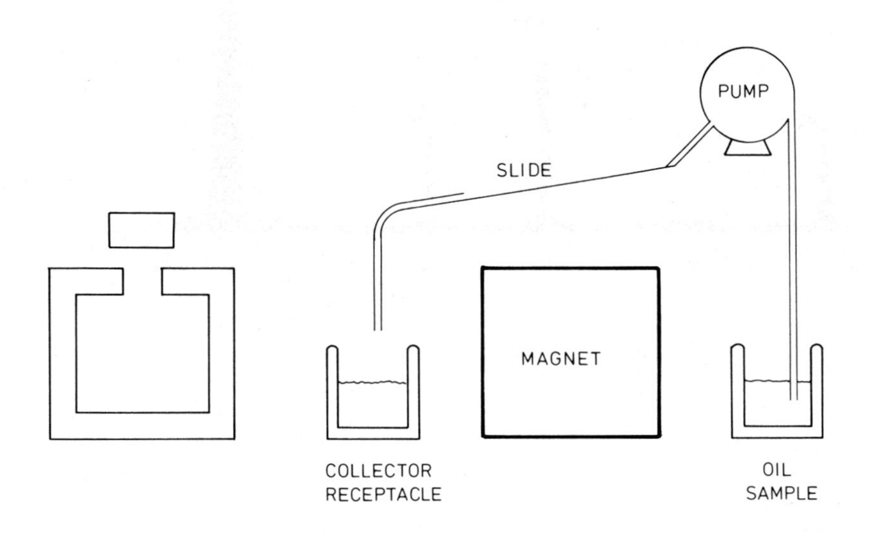

Fig.11 Schematic diagram of 'Ferrograph' (the strength of the magnetic field is greatest at the bottom of the slide.)

The monitoring of wear particles on the Ferrograph, Fig.11, on the other hand, is a technique in which most things are occluded. One of its most valuable characteristics stems from the fact that it does not see every particle and, in fact, ignores everything except wear metal particles in the oil. The separation and analysis of wear particles for size distribution, type of metal, physical shape, crystal structure etc. is often a much more sensitive indicator of the wear situation than a direct view of the worn surface. Characteristic debris from modes of failure have been discussed by Scott in a previous chapter on wear.

All wear debris analysis techniques are particle size dependent and any particular technique is sensitive only to a specific range of particle sizes. Fig.12 summarises the efficiency of these techniques as a function of particle size [9].

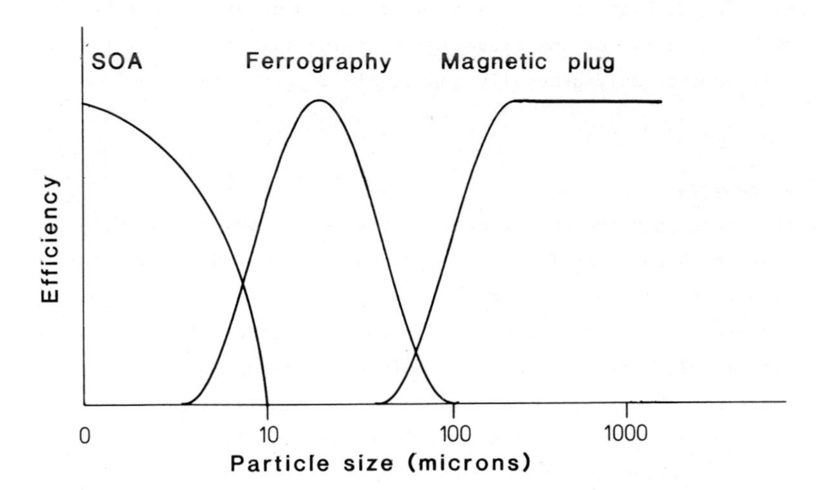

Fig.12 Efficiency of various sensors as a function of particle size

The cleanliness requirements of hydraulic systems have become more critical in recent years. Systems employing electrohydraulic servo valves in numerically controlled machine tools, high pressure systems where pump or valve clearances may be as small as 0.5 micrometre are particular examples. Analysing particulate contamination for these cases are usually carried out by particle count methods utilising either microscopes or automatic counters.

Microscope counting methods (ASTM F312, F313, ARP 598A, IP 275), are procedures which are considered unsuitable for particles smaller than 5 microns. These methods size and count statistically particles retained on a membrane surface after the fluid sample has been filtered. Apart from counting being very time consuming, there can be variation when different people count the same slide. Automatic counters are however now available utilising scanning computers and TV screens.

Other instruments using the principle of light interception count particles suspended in a liquid from either a sample bottle or direct from a hydraulic system. As these instruments operate on the blockage of light principle, they measure the projected area of a particle and then record the diameter of a circle of equivalent area. Unfortunately they cannot differentiate between solid particles and air bubbles.

The typical colour of contamination in any given hydraulic system remains fairly constant. The darkness of the particulate discolouration of a filter is therefore a rough indication of the cleanliness of the test fluid. This Patch Test procedure is however only generally applicable to gross levels of contamination.

18.5.4 Trends Analysis

This is little more than the data logger with a memory and trend analysis capability. The most elementary form is the handwritten watch log which is inspected and analysed. As in practice, real use is only made of the log after a failure has occurred, the modern tendency is to use sensors to input to a general data system which scans the monitor points and produces a regular print-out. In more advanced form, a computer is capable of combining many of the inputs and establishing trends which, by mathematical modelling, can be associated with the changes which would occur following specific defects and consequently when the system is interrogated, it can state (through a logic process) the plant and component which is deteriorating; with preset-limits such a system will even produce a statement of the 'unexpired life' remaining. Such advanced trends monitors are more commonly known as 'performance' monitors.

18.6 DETERIORATION LIMITS

Most classification societies and bodies of a similar nature have established specifications and codes of practice which define the deterioration limits which should be allowed before corrective action is taken. Typically, limits which have been established for deterioration identified by vibration methods include:

VDI Code of Practice 1056 (October 1964)
DIN 45665 (November 1967)
BS 4675: 1971
ISO 2372/3

It does seem that for vibration limits, most people in general err on the side of considerable caution although in practice, the limit must depend upon a large range of environmental factors specific to each installation.

It is only by experience, both personal and that of other that active decision limits can be reached. In the whole field of condition monitoring, developments are occurring at such a rapid speed that only an organisation such as the UK Mechanical Health Monitoring Group through its regular seminar/ symposia and courses is it possible to acquire all the information - known to consultants - needed to implement and manage an on-condition maintenance system.

REFERENCES

1 Collacott,R.A. 'Mechanical Fault Diagnosis and Condition Monitoring' Chapman & Hall, London 1977.
2 Venton,A.D.F. and Harvey,B.F. 'Reliability assessment in machinery system design' Proc. I.Mech.E. 1973.
3 Bridges,D.C. 'The application of reliability to the design of ship's machinery' Trans. I.Mar.E. 86, Part 6 1974.
4 Davies,A.E. 'Principles and practice of aircraft powerplant maintenance' Trans. I.Mar.E. 85 Part 6 1973.
5 Collacott,R.A. 'Data Sources for Reliability Statistics' UKM Publications Ltd., 92 London Road, Leicester LE2 0QR 1976.
6 Wilkinson,H.C. and Kilbourn,D.F. 'The design of ship's machinery installations' Shipping World and Shipbuilder, August 1971.
7 Mathieson,Tor-Chr. 'Reliability engineering in ship machinery plant design' Report IF/R.12 University of Trondheim,N.I.T. 1973.
8 Collacott,R.A. 'Vibration Monitoring and Diagnosis' George Godwin Limited 1978.
9 Pocock,G. 'Introduction to Ferrography' Symposium on Ferrography, British Inst. Non Destructive Testing, London 1979.

19 THE TRIBOLOGY OF METAL CUTTING

E. M. TRENT

Department of Industrial Metallurgy, University of Birmingham,
Birmingham B15 2TT

19.1 INTRODUCTION

The efficiency of metal cutting operations is very largely controlled by the behaviour of the work material and the tool material at the interface between them near the cutting edge of the tool. Metal cutting is carried out on all the major classes of metals and alloys produced commercially. The machining operations such as boring, drilling, tapping, turning, milling, planing and sawing are very varied in character. Objects as small as instrument parts or as large as industrial boilers are machined to remove excess material and to generate the necessary shapes with the required precision, but, however varied, all machining operations have certain features in common.

19.2 METAL CUTTING PHENOMENA

All machining operations involve the use of one or more tools of a wedge shape with a cutting edge, which remove a thin layer from the surface of a larger body as shown in Fig.1. The tools are always moved asymmetrically with respect to the wedge angle so that the layer removed - the "chip" - bears against and moves over one surface of the wedge, known as the "rake face" of the tool. The tool is so shaped that the freshly cut metal surface does not rub against the other face of the wedge - the "clearance face" or "flank" of the tool. (Fig.1).

The layer of metal removed is first plastically deformed by a shearing action, roughly along a plane A-B in Fig.1., extending from the tool edge A to the position B where the chip separates from the undeformed work material. This involves severe shear strain - usually a natural strain of at least 2 but often very much higher as can be seen from Fig.2. which shows a section through the forming chip during the cutting of copper. The shear plane is under high compressive stress and the tri-axial stress condition is such that with most ductile metals and alloys the high shear strain can be sustained without fracture so that a continuous chip is formed, as with the example of copper. With metals and alloys of low ductility, however, or under cutting conditions where the compressive stress on the shear plane is low, the chip may be broken into

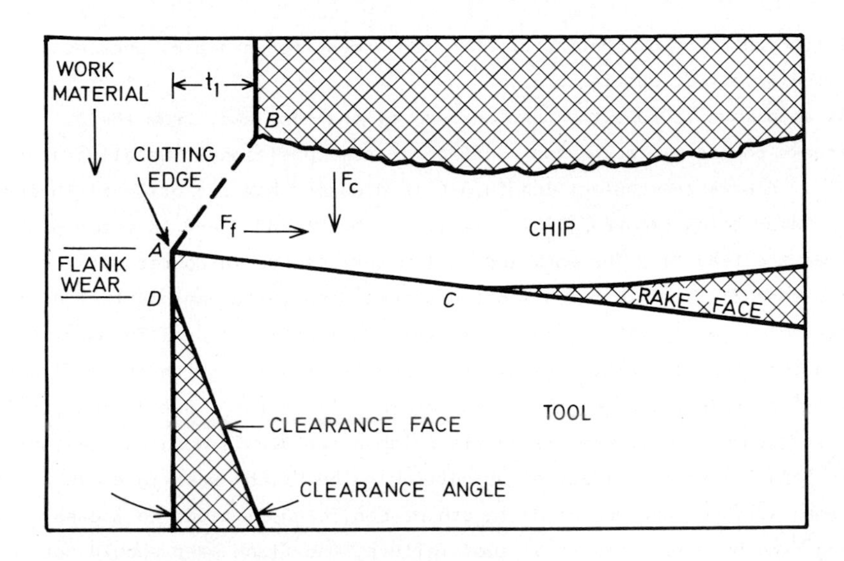

Fig.1 Features of Metal Cutting

Fig.2 Section through chip and bar of high purity copper after "quick-stop". Machined at 122m min^{-1}.

small fragments. A discontinuous chip is formed, for example, when cutting cast iron or a free cutting brass.

The chip moves across the rake surface of the tool away from the cutting edge and breaks contact with the tool surface at some position C, usually rather ill-defined. Always the contact length A-C is greater than the original thickness of the layer being removed (t_1 in Fig.1.) - the "feed" - A-C is often 5 or even 10 times the feed t_1. The work and tool materials are in contact at the cutting edge and usually for a short distance down the clearance face or flank of the tool. The length of contact in this region A-D is usually much shorter than on the rake face A-C. The clearance angle (Fig.1.) which is usually about 6° to 15° restricts the length of contact on this surface, but during cutting the most common form of wear is one in which a surface is worn on the tool nearly parallel to the direction of cutting - the "flank wear land" in Fig.1. This worn surface, and hence the length of contact A-D, tends to increase with cutting time but, to avoid total tool failure, the flank wear should not be allowed to become too large before the tool is reground or replaced.

19.3 CONDITIONS AT THE TOOL-WORK INTERFACE

19.3.1 Tool Forces and Stresses

The forces acting on the tool are (1) that required to shear the work material over the area of the shear plane A-B (Fig.1.) and (2) that required to move the chip across the tool rake face contact region A-C. The force to move the work material over the flank A-D is small compared with the other forces and can be neglected in a first approximation. Tool dynamometers have been developed and used to measure the forces acting on the tool in two directions - in the direction of cutting Fc and in the direction of the feed Ff. In most cutting operations the forces acting on the tool vary from a few kilograms to a few hundred kilograms. The cutting force Fc acts nearly normal to the rake face of the tool and exerts a largely compressive stress on this surface. The feed force Ff is almost always smaller than the cutting force (typically 40-60% of the cutting force) and exerts a shearing stress on the tool surface.

While the forces acting on the tool can be measured with accuracy, even the mean value of the stress acting on the contact area between work material and tool is difficult to determine because the area of contact is difficult or impossible to measure exactly. The stresses are not evenly distributed on the contact area and it is not easy to determine the stress distribution on this area. The general character of the stress distribution on the rake face of a cutting tool, however, is now generally accepted to be that suggested by Zorev [1] and shown diagrammatically in Fig.3. The compressive stress acting on the rake face is at a maximum at or close to the cutting edge and diminishes to zero at the end of the contact area. The maximum compressive stress near the edge

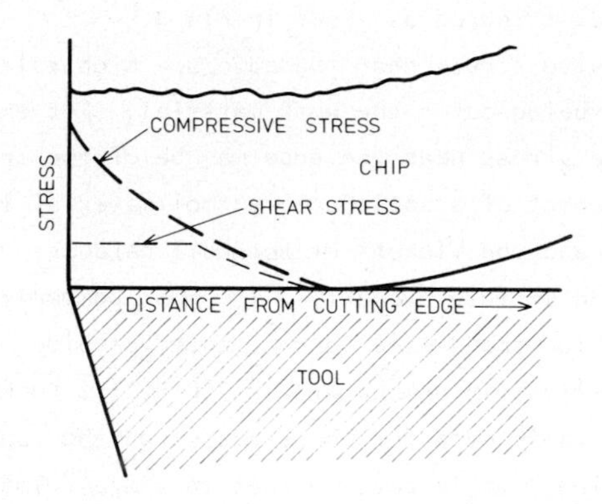

Fig.3 Stress distribution in cutting tool (after Zorev)

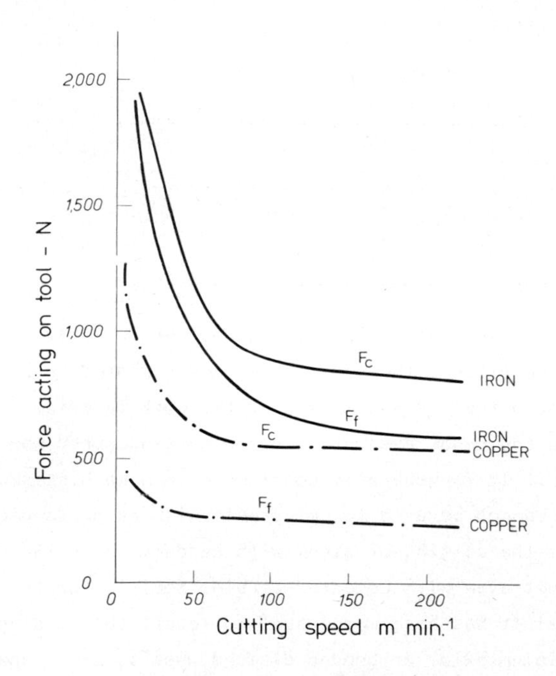

Fig.4 Forces acting on a tool as function of cutting speed.
Feed 0.25mm/rev. Depth of cut 1.25mm.

is often twice the mean stress on the area of contact. The shear stress on the rake face is more uniformly distributed as shown in Fig.3.

The values of the compressive stress near the edge are high relative to the yield stress of the material being cut - the work material. For example, in cutting steel the compressive stress near the edge may be of the order of 1500 N mm^{-2}. A major requirement of a satisfactory tool material is thus high yield stress in compression, and the Vickers or Rockwell hardness values are usually taken as an indication of this property. The most commonly used cutting tool materials are hardened high speed steel and cemented carbide. The minimum hardness of cutting tool materials in common use is 750 HV (62 Rockwell C).

Although there are few reliable data for the stresses on the contact area of tools in real cutting operations, it is certain that they are related to the yield stress of the work material. Approximate values for the mean stress acting normal to the rake face of a turning tool under a standard set of cutting conditions are shown in Table 19.1 for different work materials.

TABLE 19.1 Mean compressive stress on contact area

Work material	Compressive Stress N mm^{-2}
Iron	340
Steel (medium carbon)	770
Titanium	570
Copper	310
70/30 brass	420
Lead	14

For the cutting of materials of very high yield strength, particularly heat treated steels and nickel base alloys, the usual steel and cemented carbide tools may be inadequate because the stress imposed by the work material is high enough to deform the cutting tool edge even at very low cutting speed where the tool edge temperature is low. It is generally considered in a machine shop that high speed steel tools cannot be used to cut steels with hardness higher than 350 HV (36 Rc) and that the cutting of steel with hardness over 550 HV (53 Rc) becomes very difficult even with cemented carbide tools. For the machining of fully hardened steel it has been more usual to resort to grinding using silicon carbide, aluminium oxide or bonded diamond wheels, or to shape by electro discharge machining (EDM) or electro chemical machining (ECM). Recently the introduction of new cutting tool materials with still higher yield strength - including compacted polycrystalline diamond and cubic boron nitride tools - has made the cutting of fully hardened steels, higher strength nickel-based alloys, and other very hard materials, a more feasible proposition for industrial shaping operations.

19.3.2 Cutting Speed

One of the most important parameters in metal cutting is the velocity at which the work material passes the cutting edge - the cutting speed. This varies greatly in industrial operations from almost zero, for example near the centre of a drill, to 300 m min^{-1} or even higher. Rather to their surprise research workers measuring tool forces found that, in general, these forces do not increase as the cutting speed is raised. In many cases the forces decrease, particularly in the speed range up to 65 m min^{-1} as shown for example in Fig.4. This has been shown to be true for a wide range of work materials and cutting conditions. The forces decrease mainly because the area of contact between tool and work decreases as cutting speed is raised. Although there have been no very detailed studies of the stress acting on the tool surface as a function of cutting speed, there is no evidence to indicate that the stress acting on the tool is raised as cutting speed is increased.

Cutting speed is of particular importance in relation to the economics of machining. The cost of machining operations is reduced by increasing the rate of metal removal, and the main incentive to the development in machining in the last hundred years has been the reduction of the very high costs by the use of new machines and tools capable of machining at increased rates. In the cutting of high melting point metals and alloys the life of the cutting tool becomes progressively shorter as the cutting speed is raised until the cost of replacing worn out tools more than outweighs the advantages of higher speed. It has been the ability of the cutting tool to withstand the conditions at the tool edge which has limited the rate of machining of steel and cast iron. The development and commercial use first of high speed steels and then of cemented carbides has enabled cutting speeds to be raised by a factor of about 20 times compared with carbon steel tools and there are still many operations in which tool life is the factor limiting the rate of metal removal.

In general as the cutting speed is raised neither the forces acting on the tool nor the stresses on the area of contact are increased. The energy expended in metal cutting, however, increases approximately in proportion to the cutting speed, if other conditions remain constant. This energy is converted into heat near the cutting edge, and raises the temperature of the tool, reducing its yield stress and increasing the rate of tool wear. It is this rise in temperature which limits the ability of the tools to withstand increasing cutting speed. The generation of temperatures in metal cutting must now be considered.

19.3.3 Heat in Metal Cutting

In metal cutting energy is expended into two main regions (1) along the shear plane A-B (Fig.1.) where the work material is sheared to form the chip, and (2) at the rake surface of the tool where the chip is moved across the contact area.

The energy expended in shearing the work material to form the chip mainly results in raising the temperature of the chip and almost all of this heat is carried out of the system when the chip breaks contact with the tool. Since any one element of the chip is in contact with the tool for only a very short time - typically a few milli-seconds - only a small proportion of this heat could be conducted into the tool under the most favourable conditions. It is probable, as will be shown, that all the heat in the body of the chip is carried out of the system in most cases. A small proportion of the heat generated on the shear plane is conducted back into the body of the workpiece. The energy expended on the shear plane is normally the largest part of the total energy of cutting - often of the order of 75 to 80% of the total. The temperature of the chip is often raised to 200-350°C when cutting steel or other high melting point materials.

It is the smaller portion of the total energy of cutting - that expended in moving the chip over the tool - which is responsible for the generation of high temperatures at the tool/work interface and the conditions at this interface must therefore be considered. The high compressive stress normal to the tool rake surface has already been emphasised. The mean stress on the contact area is always much higher than the stresses normally encountered at moving interfaces in engineering systems. The very high stresses alone would result in the area of real contact between the two surfaces being a much higher porportion of the apparent contact area than is usual for sliding surfaces, and would tend to promote seizure. Other factors are also favourable to seizure. The tool is continually cutting into clean metal, being brought into contact with surfaces free from oxide or other layers which inhibit the extension of contact areas in many sliding situations. The clean work material flows continuously over the tool surface in one direction, sweeping away oxide or other layers initially present on the tool, which have little chance to re-form. Relatively high temperatures generated at the interface increase atomic activity and this also tends to promote seizure.

19.3.4 Seizure of the Tool/Work Interface

When these conditions are considered, it is not surprising to find that seizure between the tool and work surfaces is commonly observed on cutting tools [2]. The process of friction welding is often carried out under conditions less severe than those encountered in metal cutting. For example, sound joints can be made by friction welding at pressures of 75 N mm^{-2} and peripheral speeds of 50 m min^{-1}, whereas in metal cutting stresses on the contact area of 750 N mm^{-2} occur and higher speeds are often used. That seizure occurs at the tool/work interface is confirmed by numerous metallographic observations, of which three examples will be given here.

Fig.5. shows a polished and etched section through the rake face of a cemented carbide tool and adhering work material after cutting steel at 100 m min^{-1}. It shows the work material in contact with the tool surface not just at the tops of the asperities but at all the hills and valleys of the surface on a micro scale. To say that these surfaces are seized together means that sliding as normally conceived, with the two surfaces separated by a fluid film or in contact only at the asperities, is not possible. The two surfaces are mechanically interlocked and/or metallurgically bonded over the whole or a large part of the interface. That metallurgical bonding is often involved is shown by examples, such as those shown in Figs.6 and 7, in which, after stopping cutting by propelling the tool rapidly from the cutting position, the chip remains firmly adherent to the tool or separates from the tool at some position remote from the interface, leaving a layer of work material welded to the tool surface.

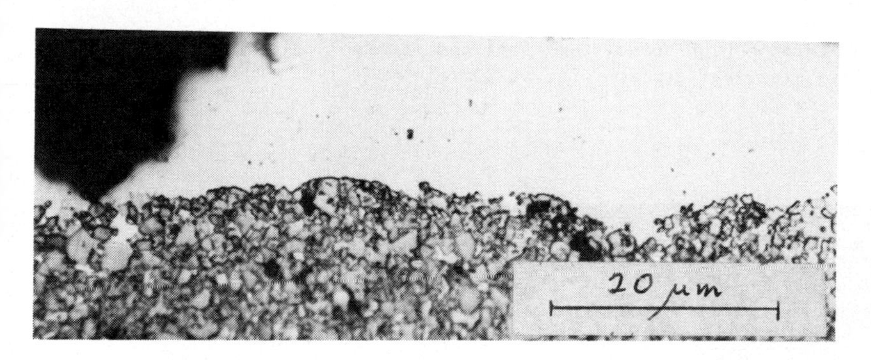

Fig.5 Section through carbide tool with adherent work material (white) after cutting steel. Shows seizure conditions at interface.

Seizure is normally thought of as a condition where a mechanism ceases to function, as when a bearing seizes, but in metal cutting the seized area is small, there is adequate power to continue cutting and the tool is strong enough to resist the stresses imposed by seizure conditions. Movement continues by shear in the work material in a region adjacent to the tool surface. This gives rise to two main sorts of conditions near the interface. The first layers of work material seized to the tool are severely work hardened and shear is then transferred to the next layers. In this way a body of "dead metal" may be built up, adherent to the tool material, which persists for long periods of cutting. Fig.8 shows an example of this feature, known as a "built-up edge". The built-up edge reaches a stable state and the size and shape depending on the work material and the cutting conditions. It is a dynamic structure with

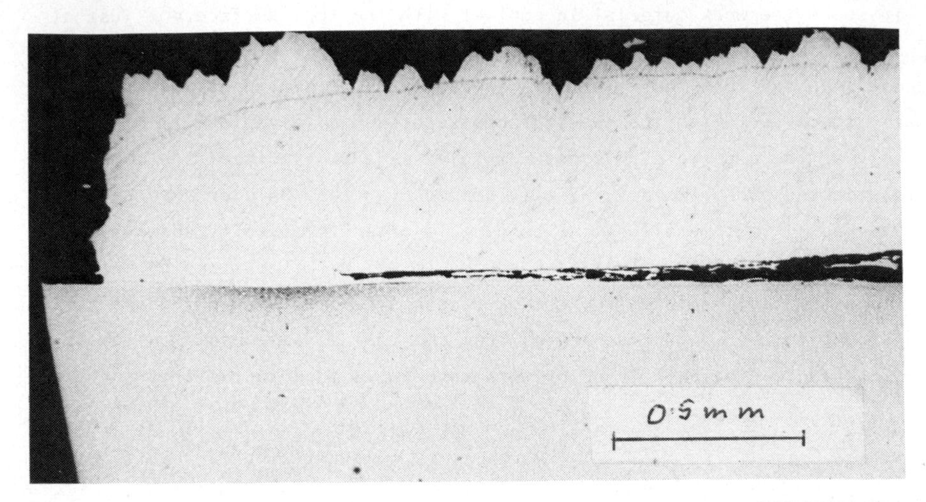

Fig.6 Section through high speed steel tool and adherent chip (austenitic stainless steel) after machining at 30m min^{-1}.

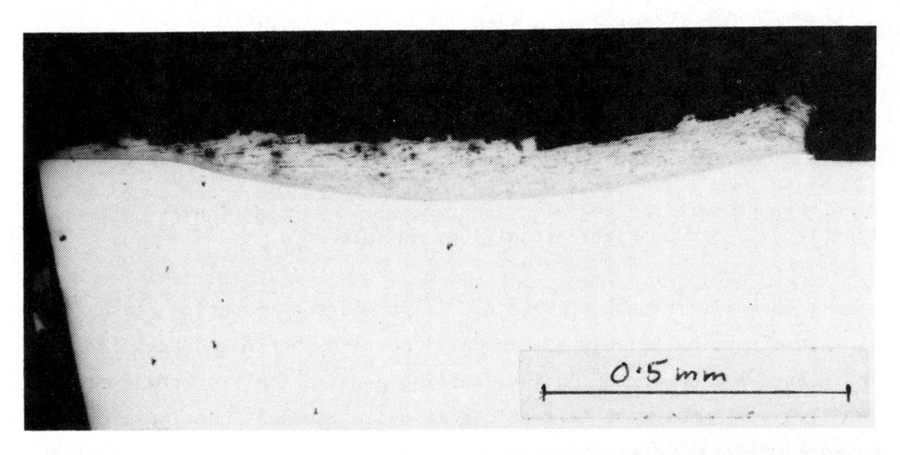

Fig.7 Section through high speed steel tool and adherent chip fragment after cutting low carbon steel at 107m min^{-1}.

fragments being continuously added and broken away. The shearing action leading to chip formation may take place at a distance of 300 μm or more from the tool surface.

A built-up edge is often formed when cutting alloys containing more than one phase, such as steel, cast iron and alpha-beta brass [3]. With such materials a built-up edge occurs when cutting at relatively low speeds, but disappears

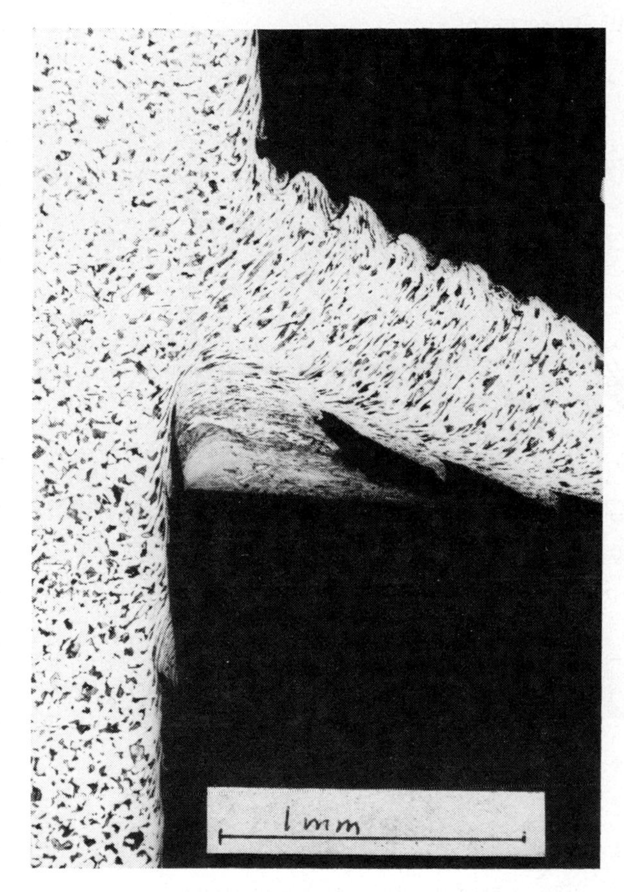

Fig.8 Built-up edge formed during cutting low carbon steel at 15m min^{-1}.

when speed or feed are raised. When cutting pure metals and solid solutions at almost any speed, and when cutting two phase alloys in the higher range of speeds, seizure conditions are observed to exist at most of the interface, but the built-up edge is absent. Movement of the work material takes place by shear concentrated into a very thin layer adjacent to the tool surface usually of the order of 25-50 μm in thickness. An example of such a layer is seen in Fig.2. for the cutting of copper and a layer when cutting a low carbon steel is seen at high magnification in Fig.9. In this layer the work material behaves more like a very viscous liquid than a normal metal and the layer is termed a "flow-zone". In the flow-zone the rate of shear strain is extremely high - 10^4 to 10^5 per second - and the amount of strain is so extreme that original structural features (such as pearlite and ferrite in steel) are completely destroyed. There is good evidence that, within the flow-zone, dynamic recovery and/or recrystallisation are taking place and the behaviour of the material is akin to its behaviour in hot working processes.

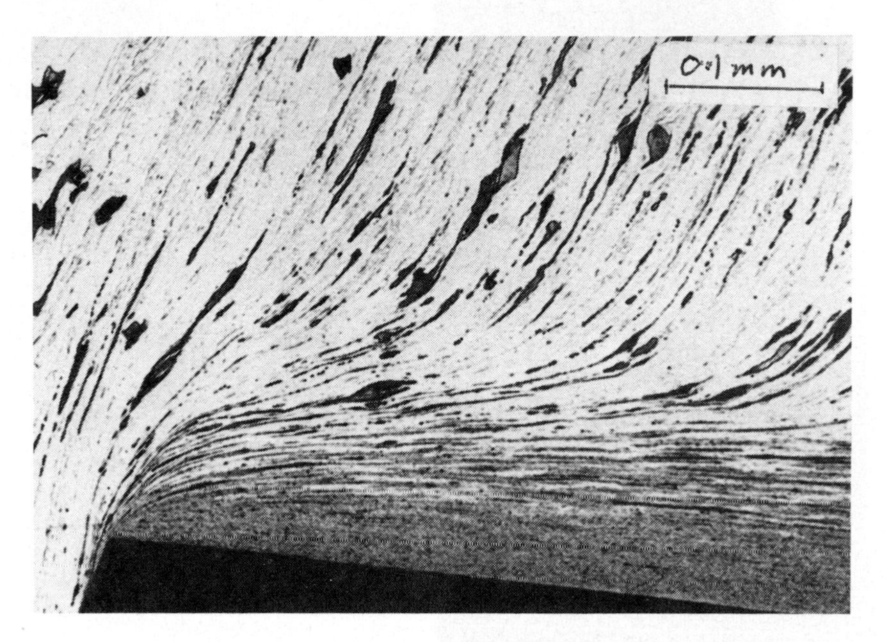

Fig.9 Flow zone at under surface of chip, adjacent to rake face of tool, formed during cutting low carbon steel at 63m min^{-1}.

19.3.5 Cutting Tool Temperatures

The energy expended in deforming the work material in the flow-zone per unit volume of metal deformed, is much higher than on the shear plane and the temperatures generated in the flow-zone are therefore higher. The contact between the flow-zone and the tool is very good. As has been demonstrated there is continuous metallic contact in many cases and heat flows readily across the boundary to heat the tool. It is the heat generated in the flow-zone at the interface between tool and work material which is the main heat source raising the temperature of the tool and creating the conditions under which cutting tools are worn.

The regions of the tools which are heated to high temperature are very localised and within these regions temperature gradients are very steep, but it is possible to study the temperature distribution in some detail for certain conditions of cutting by observation of the changes in structure or hardness of steel tools in those parts of the tools heated by cutting action above their tempering temperature. Fig.10 shows, for example, the temperature gradients in a high speed steel tool used to cut a low carbon steel at a speed of 76m min^{-1} at a feed of 0.25 mm per rev. This is characteristic of the type of temperature distribution found to occur in tools used to cut steel under conditions where a flow-zone occurs at the interface. Fig.10 shows that the temperature near the tool edge was relatively low - in this case under 650°C - but there is a high temperature region just over 1mm from the edge in the direction of chip flow, where, in this example, the temperature at the interface was over 800°C. It is fortunate that in cutting steel and many other alloys at high speed the region of highest temperature is at a distance from the edge where the compressive stress on the tool is a maximum (Fig.3).

As the cutting speed is raised the maximum temperature on the rake face of the tool increases rapidly, while the temperature near the edge is increased more slowly. The yield stress of the tool material decreases with rising temperature and, as cutting speed is raised, the temperature at the edge may reach a value where the yield stress of the tool is reduced below the compressive stress exerted by the work material. The tool edge is then plastically deformed and this leads to a rapid rise in the rate of heat generation at the cutting edge. The tool then fails catastrophically usually within a few seconds. This is the main mechanism which sets the upper limit to the rate of metal removal which can be achieved with high speed steel tools (and, at a higher level of speeds, with cemented carbide tools) when cutting steel and other high melting point alloys.

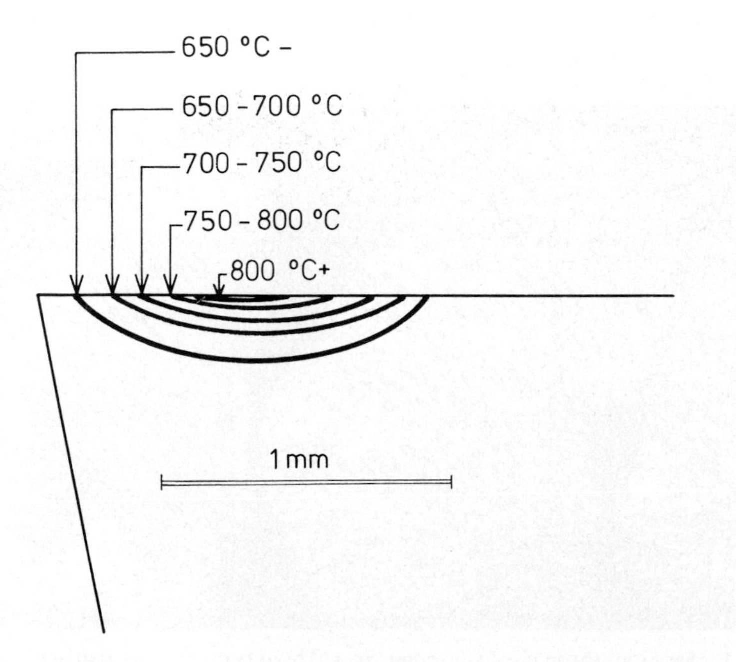

Fig.10 Temperature gradients in tool used to cut low carbon steel at 76m min^{-1}, 0.25mm/rev feed.

19.3.6 Sliding at the Tool/Work Interface

Many of the peculiar and characteristic features of machining operations arise from the unusual feature of seizure at the interface between tool and work material, but these conditions do not exist under all cutting conditions and on all parts of the contact area. The model of seizure which has been given is too simplified and must be corrected.

At very low speeds seizure may not occur. Similarly at the periphery of the contact region, even at high rates of metal removal, there is good evidence to show that sliding takes place at the interface by a type of stick-slip process. Thus a section through the outer edge of a steel chip often shows a segmented chip with a periodic structure at the interface (Fig.11.) indicating a stick-slip action. The centre of the same chip shows a flow-zone demonstrating seizure at this part of the interface (Fig.12). That sliding occurs in these peripheral regions may be attributed to two main factors (1) lower compressive stress near a free surface of the chip and (2) access of atmospheric oxygen to the interface at this position, reducing the tendency to metallic bonding.

Since the mechanisms of wear may be very different under conditions of seizure and sliding, it is useful to have in mind a model of the regions where seizure and sliding occur most usually on a cutting tool. Fig.13 shows such a map for a simple turning tool.

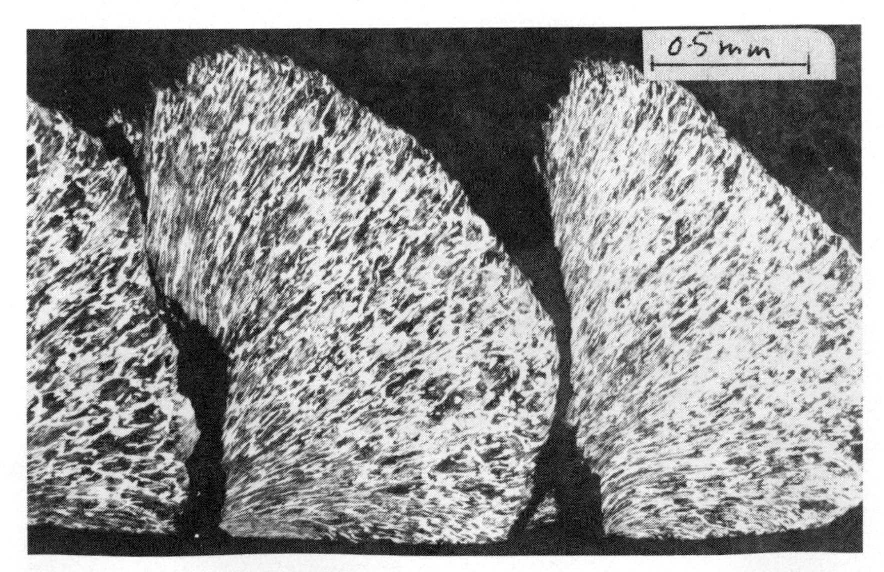

Fig.11 Section through outer edge of chip after cutting medium carbon steel at high speed. Shows stick-slip action at interface.

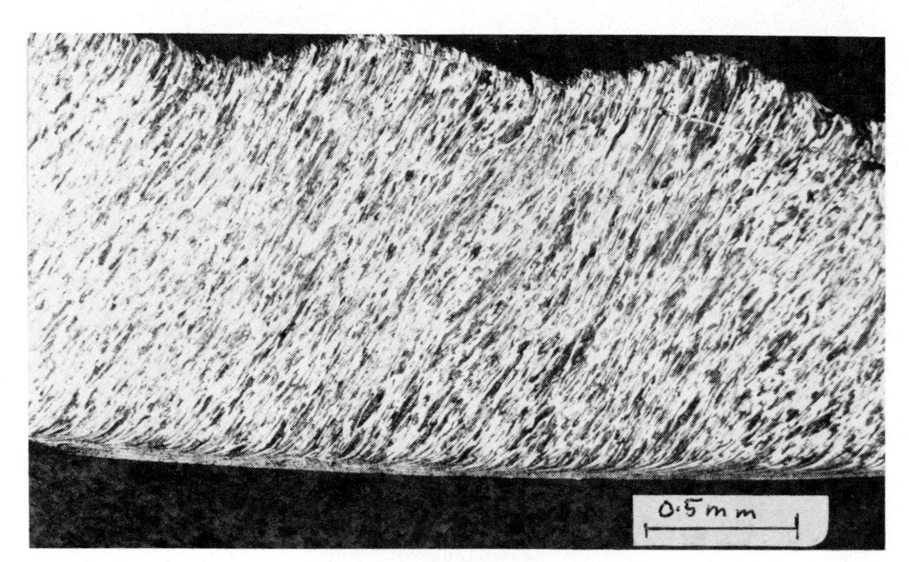

Fig.12 Section through centre of same chip as Fig.11. Shows flow-zone at interface characteristic of seizure.

19.4 CUTTING TOOL WEAR

While the upper limit to the rate of metal removal when cutting steel or other high melting point alloys is determined by the ability of the tool to withstand the cutting stresses at elevated temperatures, the life of the tool decreases as the cutting speed is raised before this limit is reached. At lower speeds the shape of the tool is changed by one or more of a number of different wear mechanisms until it can no longer cut efficiently. Fig.14 shows diagrammatically on a model turning tool the location of the chief wear features observed.

"Flank wear" on the clearance face of the tool often increases steadily with time of cutting until, when a critical amount of wear is reached, the temperature on this surface starts to rise rapidly and tool failure may be sudden. The critical amount of flank wear varies under different conditions but it may be between 0.4 and 1.5 mm. To avoid complete failure, which may be expensive, tools are normally reground or replaced before the critical wear is reached. Flank wear may occur at any cutting speed but the wear rate increases with

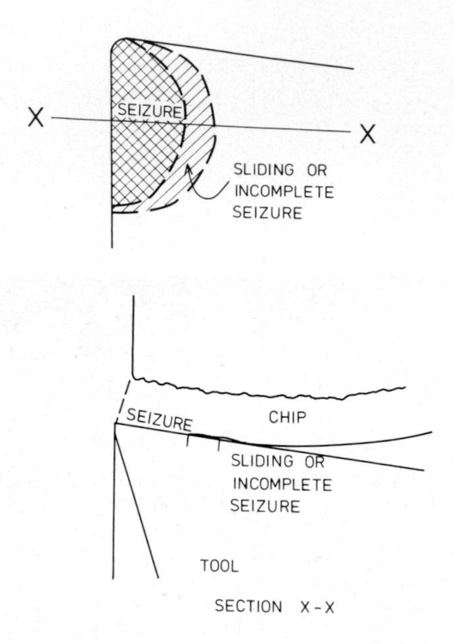

Fig.13 Diagram of turning tool showing regions of seizure and of sliding at the tool/work interface

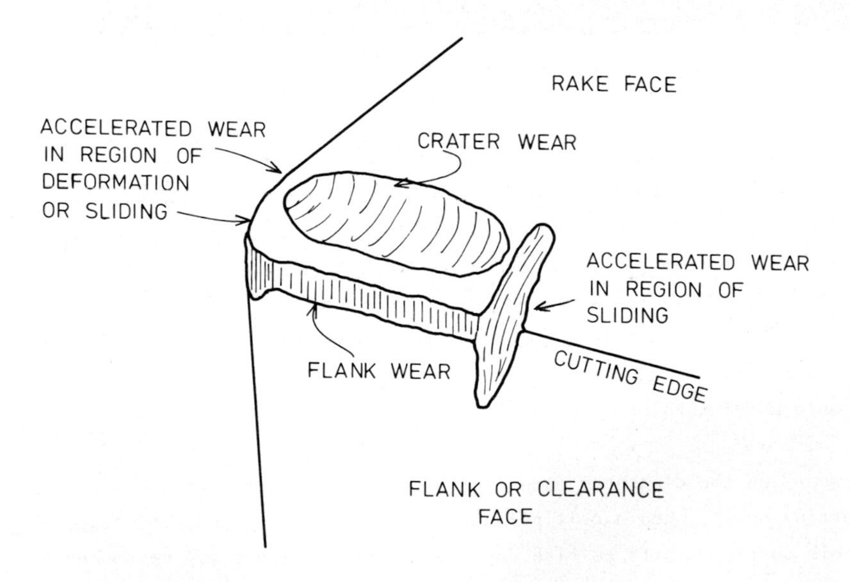

Fig.14 Diagram showing wear features on turning tool

speed as the ultimate limit for the tool material is approached. In the region of high speed cutting the rate of flank wear and the tool life often follow the relationship given by Taylor [4] for tool life in relating to cutting speed

$Vt^n = C$

V - cutting speed

t - cutting time to failure or to some standard amount of wear

n and C - constants for a given tool and work material

"Crater wear" is the term used for a groove or crater worn on the rake face of the tool, usually at some distance from the cutting edge (Fig.14). Cratering wear is characteristically observed on tools used at high cutting speeds and the rate of crater wear increases as the cutting speed approaches the ultimate limit for the tool material. As the crater becomes deeper it weakens the tool edge and may lead to fracture of the edge and tool failure.

"Flank wear" and "crater wear" are descriptive terms and the words do not imply distinct wear mechanisms. The mechanisms of wear will now be discussed. Where seizure conditions occur at the tool/work interface at least four different mechanisms of wear have been observed and these will be considered first. Wear under conditions of sliding at the interface will be considered separately.

19.4.1 Abrasion

The abrasive action of hard phases in the work material, such as oxides or carbides may contribute to the wear of cutting tools. Abrasion is, however, probably not a major cause of wear under seizure conditions unless the particles of the hard phases are large, e.g., greater than 40 μm, or present in very high concentrations, as they may be for example on the surface of castings. Even with high speed steel tools the abrasive action of dispersed, fine hard particles is probably small because, under seizure conditions they rarely impinge on the tool surface in such a way as to remove tool material. With harder tool materials such as carbides or diamond, few if any particles in the work material are harder than the tools and abrasive action is less likely. The hardness of cutting tool materials is of more significance as a measure of their ability to withstand high compressive stress than as a measure of their resistance to abrasion.

19.4.2 Surface Shearing

When cutting higher melting point metals at high speeds, the interface temperature, particularly on the rake face of the tool (Fig.10), may be very high, so that the yield stress of the steel tool is reduced to a very low value in a small volume of metal at the interface. Thin layers of the tool material may then be sheared away by the work material bonded to the tool surface. Fig.15

shows an example of this wearing action in which a crater is being worn on the rake face of a high speed steel tool when cutting carbon steel at high speed. This wear mechanism is usually observed only where the interface temperature is above 800°C on steel tools. It is one of the mechanisms of wear responsible for crater wear on steel tools and also for the final stages of flank wear just before complete tool failure. This mechanism causes rapid tool wear. It has not been observed to occur on cemented carbide tools.

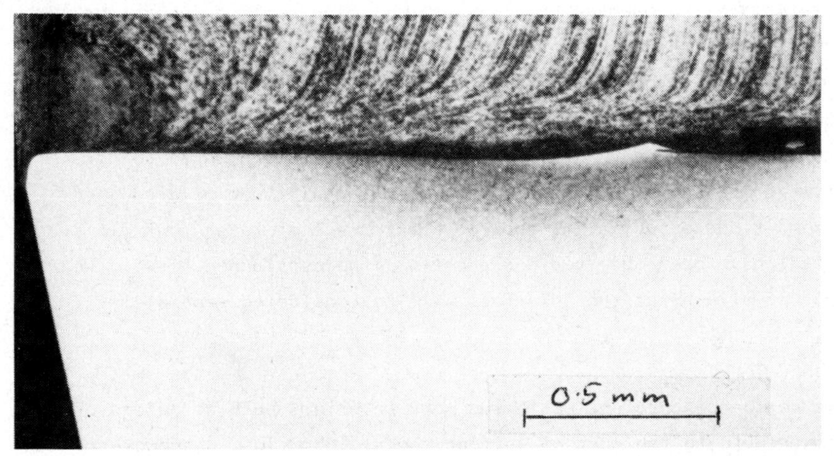

Fig.15 Section through tool used to cut low carbon steel at high speed. Showing formation of crater on rake face by shearing of high speed tool in region of high temperature.

19.4.3 Diffusion and Interaction

Under conditions of cutting where the tool and work surfaces are metallurgically bonded, the tool shape can be changed by a process of diffusion and interaction between the two materials. In the simplest situation atoms from the tool material may diffuse into the work material flowing over the surface and be carried away by it - i.e., the tool material is dissolved into the work material by a process of the same character as that of a block of salt being dissolved by a stream of water running over its surface. Diffusion is a highly temperature dependent process and diffusion wear occurs at an appreciable rate only at relatively high cutting speeds where the interface temperature is high. When cutting steels with high speed steel tools wear by a diffusion mechanism is probably significant only where the interface temperature exceeds 650°C. It has been shown that, even at moderately high cutting speeds of 25-30 m min^{-1} when cutting steel, temperatures of 700°C and over often occur at parts of the interface, and there is a rapid rise in temperature with further increments in speed.

Simple diffusion and a variety of interactions depending on the chemical compositions and metallurgical structures of the tool and work materials, occur across the interface. It is probable that wear based on atomic interactions is the most important mechanism changing the shape of high speed steel, cemented carbide or diamond cutting tools in machining the higher melting point metals and alloys at high speeds.

Fig.16 shows a section through the cratered surface of a high speed steel tool with adhering work material. The wear process was one of diffusion; the tool surface shows no signs of plastic deformation. There is good evidence that wear by diffusion is responsible for most crater and flank wear on high speed steel tools where the interface temperature is above 650°C but below the temperature required for the shearing action.

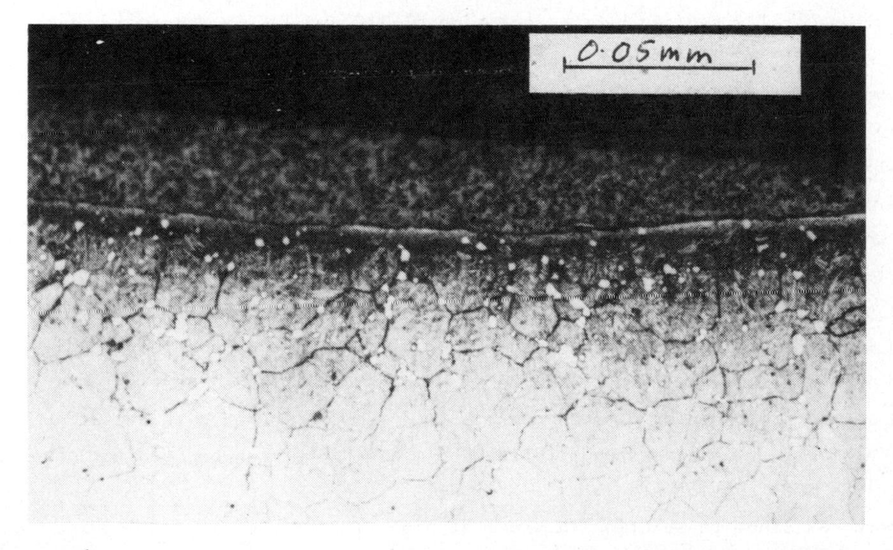

Fig.16 Section through high speed steel tool in worn crater after cutting low alloy steel at 18m min^{-1} for 38 minutes. Interface characteristic of diffusion wear

Cutting tool materials have been developed empirically and most have been developed for the machining of steel since this is the main market for cutting tools [5]. The first cemented carbides produced were alloys of tungsten carbide and cobalt (WC-Co). These are very successful for the cutting of non-ferrous metals and cast iron at speeds much higher than can be achieved with steel tools because the WC-Co alloys have higher yield stress and can resist the shearing action and deformation at high temperatures, and also because the WC-Co alloys do not react with these work materials at the high cutting speeds.

For the cutting of steel however, the WC-Co alloys are not so successful, since rapid cratering wear in particular on the rake face of the tools gives a very short life for tools used at speeds not much higher than those used with high speed steel tools. Cemented carbide tools were successful for cutting of steel at higher speeds only when a small proportion (5 to 20%) of TiC or TaC was added to the WC-Co alloys. These "steel cutting grades" of cemented carbide are able to machine steel at more than three times the speed with WC-Co alloys because the rate of wear, particularly the crater wear on the rake face, is so much lower. It is now known that their success must be attributed to the much lower rate of solution of TiC or TaC in steel at elevated temperatures, compared with that of WC. The rate of cratering in particular is affected because it is on the rake face of the tool, away from the cutting edge, that the highest temperatures at the tool/work interface are located (Fig.10). The rate of flank wear also is generally reduced by the inclusion of TiC and TaC in the tool materials when cutting steel [5].

In a more recent development the rate of wear has been decreased still further by coating cemented carbide tools with thin layers of solution resistant materials. The substances used for the coatings are TiC, TiN or HfN, which are deposited on the surface of the tools as layers 5 to 10 μm thick by a process known as chemical vapour deposition ("CVD"). These carbides and nitrides are too brittle to be used by themselves as cutting tools, but as thin layers with very fine grain size supported by the tougher cemented carbide substrate they withstand well the stresses of machine shop operations, and they are less readily dissolved in the steel flowing over the tool surfaces. With these coated tools tool life may be prolonged by a factor of 1.5 to 3 times compared with uncoated cemented carbides, or cutting speeds 30 to 60 m min^{-1} higher can be used for the same tool life when cutting steel.

Even more resistant to solution in steel at high temperatures is aluminium oxide. Al_2O_3 (alumina) has a high yield stress at high temperature and tool tips are sintered or hot pressed from alumina powder. Al_2O_3 tool tips (known as "ceramic tools") of high density and fine grain size can be used for cutting steel and cast iron at speeds as high as 600-700 m min^{-1} with very low rates of wear. These tools, however, are lacking in toughness and their use in cutting steel is restricted for this reason to a very small proportion of machine shop operations on steel. They can be more widely used for cutting cast iron at very high speed. Recently cemented carbide tools have been put on the market with thin coatings of Al_2O_3 deposited by CVD and these are being assessed for the cutting of both steel and cast iron.

Reaction between tool and work material also limits the speeds used when cutting steel and nickel-based alloys with diamond tools. Cubic boron nitride is a synthetic material made by the same type of ultra high pressure process used for the production of synthetic diamond. It has a similar structure to diamond and, although it is less hard, it can be used in cutting steels and nickel-based alloys at higher speeds because it reacts with them less readily than does diamond at elevated temperatures [6].

Thus, under the seizure conditions which prevail in many metal cutting operations, the life of the cutting tools is often controlled by processes of diffusion and interaction between tool and material at high interface temperatures when cutting materials of high melting point at high speeds.

19.4.4 Attrition

If, under conditions of seizure, the temperature-dependent wear processes of shearing and diffusion were the only ones responsible for changing the tool shape, tool life might be expected to be almost infinite at low cutting speed. At low speeds, however, tools are frequently worn by another mechanism which can be called "attrition". Sections through the edge of tools used at low speeds often show that the tool has been worn by a mechanism involving breaking away from the tool surface of fragments of microscopic size - as shown for example in Fig.17. for a high speed steel tool. Such a wear mechanism has been observed with almost all classes of tool material after cutting at low speed, and as the speed is raised it becomes less important. In many cases both diffusion and attrition wear are observed on the same worn surfaces.

Attrition wear seems to be most severe when the machine tool lacks rigidity, when vibration occurs or when there are pronounced irregularities in the work material. It involves an intermittent action in which small fragments of the tool are torn away to leave characteristically rough worn surfaces. High speed steel tools are more resistant to attrition than are cemented carbides and the life of high speed steel tools is often longer at low cutting speeds than that of cemented carbides for this reason. Twist drills for example are most commonly made from high speed steels, not only because they are cheaper but because, in many applications the life is longer than that of cemented carbide drills and performance is more consistent. When cemented carbide tools are used

Fig.17 Section through cutting edge of high speed steel tool after cutting steel at 20m min^{-1}. Worn surface characteristic of attrition wear

in the low speed range where attrition wear is dominant, the rate of wear is very dependent on the carbide grain size. Fine grained cemented carbides are much more resistant to attrition wear than coarse grained ones, and WC-Co alloys are more resistant than the steel cutting carbide grades and are often used to cut steel at low speed for this reason.

Fig.13 shows diagrammatically those parts of a turning tool where seizure and sliding are most likely to occur during cutting. Rather frequently more rapid wear is observed in the sliding regions than at the seized surfaces. For example, Fig.18 shows a deep tongue of wear on a tool in the sliding wear region where the outer edge of the chip crossed the cutting edge of the tool. The wear rate was many times higher at this position than in the adjacent seized region. High rates of wear are often observed also at the nose of the tool in turning operations, where the other edge of the chip crosses the cutting edge.

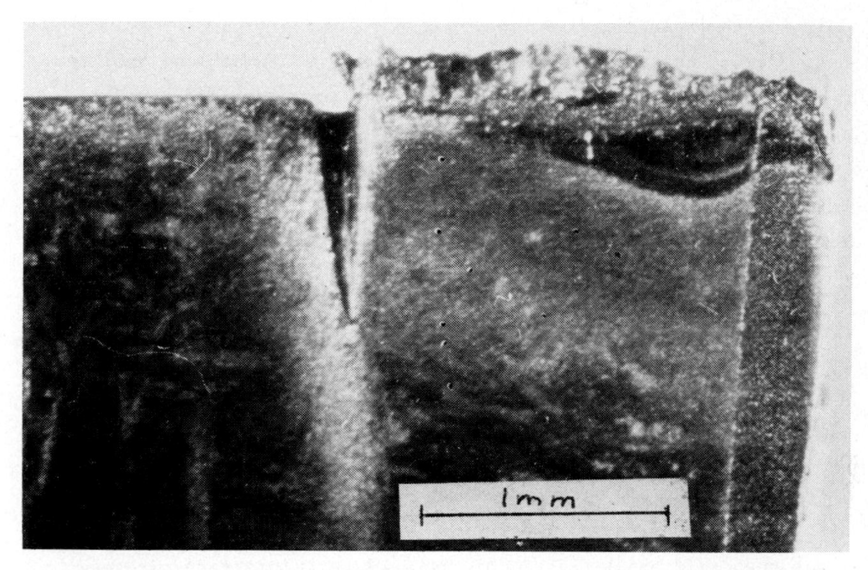

Fig.18 Clearance face of tool used to cut low carbon steel. Shows adherent work material and deeply worn groove at outside of cut, characteristic of wear by sliding action

Such increased wear rate in the sliding regions does not always occur, and the conditions influencing the severity of wear have not yet been studied in sufficient detail. It seems most probable that the wear in the sliding regions involves reaction between the tool and work surfaces and oxygen of the atmosphere, which has access to this part of the interface. One possible wear mechanism is the formation of oxide layers on the tool and the removal of these periodically by the stick-slip action of the sliding chip (Fig.11). The worn surfaces in the sliding regions are normally very smooth and this sort of wear on carbide tools may be almost as rapid as on steel tools, so that abrasion by hard particles, although possible in this region, is probably not the main mechanism of wear in most operations. The rate of sliding wear may be influenced by the use of cutting lubricants.

19.5 COOLANTS AND LUBRICANTS

A discussion of tribology in metal cutting would not be complete without considering the influence of coolants and lubricants. The cutting tool is often flooded with a fluid described either as a coolant or as a lubricant. Two main types are used - water-based fluids containing oil and other additives in suspension or solution and mineral oils with or without extreme pressure additives (mainly sulphur and chlorine containing substances). The function of the water-based fluids is mainly as a coolant with the oil present mainly to prevent corrosion, while the oil based fluids have relatively poor cooling capacity and are used mainly for their lubricant action.

Coolants are often necessary to reduce the temperature of machine, tool and workpiece in order to promote efficient operation and to maintain dimensional tolerances. Practical experience shows that a strong flow of coolant can act to increase tool life when cutting at high speeds, or to permit the use of higher cutting speeds. The direct influence of a coolant on the maximum temperature generated at the tool/work interface is usually rather slight [7]. The temperature is generated in the flow-zone at the rake face (Fig.10). Coolant cannot prevent the generation of heat at this surface and can act only to steepen the temperature gradients and reduce the volume of tool material heated to high temperature, but it can have little influence on the maximum temperature on the rake face. Acting near the cutting edge the coolant can be more effective and the major cooling effect is probably that of reducing the temperature near the edge, thus increasing the yield strength of the tool to prevent local deformation, and also reducing the rate of diffusion wear on the flank of the tool.

From what has been said about seizure at the tool/work interface, it seems unlikely that any lubricant, in either gaseous or liquid form, can penetrate to that part of the interface where seizure occurs. Lubricants can, however, act effectively in the peripheral regions where sliding occurs at the interface. By penetrating from the peripheral regions they may be able to restrict the area of complete seizure and thus to reduce the forces acting on the tool. Force measurements have shown reductions in both cutting force and feed force caused by the use of coolants and lubricants at low cutting speed. Many tests have indicated that the influence of lubricants is greatest at speeds below 30 m min^{-1} while they have very little lubricating effect over 60 m min^{-1}. The most effective lubricants in metal cutting are those with extreme pressure additives, suggesting that successful lubrication involves the formation of easily sheared surface layers when the lubricants come into contact with freshly exposed metal surfaces on the work material.

When a built-up edge is formed, coolants and lubricants are often effective in greatly reducing its size. A large built-up edge is often responsible for very poor surface finish and one of the most important functions of a cutting lubricant is to improve the surface where this is a requirement of the surface being machined. Often, in this respect, water-based lubricants, and even water itself, are effective in reducing the size of the built-up edge.

For many cutting operations lubricants are essential. At high speeds tool life may be improved by coolant action, but the use of lubricants at lower speeds to improve surface finish is not always effective in reducing the rate of wear. In fact tool wear rate is often increased by the action of lubricants in the sliding regions where it can penetrate. This acceleration of wear may occur with both steel and cemented carbide tools. High rate of wear in regions of sliding at the interface has already been discussed. In many engineering mechanisms a major function of lubricants is to prevent seizure between moving parts. In the case of metal cutting operations, the elimination of seizure is not the objective of the use of lubricants. The elimination of seizure in many cases could result in a disastrous increase in the rate of tool wear. In most cutting operations seizure between tool and work material is a normal and desirable condition, rather than a hazard to be avoided. The main useful functions of cutting fluids are to reduce temperature and thus increase cutting efficiency, to reduce cutting forces, to improve surface finish, and to help clear chips away from the cutting tool.

There is one way in which seizure between tool and work material can be modified, if not eliminated, to the advantage of certain cutting operations, and this is to include within the work material a phase which may interpose itself between tool and work during the cutting operation. Such a substance applied from within the work material can renew interfacial layers as they are swept away by the flow of the work material over the tool. Under certain conditions such substances can be described as performing the functions of an internal lubricant and are generally known as free-machining additives.

Manganese sulphide in steel may act in this way and, when the steel cutting grades of carbide are used, the sulphides from the steel will often form an intermediate layer at the seized tool/work interface. Certain calcium-aluminium silicates in steel, which like manganese sulphide are plastically deformed when the steel is sheared, also function in the same way as the sulphides. Silicate layers are formed on steel-cutting carbide tools at the interface when cutting at high speeds and are often extremely effective in reducing tool wear rate and permitting the use of higher cutting speeds.

Sulphides in copper also tend to form a thin layer at the tool/work interface where their main function seems to be to reduce the area of contact on the tool rake face and thus greatly to reduce the cutting forces which are extremely high when cutting high-conductivity copper. The addition of lead to brass results in an accumulation of lead at the brass/tool interface under many conditions of cutting. The greatest benefit which the lead confers is the breaking up of the brass chips into small fragments which are easily disposed of in high speed automatic machining.

Of all the aspects of machining the functions of cutting fluids as coolants and lubricants have probably received the least attention from scientific research. There is much to be learned concerning the ways in which they act to achieve the results for which they are used by practical machinists.

REFERENCES

1 Zorev,N.N., International Research in Production Engineering (1963), A.S.M.E., Pittsburgh, p.42.
2 Trent,E.M., I.S.I. Report No.94., 1967, p.11.
3 Williams,J.E. and Rollason,E.C., J., Inst., Met., (1970), 98, 144.
4 Boothroyd,G., "Fundamentals of Metal Machining and Machine Tools", McGraw-Hill (1975).
5 Trent,E.M., "Metal Cutting", 1977, Butterworths.
6 Hibbs,L.E. and Wentorf,R.E. Jr., 8th Plansee Seminar (1974), Paper No.42.
7 Smart,E.F. and Trent,E.M., Proc., 15th Int., Conf., M.T.D.R., (1975) 187.

20 ROLE OF LABORATORY TEST MACHINES

PROFESSOR F.T. BARWELL, M.H. JONES
Department of Mechanical Engineering, University College of Swansea.

20.1 THE EXPERIMENTAL METHOD

The establishment of facts and the understanding of relationships between physical phenomena can only be based on experience, the experimental method should therefore consist of structuring experience so as to produce unambiguous answers to certain questions. For structured experience or experiment to have meaning, it must reproduce the circumstances surrounding the occurrence of the phenomena under study. Otherwise the results, though perhaps interesting, will be irrelevant to the purpose of the investigation.

Tribological investigations are best carried out under service conditions on full scale apparatus but this is seldom possible. Information may often be required in advance of the construction of the machine involved, measurements may not be possible in the operational environment (in space for example) and full scale testing may be too costly especially when tests to destruction are required. Laboratory test methods have therefore been evolved which serve a number of vital functions in relation to the practice of engineering.

The testing of lubricants by measurement of their physical and chemical properties is well developed (See Standard Methods of Testing Petroleum and its Products, Institute of Petroleum and ASTM) but the complexity of the requirements of many engineering applications is such that satisfactory performance cannot always be predicted from such tests and it is necessary to simulate service conditions. Thus, for a product to be approved for supply to the U.S. Army for lubrication of I.C. engines, it has to be submitted to a series of engine tests and has to satisfy certain specified criteria. Once approval has been given, bulk supplies may be accepted on the basis of sufficient physical and chemical testing to ensure consistency of constitution.

Many industrial processes involve a considerable quantity of raw material and full scale testing may be inconvenient or expensive. It may not always be possible to obtain access to the interacting surfaces of interest and it may be desired to explore the effect of a variable over a greater range than would be possible with existing machinery. In these circumstances, the construction of special laboratory machines may be essential.

Tribological situations can generally be reduced to the consideration of interacting surfaces of given composition and shape with appropriate loading and relative motion together with lubricant and environment, and it is therefore attractive to provide special test machines which present no apparent resemblance to any practical machine, but which reproduce sufficiently accurately the tribological conditions involved. A number of machines are available commercially which purport to do this and provided care is taken to ensure the relevance of the test conditions to the application, they may produce much useful information.

Finally where an effort is being made to expand knowledge of the fundamentals underlying tribological action, special equipment is necessary to extend the range of observation and to isolate particular variables. It is often found that apparatus devised for fundamental research is particularly useful for applied investigations.

20.2 LUBRICANT TYPE APPROVAL TESTING MACHINES

20.2.1 Engine Tests

A great deal of lubricant testing is carried out in order to satisfy the relevant specifications for engine oil, typical examples are DEF-2101-D in the United Kingdom and MIL-L-46152 in the United States.

TABLE 20.1 Summary of Typical Engine Tests

Specification	Engine Test(s)	Properties under test
DEF-2101-D	Petter W1 - gasoline	Oil oxidation, bearing corrosion, lacquer formation
	Petter AV1 - diesel	Detergency and high temperature stability
	Oldsmobile V-8 sequence II B - gasoline	Low temperature rusting and deposits
MIL-L-46152	Ford V-8 sequence VC - gasoline	Build up of deposits due to intermittent low temperature operation
	Oldsmobile V-8 sequence III C - gasoline	High temperature oxidation
	CRC L-38 (CLR) - gasoline	Bearing corrosion and shear stability of multigrades
	Caterpillar TH - diesel	Ring sticking wear and accumulation of deposits

Fig.1 Petter Av1 Rig

Fig.2 Caterpillar rig

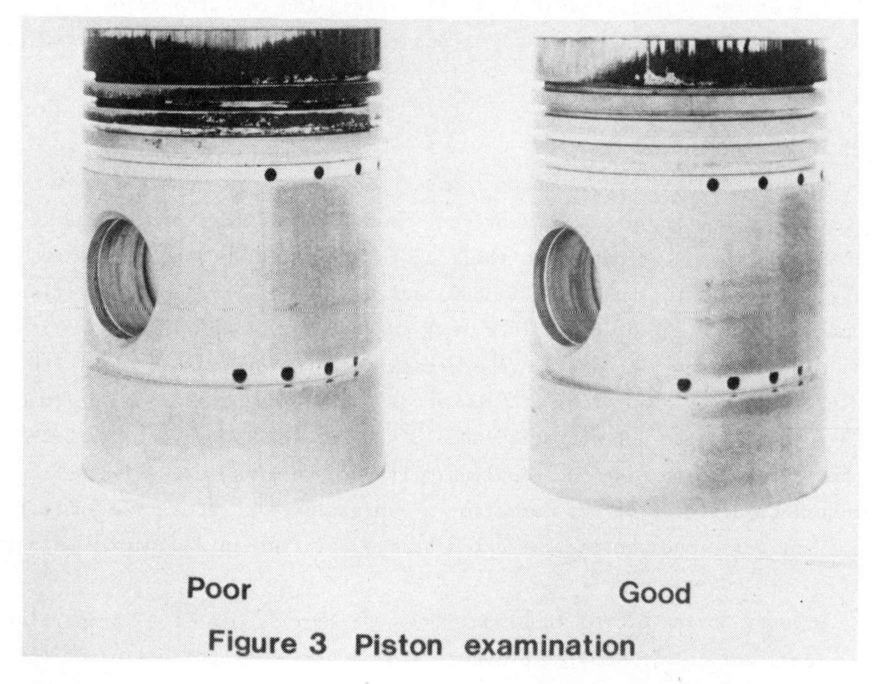

Fig.3 Piston examination

The main criterion of acceptance of a lubricant which has been subjected to type approval testing is the condition of the piston after test. This is rated by a panel of experts who assess such factors as the freedom of the rings, a minimum of scuffing of the piston crown and a minimal amount of carbon in the top ring groove. There should also be no carbon deposit in the lower ring grooves and the piston skirt should be entirely free of lacquer.

As will have been gathered from previous chapters, the action of lubricant additives is complex and there may be significant interactions of an undesirable nature. For example, some additives lead to corrosion of copper-lead bearings and in petrol-engines leaded-fuels may affect the nature of piston deposits ("grey paint"). Additional tests are therefore required which were originally carried out for 36 hours in a four cylinder Chevrolet engine. The copper-lead bearings of the test engine were weighed and examined in order to assess corrosion and the pistons examined to assess any deposit. A Petter W1. spark-ignition engine has been substituted for the Chevrolet engine in the U.K. and an engine, the CRC L-38, has been specially developed for lubricant approval tests in the U.S.A.

Although it may appear at first sight that the use of an actual engine is a straightforward means for assessing the quality of an oil, the test conditions are very critical. The composition of the fuel, the timing of the valves, the conditions of the injection system all may affect the performance of a lubricant. In particular, the exhaust arrangements can markedly affect results [1].

20.2.2 Gear Tests

The introduction of hypoid gears in back axles was made possible only by the application of extreme pressure lubricants and the maintenance of adequate lubricant quality is essential to their successful functioning. Tests to qualify lubricants to the U.K. (CS 3000) and U.S.A. (MIL-L-2105C) specifications are carried out on actual gears in a test arrangement of the type shown in Fig.4. The load carrying and extreme pressure characteristics of gear lubricants in axles under conditions of high-speed, low-torque operation, followed by low-speed, high-torque operation should satisfactorily prevent the occurrence of gear tooth ridging, rippling, pitting, welding, excessive wear, or other surface distress or the formation of objectionable deposits when tested on both untreated and phosphate treated gear assemblies in accordance with the specified procedures.

The primary action of the lubricant additive is to confer E.P. properties on the lubricant. This implies prevention of scuffing by activation of the chemically active additives by the occurrence of instantaneous temperature

Fig.4 Low speed high torque test rig hypoid gear and dynamometer installation

Fig.5 Ridging type gear failure

Rippling

Fig. 5a

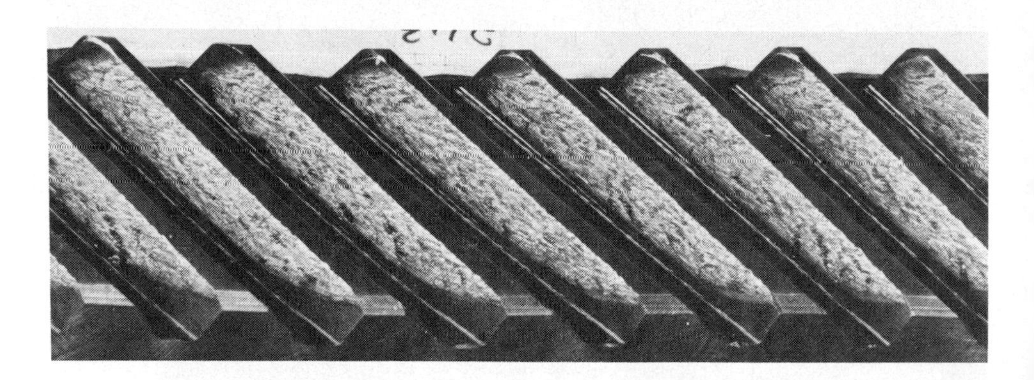

Scuffing

Scoring

Fig.5 Examples of gear damage

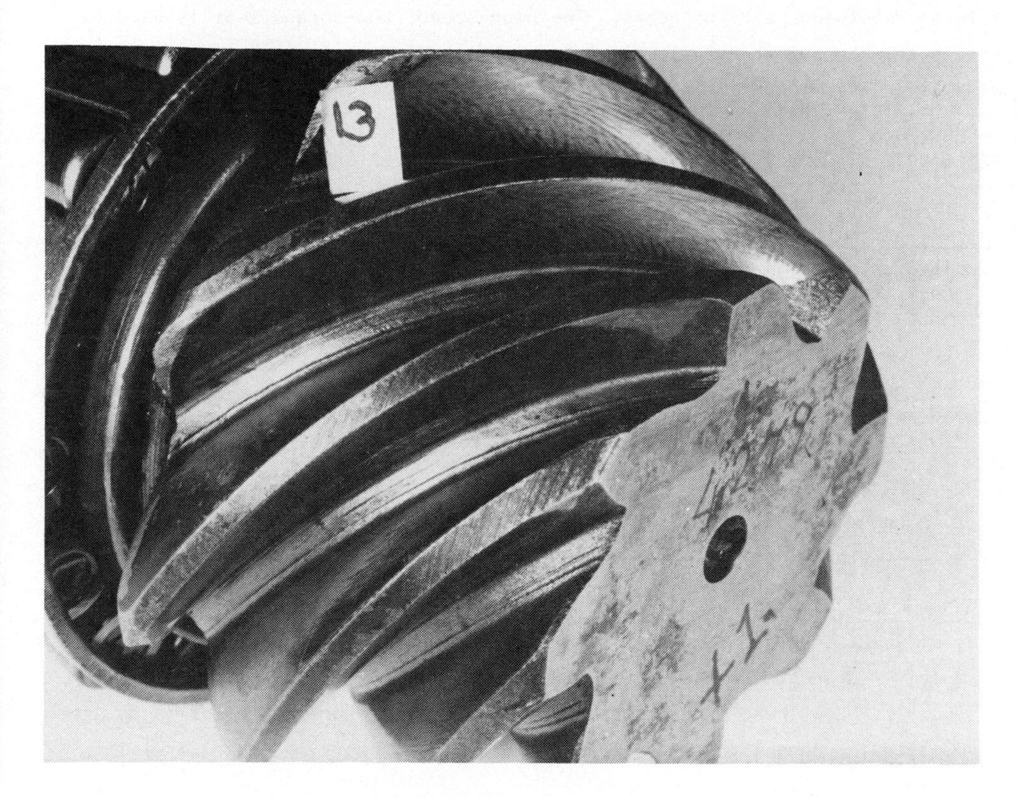

Fig.6 Rippling type gear failure

Fig.7 I.A.E. high speed gear rig

"flashes" between pairs of gears. The High-speed, Low-torque test is used to assess this behaviour using equipment which embodies an automobile rear-axle assembly. Evidence of failure is provided by the initiation of scuffing.

TABLE 20.2 Summary of the Full Axle Test of MIL-L-2105C and CS 3000

Test	Equipment	Requirements
Moisture Corrosion 7 days	CRC L-33 Federal method 5326	Maximum of 5% vapour phase corrosion of cover plate, no corrosion of functional parts.
High Speed Shock Load Test	CRC L-42 Federal method 6507 IP.234/69	Score prevention equal to or better than RG 110 or CRC 10/90 reference oil.
Low Speed High Torque Test	CRC L-37 Federal method 6506 IP 232 Proc. B.	No gear surface distress or deposit formation allowable.

When chemically active additives are used, there is always the risk that they will act under conditions and in a manner which is not desired. For example, "ridging" type failure may occur as in Fig.5 and "rippling" as in Fig.6. The "Low-speed High-torque" test is used to ensure that lubricants are formulated to prevent these types of failures from occurring under service conditions.

Lubricants for use in spur gears are usually tested for their anti-scuffing properties in machines which embody means for power circulation. One of the original machines in this class was designed by Mansion and is known as the I.A.E. machine. The gears under test are geared together, supported on two parallel shafts, and loaded by the "locking-up" of a predetermined torsional strain in the shafts. The gear sets are assembled in a "back-to-back" relationship so that power circulates continuously and the driving motor is only required to make up the losses occurring within the system.

Another machine which employs the power circulating principle is the "Ryder" machine. In this design one set of gears acts as test section and the other is of the single-helical configuration. Load is then applied by means of an axial force which is converted in the helical-gears to a torque tending to twist the shaft thereby loading the test gears.

The F.Z.G. Machine (Fig.8) also employs involute gears on the power-circulating principle.

All the above machines require gears as test pieces and, because each test must be taken to destruction, the test procedures are very expensive. There is a strong tendency therefore to employ disc-machines wherein the relative amount

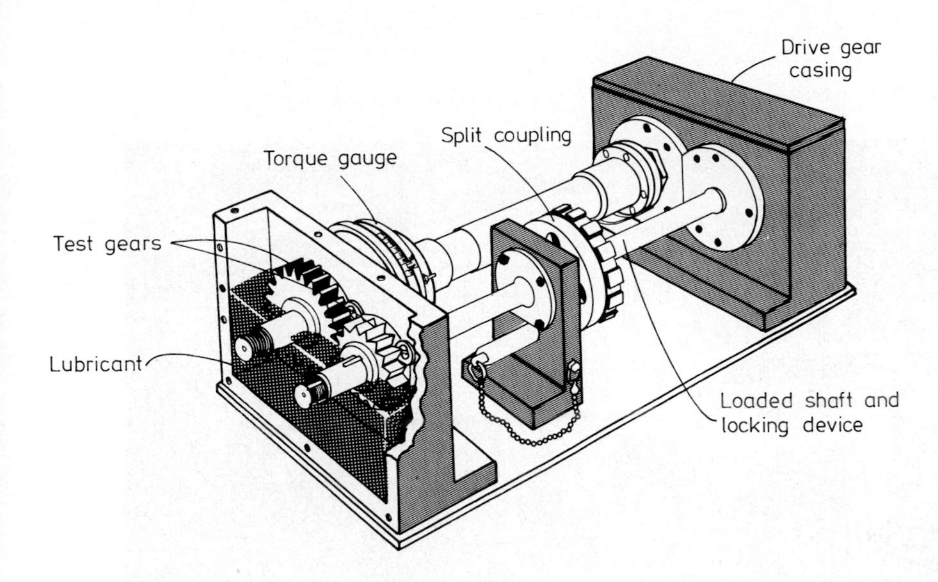

Fig.8 FZG Gear oil test

of rolling and sliding can be adjusted to correspond with the events in the meshing cycle of a spur gear.

20.2.3 Simulation of Industrial Situations

Many industrial processes requiring lubrication do not correspond to the conditions of the type approval test of lubricants, and it is necessary to investigate the existing tribological situation taking into account the nature of the interacting materials, the industrial environment and the applied forces and motions.

In this circumstance, it is frequently necessary to construct a special machine to reproduce within the laboratory the specific conditions surrounding the industrial problem.

An example of such a machine is shown in Fig.9 [2]. This consists of a machine designed and constructed at the Swansea Tribology Centre to simulate the conditions determining the action of the sideguides of rolling mills. The sideguides, normally made of bronze, are acted on by the edges of the steel strip undergoing cold-rolling which they restrain from unwanted sideways movement. They are subjected to much wear (Fig.9c), and moreover, the non-ferrous material transferred from the guide to the steel was objectionable to the customers of the material.

Accordingly a guide member was fitted with strain-gauges and thermistors and subjected to normal operation so that the forces acting on the guide could be determined together with the operational temperature. The machine illustrated

Fig.9(a) Rolling Mill with side guides

in Fig.9(b) was then designed to embody a continuous steel belt to represent the cold strip being rolled in the full scale apparatus and a member which resembled a side guide and which was forced against the edge of the strip with a force which was determined on the basis of the strain-gauge measurements.

After a series of tests had confirmed that the wear mode obtained in the laboratory corresponded exactly with that observed in practice, a series of alternative materials was investigated. It was concluded that the substitution of nodular cast-iron for the bronze presented several advantages as follows:

(a) The rate of wear was reduced

(b) The material was less expensive to procure

(c) The contamination of the product with non-ferrous material was avoided.

This example demonstrates the possibility of study of industrial problems using laboratory methods.

20.3 ANALYSIS INTO SYSTEM ELEMENTS AS THE BASIS FOR SELECTION OF LABORATORY TESTS

Tribological situations encountered in different machines may present certain features in common insofar as they all embody surfaces in relative motion.

Fig.9(b) Experimental Rig for Simulating Side Guide Wear

Fig.9(c) Simulated worn side guide

However, it will be apparent that the nature and configuration of surfaces may vary widely as between one machine and another as will the modes of force and motion. Therefore there can be no single laboratory rig which can represent all tribological situations. However, it is also true that many machines possess features which are sufficiently alike to justify the construction of test machines for the purpose of evaluating lubricants or materials of construction for use therein.

Great care is necessary however to analyse precisely the nature of the tribological situations involved in order to select the correct conditions to be applied in the test machine. A scientific attitude is necessary and meaningless jargon such as "film-strength" or "lubricity" should be avoided. (An example of allowing jargon to take over is given by the investigator who subjected to an E.P. test a lubricant required for a process imposing high pressure at very low speed).

The first step in the analytical process must be to characterize the machine element under study with respect to its position in the kinematic chain; is it a "Higher Pair" or a "Lower Pair"? Of the lower pairs, the sliding pair, Fig. 10(a) is encountered on machine slideways and the "Revolute Pair" Fig.10(b) is the basis of the majority of bearings. Apart from the "screw" pair (which is a combination of the sliding and revolute pairs) all other pairs are higher pairs and are so characterised because they must accommodate relative motion which is partly sliding and partly turning.

From the point of view of Tribology, the important distinction between lower and higher pairs is that the former allow contact to be made throughout the full extent of a surface, whereas the latter only allow "point" or "line" contact. Some common examples of higher pairs are shown in Fig.10(d-g). Thus lower pairs are said to be "conformal" and higher pairs are "counterformal". It will be apparent that the proformal difference between dispersed and concentrated contacts will be reflected in the design of the interacting components, in their materials of construction and in the properties required of any lubricants applied. [3]. Fig.11 illustrates the classification of some common machine elements into conformal and counterformal configurations as well as indicating the mode of damage characteristic of each application.

To be successful, laboratory testing machines must be simple and must employ easily manufactured test pieces. They must however reproduce the conditions of thermal and stress intensity to which it is anticipated that the lubricated system will be subjected in service. The first broad classification must be into lower and higher pairs.

There are relatively few simple machines available commercially for testing lubricants in lower pairs because, as far as lubrication proper is concerned,

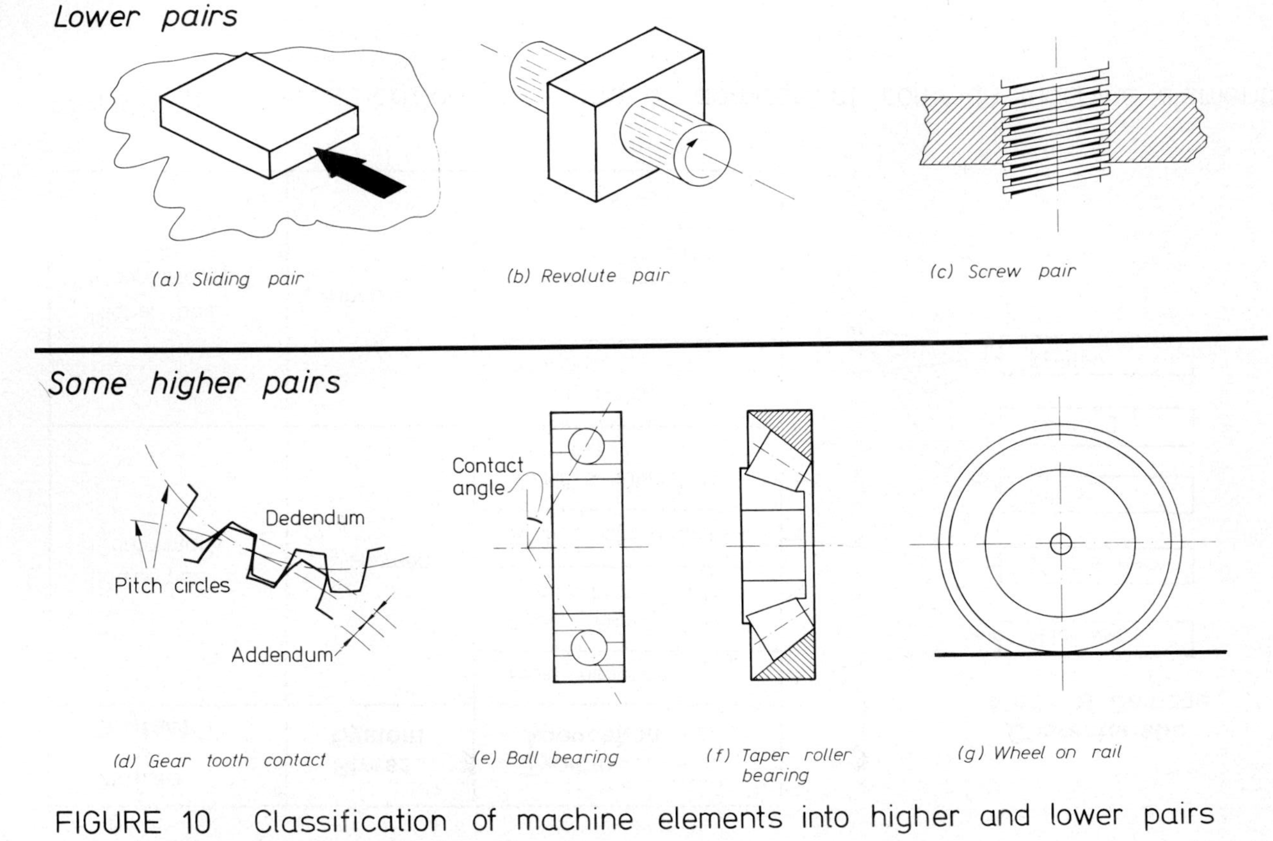

FIGURE 10 Classification of machine elements into higher and lower pairs

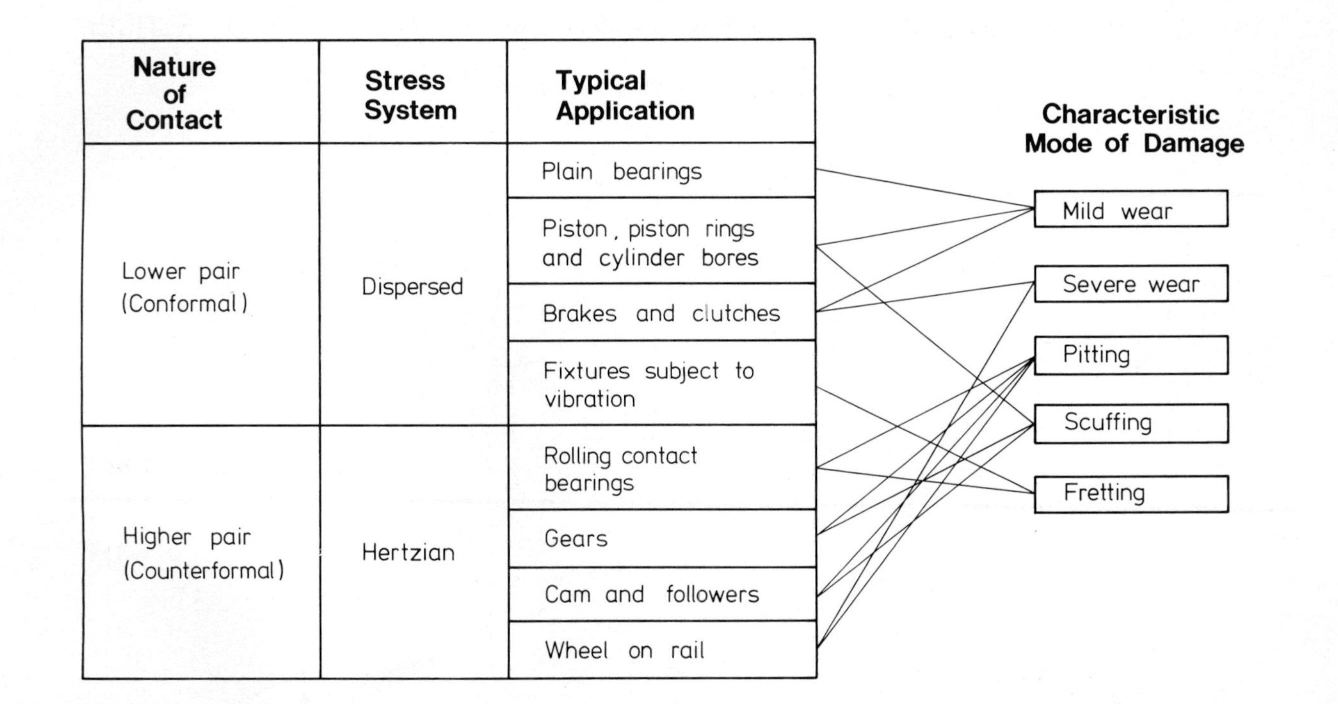

FIGURE 11 Classification of mode of damage of common machine elements

this is usually hydrodynamic and the only property required of the lubricant is its viscosity which can easily be measured by physical methods. The important properties of the lubricant in these applications are related to such factors as detergency or anti-corrosion and full scale tests are necessary to safeguard against undesirable interactions.

For testing material-lubricant combinations under conditions of pure sliding, a very convenient arrangement consists of three cylindrical pins which are held in an upper member with their axes parallel and the assembly is loaded against the side of an annular test piece which is rotated about its generating axis. The "Cygnus" machine is of this type. Both the Almen (Fig.13) and the Falex (Fig.13) machines are basically line-contact machines but a certain degree of conformity usually arises from wear of the test pieces. These two machine types present the common feature that two identical test pieces are forced against a rotating member from diametrically opposed directions, thus balancing the force on the spindle. The Almen test pieces form segments of a cylinder having an internal radius some 1.5×10^{-4}m (0.006 in) greater than that of the spindle which is 0.35 mm (0.25 in) in diameter. The corresponding test pieces of the "Falex" are "V" shaped. Loading on the "Almen" is hydraulic and the "Falex" mechanical. Results are difficult to evaluate because of the loading methods usually employed [4].

As regards higher pairs, there is a wide variety of machines which apply sliding under counterformal conditions. These have generally been developed to assess the quality of gear lubricants with particular reference to the inhibition of scuffing. One of the most successful of these has been the Shell (Stanhope-Seta) four ball machine (Fig.14). The problem of obtaining test pieces which are of uniform finish and metallurgical condition has been solved by using balls selected in batches from commercial production. Three of these balls are clamped to form a nest into which the fourth ball (held in a chuck) is forced and rotated. The balls are 12.7 mm (0.5 in) in diameter and the rotational speed of the spindle to which the chuck is attached is 1450-1500 rev min^{-1}. Load may be applied in increments up to 800 kg. The machine may be used in several procedures, the most common of which is the "Wear-scar diameter method", [5]. A number of tests (usually 20) are each made on a fresh set of balls, load being increased between each test the duration of which is one minute. The diameter of the wear scar apparent after each test is measured and plotted against load on logarithmic paper. During the tests at the lower loads wear is negligible but sufficient rubbing action takes place to leave a measurable mark which when measured and plotted, gives rise to a straight line which closely parallels that obtained by calculating the diameter of the Hertzian contact between the loaded balls. With most lubricants a load is reached at which a sharp rise in

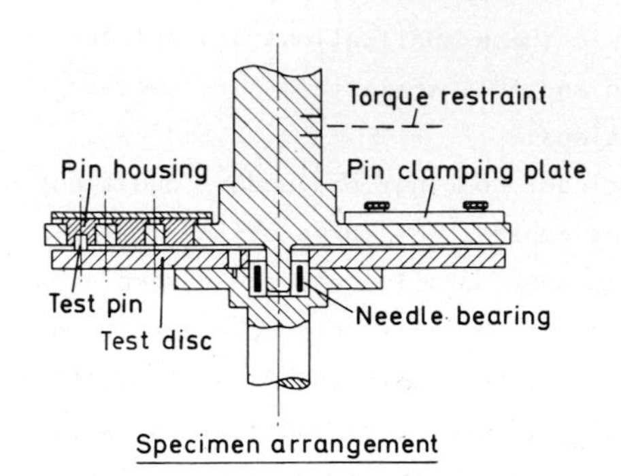

FIG.12 CYGNUS FRICTION AND WEAR TEST MACHINE

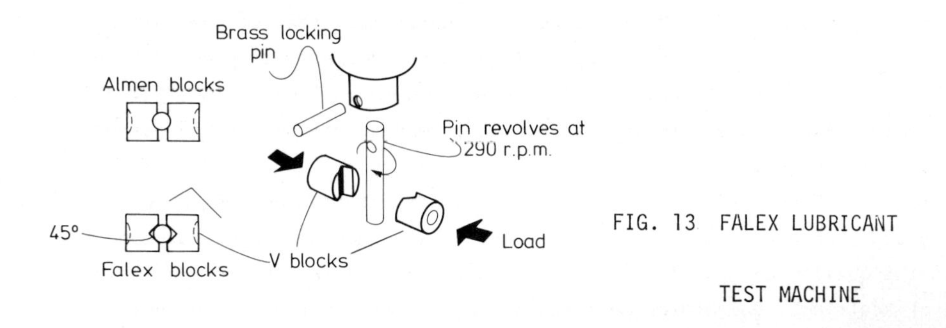

Exploded view of pin and V blocks

FIG. 13 FALEX LUBRICANT TEST MACHINE

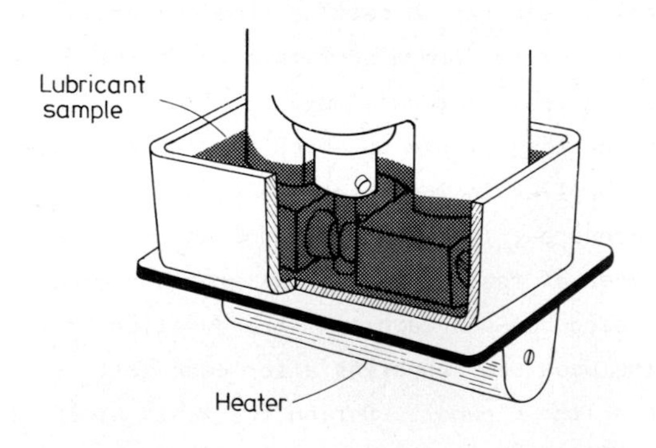

Cutaway view throughout sample pan

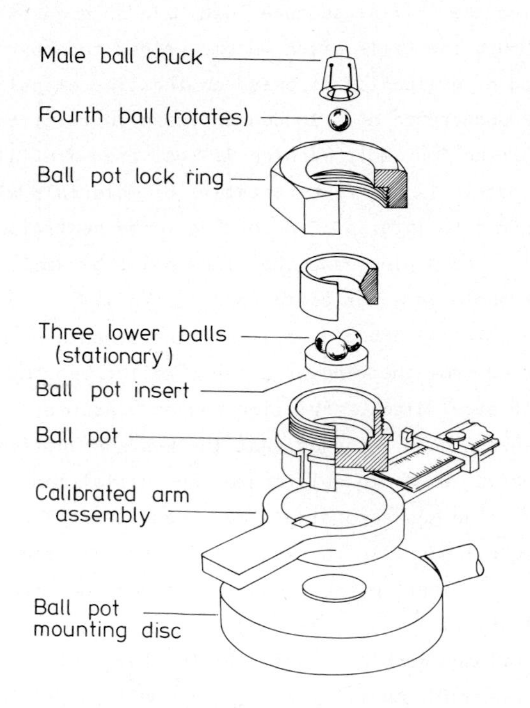

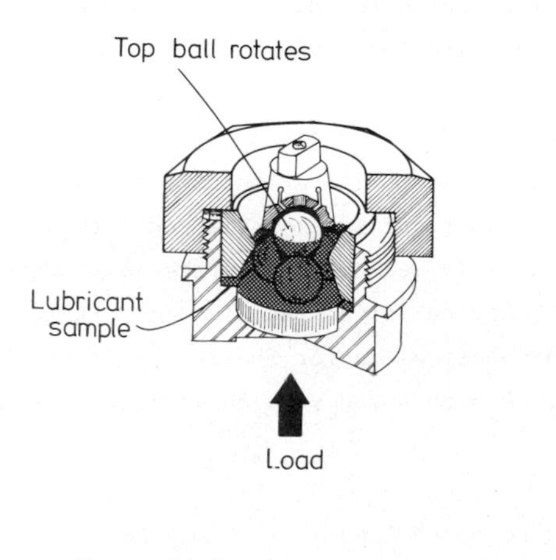

Ball and flats

Four ball machine

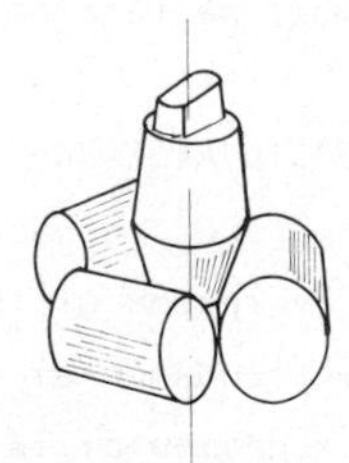

Cone and cylinder

FIGURE 14

diameter occurs. This value is usually sufficiently definite to characterise a lubricant and is called the initial seizure load. In some cases however, recovery may occur although the balls often become welded together at the higher loads. Another method of evaluation is based on the time elapsing under a constant load before the occurrence of a sudden increase in friction.

One disadvantage of the four ball machine is that the material of the test pieces, ball bearing steel, is not representative of materials with which the lubricant may be required to interact. Accordingly the central ball may be replaced by a conically ended piece and the fixed balls by small cylinders arranged to form a triangle in a horizontal plane, Fig.14.

The Timkin Machine (Fig.15) embodies a line rather than a point contact. The rotating element, formed from the ring of a taper roller bearing acts on a stationary rectangular steel block. Friction can be measured.

Most testing machines are so arranged that the wearing surfaces are in continuous or repeated contact so as to obliterate the initial manifestations of surface failure. A machine was accordingly designed at M.E.R.L. (now N.E.L.) which consisted of two cylinders which, in addition to rotation, could be traversed one relative to the other so that the contact zone was made to continually embody fresh material (Fig.16).

In all the aforementioned machines, only sliding takes place whereas in many mechanisms, involute gears for example, rolling as well as sliding can take place. Disc machines such as the "Amsler" (also the Merritt and S.A.E. machines) have discs which are loaded edgewise to provide various combinations of rolling and sliding (Fig.17).

Pitting type failure is usually associated with rolling contact and can be simulated in the laboratory [6] using a simple modification of the four ball machine (Fig.18). Instead of the three lower balls being clamped into place they are allowed to rotate within a specially designed ball race [6]. Where it is desired to investigate the pitting behaviour of a special steel, it is possible to substitute a conically ended test piece for the central ball although the three free balls must be retained. A high speed version is also available.

20.4 EQUIPMENT FOR BASIC RESEARCH

The investigations pursued in basic research laboratories may appear to be unrelated to practice by reason of the artificial conditions often imposed in order to elucidate some fundamental relationship. The closing of the communication gap between the fundamental investigator and the engineer whose practice lies in manufacturing industry for example, has been made easier by the development of the practice of modelling complex systems using computers. Thus in Fig.19 the rectangle marked A represents a real machine-element subjected to the environmental conditions and loading within the industrial environment. Its

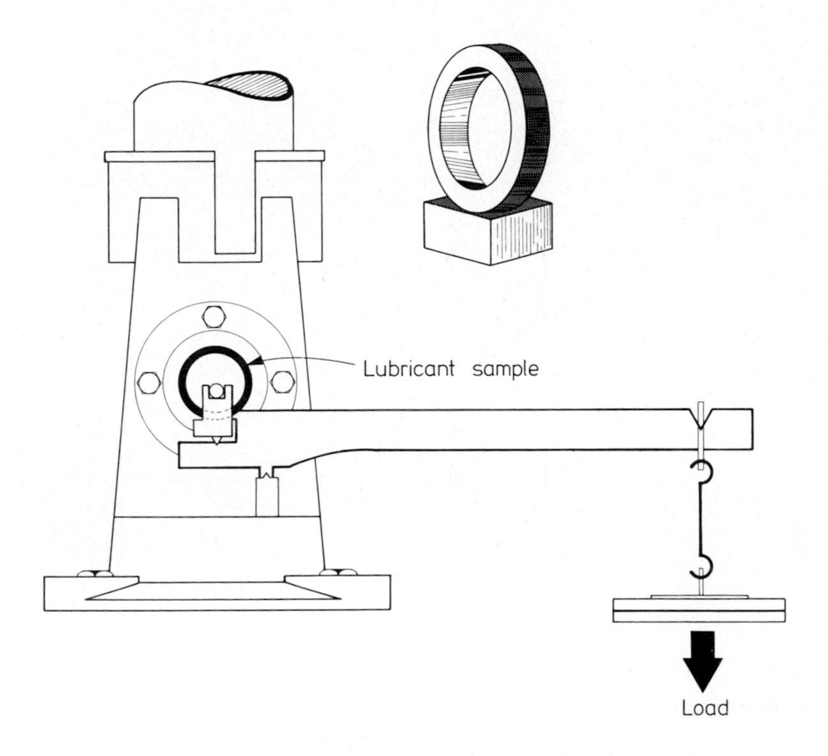

Fig.15 Timken test

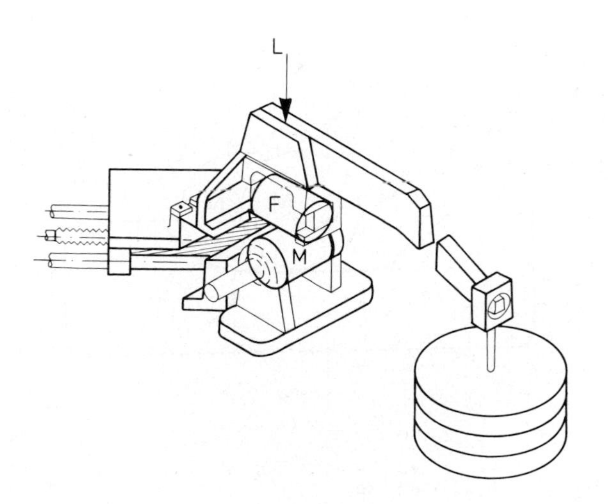

Fig.16 N.E.L. crossed cylinder machine

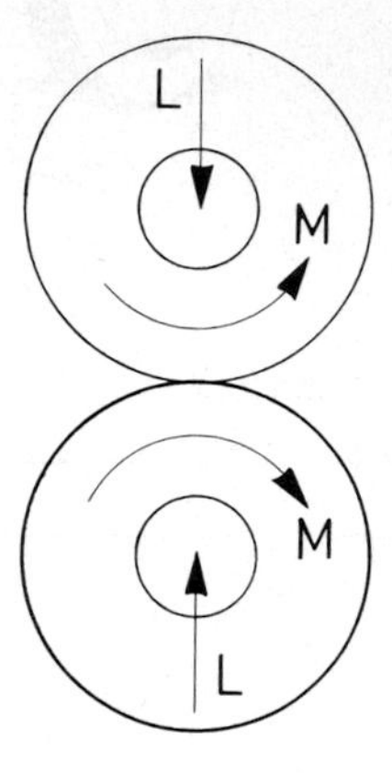

Fig.17 Amsler machine

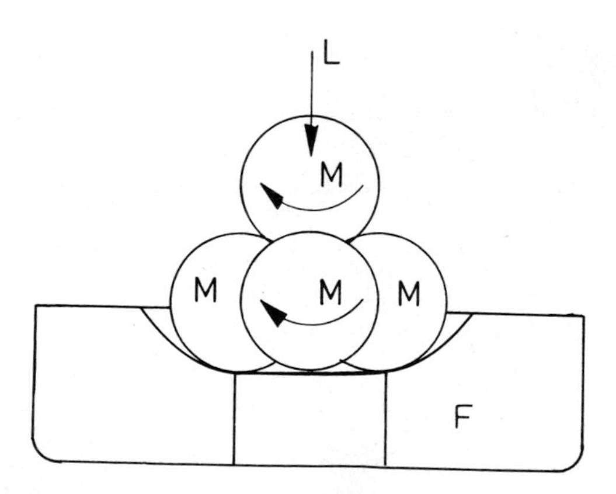

Fig.18 Rolling four ball machine

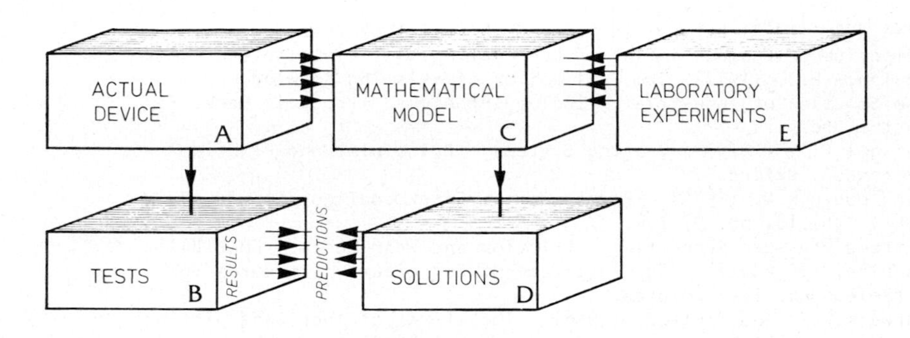

Fig.19 The role of the computer model in relating basic research to engineering practice

response to these conditions can be determined by a methodical series of tests, results of which are represented by rectangle B. Assuming that not all is well or that some development in product or production capacity is required, it may be necessary to formulate some predictions of behaviour lying outside the range of previous experience. The system can then be represented by a series of equations linked together to form a mathematical model of the machine as indicated by rectangle C. Specifications for applied conditions and loadings would be fed in and the computer would produce solutions to the equations represented by rectangle D. These solutions would provide predictions of the magnitude of output quantities which are represented by the arrows pointing to the left from D to B. The arrows pointing to the right from B to D represent the results of actual test. If there is agreement between measured outputs and calculated predictions, the model can be said to be complete. If there is divergence, the model must be refined and adjusted until acceptable agreement results. Once this agreement has been attained, the model may be used to predict the behaviour of an infinite variety of machine elements and applied conditions provided they lie within the range of equations embodied in the model.

Attempts to construct realistic mathematical models however, frequently reveal gaps in understanding of the physical system involved which can only be filled by carefully controlled laboratory experiments such as those which were necessary to elucidate the nature and operational relationships of elastohydrodynamic lubrication (as indicated at E). Such experiments usually form the basis of advanced test methods made necessary by developing practice.

REFERENCES

1 Cree,J.C. (1953) Caterpillar L1 and Chevrolet L4 test process, Symposium on Engine Testing of Lubricating Oil. Institute of Petroleum.

2 Roylance,B.J. (1977) The Application of Existing Knowledge the Solution of Industrial Tribology Problems. Proc. I. Mech. E. In the Press.

3 Barwell,F.T. (1979) 'Bearing Systems - Principles and Practice' Clarendon, Oxford.

4 Meckleburg,K.R. (1975) 'Forces in the Falex configuration'. Trans. ASLE., Vol.18, pp. 97-104.

5 Extreme Pressure Properties: Friction and Wear Tests: Four Ball Machine. IP 239/77. The Institute of Petroleum. Standards for Petroleum and its products.

6 Barwell,F.T. and Scott,D. (1956) 'The effect of lubricant pitting failure of ball bearings' Engineering, Vol.182, pp. 9-12.

GLOSSARY

Terms and Definitions

Abrasion - Wear by displacement of material caused by hard particles or hard protuberances.

Absolute Viscosity - see viscosity.

Additive - A material added to a lubricant for the purpose of imparting new properties or of enhancing existing properties.

Adhesive Wear - Wear by transference of material from one surface to another during relative motion, due to the process of solid-phase welding.

Anti-Wear Additive - An additive used to reduce wear.

Area of Contact - The area of contact between two solid surfaces is described in two ways.

(i) Apparent Area: the area of contact defined by the boundaries of the macroscopic interface of the bodies

(ii) Real Area: the sum of the local areas transmitting interfacial force directly between the bodies.

Asperities - The small scale irregularities on a surface.

B_{10} Life - see rating life.

Babbitt Metal - A non-ferrous bearing alloy, either tin or lead based consisting of various amounts of copper, antimony, tin and lead.

Base Stock (oil) - Refined petroleum oil used in the production of lubricants and other products. The base stock may be used alone or blended with other base stocks and/or additives.

Bearing - A support or guide by means of which a moving part is positioned with respect to the other parts of a mechanism.

Bearing Area - The projected bearing load carrying area when viewed in the direction of the load.

Beilby layer - An amorphous layer of deformed metal and oxide particles formed by polishing.

Blending - The process of mixing mineral oils to obtain desired viscous properties.

Boundary Lubrication - A condition of lubrication in which the friction and wear between two surfaces in relative motion are determined by the properties of the surfaces, and by the properties of the lubricant other than bulk viscosity.

Brinelling - Indentation of the surface of a solid body by repeated local impact or impacts, or by static overload.

Cavitation Erosion - Wear of a solid body moving relatively to a liquid in a region of collapsing vapour bubbles which cause local high impact pressure or temperatures.

Centre Line Average (CLA) - An English measure of surface topography representing the average departure of a line profile of the surface from the centre line.

Channeling - The tendency of a grease to form a channel by working down a bearing or distribution system, leaving shoulders to act as a reservoir and seal.

Clearance Ratio - In a bearing, the ratio of radial clearance to shaft radius.

Coefficient of Friction - The ratio obtained by dividing the tangential force resisting motion between two bodies by the normal force pressing these bodies together.

Composite Bearing Material - A solid material composed of a continuous or particulate solid lubricant phase dispensed throughout a load bearing matrix to provide continuous replenishment of solid lubricant films as wear occurs, and effective heat transfer from the friction surface.

Corrosion Inhibitor - Additives for protecting lubricated surfaces against chemical attack. They may be polar compounds wetting the metal surface preferentially, or they may absorb the water to form a water-in-oil emulsion - only the oil touches the metal. Some corrosion inhibitors combine chemically with the metal to give a non-reactive surface.

Corrosive Wear - A process in which chemical or electrochemical reaction with the environment predominates.

Cutting Fluid - A fluid applied to a cutting tool to assist in the cutting operation by cooling, lubricating or other means.

Detergent Additives - compounds which, when blended with lubricating oils, disperse the deterioration products from the fuel and lubricant, especially those formed under high temperature conditions, and, thus, minimise the formation of deposits liable to cause piston-ring sticking or other trouble.

Dispersant Additives - compounds which, when blended with lubricating oils, maintain the products of combustion from the fuel in a finely dispersed state, and thereby minimise sludge formation and filter blocking, particularly in gasoline engines operating under cold conditions.

Drop point - temperature at which a drop of grease or other petroleum product first detaches itself from the main bulk of material when a sample is steadily heated under prescribed conditions.

Duty Parameter - A dimensionless number which is used to evaluate the performance of bearings.

Dynamic Viscosity - see: Viscosity.

Eccentricity Ratio - In a bearing, the ratio of the eccentricity to the radial clearance.

Elasto-hydrodynamic Lubrication - A condition of lubrication in which the friction and film thickness between two bodies in relative motion are determined by the elastic properties of the bodies, in combination with the viscous properties of the lubricant at the prevailing pressure, temperature and rate of shear.

Embeddability - The ability of a bearing material to embed harmful foreign particles and reduce their tendency to cause scoring or abrasion.

Emulsion - A dispersion of globules of one liquid in another in which it is insoluble.

EP (Extreme pressure) Additive - A chemical substance containing one or more elements, especially sulphur, chlorine or phosphorus, able to react with metal surfaces to give inorganic films of high melting point. The presence of these films hinders welding and seizure and thus prevents scuffing and scoring, particularly in gears operating under high load conditions.

Erosion - Erosive wear is loss of material from a solid surface due to relative motion in contact with a fluid which contains solid particles.

Fatty Acids - Long chain organic acids which occur naturally as their glyceride esters in animal and vegetable oils and fats.

Filler - A substance such as lime, talc, mica and other powders, added to grease to increase its consistency or to an oil to increase viscosity.

Flash Point - The lowest temperature at which the vapour of a lubricant can be ignited under specified conditions.

Flash Temperature - The maximum local temperature generated at some point of close approach in a sliding contact.

Flexure Pivot - A type of bearing for limited movement in which the moving parts are guided by flexure of elastic members rather than by rolling or sliding surfaces.

Fretting - The removal of extremely fine particles from bearing surfaces due to the inherent adhesive forces between the surfaces particularly under the condition of small amplitude vibration.

Fretting Corrosion - A form of fretting in which chemical reaction predominates.

Friction - The resisting force tangential to a common boundary between two bodies when, under the action of an external force, one body moves or tends to move relative to the surface of the other.

Friction Polymer - An amorphous organic deposit which is produced when certain metals are rubbed together in the presence of organic liquids or gases.

Galling - A severe form of scuffing associated with gross damage to the surfaces or failure. The use of this form should be avoided.

Grease - A lubricant composed of an oil thickened with a soap or other thickener to a semi-solid or solid consistency. A lime-based grease is prepared from a lubricating oil and Calcium soap. Sodium, Barium, Lithium and Aluminium based greases are also used.

Hydraulic Fluid - A fluid used for transmission of hydraulic pressure or action, not necessarily involving lubricant properties. May be oil, water or synthetic (fire resistant) liquids.

Hydrodynamic Lubrication - A system of lubrication in which the shape and relative motion of the sliding surfaces causes the formation of a fluid film having sufficient pressure to separate the surfaces.

Hydrostatic Lubrication - A system of lubrication in which the lubricant is supplied under sufficient external pressure to separate the opposing surfaces by a fluid film.

Initial Pitting - Surface fatigue occurring during the early stages of operation of gears, associated with removal of highly stressed local areas and running-in.

Journal - That part of a shaft or axle which rotates or oscillates relatively to a radial bearing.

Kinematic Viscosity - See: Viscosity.

L_{10} life - See: Rating Life.

Lacquer - Hard, lustrous, varnish-like, oil insoluble deposit which tends to form on the pistons and cylinders of internal combustion engines.

Load Carrying Capacity - The maximum load that a sliding or rolling system can support without failure or the wear exceeding the design limits for the particular application.

Lubricant - Any substance interposed between two surfaces in relative motion for the purpose of reducing the friction or wear between them.

Mild Wear - A form of wear characterised by removal of material in very small fragments.

Non-Newtonian Viscosity - The apparent viscosity of a material in which the shear stress is not proportional to the rate of shear.

Oil - A liquid of vegetable, animal, mineral or synthetic origin feeling slippery to the touch.

Oiliness - That property of a lubricant that produces low friction under conditions of boundary lubrication. The lower the friction, the greater the oiliness.

Oil Mist (Fog) - An oil atomised with the aid of compressed air and then conveyed by the air in a low-pressure distribution system to multiple points of lubricant application.

Pitting - Any removal or displacement of material resulting in the formation of surface cavities.

Plain Bearing - Any simple sliding type of bearing as distinguished from fixed-pad, pivoted-pad or rolling-type bearings.

Porous Bearing - A bearing made from porous material such as compressed metal powders, the pores acting either as reservoirs for holding, or passages for supplying lubricant.

Pour Point - The lowest temperature at which a lubricant can be observed to flow under specified conditions.

PTFE - Polytetrafluorethylene, a polymer having outstanding low-friction properties over a wide temperature range.

PV Factor - The product of bearing pressure and surface velocity.

Rating Life - The fatigue life in millions of revolutions or hours at a given operating speed which 90 per cent of a group of substantially identical rolling element bearings will survive under a given load. The 90 per cent rating life is frequently referred to as "L_{10}-life" or "B_{10}-life".

Redwood Viscosity - A commerical measure of viscosity expressed as the time in seconds required for 50 cubic centimeters of a fluid to flow through a tube of 10 mm length and 1.5 mm diameter at a given temperature.

Root Mean Square Height (RMS) - An American measure of surface topography representing the average departure of a line profile of the surface from a mean line.

SAE - Society of Automotive Engineers.

Saybolt Viscosity - A commerical measure of viscosity expressed as the time in seconds required for 60 cubic centimeters of a fluid to flow through the orifice of the Standard Saybolt Universal Viscometer at a given temperature under specified conditions.

Scoring - The formation of severe scratches in the direction of sliding.

Scratching - The formation of fine scratches in the direction of sliding.

Scuffing - Localised damage caused by the occurrence of solid-phase welding between sliding surfaces, without local surface melting.

Severe Wear - A form of wear characterised by removal of material in relatively large fragments.

Soap - In lubrication, a compound formed by the reaction of a fatty acid with a metal or metal compound.

Solid Lubricant - Any solid used as a powder or thin film on a surface to provide protection from damage during relative movement, and to reduce friction and wear.

Spalling - Separation of particles from a surface in the form of flakes.

Stick-Slip - A relaxation oscillation usually associated with decrease in the coefficient of friction as the relative velocity increases.

Synthetic Lubricant - A lubricant produced by synthesis rather than by extraction or refinement.

Thin Film Lubrication - A condition of lubrication in which the film thickness of the lubricant is such that the friction between the surfaces is determined by the properties of the surfaces as well as by the viscosity of the lubricant.

Total Acid Number (TAN) - The quantity of base, expressed in terms of the equivalent number of milligrams of potassium hydroxide that is required to neutralise all acidic constituents present in 1 gram of sample.

Total Base Number (TBN) - The quantity of acid, expressed in terms of the equivalent number of milligrams of potassium hydroxide that is required to neutralise all basic constituents present in 1 gram of sample.

Varnish - A deposit resulting from the oxidation and/or polymerisation of fuels, lubricating oils, or organic constituents of bearing materials.

Viscosity - That bulk property of a fluid, semi-fluid or semi-solid substance which causes it to resist flow.
Viscosity is defined by the equation

$$\eta = \tau / \frac{dv}{dx}$$

τ is the shear stress, v the velocity, ds the thickness of an element measured perpendicular to the direction of flow; dv/ds is known as the rate of shear. Viscosity in the normal, that is Newtonian sense is often called dynamic or absolute viscosity. Kinematic or static viscosity is the ratio of dynamic viscosity to density at a specified temperature and pressure.

Viscosity Index (VI) - Arbitrary scale used to show the magnitude of viscosity changes with temperature in lubricating oils and other products.

Wear - The removal of material from surfaces in relative motion, normally by abrasion, adhesion or corrosion.

Wedge Effect - The establishment of a pressure wedge in a lubricant.

Wettability - A term used to indicate the ease with which a lubricant will spread or flow over a bearing surface.

ZDDP - Initials for zinc dialkyl-dithiophosphate, which is widely used as an extreme pressure agent. It is also an effective oxidation inhibitor but should not be used in mechanisms with silver bearings.

Acknowledgement. Permission to quote terms and definitions received from The Institute of Petroleum and the OECD.

AUTHOR INDEX

Numbers underlined give the page on which the complete reference is listed, other numbers refer to the page number on which the author (or his work) is mentioned in the text.

SUBJECT INDEX